4th International Conference on Competitive Materials and Technological Processes 2016 (IC-CMTP4)

IOP Conference Series: Materials Science and Engineering
Volume 175

Miskolc, Hungary
3 - 7 October 2016

ISBN: 978-1-5108-3677-8
ISSN: 1757-8981

Printed from e-media with permission by:

Curran Associates, Inc.
57 Morehouse Lane
Red Hook, NY 12571

Some format issues inherent in the e-media version may also appear in this print version.

Printed by Curran Associates, Inc. (2017)

For permission requests, please contact the Institute of Physics
at the address below.

Institute of Physics
Dirac House, Temple Back
Bristol BS1 6BE UK

Phone: 44 1 17 929 7481
Fax: 44 1 17 920 0979

techtracking@iop.org

Additional copies of this publication are available from:

Curran Associates, Inc.
57 Morehouse Lane
Red Hook, NY 12571 USA
Phone: 845-758-0400
Fax: 845-758-2633
Email: curran@proceedings.com
Web: www.proceedings.com

Table of contents

Volume 175

4th International Conference on Competitive Materials and Technology Processes (IC-CMTP4)

3 March to 7 October 2016, Miskolc, Hungary

Accepted papers received: 3 February 2017
Published online: 15 February 2017

Preface

Plenary and Keynotes

Advanced Materials for Bio- and Medical Applications

Advanced Materials for Extreme Applications

Biomaterials Derived Ceramics and Composites

Hetero-Modulus and Hybrid Materials

Materials with Extreme Dynamic Strength

Membranes and Catalysts

Minerals for Environmental and Medical Application

Nanomaterials for Environment and Health

Novel Synthesis and Processing Technology

Phase Diagram as a Tool of Materials Science

Processing, Properties and Applications of Ceramics

Testing and Characterization of Materials - Methods, Equipment and Errors

Mechanical Properties and Processing Technology of Advanced Materials

The 1st International Symposium on Intelligent and Smart Materials

The 2nd International Symposium on Innovative Construction Materials

The 2nd International Symposium on Innovative Carbon and Carbon Based Materials

IOP Conf. Series: Materials Science and Engineering **175** (2017) 011001 doi:10.1088/1757-899X/175/1/011001

ic-cmtp4

THE 4TH INTERNATIONAL CONFERENCE ON COMPETITIVE MATERIALS AND TECHNOLOGY PROCESSES

October 3-7, 2016 / Miskolc-Lillafüred, Hungary

Organized by the International Scientific Advisory Board (**ISAB**),

International Organizing Committee (**IOC**) and

Session Chairs

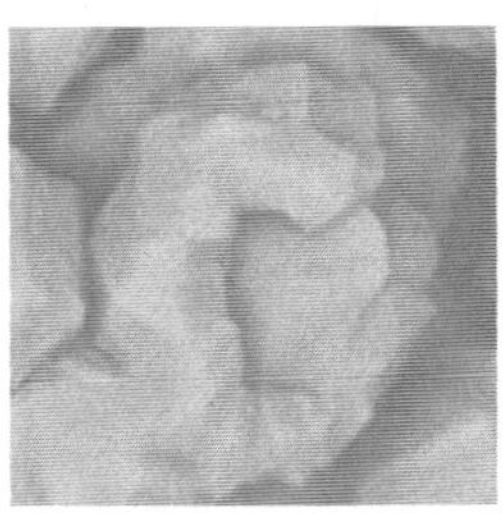

IOP Conf. Series: Materials Science and Engineering **175** (2017) 011001 doi:10.1088/1757-899X/175/1/011001

Abstract

The competitiveness is one of the most important component of our life and it plays key role in efficiency both of organizations and societies. The more scientific supported and prepared organizations develop more competitive materials with better physical, mechanical, chemical and biological properties and the leading companies are applying more competitive equipment and technology processes.

The aims the 4th International Conference on Competitive Materials and Technology Processes (ic-cmtp4) and the Symposiums **is-ism1**, **is-icbm2** and **is-icm2** are the followings:

- Promote new methods and results of scientific research in the fields of material, biological, environmental and technology sciences;
- Change information between the theoretical and applied sciences as well as technical and technological implantations.
- Promote the communication between the scientist of different nations, countries and continents.

Among the major fields of interest are innovative materials with increased physical, chemical, biological, medical, thermal, mechanical properties and dynamic strength; including their crystalline and nano-structures, phase transformations as well as methods of their technological processes, tests and measurements. Multidisciplinary applications of material science and technological problems encountered in sectors like ceramics, glasses, thin films, aerospace, automotive and marine industry, electronics, energy, construction materials, medicine, biosciences and environmental sciences are of particular interest.

In accordance to the program of the conference ic-cmtp4, and its symposiums **is-ism1, is-icbm2** and **is-icm2** we have received more than **450** inquires and registrations frome **46** countries of **5** continents. Finally from them more than **240** abstracts were accepted for presentation, including the **10** PLENARY lectures. Scientists and researchers have arrived to Miskolc-Lillafüred (Hungary) from **41** countries of Asia, Europe, Africa, North and South America and Australia.

Together with co-authors in this book are presented abstracts from more than **700** scientists and researchers..

Prof. Dr. László A. GÖMZE
chair, ic-cmtp4

IOP Conf. Series: Materials Science and Engineering **175** (2017) 011001 doi:10.1088/1757-899X/175/1/011001

IOP Conf. Series: Materials Science and Engineering **175** (2017) 011001 doi:10.1088/1757-899X/175/1/011001

THE SESSION CHAIRS AND SYMPOSIUM CHAIRS

***Session 1.* Advanced Materials for Bio- and Medical Applications**
Prof. Dr. Rainer GADOW (Germany)
MD. Prof. Dr. Alexandr L. URAKOV, (Russia)
***Session 2.* Advanced Materials for Extreme Applications**
Dr. Ludmila N. GÖMZE, (Hungary)
***Session 3.* Advanced Nanomaterials with Predesigned Properties**
Prof. Dr. Leonid KOTOV, (Russia)
***Session 4.* Biomaterials Derived Ceramics and Composites**
Prof. Dr. Bojja SREEDHAR, (India)
***Session 5.* Glasses, Coatings and Related Materials**
Prof. Dr. Nikolay NIKONOROV, (Russia)
***Session 6.* Hetero-Modulus and Hybrid Materials**
Prof. Dr. Tohru SEKINO, (Japan)
Session 7. Light-Weight Metals and Alloys
Prof. Dr. Jean-Marie DREZET, (Switzerland)
Prof. Dr. Michel PEREZ, (France)
***Session 8.* Materials with Extreme Dynamic Strength**
Prof. Dr. László A. GÖMZE, (Hungary)
***Session 9.* Membranes and Catalysts**
Dr. János LAKATOS, (Hungary)
***Session 10.* Minerals for Environmental and Medical Application**
Prof. Dr. Saveiro FIORE, (Italy) and Prof. Dr. Olga KOTOVA (Russia)
***Session 11.* Nanomaterials for Environment and Health**
Miss Saara HEINONEN, (Finland)
***Session 12.* Novel Synthesis and Processing Technology**
Prof. Dr. Sergey N. KULKOV, (Russia)
***Session 13.* Phase Diagram as a Tool of Materials Science**
Prof. Dr. Vasily LUTSYK, (Russia) and Prof. Dr. Jean-Claude TEDENAC, (France)
***Session 14.* Polymer Derived Ceramics**
Prof. Dr. Michael SCHEFFLER, (Germany)
***Session 15.* Processing, Properties and Application of Ceramics**
Dr. Nicolas TESSIER-DOYEN, (France) and Dr. Mária Maros B, (Hungary)
***Session 16.* Testing and Characterization of Materials – Methods, Equipment and Errors**
Dr. Uwe BECK, (Germany) and Sergei N. KULKOV, (Russia)
***Session 17.* Mechanical Properties and Processing Technology of Advanced Materials**
Prof. Dr. Shigeru NAGASAWA, (Japan) and Sergei N. KULKOV, (Russia)
***is-ism1.* The 1st International Symposium on Intelligent and Smart Materials**
Prof. Dr. Madhusudana RAO, (India)
***is-icm2.* The 2nd International Symposium on Innovative Construction Materials**
Prof. Dr. Tomasz SADOWSKI (Poland), Dr. Adorján BOROSNYÓI (New Zealand), Dr. Daman K. PANESAR, (Canada), Prof. Dr. Emad M.M. EWAIS, (Egypt) and Prof. Dr. László A. GÖMZE (Hungary
***is-icbm2.* The 2nd International Symposium on Innovative Carbon and Carbon Based Materials**
Prof. Dr. Vasily LEBEDEV, (Russia), Prof. Dr. Igor L. SHABALIN, (United Kingdom) and Prof. Dr. Joachim BOLLMANN, (Germany)

The financial organizer of the conference: **IGREX Engineering Service Ltd** *(Hungary)*

Some of the presented works in the ic-cmtp4 conference were selected by the Editorial Board of scientific journal **Építőanyag-JSBCM** and were published in it as the followings:

M Khramchenkov and E Khramchenkov 2016 *Építőanyag-JSBCM* **68** (2) 34 http://dx.doi.org/10.14382/epitoanyag-jsbcm.2016.6
A Isaev, V Grechishnikov, P Pivkin, Y Ilyukhin, M Kozochkin and P Peretyagin 2016 *Építőnyag-JSBCM* **68** (2) 46 http://dx.doi.org/10.14382/epitoanyag-jsbcm.2016.8
V Lutsyk, A Zelrnaya, A Zyryanov and E Nasrulin 2016 *Építőanyag-JSBCM* **68** (2) 52 http://dx.doi.org/10.14382/epitoanyag-jsbcm.2016.9
S P Buyakova, E S Kalatur, A S Buyakov and S S Kulkov 2016 *Építőanyag-JSBCM* **68** (3) 70 http://dx.doi.org/10.14382/epitoanyag-jsbcm.2016.12
I Zhukov, S P Buyakova and S S Kulkov 2016 *Építőanyag-JSBCM* **68** (3) 74 http://dx.doi.org/10.14382/epitoanyag-jsbcm.2016.13
A Urakov, N Urakova, A Reshetnikov, A Kasatkin, M Kopylov and D Baimurzin 2016 *Építőanyag-JSBCM* **68** (4) 110 http://dx.doi.org/10.14382/epitoanyag-jsbcm.2016.19
A Vereschaka, A Kutin, N Sitnikov, G Oganyan and O Sharipov 2016 *Építőanyag-JSBCM* **68** (4) 114 http://dx.doi.org/10.14382/epitoanyag-jsbcm.2016.20
G M Rodriguez Torrez, J Zarate Medina and M E Contreras Garcia 2016 *Építőanyag-JSBCM* **68** (4) 120 http://dx.doi.org/10.14382/epitoanyag-jsbcm.2016.21
M E Abad-Javier, M Cajero-Juárez and M E Contreras Garcia 2016 *Építőanyag-JSBCM* **68** (4) 124 http://dx.doi.org/10.14382/epitoanyag-jsbcm.2016.22

In the name of the International Scientific Advisory Board I would like congratulate the above authors that they papers were selected by editors of journal **Építőanyag-JSBCM.**

ACKNOWLEDGEMENT

In the name of ic-cmtp4 Conference Board I would like acknowledge and say many thanks to every scientists and colleagues we have supported us in our organizing work. Especially many thanks for their support to societies, scientific journals and organizations like:

AIPEA – International Association for the Study of Clays
Applied Rheology – International Sci. Journal on Rheology (***Switzerland***)
biomat.net – The Biomaterials Network (***Portugal***)
cfi – ceramic forum international (***Germany***)
ECA – European Crystallographic Association
Építőanyag – Journal of Silicate Based Composite Materials (***Hungary***)
GFC – Groupe Francais de la Céramique (***France***)
ICG – International Commission of Glass
Igrex Ltd. (***Hungary***)
ME – Miskolci Egyetem / University of Miskolc (***Hungary***)
refractory WORLDFORUM – Scientific Journal (***Germany***)
RAS – The Russian Academy of Sciences
SB-RAS Institute of Strength Physics and Materials Science (***Russia***)
SBAO – Society for Biomaterials and Artificial Organs (***India***)
SZTE – Scientific Society of Silicate Materials (***Hungary***)
TSU – Tomsk State University (***Russia***)

Many thanks also to colleagues of **ISAB**, **IOC** and to **session** and **symposium chairs** for their support and help in organization work and in successful transaction of the 4th International Conference on Competitive Materials and Technology Processes (**ic-cmtp4**).

I would like to say many thank personally to **Prof. Dr. Tohru Sekino** (The University of Osaka), **Prof. Dr. Sergey N. Kulkov** (Tomsk State University), **Prof. Dr. Tomasz SADOWSKI** (Lublin University of Technology), Md. **Prof. Dr. Alexandr L. Urakov**, (Izhevsk State Medical Academy), **Prof. Dr. Rainer Gadov**, (University of Stuttgart), **Prof. Dr. Shigeru NAGASAWA** (Nagaoka University of Technology), **Prof. Dr. Olga KOTOVA** (RAS Institute of Geology) and to **Prof. Dr. Takeshi TOYAMA** (Nihon University) for their excellent organization works in Japan, Russia, Asia and in European Union.

Prof. Dr. László A. Gömze
chair of conference board
University of Miskolc (Hungary)

IOP Conf. Series: Materials Science and Engineering **175** (2017) 011002 doi:10.1088/1757-899X/175/1/011002

Peer review statement

All papers published in this volume of *IOP Conference Series: Materials Science and Engineering* have been peer reviewed through processes administered by the proceedings Editors. Reviews were conducted by expert referees to the professional and scientific standards expected of a proceedings journal published by IOP Publishing.

Rheological principles of development hetero-modulus and hetero-viscous complex materials with extreme dynamic strength

L A Gömze[1,2,3] and **L N Gömze**[2]

[1]University of Miskolc, Institute of Ceramics and Polymer Engineering; Miskolc, Hungary
[2]IGREX Engineering Service Ltd, Igrici; Hungary
[3]Tomsk State University, Tomsk, Russia

E-mail:femgomze@uni-miskolc.hu and igrex2009@yandex.ru

Abstract. Materials with different crystalline and morphological compositions have different chemical, physical, mechanical and rheological properties, including wear protection, melting temperature, module of elasticity and viscosity. Examining the material structures and behaviors of differentceramic bodies and CMCs under high speed collisions in several years the authors have understood the advantages of hetero-modulus and hetero-viscous complex material systems to absorb and dissipate the kinetic energy of objects during high speed collisions. Applying the rheo-mechanical principles the authors successfully developed a new family of hetero-modulus and hetero-viscous alumina matrix composite materials with extreme mechanical properties including dynamic strength. These new corundum-matrix composite materials reinforced with $\mathbf{Si_2ON_2}$, $\mathbf{Si_3N_4}$, **SiAlON** and **AlN** submicron and nanoparticles have excellent dynamic strength during collisions with high density metallic bodies with speeds about 1000 m/sec or more. At the same time in the alumina matrix composites can be observed a phase transformation of submicron and nanoparticles of alpha and beta silicone-nitride crystals into cubic**c-**$\mathbf{Si_3N_4}$diamond-like particles can be observed, when the high speed collision processes are taken place in vacuum or oxygen-free atmosphere.

Using the rheological principles and the energy engorgement by fractures, heating and melting of components the authors successfully developed several new hetero-modulus, hetero-viscous and hetero-plastic complex materials. These materials generally are based on ceramic matrixes and components having different melting temperatures and modules of elasticity from low values like carbon and light metals (**Mg**, **Al**, **Ti**, **Si**) up to very high values like boride, nitride and carbide ceramics.Analytical methods applied in this research were scanning electron microscopy, X-ray diffractions and energy dispersive spectrometry. Digital image analysis was applied to microscopy results to enhance the results of transformations.

Keywords: ceramics, composites, diamond-like, elasticity, hetero-modulus, nanostructure, rheology, strength, viscosity

1. Introduction

In the last 10-15 years more and more ceramic materials and ceramic matrices composites (CMC) are developed and used in the leading and sophisticated industries [1-5].In our days the relatively low densityceramics and CMC materials are willingly used by industry because of their excellent properties such as hardness [6-11], wear resistance [12-15], toughness [16-18] mechanical [19-25] and dynamic strength [26-32]. In our days several methods and technologies are used to produce ceramic and ceramic based composite items with required physical and mechanical properties [33-67]. Thanking to the several years research and investigation in development and processing of high-tech ceramics and ceramic matrix composites the following scientists have achieved considerable new results:

- S. Hampshire's group in study and development of **Si_3N_4** and **SiAlON** [68-70],
- S N Kulkov's and S P Buyakova's group in **ZrO_2, $ZrWO_8$** and zirconia based CMCs [71-77],
- E V Zharikov's group in development of carbon nanotubes (**CNT**) and their application for reinforce ceramic nanocomposites [78-80],
- K Niihara's group in high-pressure synthesis and in development and application of boron nitride (**BN**) nanocomposites and nanotechnology [81-84].

The papers of S G Chuklina, A I Pylinina and I I Mikhalenko [85, 86] also well represent the wide range application opportunities of ceramic materials and ceramic based composites especially the tetragonal **ZrO_2** ceramics. In accordance to fracture and crack propagation in alloys, composites and porous materials T Sadowski with his colleagues have achieved remarkable new results. In 2004 [87] and in 2007 [88] they examined the crack length after thermal shock in functionally graded materials strip. In 2008 they described the damage state in porous ceramics under compression [89] and in 2012 successfully performed numerical analysis of the most loaded places in turbine blades under real working conditions [90]. In 2014 they described the fatigue [91] and dynamic [92] responses of composite materials and in 2015 the influence of geometrical parameters [93].

The rheological parameters and behaviors of materials are playing important roles not only during high speed collisions but in technology processes of polymers [94] and construction raw materials [95, 96]. Rheology is important also in physical, chemical and mechanical characterization of materials like porous ceramics [74] or complex materials [97] and living organism [98]. The authors of this paper also have several years experiments in development of alumina based ceramics and composite materials with increased mechanical properties [23, 27, 28, 41, 46, 99, 100] and in investigation and measurement of rheological properties of these complex materials [101-107].

On the basis of rheological principles, analysis of high speed collisions and experiments in development ceramic materials and ceramic matrix composites **our aims are** *to develop ceramic based low density hetero-modulus, hetero-viscous and hetero-plastic materials for armor shells to save transport equipments against high speed (**v>1000 m/s**) collisions with flying objects having different densities and speeds.*

2. Materials and experiments

Before choosing the materials and starting the experiments we have compared the modulus of elasticity and melting temperature of the most common technical materials including polymers, metals and ceramics (Fig. 1). Generally the polymers have very low melting temperatures and Young's modulus, meanwhile the oxide and non-oxide ceramics have much higher modulus of elasticity and melting temperatures than metals and alloys. It is well known that most of the oxide and non-oxide ceramics and ceramic matrix composites have excellent mechanical properties including hardness and mechanical strength. Inspite of the high values of their compression strengths they do not have large enough dynamic strength, thanking to their microstructures with relatively large crystals of high rigidness and strong inclination to nicks, pitting and rigid fractures.

The advantages of material structures built from components having different modules of elasticity (Figure 1.b) and similar melting temperature were first described by D.P.H. Hasselman, P.F. Betzher and K.S. Mazdiyashni [108]. Actually the material structures built from components having different modulus of elasticity are capable dissipate the mechanical stresses better than others, and hereby stop the crack propagation under excess mechanical loadings and increase dynamic strength.

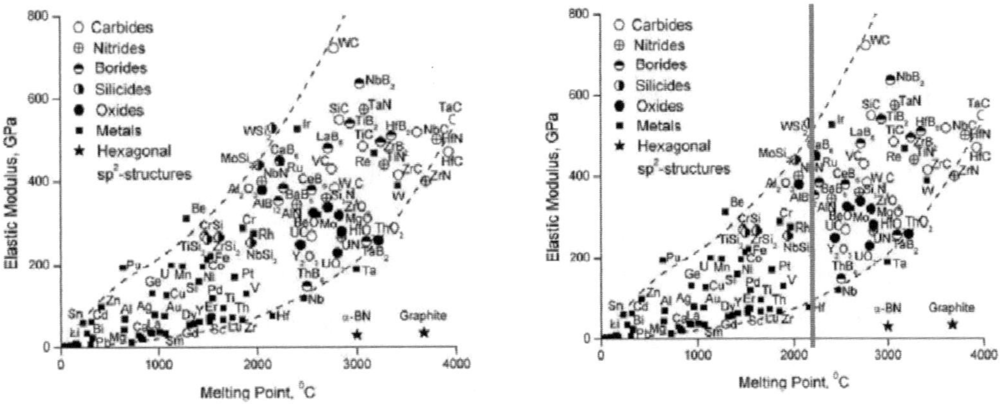

Figure 1. The melting temperatures and modulus of elasticity of most common technical materials

2.1. Materials

To create hetero-modulus composite materials with extreme dynamic strength alumina powders of different purity were used with grain size distribution from 50 nm milled powders to 100 μm large atomizer powders. The $\mathbf{Al_2O_3}$ powders were polluted with different **m%** of $\mathbf{SiO_2}$, $\mathbf{TiO_2}$, **MgO**, **CaO**, thallium oxides and metallic **Al** powders. The specimens were compacted using uniaxial pressing and sintered in $\mathbf{N_2}$ atmosphere at maximum temperature of **1600°C**. Thanking to the reactive sintering in $\mathbf{N_2}$ atmosphere the ceramic items were reinforced with **SiAlON**, **α-**$\mathbf{Si_3N_4}$ and **β-**$\mathbf{Si_3N_4}$. These quadratic thin alumina matrix composite plates were tested under **v > 1000 m/s** high speed collisions with high density flying metallic objects of homogeneous and inhomogeneous densities. The typical damages of these ceramic composite plates after the collisions are shown in Figure 2.

Thickness of ceramic plates, L = 4 mm Thickness of ceramic plates, L = 3 mm

Figure 2. Typical destructions of ceramic composites during high speed collisions

When the $\mathbf{Al_2O_3}$ corundum based ceramic plates were reinforced with **SiAlON**, **α-**$\mathbf{Si_3N_4}$ and **β-**$\mathbf{Si_3N_4}$ submicron and nano particles and tested under **v > 1000 m/s** high speed collisions with flying high density metallic objects we have observed the phase transformation of **α-**$\mathbf{Si_3N_4}$ and **β-**$\mathbf{Si_3N_4}$ into cubic phase of high density diamond-like **c-Si3N4** particles (Figure 3.a) [104, 109, 110, 111]. Thanking to the rapid and great local heating the melting of metallic parts also was observed (Figure 3.b).

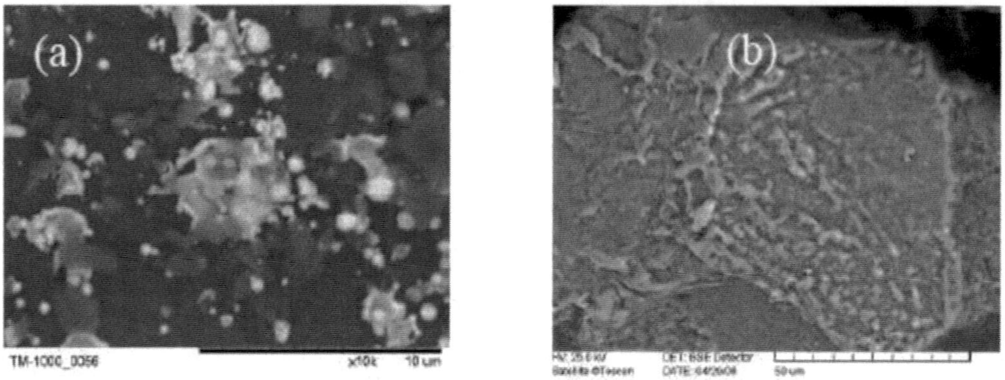

Figure 3. Typical phase transformation and melting observed during high speed collisions

2.2 Mechanical analysis of high speed collisions of objects

The energy engorgement of kinetic energy of high speed collision ($\mathbf{W_K}$) by armor shells can be determined as:

$$W_K = W_E + W_H + W_P + W_{PL} + W_S + W_V \qquad [Nm] \qquad (1)$$

where:

$\mathbf{W_E}$: energy engorgement through elastic deformation [Nm]
$\mathbf{W_H}$: energy engorgement through heating and phase transformation [Nm]
$\mathbf{W_P}$: energy engorgement through fracture indicated by pressure stress [Nm]
$\mathbf{W_{PL}}$: energy engorgement through plastic deformation [Nm]
$\mathbf{W_S}$: energy engorgement through fracture indicated by shear stress [Nm]
$\mathbf{W_V}$: energy engorgement through viscous deformation [Nm]

The kinetic energy of high speed impacting (flying) body with homogeneous density can be described as:

$$W_K = \frac{m \cdot u^2}{2\,g} = \frac{u^2}{2\,g} \cdot \rho \cdot V \qquad [Nm] \qquad (2)$$

and with inhomogeneous density can be described as:

$$W_K = \frac{u^2}{2\,g} \sum_{i=1}^{n} m_i = \frac{u^2}{2\,g} \sum_{i=1}^{n} \rho_i \cdot V_i \qquad [Nm] \qquad (3)$$

where:

$\mathbf{g}$= 9.81 m/s^2
i=1, 2, …, n: the number of different density components of flying object
$\boldsymbol{\rho}$: the density of flying object material [kg/m^3]
$\boldsymbol{\rho_i}$: density of the "i-th" component of flying object [kg/m^3]
m: mass of flying object [kg]
$\mathbf{m_i}$: mass of the "i-th" component of flying object [kg]
u: speed of flying object at the moment of collision [m/s]
V: volume of flying object [m^3]
$\mathbf{V_i}$: volume of "i-th" component of flying object [m^3]

In case when the impacting body has homogeneous density and the armor shells (ceramic plates) are made from materials having only one (constant) value of Young's modulus the engorgement of kinetic energy of flyingobject can be described as:

$$\frac{u^2}{2\,g} \cdot \rho \cdot V = \frac{R_P^2}{2E} \cdot A_1 \cdot l_1 + \frac{R_S^2 \cdot (\nu+1)}{E \cdot \cdot \nu} \cdot A_2 \cdot l_2 + W_H \qquad [Nm] \qquad (4)$$

where:
$\mathbf{v_1}$: the Poisson ration
$\mathbf{A_1}$ and $\mathbf{A_2}$: surfaces of fractures [m^2]
$\mathbf{l_1}$ and $\mathbf{l_2}$: deep and "movement" of fractures [m]
$\mathbf{R_p}$ and $\mathbf{R_S}$: the pressure and shear strength of ceramic plates (tiles) [N/m^2]
$\mathbf{W_H}$: energy engorgement through heating and recrystallization of flying object particles and alumina based ceramic materials surrounding the collision [Nm]

In case when the impacting body has inhomogeneous density and the armor shells (ceramic plates) are made from materials having only one value of Young's modulus the engorgement of kinetic energy of flyingobject can be described as:

$$\frac{u^2}{2\,g} \sum_{i=1}^{n} \rho_i \cdot V_i = \frac{R_P^2}{2E} \sum_{i=1}^{n} A_{1i} \cdot l_{1i} + \frac{R_S^2 \cdot (\nu+1)}{E \cdot \cdot \nu} \cdot \sum_{i=1}^{n} A_{2i} \cdot l_{2i} + W_H \qquad [Nm] \qquad (5)$$

where:
$\mathbf{v_1}$: the Poisson ration
$\boldsymbol{\rho_i}$: density of the "i-th" component of flying object [kg/m^3]
$\mathbf{A_{1i}}$ and $\mathbf{A_{2i}}$: surfaces of fractures caused by "i-th" density component of flying object [m^2]
i=1, 2, … n: the number of different density components of flying object
$\mathbf{l_{1i}}$ and $\mathbf{l_{2i}}$: deep and "movement" of fractures, caused by "i-th" density component of flying object [m]
$\mathbf{R_p}$ and $\mathbf{R_S}$: the pressure and shear strength of ceramic plates (tiles) [N/m^2]
$\mathbf{V_i}$: volume of "i-th" component of flying object [m^3]

In case when the flying object (impact body) has homogeneous density and the armor shells (ceramic plates) are made from materials having several values of Young's modulus the engorgement of kinetic energy of falling object can be described as:

$$\frac{u^2}{2\,g} \cdot \rho \cdot V = \sum_{j=1}^{N} \frac{R_{Pj}^2}{2E_j} \cdot A_{1j} \cdot l_{1j} + \sum_{j=1}^{N} \frac{R_{Sj}^2 \cdot (\nu_j+1)}{E_j \cdot \nu_j} \cdot A_{2j} \cdot l_{2j} + W_H \qquad [Nm] \qquad (6)$$

where:
$\mathbf{A_{1j}}$ and $\mathbf{A_{2j}}$: surface of fractures of "j-th" Young's modulus component of ceramic body [m^2]
$\mathbf{E_j}$: the Young's modulus of the "j-th" component of ceramic body [N/m^2]
j=1, 2, …, n: the number of different Young's modulus component of ceramic body
$\mathbf{l_{1j}}$ and $\mathbf{l_{2j}}$: deep and "movement" of fractures of "j-th" Young's modulus component of ceramic body [m]
$\mathbf{R_{Pj}}$ and $\mathbf{R_{Sj}}$: the pressure and shear strength of "j-th" Young's modulus component of ceramic body [N/m^2]
$\mathbf{v_j}$: the Poisson ration of "j-th" Young's modulus component of ceramic body [m^3]

In case when the flying object (impacting body) has inhomogeneous density and the armor shells (ceramic based composite plates) are made from materials having several values of Young's modulus the engorgement of kinetic energy of high speed collision can be described as:

$$\frac{u^2}{2\,g} \sum_{i=1}^{n} \rho_i \cdot V_i = \sum_{j=1}^{N} \frac{R_{Pj}^2}{2E_j} \cdot A_{1j} \cdot l_{1j} + \sum_{j=1}^{N} \frac{R_{Sj}^2 \cdot (\nu_j+1)}{E_j \cdot \nu_j} \cdot A_{2j} \cdot l_{2j} + W_H \qquad [Nm] \qquad (7)$$

where:
$\mathbf{v_j}$: the Poisson ration of "j-th" Young's modulus component of ceramic body [m^3]
$\boldsymbol{\rho_i}$: density of the "i-th" component of flying object [kg/m^3]
$\mathbf{A_{1j}}$ and $\mathbf{A_{2j}}$: surface of fractures of "j-th" Young's modulus component of ceramic body [m^2]
$\mathbf{E_j}$: the Young's modulus of the "j-th" component of ceramic body [N/m^2]

i=1, 2, … n: the number of different density components of flying object
j=1, 2, …, N: the number of different Young's modulus component of ceramic body
$\mathbf{l_{1j}}$ and $\mathbf{l_{2j}}$: deep and "movement" of fractures of "j-th" Young's modulus component of ceramic body [m]
$\mathbf{R_{Pj}}$ and $\mathbf{R_{Sj}}$: the pressure and shear strength of "j-th" Young's modulus component of ceramic body [N/m^2]
$\mathbf{V_j}$: volume of "j-th" component of flying object [m^3]

3. Results and discussion

Metallic flying objects having inhomogeneous densities were used for high speed collision tests. Examining the collision process and the ceramic composite armor shells in the place and surrounding of the hits a considerable phase transformation could be observed both in $\mathbf{Al_2O_3}$ and in $\mathbf{Si_3N_4}$ components as it is shown in Figure 4.

Figure 4. Phase transformation of ceramic components of armor shells

Similar phase transformation could be observed in the metallic components (Figure 5) of the high speed flying objects.

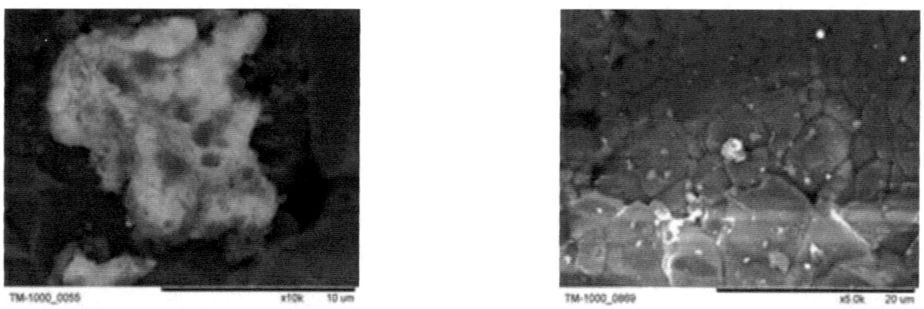

Figure 5. Melting and phase transformation of metallic components of collisions

On the basis of Figures 4 and 5 the thermic part of the energy engorgement of high speed collision [27] can be described as:

$$W_H = W_{HS} + W_{RC} + W_{RM} \qquad [\mathrm{Nm}] \qquad (8)$$

where:
$\mathbf{W_H}$: energy engorgement through heating [Nm]
$\mathbf{W_{HS}}$: energy engorgement through heating and melting the surrounding of the collision and fall [Nm]
$\mathbf{W_{RC}}$: energy engorgement through recrystallization of ceramic particles surrounding the collision and fall [Nm]
$\mathbf{W_{RM}}$: energy engorgement through spraying and recrystallization of falling metallic body [Nm]

The melting of the high speed metallic components with inhomogeneous density gave the idea to create armor shells from components having the same or similar module of elasticity meanwhile their melting temperatures are different (Fig. 6).

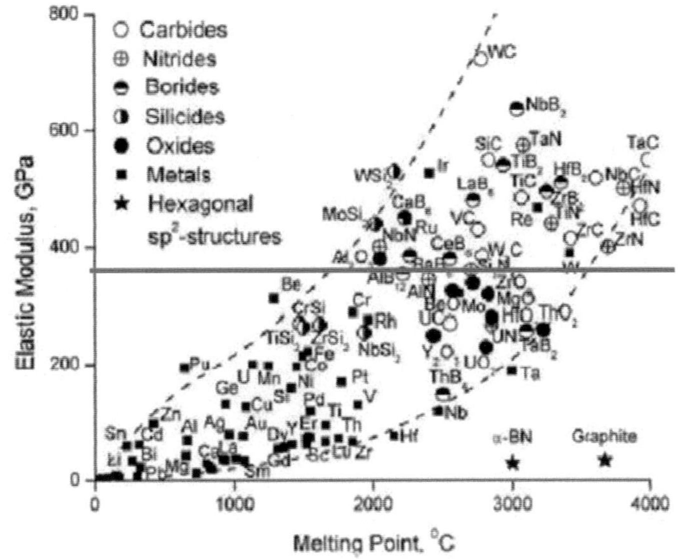

Figure 6. The armor shells components have different melting temperatures

Thanking to the different melting temperatures of components this material structure has good ability to absorb not only elastic energy but absorb and dissipate the viscous energy during crack propagation. Although these materials have a certain "self-healing" properties thanking to the phase transformation (crystallization) of melted components during cooling after the hits.

Finally we decided to develop self-healing armor shells from ceramic-metal composites from hetero-modulus and hetero-melting components. These composite material structures are built from particles having different Young's modulus, different viscosity and plasticity. For this we developed special high porosity (Fig. 7.a) or foam-like (Fig. 7.b) ceramic composites first and after impregnated them with different metals and alloys (Fig.8).

a, b,

Figure 7. The developed new porous (a) and foam (b) ceramic composite structure

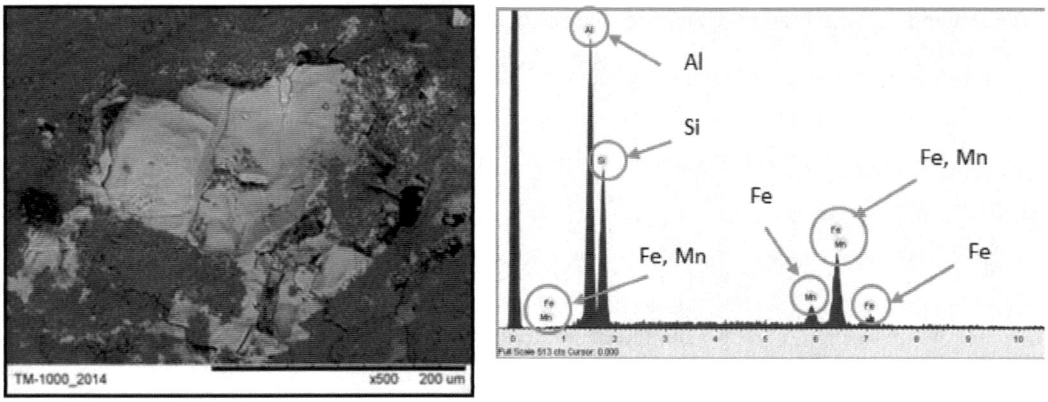

Figure 8.The developed light weight ceramic-metal armor shells

This kind of material structure can be mechanically and rheologically characterized as hetero-modulus, hetero-viscous and hetero-plastic complex materials in the moments of high speed collisions. The energy engorgements of this kind of materials during high speed collisions can be described as:

$$\frac{u^2}{2g}\sum_{i=1}^{n}\rho_i \cdot V_i = \sum_{j=1}^{N}\frac{R_{Pj}^2}{2E_j}\cdot A_{1j}\cdot l_{1j} + \sum_{j=1}^{N}\frac{R_{Sj}^2\cdot(\nu_j+1)}{E_j\cdot\nu_j}\cdot A_{2j}\cdot l_{2j} + \sum_{k=1}^{K}\eta_k\cdot\frac{A_{3k}\cdot l_{3k}}{t_k} + W_H \qquad (9)$$

where:

$\mathbf{v_j}$: the Poisson ration of "j-th" Young's modulus component of ceramic based armor body [m^3]
$\boldsymbol{\rho_i}$: density of the "i-th" component of flying object [kg/m^3]
$\mathbf{A_{1j}}$ and $\mathbf{A_{2j}}$: surface of fractures of "j-th" Young's modulus component of the ceramic based composite armor plate [m^2]
$\mathbf{E_j}$: the Young's modulus of the "j-th" component of the ceramic composite armor plate [N/m^2]
i=1, 2, …, n: the number of different density components of flying object
j=1, 2, …, N: the number of different Young's modulus components of the ceramic based composite armor plate
k=1, 2, …, K: the number of different viscosity components of the ceramic based composite armor
$\mathbf{l_{1j}}$ and $\mathbf{l_{2j}}$: deep and "movement" of fractures of "j-th" Young's modulus component of the ceramic based composite armor plate [m]
$\mathbf{R_{Pj}}$ and $\mathbf{R_{Sj}}$: the pressure and shear strength of "j-th" Young's modulus component of the ceramic based composite armor plate [N/m^2]
$\mathbf{V_i}$: volume of "i-th" component of flying object [m^3]

4. Conclusions

During this research the Authors successfully found some good pore forming additives which could decrease considerably the wetting angle between the ceramic matrices and the impregnated metals.

Thanking to the different values of heat conductivity and melting points of components these low density armor shells have not only excellent hardness, wear resistance and dynamic strength but good ability to self-healing and extra high tolerance to damage.

Acknowledgements

The authors would like acknowledge to IGREX Engineering Service Ltd for their financial, technological and materials to continue this research during several years. Thanking to their long term selfless financial and technological support all developed new materials with extreme dynamic strength and developed know-how belong to IGREX Ltd.

References

[1] K Katahira, y Watanabe, H Ohmori and T Kato 2002 *Int. Journal of Machine Tools and Manufacture* **42** 1307
[2] C B Carter and M G Norton 2007 Ceramic Materials: Science and Engineering *Springer* 1-716
[3] E S Lukin, N A Makarov, A I Kozlov, N A Popova, E V Anufrieva, M A Vartanyan, Yu A Kozlov, M N Safina and D O Lemeshev 2008 *Steklo i keramika***65** (10) 27
[4] A Morikawa, T Suzuki, K Kikuta, A Suda and H Shinjo 2009 *Építőanyag-JSBCM***61** (1) 2 http://dx.doi.org/10.14382/epitoanyag-jsbcm.2009.1
[5] E Rocha-Rangel, D Hernández-Silva, E Terés-Rojas, E Martines-Franco and S Diaz-De La Torre 2010 *Építőanyag-JSBCM***62** (3) 75 http://dx.doi.org/10.14382/epitoanyag-jsbcm.2010.15
[6] S D Skrovanek and R C Bradt 1979 *J. Am. Ceram. Coc.***62** (3-4) 215
[7] G R Anstis, P Chantikul, B R Lawn, D B Marshall 1981 *J. Am. Ceram. Soc.* **64** 533
[8] R W Rice, C C Wu and F Borchelt 1994 *J. Am Ceram Soc.* **77** (10) 2539
[9] A Krell and B Plank 1995 *J. Am. Ceram Soc.* **78** (4) 1118
[10] B Venkataraman and G Sundarajan 1996 *ActaMaterialica***44** (3) 451
[11] A Gubernat, L Stobierski and P Labaj 2007 *J. Europ. Ceram. Soc.***27** 781
[12] T Senda, E Yasuda, M Kaji and B C Bradt 1999 *J. Am Ceram. Soc.***82** (6) 1505
[13] J Szépvölgyi 2004 *Építőanyag-JSBCM***56** (3) 86 (in Hungarian) http://dx.doi.org/10.14382/epitoanyag-jsbcm.2004.1
[14] S Kulkov and N Savchenko 2008 *Építőanyag-JSBCM***60** (3) 62 http://dx.doi.org/10.14382/epitoanyag-jsbcm.2008.10
[15] N I Ershova and I Y Kelina 2009 *Építőanyag-JSBCM***61**(2) 34 http://dx.doi.org/10.14382/epitoanyag-jsbcm.2009.6
[16] A Muchtar and L C Lim 1998 *ActaMaterialica***46** (5) 1683
[17] V A Izhevskyi, L A Genova, J C Bressiani and A H Bressiani 2000 *Cerâmica***46** 297
[18] A Zerr, R Riedel, T Sekine, J E Lowther, W Y Ching and I Tanaka 2006 Adv. Mater. **18** 2933
[19] P Chantikul, S J Bennison and B R Lawn 1990 *J. Am. Ceram. Soc.***73** (8) 2419
[20] J Seidel, N Claussen and J Rödel 1995 *J. Eur. Ceram. Soc.* **15** 395
[21] J B Wachtman, W R Cannon and M J Matthewson 2009 Mechanical properties of ceramics *2nd Edition Wiley* pp1-496
[22] R W Rice 1997 *J. Mater. Sci.* **32** 1973
[23] J Csányi and L A Gömze 2001 *Építőanyag-JSBCM***53** (2) 66
[24] A G Tkachev and O N Tkacheva 2009 *Steklo i keramika***66** (2) 15
[25] Y T O, J B Koo, K J Hong J S Park and D C Shin 2004 *Mater. Sci and Eng. A* **37** (4) 191
[26] H He, T. Sekine, T. Kobayashi and H. Hirosaki 2000 *Phys. Rev. B***62** (17) 11412
[27] L A Gömze and L N Gömze 2009 *Építőanyag-JSBCM,***61** (2) 38 http://dx.doi.org/10.14382/epitoanyag-jsbcm.2009.7
[28] L A Gömze and L N Gömze 2011 *IOP Conf. Ser.: Mater. Sci. Eng.***18** 082001 http://dx.doi.org/10.1088/1757-899X/18/8/082001
[29] S A Kinelovskii and K K Maevskii *J. Appl. Mech. Tech. Phys.* 2013 **54** (4) 524
[30] A Ainabayev, M Arkhipov, A Baideldinova, K Omarova and G Ksandopulo 2013 *IOP Conf. Ser.: Mater. Sci. Eng.***47** 012044 http://dx.doi.org/10.1088/1757-899X/47/1/012044
[31] V V Yakushev, A N Zhukov, A V Utkin, A I Rogacheva and V A Kudakina 2015 *Combustion, Explosion and Shock Waves* **51** (5) 603
[32] A Baideldinova, Ksandopulo G, Mukhina L 2016 *IOP Conf. Ser.: Mater. Sci. Eng.***123** 012032 http://dx.doi.org/10.1088/1757-899X/123/1/012032
[33] C Csizi 2002 *Építőanyag-JSBCM* **54** (2) 49 (in Hungarian) http://dx.doi.org/10.14382/epitoanyag-jsbcm.2002.10
[34] M Nagy, L Kotsis, É Makó, A Vértes and Z Klencsár 2002 *Építőanyag-JSBCM***54** (3) 70 (in Hungarian) http://dx.doi.org/10.14382/epitoanyag-jsbcm.2002.13
[35] Z Shen, M Johnson, Z Zhao and M Nygren 2002 *J. Am. Ceram Soc.***85** 1921

[36] L Hoffman, Z Jőzsa and R Nemes 2003 *Építőanyag-JSBCM***55** (1) 13 (in Hungarian) http://dx.doi.org/10.14382/epitoanyag-jsbcm.2003.3
[37] Z Károly, J Szépvölgyi and Z Farkas 2003 *Építőanyag_JSBCM***55** (2) 46 (in Hungarian) http://dx.doi.org/10.14382/epitoanyag-jsbcm.2003.9
[38] A Krell, P Blank, H Ma, T Hutzler and M Nebelung 2003 *J. Am. Ceram Soc.***86** (4) 546
[39] N Hegman, P Szücs and J Lakatos 2003 *Építőanyag-JSBCM***55** (3) 101 (in Hungarian) http://dx.doi.org/10.14382/epitoanyag-jsbcm.2003.18
[40] J Szépvölgyi and Z Károly 2004 *Építőanyag-JSBCM***56** (1) 12 (in Hungarian) http://dx.doi.org/10.14382/epitoanyag-jsbcm.2004.2
[41] J Csányi, L A Gömze and I Z Kövér 2004 *Építőanyag-JSBCM***56** (3) 103 (in Hungarian) http://dx.doi.org/10.14382/epitoanyag-jsbcm.2004.13
[42] M C Santoa, A R A Nagueira and I A Nogrega 2005 *J. Braz. Chem. Soc.***16** 188
[43] T Rádóczy and K Kovács 2006 *Építőanyag-JSBCM***58** (2) 34 (in Hungarian) http://dx.doi.org/10.14382/epitoanyag-jsbcm.2006.8
[44] B N Kim, K Hiraga, K Morita and H Yoshida 2007 *Scripta Mater.***57** (7) 607
[45] G Varga 2007 *Építőanyag-JSBCM***59** (1) 6 http://dx.doi.org/10.14382/epitoanyag-jsbcm.2007.2
[46] J Tamásné Csányi and L A Gömze 2008 *Építőanyag-JSBCM***60** (1) 15 http://dx.doi.org/10.14382/epitoanyag-jsbcm.2008.4
[47] J Wang, C Y Huang and Y C Wu 2009 *Ceramics International***35** 1467
[48] K Katahira and H Ohmori 2009 *Építőanyag-JSBCM***61** (4) 114 http://dx.doi.org/10.14382/epitoanyag-jsbcm.2009.21
[49] J G Chigvinadze, J V Acrivos, S M Ashimov, D Gulamova, T V Machaidze and D Uskenbaev 2009 *Építőanyag-JSBCM***61** (4) 104 http://dx.doi.org/10.14382/epitoanyag-jsbcm.2009.19
[50] J G Miranda-Hernández, E Rocha-Rangel and S D de la Torre 2010 *Építőanyag-JSBCM***62** (1) 2 http://dx.doi.org/10.14382/epitoanyag-jsbcm.2010.1
[51] S N Kulkov and M V Grigoriev 2010 *Építőanyag-JSBCM***62** (3) 66 http://dx.doi.org/10.14382/epitoanyag-jsbcm.2010.13
[52] R Naslain 2011 *IOP Conf. Ser.: Mater. Sci. Eng.***18** 082002 http://dx.doi.org/10.1088/1757-899X/18/8/082002
[53] O Khasanov, U Reichel, E Dvilis and AKhasanov 2011 *IOP Conf. Ser.: Mater. Sci. Eng.***18** 082004 http://dx.doi.org/10.1088/1757-899X/18/8/082004
[54] L Kozielski, M Adamczyk and J Erhart 2011 *IOP Conf. Ser.: Mater. Sci. Eng.***18** 092001 http://dx.doi.org/10.1088/1757-899X/18/9/092001
[55] V Reschke, A Laskowsky, M Kappa, K Wang, R K Bordia and M Scheffler 2011 *Építőanyag-JSBCM***63** (3-4) 57 http://dx.doi.org/10.14382/epitoanyag-jsbcm.2011.10
[56] F A Santos, M H F V Fernandes, E Davim, D G Pinatti, D R R Lazar and C Santos 2013 *IOP Conf. Ser.: Mater. Sci. Eng.***47** 012003 http://dx.doi.org/10.1088/1757-899X/47/1/012003
[57] E S Kalatur, A V Kozlova, S P Buyakova and S N Kulkov 2013 *IOP Conf. Ser.: Mater. Sci. Eng.***47** 012004 http://dx.doi.org/10.1088/1757-899X/47/1/012004
[58] H L Chu, C L Wang, H E Lee, Y YSie, R S Chen, W S Hwang and M C Wang 2013 *IOP Conf. Ser.: Mater. Sci. Eng.***47** 012005 http://dx.doi.org/10.1088/1757-899X/47/1/012005
[59] J H Pee, Y J Kim, C Y Kim, W S Cho and H T Kim 2013 *IOP Conf. Ser.: Mater. Sci. Eng.***47** 012011 http://dx.doi.org/10.1088/1757-899X/47/1/012011
[60] L Mlinárik and K Kopecskó 2013 *IOP Conf. Ser.: Mater. Sci. Eng.***47** 012014 http://dx.doi.org/10.1088/1757-899X/47/1/012014
[61] M Takato, M Kaneko, T Nishi, K Saito and F Uchikoba 2013 *IOP Conf. Ser.: Mater. Sci. Eng.***47** 012015 2013 *IOP Conf. Ser.: Mater. Sci. Eng.***47** 012015
[62] T Konegger, P Schneider, V Bauer, A Amsüss and A Liersch 2013 *IOP Conf. Ser.: Mater. Sci. Eng.***47** 012054 http://dx.doi.org/10.1088/1757-899X/47/1/012054
[63] C Yu, C Deng, W Yuan and H Zhu 2013 *IOP Conf. Ser.: Mater. Sci. Eng.***47** 012059 http://dx.doi.org/10.1088/1757-899X/47/1/012059

[64] J Li, W J Yuan, C J Deng and H X Zhu 2013 *IOP Conf. Ser.: Mater. Sci. Eng.***47** 012060 http://dx.doi.org/10.1088/1757-899X/47/1/012060
[65] D Horkavcova, M Cerny, L Sanda, P Novak, E Jablonska, Z Zlamalova-Cflova and A Helebrant 2016 *IOP Conf. Ser.: Mater. Sci. Eng.***123** 012002 http://dx.doi.org/10.1088/1757-899X/123/1/012002
[66] X Zhang, B Zhong, L Liu, X Huang, G Wen, Y Huang and J Bollmann 2016 *IOP Conf. Ser.: Mater. Sci. Eng.***123** 012051 http://dx.doi.org/10.1088/1757-899X/123/1/012051
[67] M Sobiesiak 2016 *IOP Conf. Ser.: Mater. Sci. Eng.***123** 012052 http://dx.doi.org/10.1088/1757-899X/123/1/012052
[68] S Hampshire, H K Park, D P Thompson and K H Jack 1978 *Nature***274** 880
[69] A S Hakeem, R M A Khan, M MAl-Malki, F Patel, A I Bakare, S Ali, S Hampshire and TLaoui 2014 *Advancesin Science and Technology***89** 63
[70] A Bachar, C Mercier, ATricoteaux, A Leriche, C Follet and S Hampshire 2016 *J. Eur. Ceram. Soc.***36** (12) 2869
[71] E S Kalatur, A V Kozlova, S P Buyakova and S N Kulkov 2013 *IOP Conf. Ser.: Mater. Sci. Eng.***47** 012004 http://dx.doi.org/10.1088/1757-899X/47/1/012004
[72] E Dedova, E Klevtsova and S Kulkov 2013 *IOP Conf. Ser.: Mater. Sci. Eng.***47** 012021 http://dx.doi.org/10.1088/1757-899X/47/1/012021
[73] S N Kulkov, E S Dedova, F Pedraza and J Erdélyi 2014 *Építőanyag-JSBCM***66** (2) 35 http://dx.doi.org/10.14382/epitoanyag-jsbcm.2014.7
[74] S N Kulkov, S P Buyakova, M Chatzinikolaidou and I Kocserha 2015 *Építőanyag-JSBCM***67** (4) 155 http://dx.doi.org/10.14382/epitoanyag-jsbcm.2015.26
[75] T V Kolmakoba, S P Buyakova and S N Kulkov 2016 *IOP Conf. Ser.: Mater. Sci. Eng.***123** 012005 http://dx.doi.org/10.1088/1757-899X/123/1/012005
[76] S P Buyakova, E S Kaltur, A S Buyakov and S S Kulkov 2016 *IOP Conf. Ser.: Mater. Sci. Eng.***123** 012040 http://dx.doi.org/10.1088/1757-899X/123/1/012040
[77] I Zhukov, S P Buyakova and S N Kulkov 2016 *Építőanyag-JSBCM***68** (3) 74 http://dx.doi.org/10.14382/epitoanyag-jsbcm.2016.13
[78] E V Zharikov, K S Zaramenskikh, N A Popova, P PFaikov, L D Iskhakova, M N Gerke, S V Kutrovskaya and D S Nogtev 2011 *Glass and Ceramics* **68** 80
[79] X Devaux, S Yu Tsareva, A N Kovalenko, K S Zaramenskih, E McRae and E V Zharikov 2012 *Physica E***44** 1028
[80] E V Zharikov, T T T Duong, P P Faikov, N A Popova and D M Sovyk 2016 *Inorganic Materials: Applied Research***7** 163
[81] J Liu, Z Fu, W Wang, J Zhang, H Wang, Y Wang, S Lee and K Niihara 2014 *J. Eur. Ceram. Soc.***34** 3095 http://dx.doi.org/10.1016/j.jeurceramsoc.2014.04.004
[82] S T Nguyen at al 2015 *J. Eur. Ceram. Soc.***35** (9) 2651
[83] S Shinoda, T Aoba, T Suzuki, T Nakayama, H Suematsu and K Niihara 2017 *Jpn. J. Appl. Phys.***56** 019201 https://doi.org/10.7567/JJAP.56.019201
[84] H B Cho, T Nakayama, H Suematsu, T Suzuki, W Jiang, K Niihara E Song, N S A Eom, S Kim and Y H Choa 2016 *Composites Science and Technology***129** 205 http://dx.doi.org/10.1016/j.compscitech.2016.04.033
[85] S G Chuklina, A I Pylinina, L I Podzorova, N A Mikhailina, and I I Mikhalenko 2016 *Russian Journal of PhysicalChemistry***90** (12) 2370
[86] A I Pylinina and I I Mikhalenko 2013 *Theoretical and Experimental Chemistry***49** (1) 65
[87] T Sadowski and A Neubrand 2004 *Int. J. Fract.***127** 135
[88] T Sadowski, M Boniecki and Z Librant 2007 *Int. J. Heat and Mass Transfer***50** 4461
[89] T Sadowski and S Samborski 2008 *Comp. Mater. Sci.***43** 75
[90] T Sadowski and P Golewski 2012 *Comp. Mater. Sci.***64** 285
[91] T Sadowski, M Kneé and P Golewski 2014 *Key Engineering Materials***601** 25
[92] V Burlayenko and T Sadowski 2014 *Meccanica***49** 2617

[93] T Sadowski and P Golewski 2015 *Archives of Metallurgy and Materials* **60** (4) 2737
[94] Y Wang, D A Lakho and D Yao 2015 *Épitőanyag-JSBCM* **67** (4) 132 http://dx.doi.org/10.14382/epitoanyag-jsbcm.2015.21
[95] C Tóth and Z Soós 2015 *Épitőanyag-JSBCM* **67** (4) 126 http://dx.doi.org/10.14382/epitoanyag-jsbcm.2015.20
[96] H Hoornahad, E A B Koenders and K van Breugel 2015 *Épitőanyag-JSBCM* **67** (4) 136 http://dx.doi.org/10.14382/epitoanyag-jsbcm.2015.22
[97] A Slobodov, A Uspenskiy, S Yavshits and A Mischenko 2015 *Épitőanyag-JSBCM* **67** (4) 164 http://dx.doi.org/10.14382/epitoanyag-jsbcm.2015.28
[98] R Bansil, J Hardcastle and M Constantino 2015 *Épitőanyag-JSBCM* **67** (4) 150 http://dx.doi.org/10.14382/epitoanyag-jsbcm.2015.25
[99] S N Kulkov, S P Buyakova and L A Gömze 2014 *Épitőanyag-JSBCM* **66** (1) 2 http://dx.doi.org/10.14382/epitoanyag-jsbcm.2014.1
[100] V V Promakhov, S P Buyakova, V Illavszky, S N Kulkov and L A Gömze 2014 *Épitőanyag-JSBCM* **66** (3) 81 http://dx.doi.org/10.14382/epitoanyag-jsbcm.2014.15
[101] L A Gömze 2003 *Épitőanyag-JSBCM* **55** (4) 133 (in Hungrian) http://dx.doi.org/10.14382/epitoanyag-jsbcm.2003.23
[102] L A Gömze 2004 *Épitőanyag-JSBCM* **56** (2) 46 (in Hungarian) http://dx.doi.org/10.14382/epitoanyag-jsbcm.2004.5
[103] L A Gömze and Á Kovács 2005 *Épitőanyag-JSBCM* **57** (2) 34 (in Hungarian) http://dx.doi.org/10.14382/epitoanyag-jsbcm.2005.7
[104] L A Gömze and L N Gömze 2010 *Épitőanyag-JSBCM* **62** (4) 98 http://dx.doi.org/10.14382/epitoanyag-jsbcm.2010.18
[105] Á Egész and L A Gömze 2013 *Épitőanyag-JSBCM* **65** (4) 107 http://dx.doi.org/10.14382/epitoanyag-jsbcm.2013.20
[106] L A Gömze, L N Gömze, S N Kulkov, I Shabalin, I Gotman, E Pedraza, G L Lecomte-Nana, T Mayorova, E Kurovics and A Hamza 2015 *Épitőanyag-JSBCM* **67** (4) 143 http://dx.doi.org/10.14382/epitoanyag-jsbcm.2015.24
[107] L A Gömze 2015 RHEOLOGY, Compilation of Scientific Papers I. *Published by IGREX Ltd, Hungary* pp 1-193 **ISBN 978-963-12-3088-8**
[108] D P H Hasselman, P F Becher and K S Mazdiyashni 1980 ZeitshriftWerkstoff-tech **11** 3 82
[109] L A Gömze and L N Gömze 2011 in Thesises of *Int. Conf. XIII Khariton's Readings* Extreme States of Substance (Sarov) Russia 160
[110] L A Gömze and L N Gömze 2013 *Materials Science Forum* **729** 43
[111] L A Gömze 2016 APPLIED MATERIALS SCIENCE I., Compilation of Selected Scientific Papers *Published by IGREX Ltd, Hungary* pp 1-189 **ISBN 978-963-12-6600-9**

IOP Conf. Series: Materials Science and Engineering **175** (2017) 012002 doi:10.1088/1757-899X/175/1/012002

Scientific basis for creation of construction materials based on titanium and alumina minerals

O Kotova, A Shmakova and A Ponaryadov

Laboratory of Mineral Raw Technology, Institute of Geology Komi SC UB RAS, 167982 Syktyvkar, Russia

E-mail: kotova@geo.komisc.ru

Abstract. Currently there is a need in affordable and accessible materials with specific physical and chemical properties. Al- and Ti-minerals are excellent test objects for correlation between structure and physical properties of mineral. For example, corundum and ilmenite are related to the same structural type (rhombohedral lattice R3) and possess various physical properties. With the help of modern equipment we studied titanium- and aluminum-containing concentrates of natural raw and also products of various kinds of influences on them, which showed signs of nanostructuring. We observed methods of directional change of physical and chemical characteristics of Al- and Ti- minerals and creation of new commercial products.

1. Introduction

Al- and Ti-minerals (products of natural and technogenic processes) are considered in the applied mineralogy as geomaterials for specific technological tasks.

Aluminum and titanium, among other non-ferrous metals (copper, magnesium, lead, zinc, tin), have a great industrial importance. The cost of non-ferrous metals, as compared with black metals, is high; besides the world market has greatly increased demand for technological and environmental parameters of commercial products. All this stipulates the necessity, firstly, to obtain a deeper knowledge about Al and Ti minerals as carriers of chemical elements and useful properties, about their actual crystal structure; secondly, to understand the mechanisms of changes of the parameters of the minerals under the influence of natural and anthropogenic impacts; thirdly, to obtain new products competitive in the world market [1–4].

The main purpose of this work is the development of scientific base for the creation of construction materials based on titanium and aluminum minerals.

2. Objects and Experimental procedure

The objects of study are Al- and Ti-minerals of natural raw (bauxites, ilmenite-leucoxene ores) and their products (including those with signs of nanostructuring).

The main method of monitoring of the composition and properties of titanium ores and associated minerals, as well as the products of technological processes is a quantitative phase analysis.

The shape and size of mineral were obtained by TEM JSM-6400. Other modern methods were also used, for example, X-ray fluorescence techniques (XRF), optical-mineralogical methods (stereomicroscope MBS-10, polarizing microscope POLAM L-311), etc.

Applying X-ray small-angle scattering method to bauxites we show that the rocks include fine particles with linear size equal to 40 nm. To prove the data of small-angle scattering we tested the samples of hematite-boehmite bauxites from the Vezhayu-Vorykvinskoe deposit with the help of electron microscopy. The results of small-angle scattering comply with electron microscopy results (Figure 1).

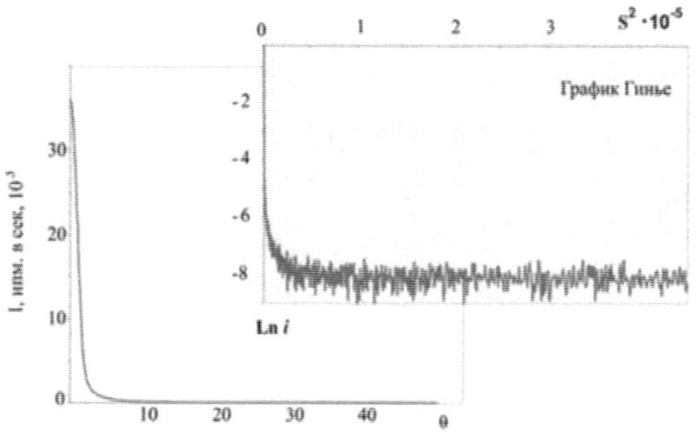

Figure 1. Dependence intensity from angle and SEM secondary electron image of aggregated particles in hematite-boehmite bauxites.

3. Results and Discussion

The development of geomaterial industry should be supported by appropriate resource base. Aluminum ores (bauxites especially) are deficit Russian mineral resource. The main reserves of bauxites in Russia are related to Iksinskoe and Kalinskoe deposits of the Northern Urals, which cannot meet the needs of the domestic industry. Vezhayu-Vorykvinskoe deposit in Timan province (Komi Republic) is promising. Russian aluminum plants operate mainly from imported and tolling raw [4, 5].

The mineral resources of titanium ores are concentrated mainly in the primary deposits of the Northwest (Kola Peninsula, Murmansk region) and Ural (Chelyabinsk region) federal districts and complex (with zirconium) placers from Northern (Komi Republic), Central, Volga and Ural federal districts. According to the content of titanium and technological properties these deposits are inferior to foreign ones.

The changeability of the ores and their further behavior at processing and enrichment is directly connected with their composition and genesis. The stages of ore transformation are often hard to determine. It is possible to get information from the study of such objects but with more pronounced transformations. We studied titanium minerals and stages of their transformation in exogenous conditions of modern coastal-marine placers (on the example of Australia) due to problems of study of buried paleoplacers (e.g. Pizhma, Russia). The complex of mineralogical and analytical methods was used to study the mineral composition, morphostructural characteristics and alteration degree of minerals from the modern coastal-marine placer Stradbroke (Eastern Australia) and Pizhemskoe paleoplacer from Middle Timan (Russia) [1].

It was established that the ore sands of Stradbroke ilmenite are predominated by ilmenite, and Pizhemskoe titanium deposit – by leucoxene (Figure 2). Paleoplacers of Pizhma represent more rebellious ores due to predominant titanium (and not only) structures, alongside with the main ore components there are other metals, which occurrence form can be different (isomorphic impurity, independent mineral phases).

The physical and chemical characteristics of Ti- and Al-minerals often do not meet the demands of consumers. The methods of directional change of these characteristics under various types of energy impacts are widely used and developed. For example, the most important obstacle for technological processes is the enrichment of bauxites by nonmagnetic ferrous oxides - goethite-hydrogoethite and hematite-hydrohematite. There is an assumption that if goethite and hematite are transformed to the ferromagnetic phases in the bauxites, then it could provide separation and selective recovery of aluminum and ferrous components, which in turn will improve the quality of aluminum concentrate and give the possibility to obtain additional - ferrous – industrial product. This is confirmed by the experience of transformation of nonmagnetic oxyhydroxides and ferrous oxides into magnetite to improve the washability of the iron ores. At present we have convincing evidences of the efficiency of irradiation of ferrous nonmagnetic materials by high-energy electrons, which results in transformation of non-magnetic and low magnetic minerals into ferromagnetic minerals [6].

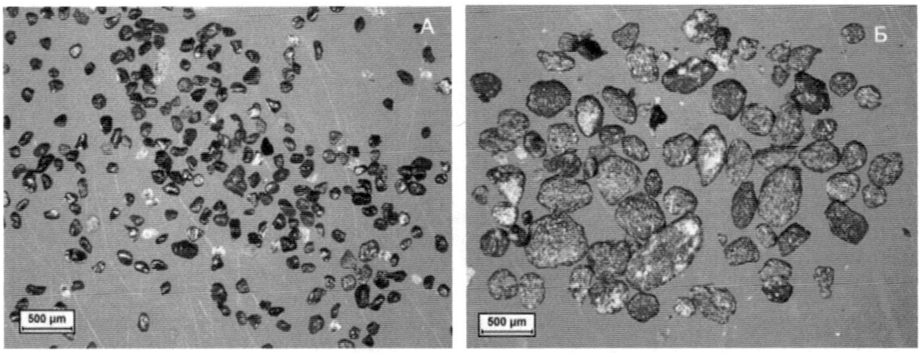

Figure 2. Titanium minerals of modern coastal-marine placer in Stradbroke Island, Eastern Australia (A) and Pizhma paleoplacer in Middle Timan (B).

Using a modern instrument base for mineralogical studies, we at the first time examined the effect of thermal and radiation-thermal influence on Middle Timan ferriferous bauxites (Figure 3) to improve their technological properties.

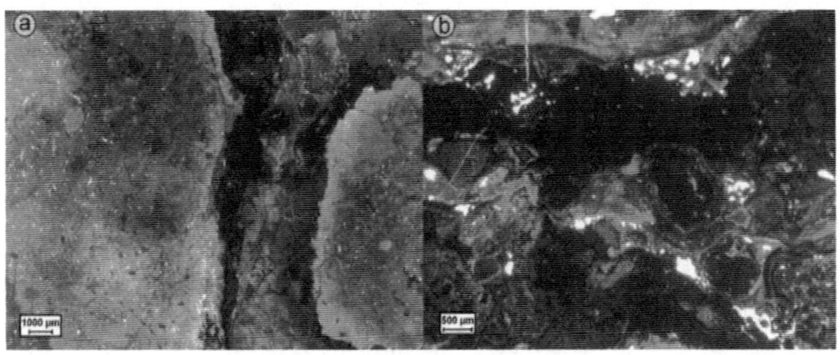

Figure 3. Image of thin sections of bauxites in plane polarized light (a) Clastic structure of boehmite-diaspore-chamosite-anatase-rutile aggregates. (b) Hematite-goethite structures in the form of concretions

We showed that after heating up to 600°C and 60 minutes-soaking a complete decomposition of goethite and boehmite occurred with formation of hematite and γ-Al_2O_3 spinelide accordingly. The magnetic properties of the bauxites reduced and the yield of magnetic fraction was lower. The radiation-thermal exposure results in forming ferromagnetic phases (Figure 4), which facilitates the separation of alumina and ferriferous components of the bauxites, and also allows solving the problem of selective extraction of phases of rare and rare earth elements. To realize this scenario it is necessary to disintegrate bauxite substance more finely.

It should be noted that titanium and aluminum ores are raw for compositional materials. In the article [3] were studied the dependence of physical properties on the structure of composite materials prepared from natural ilmenite, and alumina ceramics (Figure 5) with different content of impurities. X-ray analysis showed that in both cases the corundum structural type was predominant. The observed effect of the strong influence of impurities on the radiophysical characteristics allows synthesizing both radiotransparent and radioabsorbing materials (Figure 6) with the same crystal structure

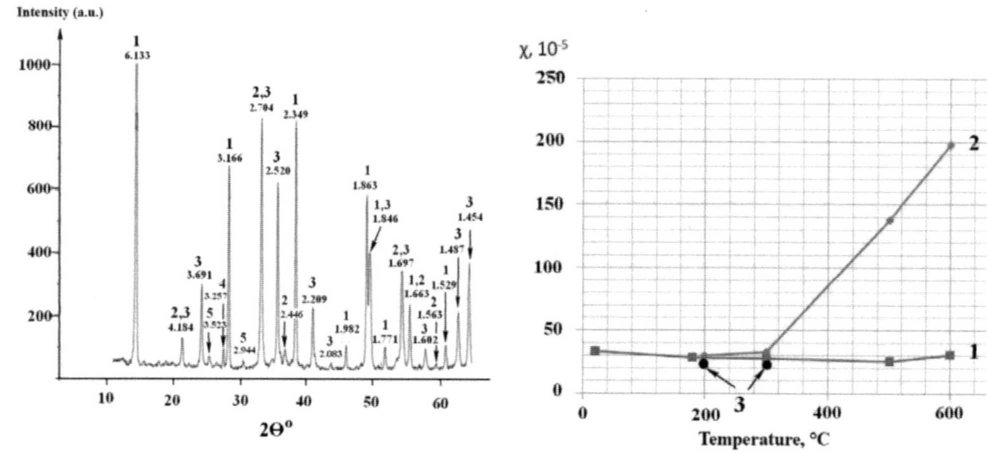

Figure 4. XRD pattern of radiation-thermal modified bauxite and thermal dependence of volumetric magnetic susceptibility revealed in different experiments: 1 — heating without irradiation, 2 — heating with irradiation, 3 — magnetic susceptibility values in the irradiation experiments after 20 min exposure at 200 and 300 °C.

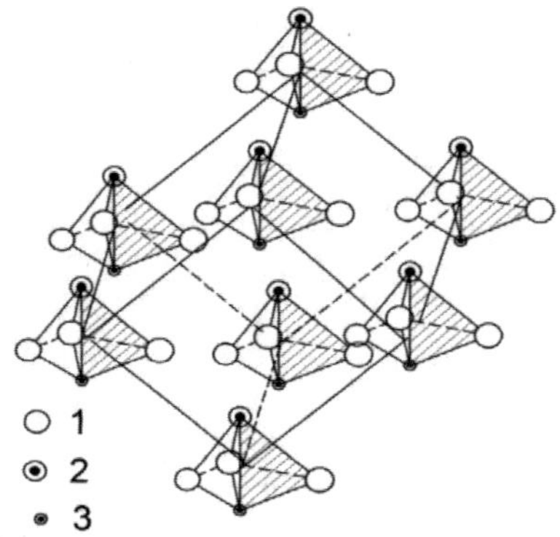

Figure 5. The unit cell of corundum (trigonal rhombohedral crystal system, space group R3c to (1 - oxygen atom, 2 - Al atom in corundum and Fe atom in ilmenite, 3 - Al (Cr, Fe) atom in corundum or Ti in ilmenite [3].

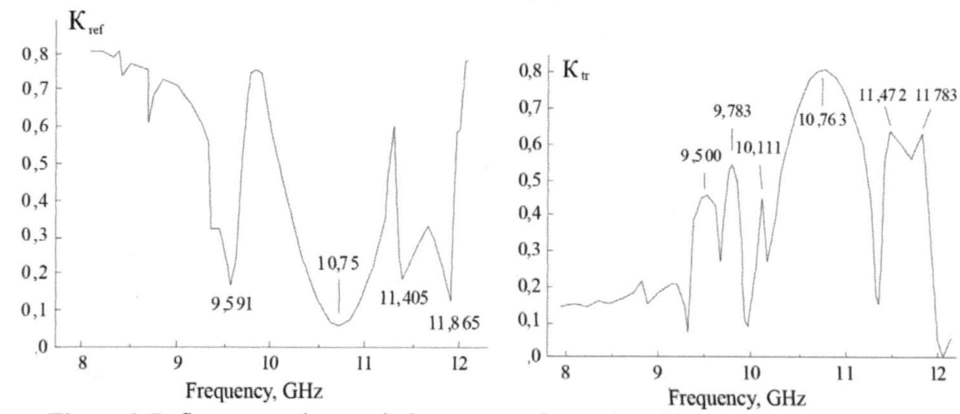

Figure 6. Reflectance and transmission spectra of ceramics with corundum structure [3].

RF properties of ceramic and composite materials based on corundum and ilmenite are characterized by the simultaneous expression of both scattering and absorption. The fine structure of reflectance spectra is characteristic for solid solutions with a complex chemical composition. Therefore to obtain radiotransparent materials it is necessary to minimize the amount of impurities. During the synthesis of radioabsorbing materials, on the other hand, it is necessary to add magnetically active impurities forming a dielectric ferrite-like structure. The materials, synthesized from natural ilmenite, can be used as EMR (electromagnetic radiation) absorbers, and on the basis of corundum ceramics containing active impurities it is possible to produce SHF (super high frequency) filters, which are EMI transmissive in the narrow frequency range.

New applications of titanium ore based on its mineralogical and processing characteristics have a real commercial interest: the list of commodity products is expanding (and consequently the prospects for production), energy costs are reducing, environmental risks are decreasing, efficiency of sustainable development of region is increasing. One of the promising directions of solution of this problem can be the synthesis of titanium dioxide nanotubes from mineral raw materials (for example from titanium ore of Pizhemskoe deposit) [7].

In recent decades one-dimensional nanostructure materials, derived from titanium dioxide, have been widely used in photocatalysis, as a base for catalysts, implants, reagents enhancing blood coagulability, batteries etc. [2, 8].

Due to a high cost of commercially available TiO_2 nanoparticles their synthesis from cheap natural raw materials by hydrothermal method becomes very important.

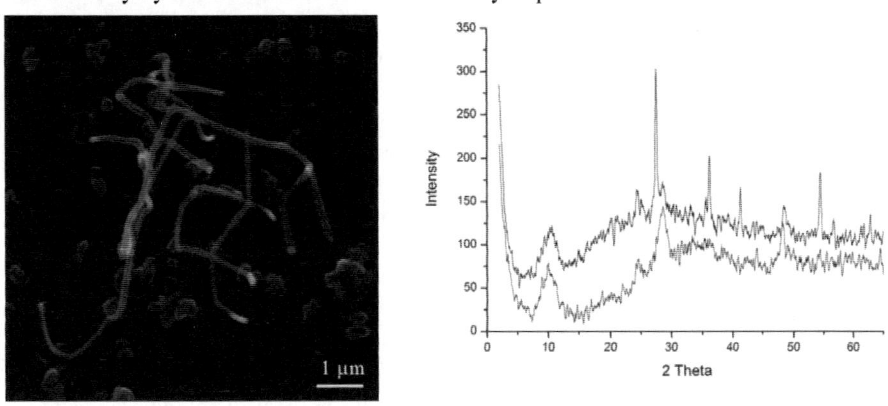

Figure 7. SEM image and XRD patterns for TiO_2 nanotubest obtained from natural leucoxene ore (Pizhemskoe deposit)

The TiO_2 nanotubes (figure 7) were obtained using simple hydrothermal methodin Laboratory of mineral raw in Institute of Geology Komi SC UB RAS. An inexpensive natural raw material – non-magnetic fraction of gravity concentrate of titanium ore of Pizhemskoe deposit – was used as a starting material. The synthesized titanium dioxide nanotubes have outer diameter 70–100 nm and length up to 4500 nm. The synthesized TiO_2 nanotubes have a large surface area that results in a good sorbent material.

4. Conclusions

The main (strategic) task of mineralogical geomaterial science is to combine the efforts of specialists in related scientific areas (geologists, mineralogists, crystallographists, chemists, physicists, engineers and others) with the aim of

- development of theoretical and experimental bases of the development of mineral resources,

- determination of fundamental laws of formation and properties of natural and technogenic structures (including nanoscale phases), their kinetic and dynamic features in technological processes,
- development of methods for directed change and decomposition of geomaterials in the process of technical functioning
- synthesis of new materials and creation of science-based technologies.

5. Acknowledgement

The authors express gratitude to the common use center Geonauka for their help in analytical work. This work was supported by the Program of UB RAS (project 15-18-5-33).

References

[1] Kotova O., Ozhogina E., Ponaryadov A., Golubeva I. Titanium minerals for new materials // *IOP Conf. Series: Materials Science and Engineering* **123** (2016) 012025.

[2] Kotova O. B., Ponaryadov A. V., Valyon J. Nanostructured mineral surface: sorption properties // *Vestnik of Institute of Geology of Komi SC, UB RAS*, 2007, No. **10**, pp. 8-10. (in Russian)

[3] Goldin B. A., Ryabkov Yu. I., Sekushin N. A., Nazarova L. Yu. Problems of radiolucent and radio absorption of ceramic and compositional materials with corundum structure (comparative analysis) // *Izvestia Komi naucnogo tsentra*, V. **3**, 2010. Pp. 66–68. (In Russian)

[4] Kotova O. B., Vakhrushev A. V. Timan Bauxites: mineralogical and technological characteristics // *Vestnik of Institute of Geology*, Komi Science Centre, UB RAS 2011, V. 195. No. **3**, pp. 12-16.

[5] Kotova O. B., Razmyslov Y. N., Rostovtsev V. I., Silaev V. I. Radioactive and thermical modification of ferrous bauxite during processing // *Obogascchenie rud,* 2016. № **4**. C. 16–22. (In Russian)

[6] Razmyslov Y. N. Energy-driven phase changes of bauxites // *Vestnik of Institute of Geology of Komi SC*, UB RAS, 2016. № **6**. C. 33–34.

[7] Kotova O. B., Ponaryadov A. V., Gömze L. A. Hydrothermal synthesis of TiO2 nanotubes from concentrate of titanium ore Pizhemskoe deposit (Russia) // *Vestnik IG Komi SC* UB RAS, 2016. № **1**. C. 34–36.

[8] Ponaryadov A. V., Kotova O. B. Synthesis and properties of nano-dispersed structures of titanium minerals on the example of anatase. // *Reports of the Academy of Sciences*, 2009, Vol 425, No. **5**, pp. 1-4. (in Russian)

IOP Conf. Series: Materials Science and Engineering **175** (2017) 012003 doi:10.1088/1757-899X/175/1/012003

Engineering estimation of time-dependent deformation characteristics as bending moment relaxation and released unfolding motion of creased paperboard

Sh Nagasawa

Department of Mechanical Engineering, Nagaoka University of Technology, 940-2188 Nagaoka, Japan

E-mail: snaga@mech.nagaokaut.ac.jp

Abstract. Paperboards are recognized to be important raw materials for packaging industry due to their advantages such as high strength-to-weight ratio, recyclability. Regarding the development of advanced packaging materials and the requirement of smart formed products, a study of sheet's response behaviour is necessary for expanding the advanced converting industry. After introducing a couple of past research works concerned crease technologies, a fundamental mechanisms of crease deformation is reviewed using the scoring depth and the folding angle of a paperboard. Since one of important forming characteristics is a time-dependent stress relaxation or time-delayed strain during a fold/unfold process, the author's experimental approaches for estimating a short term (less than 10 seconds) dynamic deformation behaviour of creased paperboard are discussed.

1. Introduction

Paperboards are recognized to be important raw materials for packaging industry due to their advantages such as a high strength-to-weight ratio, printability, recyclability. The paperboards can be converted and used as carton boxes. In order to convert a raw paperboard into packaging containers, the boards are firstly printed with color inks. Next, they are subjected to cutting and creasing processes. The objectivity of these two processes is to convert such printed boards into designed blank forms. After that, the blanks are folded and glued to obtain the containers [1].

Since coated and printed paperboards are kinds of laminated composite materials which are made of multiple plies and composed of numerous fibers, they normally exhibit complicated mechanical properties and forming behaviors. In addition, the successful forming of the paperboards tends to be more difficult than that of other monolithic resin or metallic sheet materials. Regarding the continuous development of advanced packaging materials and the requirement of smart formed products, understanding of sheet's response behaviors and their forming characteristics are necessary for the converting industry. They contribute to design suitable forming conditions and to develop smart forming technologies. A few reports for in-plane elongation of paperboard during indentation of creaser were shown by Hallady and Ulm [2] .Actual creasing range was investigated as the relationship between the depth and width by Hine [3]. Carlsson et al. [4] explained about the bending behavior of creased part as composed of two parallel beams. Beex et al., Nygards et al. and Sudo et al. [5-7] reported a sort of full model based on the through-thickness discretization of creased paperboard.

The author's group investigated various forming characteristics of a white-coated paperboard

IOP Conf. Series: Materials Science and Engineering **175** (2017) 012003 doi:10.1088/1757-899X/175/1/012003

subjected to a sort of cutting and/or creasing process. Nagasawa et al. [8-9] reported about the quasi-static folding stiffness with respect to the indentation depth of a creaser and also discussed with the crease deviation effect on the folding of creased part. However, since the paperboard is a sort of composite material made of laminated thin papers and numerous fibers, its crease-based forming or shaping behaviors are fairly complicated.

In this work, a couple of current research works concerned board forming processes are reviewed. One of important forming characteristics is a bending moment (which is composed of in-plane bending stress and out-of-plane delamination) relaxation and/or released (creep) behavior during a unfolding process. Before reviewing the time-dependent behavior of bending moment resistance and unfolded deformation, the in-plane loading behavior of paperboard is briefly reviewed.

Regarding the in-plane tensile deformation of paper, the stress relaxation by a specified forced displacement was known by several reports [10-11] and the stress was often characterized with the logarithmic reduction of elapsed time. When the elapsed time of relaxation process is sufficiently taken as long, it is known that the relaxation response tends to be non-linear [10]. Generally speaking, a generalized Maxwell model can be considered to describe the non-linear relaxation. However, when the response is primary composed of the logarithmic characteristic and a bit offset from the main part, a modified Maxwell model, called as KWW (Kohlrausch-Williums-Watts) rule, is analyzed [12]. When considering a package box making, the scoring and folding of paperboard is empirically processed in a short time. The folding duration seems to be less than a few seconds. Therefore, we may focus here on the working time as shorter than 10seconds. Apart from the in-plane uni-axial tensile deformation, the folding of creased part is complicated due to the structure of multiple plies. Recent author's approach for estimating the time-dependent relaxation of bending moment resistance and released unfolded deformation of creased part is synthetically discussed [13-15].

2. Quasi-static deformation mechanism of crease of paperboard

Figure 1 illustrates a stripped carton development by a punching process and its box form. A stable folding of paperboard is enable by keeping a smart profile of scored crease line and its inner bulge. In this review paper, the crease deformation characteristics were investigated by varying the mechanical conditions of die set.

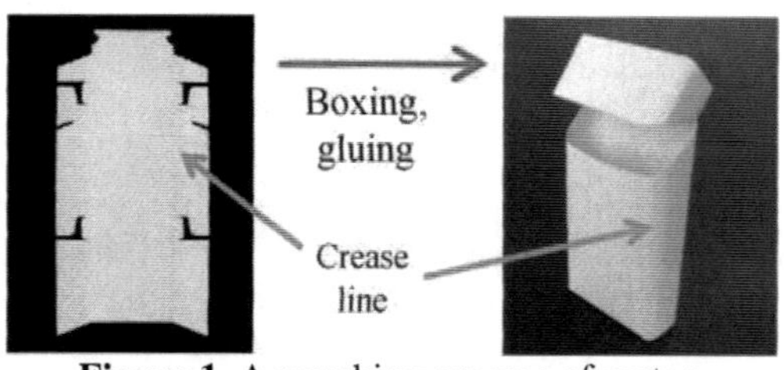

Figure 1. A punching process of carton development and folding on crease lines

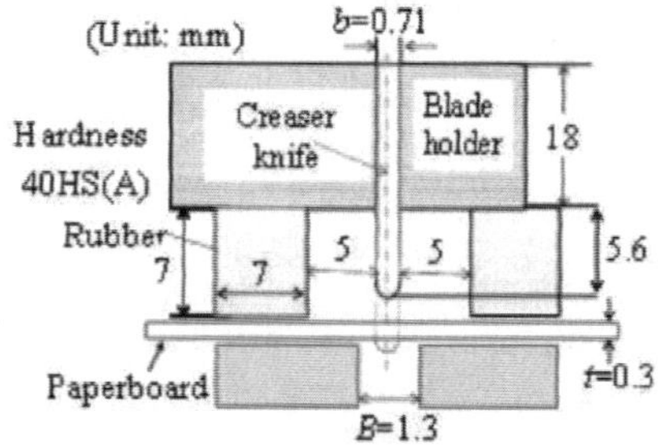

Figure 2. Layout of out-of-plane scoring by creaser

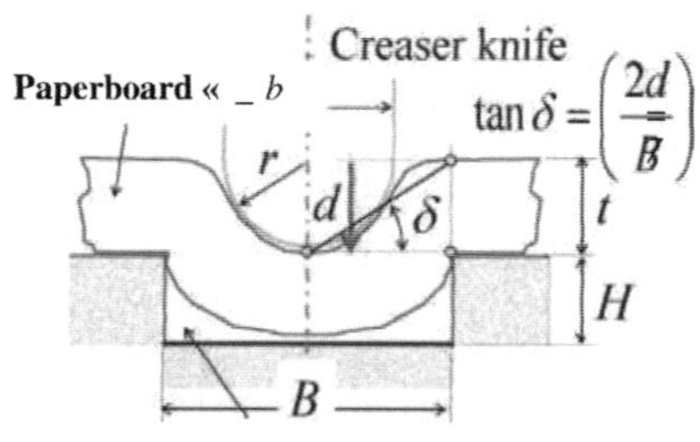

Figure 3. Schematic of scoring state and parameters

Figure 2 illustrates experimental apparatus in making the test pieces of paperboard scored by a round-edge creaser with radius of r=0.355 mm and the rubber fixture of shore hardness 40A. Here, using the paperboard thickness *t* and the thickness of creasing rule b, the groove width *B* was also empirically chosen as 2t+b = 1.3 mm. Figure 3 shows a detail of scored zone of paperboard when the creaser moves downward with the indentation depth *d.* Since the paperboard consists of multiple-plies, the in-plane sliding and de-lamination occur at the scored zone. The normalized indentation depth *y*= tan£= 2d/B were chosen as 0.0—1.0, while the feed velocity was chosen as F=0.0167 mm's"^{1}. The creaser direction angle *$* was chosen as 90° with respect to the machine direction (MD).

The test pieces were prepared as 10 pieces of rectangle-formed white-coated paperboard (the basis weight *p* =228~237 g-m^{-2}), which had a width of 15 mm, length of 60 mm, and thickness of *t*= 0.3 (0.297~0.303) mm, for each condition of chosen normalized indentation depth *y* The in-plane tensile test properties in MD, based on JIS-P-8113 (gauge length: 180 mm, width: 15 mm, feed velocity: F=0.33 mm-s^{-1}), were shown in Table 1. All the paperboards were kept in a room which had the temperature of 296 K and the humidity of 50 %RH.

Table 1. In-plane tensile properties of white-coated paperboard in machine direction

Ultimate tensile strength	σ_B	MPa	41.1 (40.2~42.7)
Proof stress at 0.2% strain	σ_y	MPa	11.1 (11.4—10.9)
Breaking true strain	ε_B	%	1.71 (1.81—1.62)
Young's modulus	*E*	GPa	5.72 (5.91—5.53)

Figure 4 shows a general view of experimental apparatus CST-J-1 [9] for repeated folding motion test and a schematic diagram of folding process at the first round fold.

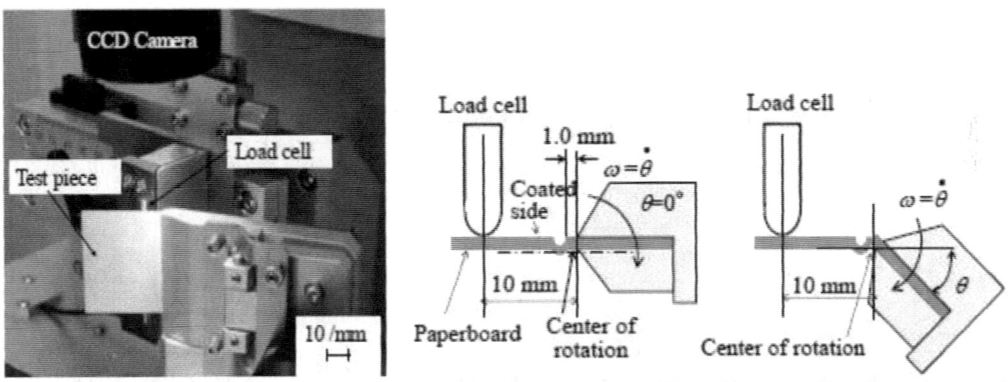

(a) General view (b) Conception of bending test (loading process)

Figure 4. Bending test apparatus

Figure 5 shows examples of relationship between a folding angle *d* and a bending line moment (resistance for the unit width) *M* when varying the tracking (maximum) angle *0,* the normalized indentation depth *y*in the cases of fold repetition n=2 and the rotation velocity w=0.2 rps. When y> 0.6, the overshoot of *M* disappeared [8, 9]. The transition of this overshoot disappearance was illustrated in Fig.6 in order to exaggerate the transition of creasing process. At the peak position of overshoot, since a certain level of long-parallel de-laminations occur in the scored zone, the in-plane bending stiffness is remarkably decreased. When the folding angle *d*passes through the peak point of overshoot, the inner bulging is promoted but the deformation of de-laminated zone is restricted with the immobile nodes.

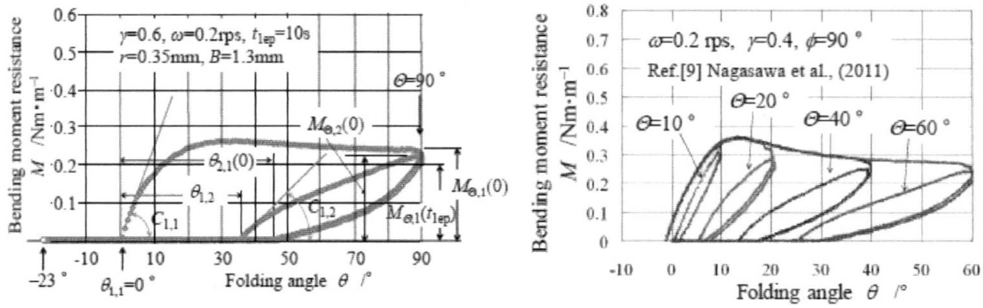

Figure 5. Relationship between bending line moment and rotation angle in case of second round folding: n=2. (a) Left: Analysis parameters on folding resistance diagram (0=90°, γ=0.6), (b) Right: Examples of bending moment responses by varying the tracking angles (0=10~60°, γ=0.4)

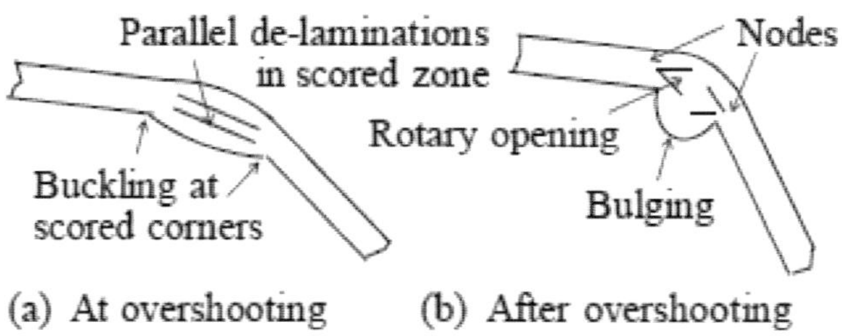

Figure 6. Transition model of creasing process

Figure 7 shows the first term gradient $C_{1,1} = \partial M|\partial\theta|_{\theta-\theta 1,1}$with respect to *y*, while Fig. 8 shows the initial released angle $\theta_{2,1(0)}$ with the tracking angle θ when varying the normalized indentation depth *y* at the first fold repetition *n=1* [9]. Seeing Fig.7, the variation tendency of gradient $C_{1,1}$ with *y* is obviously changed around *y* =0.6. This seems to be related to the disappearance of overshoot of bending moment. In Fig.8, the angle difference of $\theta - \theta_{2,1(0)}$ is a sort of spring back and $\theta_{2,1(0)}$ is the residual deformation, tentatively. This value $\theta_{2,1(0)}$ was roughly estimated as a half of tracking angle θ when varying θ and y Figure 9 shows a representative example of the relationship between the folding rotation angle *d* (up to the tracking angle 0=90° in 15 times) and the bending moment *M* when choosing the nominal shear strain *y* = 0.3, 0.8. Since the paperboard has viscous elasticity and creep characteristics during the folding process, the rotation velocity *w* of the fixture was normally set to ω = 0.2 rps and the elapsed time for returning back until the next folding (release time) was set to t_{2ep}, which was defined as the duration starting from a releasing position $\theta_{2,n}$ until reaching a restarting position $\theta_{1,n+1}$ Here, *n* =1,2,3...15 is a dummy suffix used to count the number of folding repetitions. In every folding round, the stopping time before returning back (holding time) was set to t_{1ep}= 10s at a tracking position of θ. The relationship between the first term gradient of bending moment $C_{1,n}$ and the number of folding repetitions *n* is shown in Fig.10, while Fig.11 shows the bending moment $M_{\Theta\text{-}90°}$ at t_{1ep}= 0s when choosing y=0, 0.3, 0.8. It is revealed that $C_{1,n}$ and $M_{\Theta\text{-}90°}$, tend to be a little decreased but almost stable with respect to the folding number n, when watching the range of 5<n<i5.

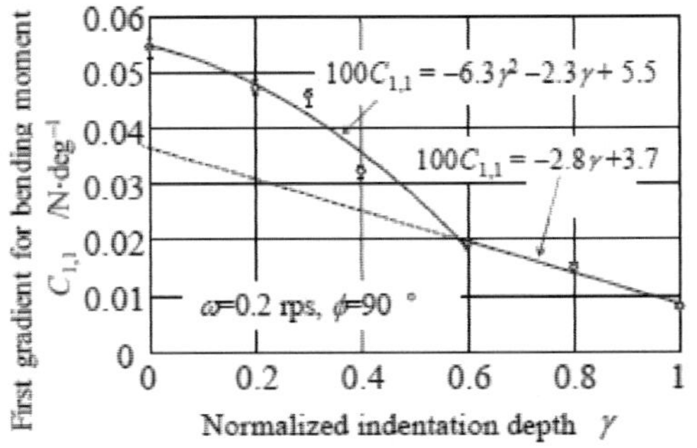

Figure7. Dependency of first gradient of bending line moment on normalized indentation depth

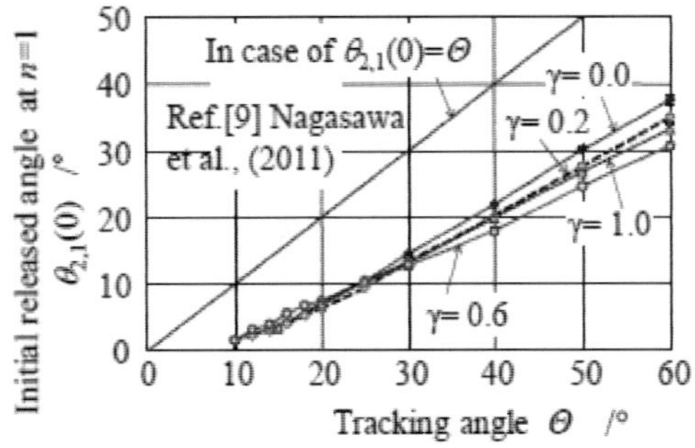

Figure 8. Dependency of initial released angle on tracking angle

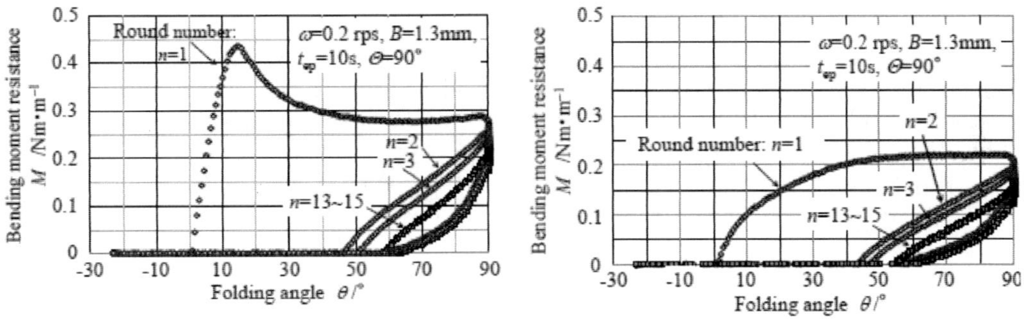

(a) In case of γ=0.3, n=1~3, 13~15

(b) In case of γ=0.8, n=1~3, 13~15

Figure 9. Response of bending moment with respect to folding angle (Ref.[14] Nagasawa et al. (2015))

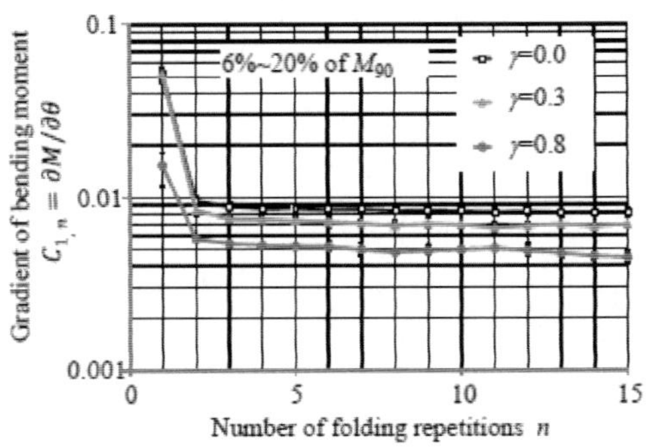

Figure 10. Effect of folding repetitions n on first term gradient moment of bending moment $C_{1,n}$

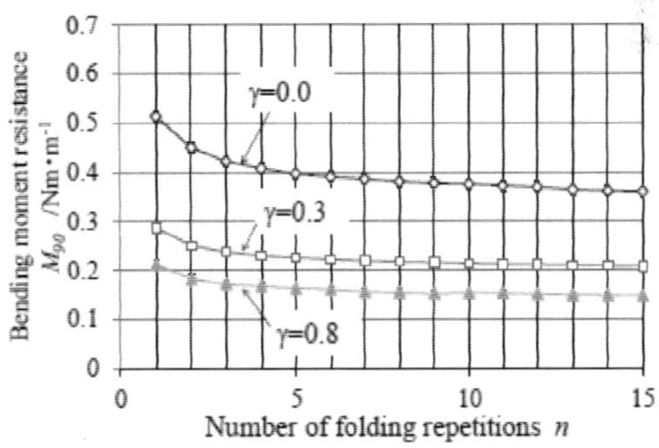

Figure 11. Effect of folding number n on bending $M00)$ at tracking position 90°

3. Relaxation of bending moment resistance at tracking position

Observing the time-dependent response of bending moment resistance M_Θ at the tracking position Θ in Fig.5 and Fig.9, the relaxation response of M_Θ with respect to the elapsed (holding) time t_{1ep} was revealed as shown in Fig.12. In order to characterize this relaxation, a linear approximation was introduced with the logarithmic term $\ln(t_{1ep})$ using Eq.(1). Two relaxation coefficients a_1, a_0 are introduced by Eq.(1). Here, $a_0$is defined as $M_\Theta(1)$ (at t_{1ep}=1s).

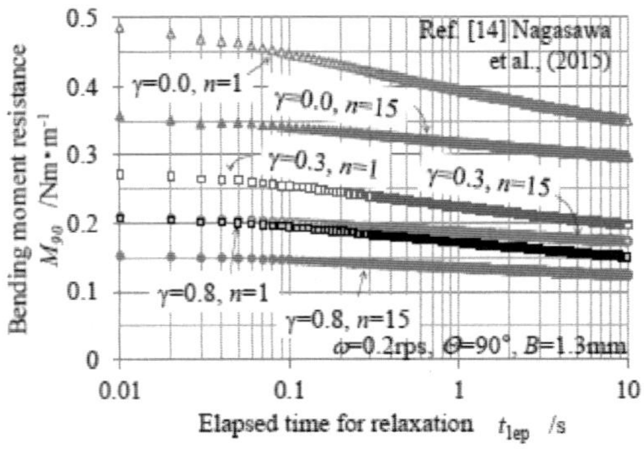

Figure 12. Relaxation of bending moment resistance with elapsed (holding) time in case of Θ =90°

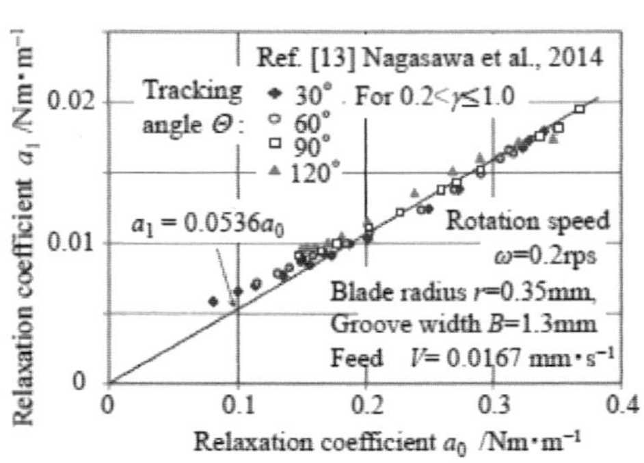

Figure 13. Relationship between coefficients a_1 and a_0 for 0.2<γ<1.0 and 30°< Θ <120° (η=1)

Figure 13 shows a result of scatter diagram of a_o, a_1 for 0.2<γ<1.0 and 30°<Θ<120° when the folding repetition was $n=1$ and the rotation velocity was ω=0.2rps. Here, the rubber fixture (shown in Fig.2) was not considered (without rubber fixture). The correlation between a_1 and a_o was arranged by using a linear approximation with Eq.(2). The exponential coefficient p_1 was estimated as 0.0536 in the case of Fig. 13 by using the least squared method. When considering the rubber fixture, the coefficients a_0, a_1 were decreased with 20~30% of non-rubber condition. However, since the values of a_0, a_1 were varied in a similar tendency with each other, the value of p_1 was 0.056. Namely, the exponent coefficient of relaxation pi is not remarkably varied by using the rubber fixture.

$$M_\Theta = -a_1 \ln (t_{1ep}) + a_0 \quad (1)$$

$$M_\Theta / a_0 = 1 + \ln(t_{1ep})^{-p_1},\ p_1=a_1/a_0 \quad (2)$$

In the case of multiple folding repetition, when considering n=i~i5, the relaxation coefficients a_0, a_1 were estimated with Eq. (2) from Fig.12, and its scatter diagram was shown in Fig.14. Here, the scoring condition γ=0.0, 0.3, 0.8 were chosen for ω=0.2rps, 6=90° and the holding time t_{1ep} was chosen as 10s while the released time t_{2ep} (the n-th round release-elapsed time until the next folding, duration starting from a released position $\Theta_{2,n}$ until reaching a restart position $\Theta_{1,n+1)}$ was chosen as 0s for every folding. The value of p_1 (=a_1/a_0) was estimated for the folding number n. This result was shown in Fig.15.
0.06

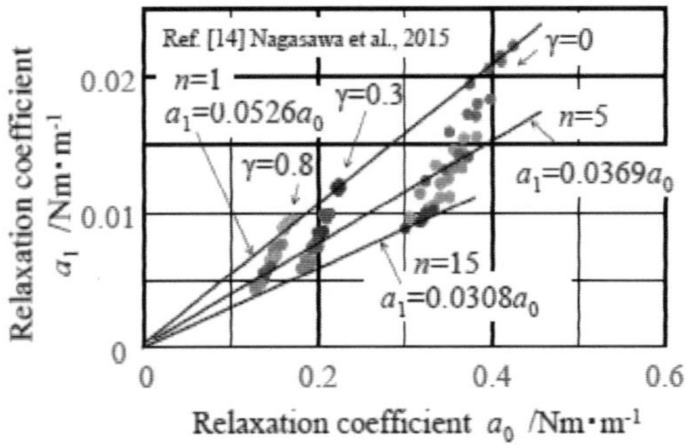

Figure 14. Relaxation coefficients a_1 and a_0 with respect to γ=0, 0.3, 0.8 for n=1~15

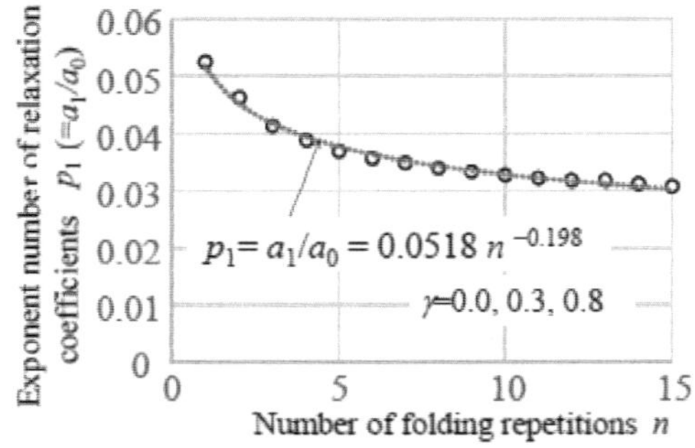

Figure 15. Variance of exponent coefficient p_1 with number of folding repetitions n

Regarding the effect of rotary velocity ω on the bending moment resistance at the tracking position, it is known that the exponential coefficient $p_1$tends to increase for ω<0.1 rps, while the value of p_1 is almost constant for $0.2<\omega<0.6$ rps. Namely, the relaxation of bending moment resistance basically depends on the preparation time of specified folding angle (tracking angle).

4. Released unfolding motion

When choosing the released time t_{2ep} as a positive value in Fig.5, the angle of folded paperboard tends to be decreased (unfolded). This is a sort of creep deformation and residual stress released state under an unloading condition. This behaviour was investigated by the use of the released angle $\Theta_{2,1}$ captured by a video camera. Figure 16 illustrates the released (unloading) process from the tracking position when the folding repetition number is n=1. The initial released angle is defined as $\Theta_{2,1}(t_{2ep}=0)$ when the reaction force of load cell becomes zero. From the previous research of relaxation at the tracking position, the time-dependent deformation of creased part seems to be characterized by the preparation duration or the rotation velocity w and the holding time t_{1ep}. Therefore, in this work, the relationship between $\Theta_{2,1}$ and t_{2ep} was investigated for ω = 0.02~0.6 rps when γ=0.6 and t_{1ep}=0s.

Figure 17 shows the initial released angle $\Theta_{2,1}$ *(0)* and Fig.18 shows the time-dependent released angle $\Theta_{2,1}$ (t_{2ep}) with respect to ω [15].

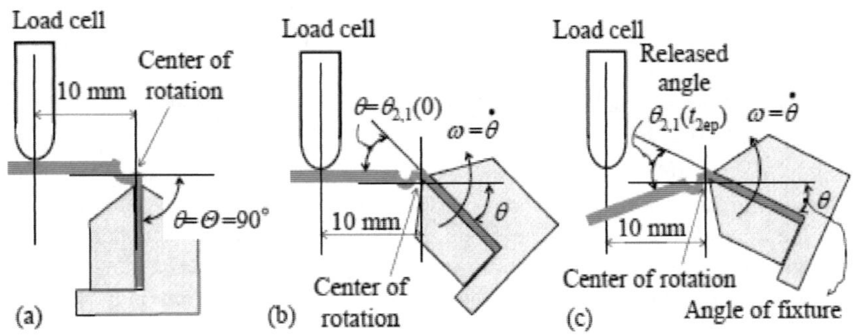

Figure 16. Schematics of unloading and relationship between released angle and fixture angle (a) State of tracking angle of 90°; (b) State of initial released angle t_{2e}=0) when reaction force of load cell becomes zero; (c) In case of unfolded state when detaching the load cell (for t_{2ep}>0).

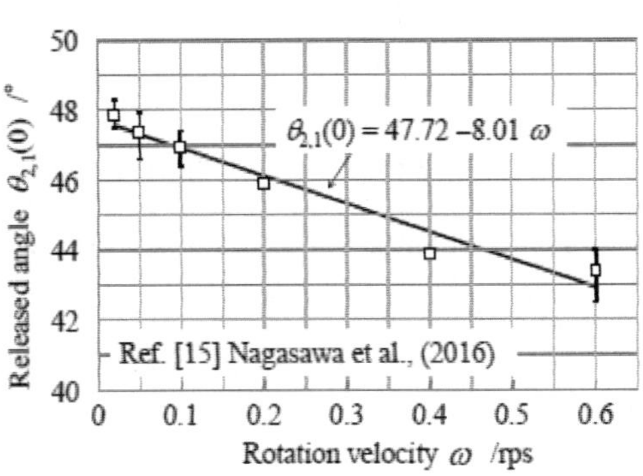

Figure 17. Dependency of initial released angle on folding velocity in case of γ=0.6, Θ=90°

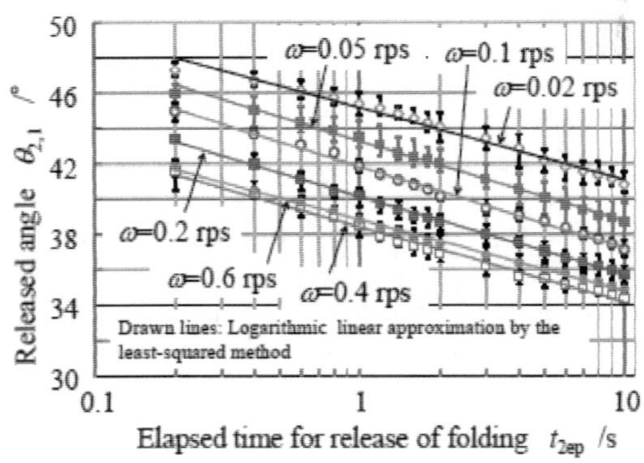

Figure 18. Released response of folding angle with respect to elapsed time after releasing in case of γ=0.6, Θ=90°

The released angle 62,1 was approximated with the logarithmic term $\ln(t_{2ep})$ as shown in Eq.(3) and Eq.(4). Two released coefficients b_1, b_0 are introduced and investigated with respect to the rotary velocity ω. Figure 19 shows the relationship between the exponent number $p_2=b_1/b_0$ and ω.

$$\Theta_{2,1} = -b_1 \ln(t_{2ep}) + b_0 \quad (3)$$

$$\Theta_{2,1} / b_0 = 1 + \ln(t_{2ep})^{-p_2}, \; p_2=b_1/b_0 \quad (4)$$

Figure19. Dependability of exponential coefficient p_2 on folding rotation velocity ω for a single folding repetition $n=1$

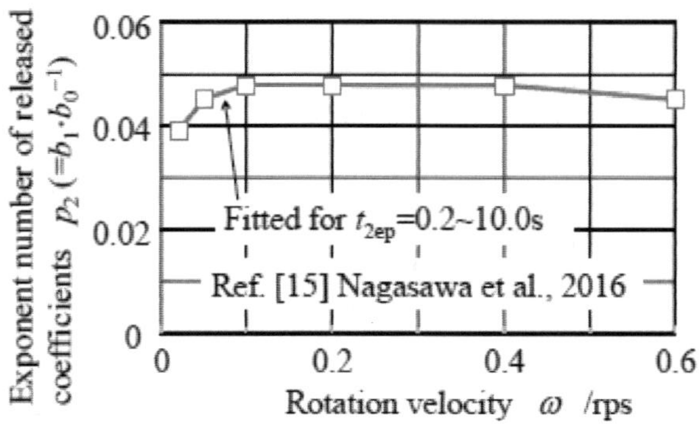

When choosing the rotary velocity as ω> 0.2 rps (1.26 rad-s), it is revealed that the recovered angle (by released) tends to be insensitive (stable) with the rotation velocity.

5. Conclusions and summary

Several folding processes of creased part of paperboard were investigated by using a creasing strength tester CST-J-1. Through this works, the followings are revealed.

(1) The relaxation of bending moment resistance at the tracking (maximum) angle Θ is characterised by a logarithmic function of holding time when varying the normalized indentation depth y (which determines the scoring condition). The bending moment gradient a_1, the bending moment of intercept a_0 and the exponential coefficient $p_1 = a_1/a_0$ are the fundamental and essential quantities, which are almost independent to the mechanical condition (normalized indentation depth and tracking angle).

(2) The creep-recovery (release) response of folded angle during returning back is characterized and approximated by a logarithmic function of elapsed time. Here, the released angle gradient b_1, the released angle of intercept b_0 and the exponent number coefficient p_2 (= $b_1 b_0^{-1}$) were introduced for characterizing the unfolding motion of creased paperboard.

References

[1] M.J. Kirwan, Folding cartons (Handbook of Paper and Papaperboard Packaging Technology), 2nd eds., M.J. Kirwan (Ed.), Wily-Blackwell (2013) [265-312].

[2] J.F. Hallady, R.W.K. Ulm, Creasing and Bending of Folding Boxboards, TAPPI Paper Trade Journal, 108(5), (1939), [36-40].

[3] D.J. Hine, Testing boxboard creasing, Modern Packaging, 8 (i959), [i22-i28].L. Carlsson, A.D. Ruvo, C. Fellers, Bending properties of creased zones of paperboard related to interlaminar defects. Journal of Material Science, 18(5), (1983), [1365-1373].

[4] L. Carlsson, A.D. Ruvo, C. Fellers, Bending properties of creased zones of paperboard related to interlaminar defects. Journal of Material Sicence, 18(5), (1983), [1365-1373]

[5] L.A.A. Beex, R.H.J. Peerlings, An experimental and computational study of laminatedpaperboard creasing and folding, International Journal of Solids and Structures, 46(24) (2009), [4192-4207].

[6] M. Nygards, M. Just, J. Tryding, Experimental and numerical studies of creasing of paperboard, International Journal of Solids and Structures, 46(11-12), (2009), [2493-2505].

[7] A. Sudo, S. Nagasawa, Y. Fukuzawa, I. Katayama, Analysis of exfolidation of laminated layers and creasing deformation of paperboard, Proceedings of Hokuriku-Shinetsu Branch Annual Conference of Japan Society for Mechanical Engineers, 047-1, (2005), [35-36] (in Japanese).

[8] S. Nagasawa, Y. Fukuzawa, T. Yamaguchi, S. Tsukatani, I. Katayama, Effect of crease depthand crease deviation on folding deformation characteristics of coated paperboard, Journal of Materials Processing Technology, 140 (2003), [157-162].

[9] S. Nagasawa, M. Nasuruddin, Y. Shiga, Bending Moment Characteristics on Repeated Folding Motion of Coated Paperboard Scored by Round-Edge Knife, Journal of Advanced Mechanical Design, Systems, and Manufacturing, 5(4), (2011) [385-394].

[10] B.D. Craven, Stress Relaxation and Work Hardening of Paper, Appita Journal, 16(2) (1962), [57-70].

[11] F. Johanson, J. Kubat, Measurements of Stress Relaxation in Paper, Svensk Papperstidning, 20(31) (1964) [822-832].

[12] K.W. Sharon, M. Etienne, L.Efraim, Stress Relaxation of a Paper Sheet under Cyclic Load: An Experimental and Theoretical Model, Material Science and Application, 1 (2010), [317-322].

[13] S. Nagasawa, Y. Shiga, Y. Fukuzawa, Effects of scoring depth, tracking angle, rubbering on bending-moment relaxation of creased paperboard, Advanced Materials Research, 939 (2014), [53-59].

[14] S. Nagasawa, S. Ozawa and Y. Fukuzawa, Effects of folding numbers, scoring depth and bending velocity on bending-moment relaxation of creased paperboard, Mechanical Engineering Journal, 2(1) (2015), 14-00346 [1-9].

[15] S. Nagasawa, S. Ozawa, Effect of bending velocity on time-dependent release behavior of creased white-coated paperboard, Mechanical Engineering Journal, 3(3) (2016), 16-00182 [1-11]

Topological Insulators: Electronic Band Structure and Spectroscopy

S Palaz[1], H Koc[2], A M Mamedov[3,4] and E Ozbay[3]

[1]Department of Physics, Faculty of Science and Letters, Harran University, 63000, Urfa, Turkey
[2]Department of Physics, Faculty of Science and Letters, Siirt University, 56000, Siirt, Turkey
[2]Nanotechnology Research Center (NANOTAM), Bilkent University, 06800, Bilkent, Ankara
[3]International Scientific Research Center, Baku State University, Baku, AZ-1148, Azerbaijan

E-mail: mamedov@bilkent.edu.tr

Abstract. In this study, we present .the results of our ab initio calculation of the elastic constants, density of states, charge density, and Born effective charge tensors for ferroelectric (rhombohedral) and paraelectric phases (cubic) of the narrow band ferroelectrics (GeTe, SnTe) pseudopotentials. The related quantities such as bulk modulus and shear modulus using obtained elastic constants have also been estimated in the present work. The total and partial densities of states corresponding to the band structure of Sn(Ge)Te(S,Se) were calculated. We also calculated the Born effective charge tensor of an atom (for instance, Ge, Sn, Te, etc.), which is defined as the induced polarization of the solid along the main direction by a unit displacement in the perpendicular direction of the sublattice of an atom at the vanishing electric field.

Keywords: ab initio calculation; elastic constant; charge density; Born effective charge tensors

1. Introduction

In recent years, the experimental and theoretical studies associated with these materials have greatly increased. The main results of these experimental and theoretical studies are that these materials show the characteristics of narrow gap semiconductors. An interesting feature of these materials is that the variation susceptibility mass with carrier concentrations in the two band region for GeTe is opposite to that observed for SnTe. That is, the mass for GeTe increases more rapidly while the mass for SnTe increases less rapidly with concentrations in the two band region [1]. The optical and electronic properties of GeTe get significantly modified depending on the change in the microscopic structure from the crystalline to the amorphous phase. Due to this characteristic, GeTe is an important basic material in phase-change alloys used in optical rewritable compact disks (CDs) and digital versatile disks (DVD) [2-5]. As a result of being narrow gap semiconductors, these materials are suitable as infrared detectors and light emitting devices [6].

In the past, some detailed works [1, 5, 6-9] have been carried out on the structural, electronic, optical, elastic, and dynamical properties of these compounds. Tsu et al. [1] described the electrical, optical, and band structure properties of GeTe and SnTe using thin films evaporated onto heated NaCl substrates to lower carrier concentrations. Okoye et al. [6] calculated the optical and electronic properties of SnTe and GeTe using the generalized gradient and the local density approximation. Shaltaf et al. [5] investigated the electronic and phonon band structure, Born effective charge and optical dielectric tensors, and elastic properties of GeTe as a ferroelectric material in its low-temperature rhombohedral phase using the density functional perturbation theory. Husnu et al. [7] investigated the structural, electronic, and optical properties of narrow-band ferroelectric compounds (Ge, Sn) Te in the ferroelectric and paraelectric phase using the using the generalized gradient approximation. Nonaka et al. [8] analyzed the crystal structure of GeTe and $Ge_2Sb_2Te_5$ meta-stable phase using the powder x-ray diffraction method. Herman et al. [9] investigated the relativistic band structure of IV-VI compounds (GeTe, SnTe, PbTe, PbSe, and PbS) using the first-order perturbation theory.

As far as we know, no ab initio general potential calculations of the elastic constants, Born effective charge tensors for ferroelectric and paraelectric phases of SnTe, the charge density for paraelectric phases of these compounds, and the elastic constants, Born effective charge tensors for paraelectric phase of GeTe have been reported. Our main goal in this study is to provide some additional information using the density functional theory. Therefore, in this work, we investigated the elastic constants, density of states and Born effective charge tensors for ferroelectric and paraelectric phases of the narrow band ferroelectrics (GeTe, SnTe). The method of calculation is given in Section 2; the results are discussed in Section 3. Finally, the summary and conclusion are given in Section 4.

2. Method of calculation

In all of our calculations that were performed using the ab-initio total-energy and molecular-dynamics program VASP (Vienna ab-initio simulation program) [10-13] that was developed within the density functional theory (DFT) [14], the exchange-correlation energy function is treated within the GGA (generalized gradient approximation) by the density functional of Perdew et al. [15]. The potentials used for the GGA calculations take into account the $4s^24p^2$ valence electrons of each Ge-, $5s^25p^2$ valence electrons of each Sn-, and $5s^25p^4$ valence electrons of each Te-atoms. The GGA calculation within the core-state model potential of A (Ge, Sn) has only four valence electrons because the $3d^{10}$and $4d^{10}$ have become part of the core. When including a plane-wave basis up to a kinetic-energy cutoff equal to 17 Ha, the properties investigated in this work are well converged. The Brillouin-zone integration was performed using special k points sampled within the Monkhorst-Pack scheme [16]. We found that a mesh of 12 x 12 x 12 k points was required to describe the elastic constants, density of states and Born effective charge tensors well. This k-point mesh guarantees a violation of charge neutrality less than 0.008e. Such a low value is a good indicator for an adequate convergence of the calculations. The unit cell of the crystal structure contains 3 GeTe and SnTe molecules (6 atoms) for ferroelectric phase and 1 GeTe and SnTe molecules (2 atoms) for paraelectric phase. When we started the calculations, we have optimized the structural properties first. The lattice parameters obtained as a result of this optimization are shown in Table 1 along with the experimental and theoretical results. The structural parameters obtained are in good agreement with the experimental and theoretical values [6, 17-18]. We used these structural properties in all of our subsequent calculations.

Table 1. The calculate equilibrium lattice parameters (a, and c) for GeTe and SnTe in Ferroelectric and Paraelectric phase

Material	**Structure**	**Reference**	a **(Å)**	c **(Å)**
GeTe	Rhombohedral (R3m)	Present	4.228	10.886
		Experimental[a]	4.156	10.663
	Cubic ($Fm\bar{3}m$)	Present	6.024	
		Theory (GGA)[b]	6.011	
		Theory (LDA)[b]	5.858	
		Experimental[c]	5.996	
SnTe	Rhombohedral (R3m)	Present	4.502	11.489
	Cubic ($Fm\bar{3}m$)	Present	6.309	
		Theory (GGA)[b]	6.404	
		Theory (LDA)[b]	6.231	
		Experimental[c]	6.327	

[a]Reference [6] ; [b]Reference [17]; [c]Reference [18]

Table 2. The calculated elastic constants (in GPa), isotropic bulk modulus (B, in GPa), shear modulus (G, in GPa), for GeTe and SnTe in Ferroelectric and Paraelectric phase

	SnTe($Fm\bar{3}m$)	**SnTe (R3m)**	**GeTe($Fm\bar{3}m$)**	**GeTe (R3m)**		
	Present	**Present**	**Present**	**Present** **Theory[a]Expt[b]**		
Elastic constant (GPa)				Clamped	relaxed	
C_{11}	135.110	76.810	170.411	98.256	115.70	
C_{12}	-0.514	16.562	10.057	112.00		
C_{13}		29.507		21.996	22.41	
C_{33}		59.712		19.36		
C_{44}	15.130	30.124	28.974	32.096	36.76	
C_{66}		42.711		27.20		
Bulk modulus (GPa)				69.003	86.51	
B_V	44.694	40.498	63.508	59.51		
B_R		40.458		38.123		
B	44.694	40.478	63.508	49.891		
Shear modulus (GPa)						
G_V	36.203	32.293	49.455	48.655		
G_R	21.952	28.314	38.915	48.027		
G	29.078	30.304	44.185	48.341	44.80	49.96
				39.535		
				35.881		
				37.708		

[a]Reference [5] ; [b]Reference [20]

3. Results and discussion

3.1. Elastic constants

We have derivated the elastic constants from the strain-stress relationship [19] for VASP calculations. The calculated elastic constants, bulk and shear moduli are given in Table 2 together with the theoretical and experimental results [5, 20]. The obtained elastic constants for GeTe in ferroelectrics phase are comparison with the theoretical values [5]. The C33 which is lower than the clamped value is higher than relaxed value. Apart from this, other elastic constants are in good agreement with the theoretical values. As can be seen from Table 2, the calculations carried out in both phases showed the presence of abrupt changes in the calculated parameters (elastic constants, bulk and Shear modulus). Unfortunately, there are no theoretical and experimental result for comparing with the present work for each phase of SnTe and paraelectric phase of GeTe. The calculated C11 and C33 elastic constant measure the a- and c- direction on resistance to linear compression, respectively. The C33 for either compounds in ferroelectric phase are lower than the C11. Therefore, the c- axis for either compounds are more compressible than the a- axis. We have defined the polycrystalline bulk (B) modulus and shear modulus (G) using the subscripts V (Voigt), R (Reuss) and H (Hill). The two approaches methods called the Voigt and the Reuss [21, 22] are used to calculate the polycrystalline modulus [23-25] using elastic constants. The bulk modulus is a measure of resistance to volume change by an applied pressure. The shear modulus is a measure of resistance to reversible deformations upon shear stress [26]. The bulk modulus given Table 2 is a good agreement with experimental and theoretical values. The calculated shear modulus for GeTe in both phases are higher than SnTe.

3.2. Electronic Properties

Band gap values obtained from the band structures of GeTe and SnTe in ferroelectric and paraelectric were calculated as 0.742 eV and 0.359 eV in the ferroelectric phase and 0.376 eV and 0.028 eV in the paraelectric phase, respectively, and the band gap character of these compounds for either of the phases was determined as direct [7]. Our results are in agreement with the results obtained in the previous calculation [5-6,27-29] (see Table 3). The change in the energy gap for the SnTe and GeTe can be understood qualitatively in terms of the difference between the relativistic effect in Ge and Sn (spin-orbital coupling) and the relativistic correction is extremely important in determining the positions of the energy bands. This change in energy gap may also be due to the presence of spontaneous polarization (change of the crystal field at the phase transition).

Table 3. Energy band gap for GeTe and SnTe in Ferroelectric and Paraelectric phase

Material	Structure	Reference	$E_g\,(eV)$
GeTe	Rhombohedral (R3m)	Present	0.742 direct
		Theory[a]	0.48 direct
		Theory[b]	0.48
		Theory[c]	0.3369
	Cubic ($Fm\bar{3}m$)	Present	0.376 direct
		Theory (LDA-GGA)[d]	0.399-0.340
		Experimental[d]	0.2
SnTe	Rhombohedral (R3m)	Present	0.359 direct
	Cubic ($Fm\bar{3}m$)	Present	0.028 direct
		Theory (LDA-GGA)[d]	0.074-0.061
		Experimental[d]	0.2

[a]Referance [5]; [b]Referance [27]; [c]Referance [28]; [d]Referance [6,29]

The total and partial densities of states for GeTe and SnTe in ferroelectric phase are illustrated in Figure 1. As you can see from these figures, the lowest valence bands occur between about -25 and -22 eV are dominated by d states. Similarly, the valence band that occur between approx. -12 and -5 eV are dominated by s states. The highest occupied valence bands results from the sp-hybridization of Ge and Te bands. The other states also contribute to the valence bands, but the values of densities of these states are so small compared to the sp-hybridization. The lowest unoccupied conduction bands just above the Fermi energy level are dominated by p states.

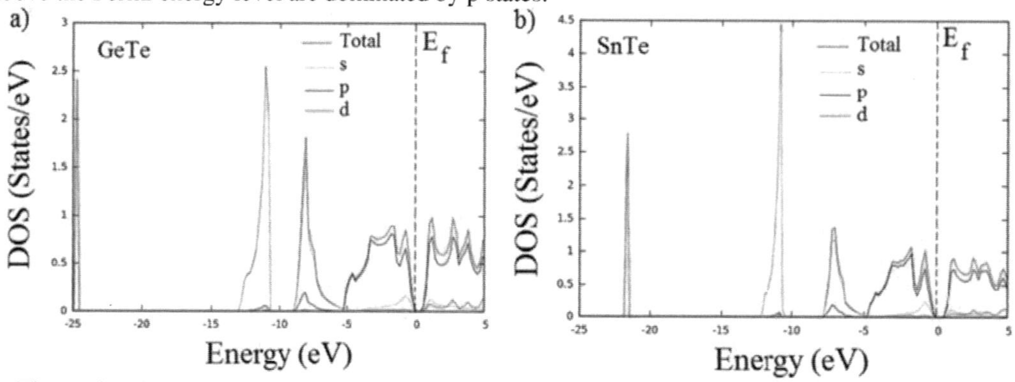

Figure 1. The total and projected density of states for a) GeTe and b) SnTe in ferroelectric phase

The two-dimensional charge density distribution is illustrated in Figure 3. Examination of the nature of chemical bonding defining the distribution of valence charges between atoms is necessary to explain the overall shape. The overall shape of charge distributions suggests the covalent bond of Ge-Te and Sn-Te. Ionicity is directly associated with the character of the chemical bond. In these figures, the shape of charge distribution shows a low electron accumulation between Ge and Sn atoms. This is a clear indication of the ionic bond. The large electron accumulation between Te atoms and their as spherical character indicates a strong directional covalent between these atoms. Therefore, the bonding of GeTe and SnTe as general can be classified as a combination of ionic, covalent, and metallic characters. In addition, evidence of metallic bonding is presented in electron density distribution map, as illustrated in the Figure 1 and Figure 2.

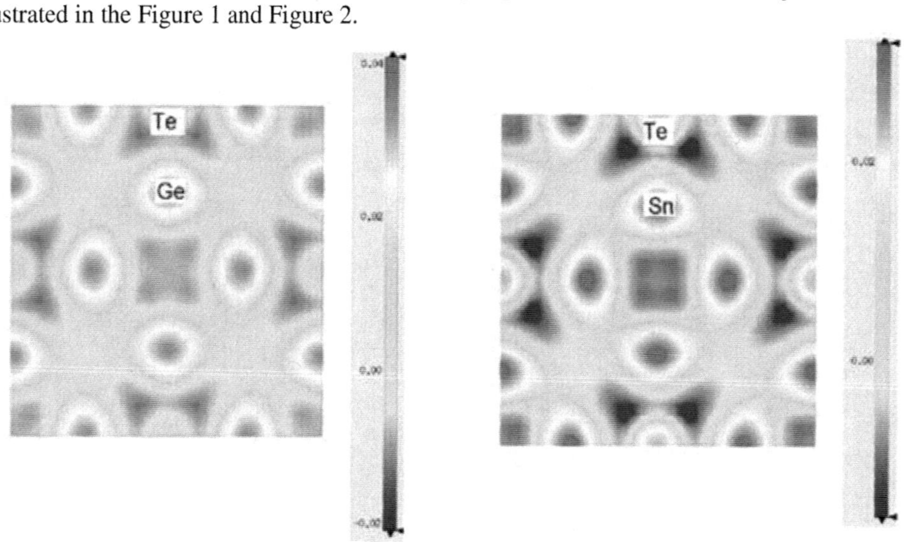

Figure 2. Charge density distribution of the valance charge of GeTe and SnTe in the 100 direction

3.3. Born effective charges

The Born effective charge (BEC) Z* is a fundamental quantity for the study of lattice dynamics, controlling the long range Coulomb part of the force constants and defined the change in polarization as a result of the displacement of an atom in the crystal [30]. The polarization value at different directions of the crystal may be different. Therefore, BEC can take different values for that depending

IOP Conf. Series: Materials Science and Engineering **175** (2017) 012004 doi:10.1088/1757-899X/175/1/012004

on the polarization. BEC especially occurs in the ionic and covalent bonded structure (strong covalent bond between atoms). The BEC tensor of an atom k is defined as

$$Z^*_{k,\gamma\alpha} = V \frac{\partial P_\gamma}{\partial_{k,\alpha}}\bigg|_{E=0} \quad (1)$$

where P_γ represents the polarization and the derivative is evaluated under the condition of vanishing macroscopic electric field. V is the unit cell volume. BEC quantities are important to study ferroelectric materials, because phase transitions in these materials hold a place from the competition of long range Coulomb interactions and short range force [31]. Until now, anomalous charges in ferroelectric materials were explained qualitatively in the framework of a bond orbital model [32]. Our study helps to clarify the debate on the real nature of the chemical bonding in narrow band ferroelectrics and bring out the role of the Ge(Sn) atom. The BEC tensors of GeTe and SnTe in cubic and Rhombohedral phase are diagonal, with one (cubic: $Z^*_{11} = Z^*_{22} = Z^*_{33}$) and two (Rhombohedral: $Z^*_{11} = Z^*_{22} \neq Z^*_{33}$) independent components. SP hybridization between Ge and Te is well known and was already pointed out from our calculation [7,33]. In this context, it seemed realistic, following [34], to focus on sp-hybridization changes to explain intuitively large anomalous contributions [35]. Results for theoretical cubic and rhombohedral structures are reported in Table 1 and 5. The first line brings together the charge of nucleus and the core electrons included in the pseudopotential. The other contributions come from the valence electron levels. Sets of bands were identified by the name of themain atomic orbital which generated this energy level in the investigated compound. Their dominant character was confirmed by partial density plots [7].

We computed the BEC tensors for the ferroelectric phase, too. In the present paper we only comment on eigenvalues of these tensors (Tables 1 and 4) that already allow a pertinent comparison with the cubic phase. The Z* eigenvalues of Ge (Sn) and Ge correspond to an eigenvector aligned along the ferroelectric axis. In the case of Te, eigenvector associated to the highest eigenvalue approximately point in the Ge(Sn) – Te direction: we identify this highest contribution as Te (∥) while the others are referred to as Te(⊥), by analogy with the cubic phase. The Z* eigenvalues of Ge and Te in rhombohedral phase are in agreement with the results obtained in the previous calculation [5].

Table 4. The calculated Born Effective charge for Ge, Sn, and Te in in Ferroelectric and Paraelectric phase

	Nominal	Z^*	Cubic ($Fm\bar{3}m$)	Rhombohedral (R3m)	
Ge	+4	Z^*_{11}	10.432	5.515	6.771[a]
		Z^*_{33}	-	3.349	4.550[a]
Te	-2	Z^*_{11}	-11.761	-5.406	-6.638[a]
		Z^*_{33}	-	-3.373	-4.479[a]
Sn	+4	Z^*_{11}	7.121	5.805	
		Z^*_{33}	-	5.251	
Te	-2	Z^*_{11}	-6.760	-5.716	
		Z^*_{33}	-	-5.281	

Reference [5]

4. Conclusion

In the present work, we conducted a detailed investigation of the elastic constants, density of states and Born effective charge tensors for ferroelectric and paraelectric phases of the narrow band ferroelectrics (GeTe and SnTe) crystals using the density functional methods. The second-order elastic constants have been calculated, and the related quantities such as bulk modulus and shear modulus have also been estimated. BECs of Ge, Sn, and Te atoms in rhombohedral and cubic phases are analyzed. The Born effective charge tensor is strongly anisotropic with a difference of 2.0 between parallel and perpendicular components. We examined the band structure as well as the change of Born effective charge tensors with atomic positions and all of our results have been discussed and compared with the available theoretical and experimental results.

References

[1] Tsu R, Howard W E and Esaki L 1968 *Phys. Rev.* **172** 779-788
[2] Ovshinsky S R 1968 *Phys. Rev.Lett.* **2** 1450-1453
[3] Libera M and Chen M 1993 *J. Appl. Phys.* **73** 2272-2282
[4] Yamada N 1996 *MRS Bull.* **21** 48-50
[5] Shaltaf R, Durgun E, Raty JY, GhosezPh and Gonze X 2008 *Phys. Rev.* B. **78** 205203
[6] Okoye CM I 2002 *J.Phys.: Condens. Matter.* **14** 8625
[7] Koc H, Simsek S, Mamedov A M and Ozbay E 2015 *Ferroelectrics* **483** 43-52
[8] Nonaka T, Ohbayashi G, Toriumi Y, Mori Y and Hashimoto H 2000 *Thin Solid Films* **370** 258-261
[9] Herman F, Kortum LR, Ortenburger IB and Dyke JP 1968 *J de Phys: Colloque.* **29** C4-62-C4-77.
[10] Kresse G and Hafner J 1993 *Phys. Rev.* B. **47** 558-561
[11] Kresse G and Furthmüller J 1996 *Comput. Mater. Sci.* **6** 15-50
[12] Kresse G and Joubert D 1999 *Phys. Rev.* B **59** 1758-1775
[13] Kresse G and Furthmüller J 1996 *Phys. Rev.* B **54** 11169-11186
[14] Hohenberg P and Kohn W 1964 *Phys Rev.* **136** A1133-A1138
[15] Perdew JP and Burke S Ernzerhof M. 1996 *Phys. Rev.Lett.* **77** 3865-3868
[16] Monkhorst HJ and Pack JD 1976 *Phys. Rev.* B **13** 5188-5192
[17] Pereira PB Sergueev I, Gorsse S, Dadda J, Müller E and Hermann RP 2013 *Phys. Status Solidi* B **250** 1300-1307
[18] Seddon T, Gupta SC and Saunders GA 1976 *Solid State Communications* **20** 69-72
[19] Page YL and Saxe P 2001 *Phys. Rev.* B. **63**174103
[20] Onodera A, Sakamoto I, Fujii Y, Mori N and Sugai S 1997 *Phys. Rev.* B. **56**7935-7941
[21] Voight W 1928 Lehrbook der kristallphysik Leipsig: Teubner
[22] Reuss A 1929 *Z. Angew. Math. Mech.* **9** 49-58
[23] Haines J, Léger JM and Bocquillon G 2001 *Annu Rev. Mater. Res.* **31**1-23
[24] Watt JP and Peselnick L 1980 *J. Appl. Phys.* **51**1525-1531
[25] Wu Z, Zhao E, Xiang H, Hao X, Lui X and Meng J 2007 *Phys. Rev.* B **76**054115
[26] Shein IR and Ivanovskii AL 2008 *J.Phys: Condens. Matter.* **20**415218.
[27] Rabe KM and Joannopoulos JD 1987 *Phys. Rev.* B **36** 6631
[28] Ciucivara A, Sahu BR and Kleinman L 2006 *Phys. Rev.* B **73** 214105
[29] Chang LL, Stiles PJ and Esaki L 1966 *IBM J. Res.* Rev. **10** 484
Esaki L and Stiles PJ 1966 *Phys. Rev. Lett.* **16** 1108
[30] Ghosez PH, Michenaund JP and Gonze X 1988 *Phys. Rev.* B **58** 6224-6240
[31] Ghosez PH, Gonze X and Michenaund JP 1996 *Europhys Lett.* **33** 713-718
[32] Harrison WA 1989 Electronic structure and the properties of solids-The physics of the chemical band (Dover Publications)
[33] Koc H, Simsek S, Palaz P, Oltulu O, Mamedov AM andOzbay E 2015 *Phys. Status Solidi* C**12**651-658
[34] Gonze X and Lee C 1997 *Dynamical matrices*, **55** 10355-10368
[35] Zhong W, King-Smith RD and Vanderbilt D 1994 *Ferroelectrics* **72** 3618-3621

IOP Conf. Series: Materials Science and Engineering **175** (2017) 012005 doi:10.1088/1757-899X/175/1/012005

Whole blood of mammalian species in the oscillating shear field: influence of erythrocyte aggregation

U Windberger[1], Ch Pöschl[1], S Peters[1], J Huber[2], R van den Hoven[3]

[1]Department Biomedical Research, Medical University of Vienna, Austria
[2]University Clinic for Ruminants, Kremesberg, Veterinary University Vienna, Austria
[3]Clinical Unit of Equine Internal Medicine, Veterinary University Vienna, Austria

Email: ursula.windberger@meduniwien.ac.at

Abstract. This is the rheologicalanalysis of mammalian blood of species with a high (horse), medium (man), and low (sheep) erythrocyte (RBC) aggregability by small amplitude oscillation technique. Amplitude and frequency sweep tests in linear mode were performed with blood from healthy adult volunteers, horses, and sheep in CSS-mode. Blood samples were hematocrit (HCT) adjusted (40%, 50%, 60%) and tested at 7°C, 22°C, and 37°C. Storage modulus (G′) increased with HCT and decreased with temperature in each species, but the gradient of this increase was species-specific. The lower dependency of G′ on the equine HCT value could be a benefit during physical performance when high numbers of RBCs are released from the spleen in the horse. In sheep, a HCT-threshold had to be overcome before elasticity of the blood sample could be measured, suggesting that the cohesive forces between RBCs, and between RBCs and plasma molecules must be very low. The frequencies for tests under quasi-staticcondition were in a narrow range around the physiologic heart rate of the species. In horse, time-dependent influences concurred at frequencies lower than 3 rad.s^{-1} probably due to sedimentation of RBC aggregates. In conclusion, elasticity of blood depends not only on the amount of blood cells, but also on their mechanical and functional properties.

Introduction

The rheology of blood is influencedby the quantitative and qualitative properties of blood cells, by the constitution of blood plasma, and by the adhesive and cohesive forces between blood cells and blood plasma. Compared toparticlesystems or dispersions, the viscosity of whole blood (WBV) is low, even forhighcellular fractionssuch as 40 – 45 %, which is the physiological hematocrit (HCT) of man and most animal species. This is the result of the unique red blood cell (RBC) properties.

RBCs have a high surface-to-volume ratio that allowsviscous deformation in narrow vessels (1). Exposure to shear forces causes a deformation from the relaxed biconcave towards folded shapes (2-4).The RBCs return to their biconcave resting shape after disappearance of the shearing force, and even exhibit a shape memory effect (5). The RBC membrane is not static, but an active ATP-driven structure. Not only the typical shape, but also dynamic fluctuationsof the cell membrane at physiological body temperature are increased in relation to metabolic and thermal energy (6). Furthermore, RBC membranes use a tank thread motion to absorb hydrodynamic stress - a process that facilitates laminar blood flow and suspension stability within the blood vessel. The tank thread frequency starts at a certain threshold (7, 8), which is dependent on the viscosity of the suspending medium (9, 10). Another dynamic feature of RBC membranes is that they enable the cells to sustain a rolling motion in shear flow, which avoids energetically costly deformation (8). It should be kept in mind that the volume concentration of RBCs changes dynamically throughout the body with the lowest HCT values being present in the microvasculature. As a result, the viscous resistance of blood flowing through narrow capillaries is only 20% higher than that of pure plasma (11).

RBCs form aggregates - called as Rouleaux - at low volume flow (12-15).These aggregates break up when shear rates increase at higher flows. There is remarkable diversity in RBC aggregability among the mammalian species (16). For instance, RBC aggregation is too low to be measured with routine techniques in cow, sheep, goat, mouse, and rat blood. In contrast, in horse and other equidae, the physiological RBC aggregation is as high as it would be in inflammatory disease in man (17, 18). RBC aggregation affects venous vascular resistance (19), and clinical cases associated with high RBC aggregation indicate specific diseases (20). The impact of RBC aggregability for in-vivo blood flow is discussed controversially. Although there is evidence that high RBC aggregation blunts the parabolic blood velocity profile (21), moderate RBC aggregation seems to promote blood flow (22, 23), at least in comparison to non-aggregating RBC suspensions. It is generally agreed that in-vivo blood viscosity rises when intravascular shear forces are low enough to allow RBC aggregation.By considering blood flow as Poiseuille flow (24), the lowest wall shear rates are calculated for the postcapillary venules. RBC aggregates have been seen with high-speed video microscopy in the venules after passing the postcapillary region (25). The physiological relevance of RBC aggregation lies in the phase separation phenomenon of composite fluids in narrow tubes reaching a critical diameter. Due to the axial migration of RBCs in the tube flow (26), a marginal cell-free layer (CFL) is formed, whereby the magnitude of the width of this layer (CFLW) is linked to RBC aggregability.This CFL reduces the endothelial shear stress, and subsequently, an intrinsic response mediated by vasoactive factors starts to adjust the vascular diameter (27,28). This in turn modulates the distribution of RBCs in subsequent vessels (29-31).

Rheological properties of blood are usually measured by viscosityas a function of shear rate. Methods to test properties of singular RBCs were recently summarized (32). Blood was characterized as a shear thinning viscoelastic fluid, showing different degree of thixotropy in relation to RBC aggregation. Typically, at low shear rates RBCs aggregate to clusters, while at high shear ratesRBCsare singularly suspended in blood plasma (33).

In contrast to rotational flow, small amplitude oscillating shear flow (SAOS) can be used to study the elastic property of fluids under quasi-static conditions. An application of this method to human blood has been described recently (34). Applying this principle to blood, a specific "texture"of blood will result from the forces between RBCs, and from the quality of the contact between RBCs and plasma molecules.

In the present study we tested human, equine and ovine blood atSAOS. As expected from the differences in RBC aggregability among these species, we observed that shear moduli varied, being highest in the horse. In fact, it was difficult to measure ovine blood, which likely is the result of the minute RBC aggregability of this species. It must be noted that yield stress of blood is lower than vascular wall stress under physiologic circumstances. However, intravascular shear rates are not fixed, but vary dynamically in the body. A decrease in blood flow either generalized due to blood loss or due to vascular dysfunction such as in septicaemia, or more localized due to vascular occlusion following intima hyperplasia or thromboembolic events can easily lower the wall shear rates to those shear forces that we set in our experimental protocol. Since blood is a viscoelastic fluid, storage modulus (G′) values were always below loss modulus (G′′)-values. A percolated structure in normal plasma could only be achieved at very high and thus unphysiologic HCT values. In such circumstances, blood behaved jellylike.

IOP Conf. Series: Materials Science and Engineering **175** (2017) 012005 doi:10.1088/1757-899X/175/1/012005

Material and Methods

Blood samples

Whole blood of healthy man, horse, and sheep were used for this study. 4 human volunteers (2f, 2m, age: 22-35 y), 4 *Warmblut* horses (2 mares, 2 geldings, age: 13-26 y), and 10 female *Milchschaf* sheep (age: 1.2-3.6 y) were used. 90 mL blood was withdrawn into EDTA tubes from each individual by venous puncture (man: V. radialis, horse and sheep: V. jugularis) with a 16 G needle connected to a vacutainer system. Samples were centrifuged at 2000 rpm for 5 minutes and blood plasma was separated. New samples were reconstituted out of RBC concentrate and autologous plasma to generated whole blood samples with HCT values of 40, 50, and 60%. The samples were carried in insulated bags to the laboratoryand kept in the fridge prior to their measurement. All measurements were finished within 6 hours following withdrawal.

The procedure was approved by the Ethics Committee of the Medical University Vienna (1892/2013) and the Austrian Federal Ministry of Science and Research (animal license number: GZ-1744/115-97/98).

Rheological protocol

SAOS measurements were performed using the stress controlled Physica MCR301 rheometer (Anton Paar, Graz, Austria). 3.5 mL of each blood sample was filled into the stainless steel double gap cylinder system and was analyzed at three different temperatures (7, 22, 37°C) starting with the lowest temperature. Isothermal amplitude and frequency sweep tests were performed.

Strain dependency of blood samples was measured at fixed frequency (10 rad s^{-1}) to compare the yield stresses between the species. In order to carry out the subsequent frequency tests in linear regime, some of the human and horse blood samples were tested at frequencies between 1 and 20rad s^{-1}. After a pre-shear interval (man and sheep: 30 s rotation at 300 s^{-1} followed by a 20 s interval at 1 s^{-1}; horse: 30 s oscillation at 0.01 Pa and 10 rad s^{-1}), increasing shear stresses of 0.001 – 10 Pa were applied by a logarithmic shear stress ramp. Yield points were estimated by the Rheoplus software (version 4.2, Anton Paar, Graz, Austria).

Frequency sweep tests were performed at 5-10mPathroughout the whole frequency range (20 - 1 rad s^{-1}). The blood samples were subjected to frequency sweep from high to low frequency only.

Descriptive statistic was performed by IBM®SPSS®Statistics (version 22).

Results

Amplitude sweep tests

Allowed frequencies for the frequency sweep tests were within the narrow frequency range between 3 and 15 rad s^{-1}. These frequencies correspond to the range of the physiologic heart rate of the species at rest (0.5 - 2.5 Hz).

The software easily defined the linear viscoelastic (LVE) range at 7°C in man and horse (as well as in sheep at high HCT: 60%), but became imprecise when the temperature increased. Especially at 37°C, the G′-values were sloping – although continuously decreasing. Occasionally, LVE at 37°C had to be determined based on G′′-values that always displayed a plateau until a certain shear stress was reached. As expected, experiments became more reproducible when HCT was high and temperature was low.

Basically, yield points and corresponding G′-values were very low in blood although traceable. G′-values increased with hematocrit and decreased with temperature. Data are provided in table 1 and figure 1. In blood samples at 40% HCT and 37°C,yield pointsare not provided due to increased sloping of G′, although G′′ showed a plateau value. Yield points appeared to be slightly higher in horsecompared to man. In sheep, yield points could be determined at 60% HCT only, however, the obtained values werethe highest among the three species.

Frequency sweep tests
G′-values increased with the HCT and decreased with the temperature (see table 2 and figure 2). At 40% HCT and 37°C the interquartile distance became larger indicating greater variance among the samples. There was no G′-G′′-crossover within the allowed frequencies. Sheep with the lowest RBC aggregation showed the lowest G′-values. It is interesting to note that G′ was lower in man than in horse at 40% HCT, but higher in man than in horse at 60% HCT at each temperature. At 50% HCT the median values of man and horse were nearly identical.

Table 1. Yield points(τ_y) of human, equine and ovine blood. Data are presented as median values and quartiles in parentheses

$\tau_y$at 7°C in mPa	40% HCT	50% HCT	60% HCT
Man	21(8/22)	33(13/46)	46(45/46)
Horse	33(20/86)	21(20/78)	72(13/185)
Sheep	-	-	464(348/465)
τ_y at 22°C in mPa	40% HCT	50% HCT	60% HCT
Man	3(1/8)	21(10/21)	21(12/39)
Horse	27(9/46)	20(10/30)	60(20/185)
Sheep	-	-	214(213/215)
τ_y at 37°C in mPa	40% HCT	50% HCT	60% HCT
Man	-	9(4/9)	9(4/18)
Horse	-	10(9/18)	54(4/189)
Sheep	-	-	97(31/99)

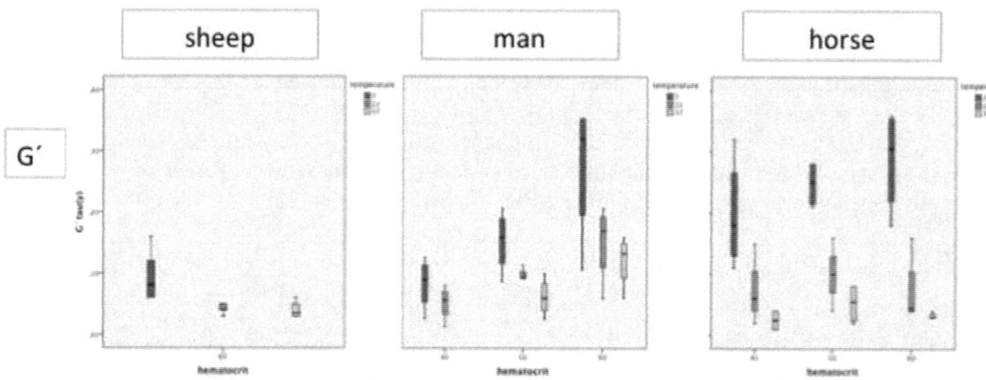

Figure 1. Amplitude sweep test: storage modulus during elastic deformationin sheep, man, and horse blood at 3 different HCT levels (40%, 50%, 60%) and 3 temperatures (blue boxes: 7°C, green boxes: 22°C, yellow boxes: 37°C). In sheep, tests could be performed at 60% HCT only.

There were only minute changes of the tanδ-value with HCT and temperature, although the loss factor decreased with the HCT and increased with the temperature in each sample. Tanδ-values at 9.38 rad s^{-1} in man were between 2.30(2.11/2.88) (measured in human blood at 60% HCT and 7°C) and 4.17(4.11/4.46) (measured in human blood at 40% HCT and 37°C).In horse tanδ-values were between 1.80(1.58/1.82) and 1.90(1.24/2.72), and in sheep tanδ-values were up to 4.56(2.98/5.38), verifying blood as viscoelastic fluid.

IOP Conf. Series: Materials Science and Engineering **175** (2017) 012005 doi:10.1088/1757-899X/175/1/012005

Table 2. Storage modulus at 9.38 rad s^{-1} in human, equine, and ovine blood samples. Data are expressed as median and quartiles in parentheses

$G'_{9rad/s}$ at 7°C in mPa	40% HCT	50% HCT	60% HCT
Man	111(92/131)	262(253/289)	441(410/534)
Horse	176(145/221)	260(238/281)	340(311/415)
Sheep	-	-	130(101/258)
$G'_{9rad/s}$ at 22°C in mPa	40% HCT	50% HCT	60% HCT
Man	51(35/67)	132(116/151)	259(225/289)
Horse	92(70/109)	132(130/139)	179(161/216)
Sheep	-	-	86(69/105)
$G'_{9rad/s}$ at 37°C in mPa	40% HCT	50% HCT	60% HCT
Man	31(19/35)	77(69/92)	171(159/181)
Horse	56(31/81)	77(66/80)	119(114/135)
Sheep	-	-	55(50/63)

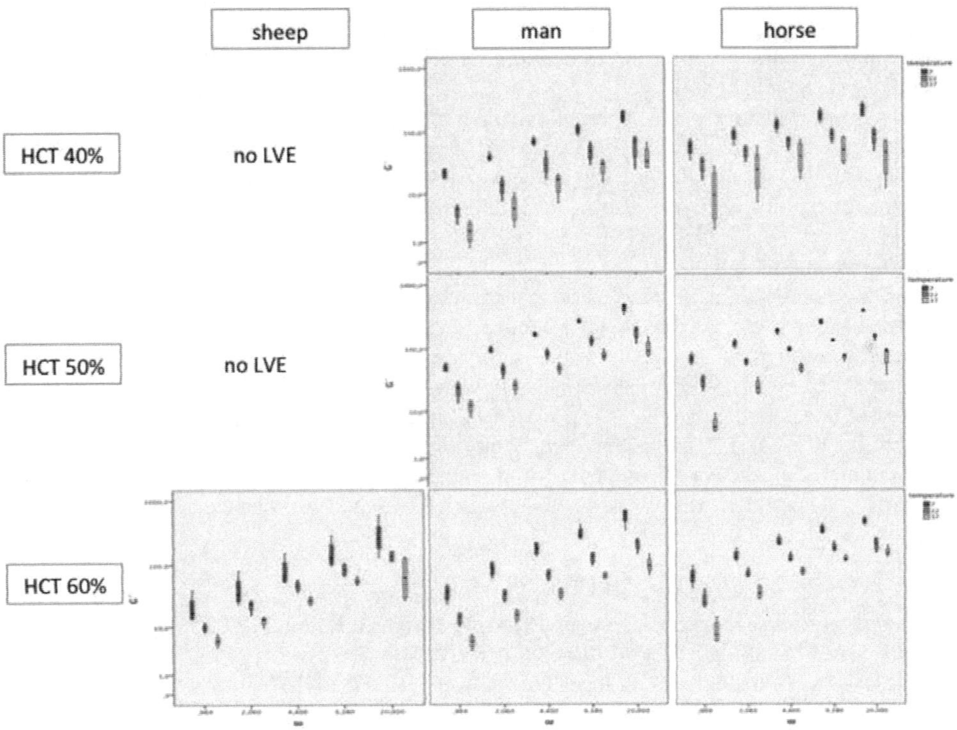

Figure 2. Frequency sweep test: storage modulus of sheep, man, and horse blood at 3 different HCT levels (40%, 50%, 60%) and 3 temperatures (blue boxes: 7°C, green boxes: 22°C, yellow boxes: 37°C)

Discussion

The present investigation shows that whole blood of different mammalian species can be measured by small amplitude oscillation. To study the influence of RBC properties on our protocol, we used three species: the horse representing the species with highest RBC aggregability of all, followed by man with intermediate, and sheep with nearly immeasurable RBC aggregability (16). Using autologous plasma, the samples were adjusted to several predetermined HCT values and tested at different temperatures. By this approach, we demonstrated that the shear moduli increased with the HCT and decreased with the temperature in each species, but the gradient of this increase was species-specific.

At a frequency analog to the physiologic heart rate, G′ was higher in horse blood than in human blood at the physiological HCT (40%), but lower in horse compared toman when the HCT was raised (60%). This implicates that the gradient of the G′-increase in relation to the cellular fraction in blood was lower in the horse. This could be of physiological relevance in this species. Resting horses possess a splenic reservoir for RBCs with a HCT of about 80%. After an exercise-induced contraction, these splenic RBCs are added to the RBCs in the circulation thereby increasing HCT to maximally 65% (35). The extra number of circulating RBCs effectively increases the blood oxygen store and the oxygen carrying capacity. However, in parallel, the blood viscosity and the elasticity of the blood suspension is elevated, as well. On the other hand, the low dependency ofG′on the HCT valuecould be a benefit during exercise.

In sheep, a HCT-threshold must be overcome before elasticity ofthe blood samplewas detected.Ovine RBCs are smaller than those of man and horse (36) and have a reduced deformability if exposed to shear stress (37). Based on their minute RBC aggregability (16), RBCs are singularly suspended in the plasma volume. Flow curves show the low shear thinning of ovine blood. Our results notice that apart from the attracting forces between the RBCs, also the forces between RBCs and the surrounding plasma should be low.

Another interesting finding of the sheep blood concerns the yield point that could not be measured at HCT of 40% and 50%. However, at HCT of 60% the yield point was higher than in man and horse. The physiological HCT of ovine blood ranges between 30 and 38% (36). A release of splenic RBCs by a sympathetic stimulus is insignificantin sheep due to the diverse composition and function of the spleen. Sheep do not exhibit high HCT values regularly. But if they do so, the yield stress of blood ishigherthan in man and horse.

In conclusion, blood is a fragile suspension that shows its best stability around the resting heart rate of the particular species. At frequencies above 15 rad $s^{-1}$and below 3 rad s^{-1}, G′-values weresignificantly sloping, showed an irregular sequence, or were even absent. We assume that the weak cohesive forcesof bloodare disrupted at high frequencies, while at low frequencies, sedimentation of RBCs in the rheometer gap may result in phase separation and shear banding during the measurement. It is logical thatsuspension stability will increase with the increase of the volume fraction of cellular elements in a composite fluid. In addition, we showon our HCT standardized samples that G′-values and yield stresses are higher when RBCs have the chance to form aggregates.

References

[1] Waugh RE and Hochmuth RM *Mechanics and deformability of hematocytes* In: *The Biomedical Engineering Handbook* ed Bronzino JD (Boca Raton, FL: CRC) pp 474-486
[2] Branemark PI and Bagge U 1977 *Blood Cells* **3**11-24
[3] Fedosov DA, Peltomäki M, Gompper G *Soft Matter* **10** 4258-4267
[4] Gaehtgens P, Schmidt-Schönbein H 1982 *Naturwissenschaften* **69** 294-296
[5] Fischer TM 2004 *Biophysical Journal* **86** 3304-3313
[6] Park YK, Best CA, Auth T, Gov NS, Safran SA, Popescu G, Suresh S, Feld MS 2010 *PNAS* **107** 1289-1294
[7] Basu H, Dharmadhikari AK, Dharmadhikari JA, Sharma S, Mathur D 2011 *Biophysical Journal* **101** 1604-1612
[8] Dupire J, Socol M, Viallat A 2012 *PNAS* **109** 20808-20813
[9] Fischer TM 2007 *Biophysical Journal* **93** 2553-2561
[10] Tran-Son-Tray R 2006 A study of the tank thread motion of red blood cells in shear flow. PhD thesis. Washington University, St. Louis, MO, USA

[11] P Gaehtgens 1981 *Drug Research* **31** 1995-1998
[12] Barshtein G Wajnblum D Yedgar S 2000 *Biophysical Journal* **78** 2470-2474
[13] Brust M, Aouane O, Thiebaud M, Flormann, Verdier C, Kaestner L, Laschke MW, Selmi H, Benyoussef A, Podgorski T, Coupier G, Misbah C, Wagner C 2014 *Scientific Reports* **4** 4348; DOI: 10.1038/srep04348
[14] Cokelet GR, Meiselman HJ 2007 *Macro- and microrheological properties of blood* In: *Handbook of Hemorheology and Hemodynamics* eds Baskurt OK, Hardeman MR, Rampling MW, Meiselman HJ (Amsterdam, Berlin, Oxford, Tokyo, Washington, DC: IOS Press) pp 45-71
[15] Johnson PC, Bishop JJ, Popel S, Intaglietta M 1999 *Biorheology* **36** 457-460
[16] Windberger U, Bartholovitsch A, Plasenzotti R, Korak HJ, Heinze G 2003 *Experimental Physiology* **88.3** 431–440
[17] Windberger U, Baskurt OK 2007 *Comparative hemorheology* In: *Handbook of Hemorheology and Hemodynamics* eds Baskurt OK, Hardeman MR, Rampling MW, Meiselman HJ (Amsterdam, Berlin, Oxford, Tokyo, Washington, DC: IOS Press) pp 267-285
[18] Lowe GDO 1998 *Clinical Blood Rheology* 1-2 (Boca Raton, FL: CRC Press)
[19] Cabel M, Meiselman HJ, Popel AS, Johnson PC 1997 *Am. J. Physiol.***272** H1020-H1032
[20] Toth K, Kesmarky G, Alexy T 2007 *Clinical significance of hemorheologic alterations* In: *Handbook of Hemorheology and Hemodynamics* eds Baskurt OK, Hardeman MR, Rampling MW, Meiselman HJ (Amsterdam, Berlin, Oxford, Tokyo, Washington, DC: IOS Press) pp 392-432
[21] Bishop JJ, Nance PR, Popel AS, Intaglietta M, Johnson PC 2001 *Am. J. Physiol. Heart Circ. Physiol* **280** H222-H236
[22] Soutani M, Suzuki Y, Tateishi N, Maeda N 1995 *Am. J. Physiol. Heart Circ. Physiol.* **268** H1959-H1965
[23] Charansonney O, Mouren S, Dufaux J, Duvelleroy M, Vicaut E 1993 *Biorheology* **30** 75-84
[24] Pantos J, Efstathopoulos E, Katritsis G 2007 *Current Vasc. Pharmacol.* **5** 113-119
[25] Kim S, Popel AS, Intaglietta M, Johnson PC 2005 *Am. J. Physiol. Heart Circ. Physiol.* **288** 584-590
[26] Bishop JJ, Popel AS, Intaglietta M, Johnson PC 2001 *Am. J. Physiol. Heart Circ. Physiol.* **281** H939-H950
[27] Moncada S, Palmer RM and Higgs EA 1981 *Pharmacological Reviews* **43** 109–142
[28] Malek K and Izumo S 1992 *Am. J. Physiol.* **263** C389–C396
[29] Ong PK, Jain S and Kim S 2012 *Microvasc. Res.***83** 118-125
[30] Hightower CM, Vazquez BY, Park SW, et al. 2010 *Wiley Interdisciplinary Reviews: Systems Biology and Medicine***3** 458-470
[31] Yalcin O, Aydin F, Ulker P, et al. 2006 *Am. J. Physiol.Heart Circ. Physiol.* **290** H765-H771
[32] G Tomaliuolo 2014 *Biomicrofluidics* **8** 051501
[33] Baskurt OK, Meiselman H 2003 *Semin. Thromb. Hemost.* **29** 435-450
[34] Tomaliuolo G, Carciati A, Caserta S, Guido S 2015 *Rheologica Acta* **5** 1-11
[35] Persson SGB 1983 *The significance of haematological data in the evaluation of soundness and fitness in the horse* In: *Equine exercise physiology* eds Snow DH, Persson SGB, Rose RJ (Cambridge, UK: Granta Editions)
[36] Kraft W, Duerr UM (eds) 2005 *Klinische Labordiagnostik in der Tiermedizin* (Stuttgart, New York: Schattauer) 6th edition pg. 59
[37] Plasenzotti R, Stoiber B, Posch M, Windberger U 2004 *Clin. Hemorheol. Microcirc* .**31** 105-111

IOP Conf. Series: Materials Science and Engineering **175** (2017) 012006 doi:10.1088/1757-899X/175/1/012006

Piezoresistive Carbon-based Hybrid Sensor for Body-Mounted Biomedical Applications

M Melnykowycz[1], M Tschudin[2] and F Clemens[1]

[1]Laboratory for High Performance Ceramics, Empa, Swiss Federal Laboratories for Materials Science and Technology, Dübendorf, Switzerland
[2]STBL Medical Research AG, Wilen, Switzerland

E-mail: mark.melnykowycz@empa.ch

Abstract. For body-mounted sensor applications, the evolution of soft condensed matter sensor (SCMS) materials offer conformability andit enables mechanical compliance between the body surface and the sensing mechanism. A piezoresistive hybrid sensor and compliant meta-material sub-structure provided a way to engineer sensor physical designs through modification of the mechanical properties of the compliant design. A piezoresistive fiber sensor was produced by combining a thermoplastic elastomer (TPE) matrix with Carbon Black (CB) particles in 1:1 mass ratio. Feedstock was extruded in monofilament fiber form (diameter of 300 microns), resulting in a highly stretchable sensor (strain sensor range up to 100%) with linear resistance signal response.

The soft condensed matter sensor was integrated into a hybrid design including a 3D printed metamaterial structure combined with a soft silicone. An auxetic unit cell was chosen (with negative Poisson's Ratio) in the design in order to combine with the soft silicon, which exhibits a high Poisson's Ratio. The hybrid sensor design was subjected to mechanical tensile testing up to 50% strain (with gauge factor calculation for sensor performance), and then utilized for strain-based sensing applications on the body including gesture recognition and vital function monitoring including blood pulse-wave and breath monitoring. A 10 gesture Natural User Interface (NUI) test protocol was utilized to show the effectiveness of a single wrist-mounted sensor to identify discrete gestures including finger and hand motions. These hand motions were chosen specifically for Human Computer Interaction (HCI) applications. The blood pulse-wave signal was monitored with the hand at rest, in a wrist-mounted. In addition different breathing patterns were investigated, including normal breathing and coughing, using a belt and chest-mounted configuration.

1. Sensor Designs and Body-Mounted Applications

The development of body-mounted sensor technologies is leading to the creation of compliant and soft sensor structures, which can augment or improve upon the use of conventional sensors currently in wide use. Soft condensed matter sensors (SCMS) can provide the ability to measure surface strains directly on the skin surface, and offer integration possibilities in clothing, or as standalone bands to be positioned at different locations.

1.1. Body-Mounted SensorApproaches

The explosion of wearable sensors and fitness trackers in the past years has been driven by the availability of low-cost MEM based sensors, specificallyinertial measurement unit (IMU) chips that include accelerometers and gyroscopes sensors that allow motion capture. Motion tracking can be accomplished with sensors placed at key linkage points on the human body, such as the wrist, elbow, and knee as well as on the torso, head etc. A rigid body model based on the human skeleton can then be used to track how a person is moving. Similarly, motion recognition and depth camera systems can be used to visually track motion. The information from both systems can then be used to animate characters in virtual environments. In 3D game and animation design, the movement of human or character 3D models are also based on rigid body motion, making it straightforward to use captured motion in designing virtual character motion. For consumer products, motion sensors can be easily integrated in devices such as smart watches or fitness trackers to track basic motions and extrapolate the activity of the person such as walking, running, sleep patterns, etc.

Motion sensors are not able to accurately determine the deformation of the body surface however. This makes it difficult to track the motion of individual muscles, where electrical impulse may be used instead. It is worthwhile to mention that electrical impulse measurements will also not allow for surface strain measurement. For strain measurements commercially piezoresistive bend sensors are available, which can be used to monitor the movement of individual fingers, or basic deflection and bending. Alternative electro-active polymer (EAP) sensors can be used to measure bending and strain as capacitive sensors mounted on the body.

In comparison to our soft body, commercial available piezoresistive sensors are still stiff and too large. In the current work, piezoresistive SCMS fibers were integrated into a compliant polymer substrate structure for body-mounted strain sensing, with applications in vital sign monitoring and gesture monitoring for Natural User Interface (NUI) design approaches.

1.2. Applications for Body-Mounted Sensors

In biomedical field, wearable sensors are already used to analyze motion, but consumer-grade devices do not provide data that can be trusted by medical practitioners for true remote medical use cases. The accuracy of heart rate monitors for example, can vary and therefore only be used to have a general understanding of how a patient's heart rate is changing. In the current work, a SCMS based on Carbon Black (CB) and a thermoplastic elastomer (TPE) was investigated for different body-mounted sensor applications, including wrist motion, blood pulse-wave, and breathing monitoring.

While the initial applications of wearable sensors have focused largely on fitness trackers, many applications exist in the biomedical and computing fields. User interaction patterns in personal computing have focused almost exclusively on the computer/mouse, as those were the input devices that were logical to build. The advent of mobile computing with the smartphone and tablet devices ushered in user interaction patters based around tapping, swiping, and pinch/zooming with the fingers. Currently the release of virtual reality devices brings another dimension to user interaction where a full 3D volume around a user may be used for interaction with a computing program. Later, smart glasses and augmented reality will also require novel interaction patters that are natural for people to perform as they move between different device types.

2. Sensor Material Production

The production of the carbon black based conductive polymer composites is documented in the previous work [1]. The SCMS fibers were extruded using a Capillary Rheometer (Rosand RH7Flowmaster). A Niggeloh 200 bar pressure transducer was used to monitor the pressure of the melt through an orifice of 0.3 mm in diameter.

3. Sensor Structural Design and Production

A primary design consideration for body-mounted sensor design is ensuring that the sensor conforms to and stays relatively fixed on the area of interest on the body. Silicone was a logical material choice for designing a structural carrier for the sensor, however silicone is also excessively deformable and with a high Poisson's Ratio, will contract when stretched over a surface, thereby reducing its surface area and creating a larger stress concentration on the body surface than may be necessary.

Sensor structures were produced with a hybrid TPE-silicone structure.Previouslya similar configuration worked well for initial investigationson textile substrate [2], but in order to better tune the mechanical properties to different applications, a hybrid sensor design approach was taken. In the current work, a compliant design approach was chosen(Figure 1a). An auxetic array was built up by 3D printing using a commercially available TPE (FilaFlex®), which was then casted in silicone together with the sensor fiber. The auxetic unit cell was chosen, since it displays a negative Poisson's Effect[3], and expands in the width direction while under tensile load. This would then counteract the contraction of the silicone material when stretched on the body.

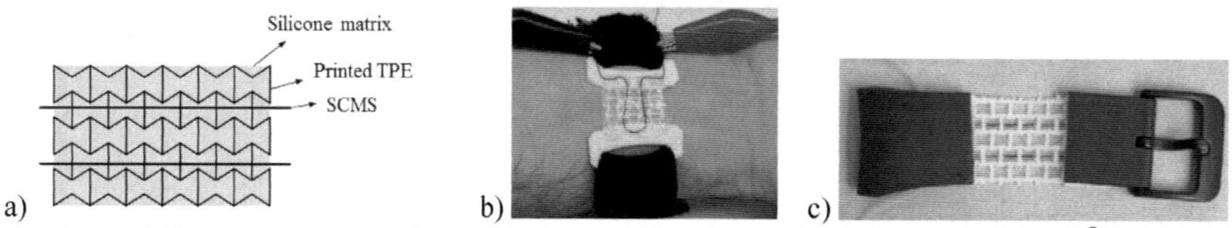

Figure 1. Textile band sensor (a), hybrid structure (b), integration in Samsung Gear S2®band (c)

For soft body dynamic detection, the sensor structures was connected to a textile band (b) or smartwatch (b). The wires of the sensor structure were connected to the blood pressure watch prototype developed by STBL Medical Research AG (Freienbach, Switzerland). Via the blood pressure watch a current was applied to the band sensor and the resulting voltage across the sensor was recorded. A 24 bit analog-to-digital converter (ADC) was used in the design to provide a high resolution measurement of the band sensor signal with a data acquisition rate of 40 Hz. The blood pressure watch prototype was connected via Universal Serial Bus (USB) connection to a laptop with a custom LabView® program which ingested and saved the sensor data from the band sensor.

4. Evaluation for the use of Natural User Interfaces

NUI is an approach to Human Computer Interaction (HCI) which attempts to use gestures to interact with computing systems that are close to the natural movements of the human body. The development of NUI interaction devices is necessary for the evolution of computing, which is moving towards virtual and augmented reality technologies that are used in a mobile and room-scale or standing user environment. In order to design a motion-capture test specifically for NUI, an analysis was made of current NUI patterns utilized in mobile computing devices including smartphones, Leap Motion® Controller, Microsoft HoloLens®, and the Myo® Armband. A primary application concept for the strain sensor is as integration of the sensor in a smartwatch band. To evaluate this concept a wristband prototype was prepared by integrating the compliant sensor with the Samsung Gear S2®smartwatch, and evaluated using the NUI gesture test described in Table 1.

A previous investigation had focused on the motion capture capability of the textile wristband, placed at different positions around the wrist, and it was shown that multiple sensors would be able to define discrete hand positions [2]. The current work sought to integrate the sensor material in a form that can be built as a wristband for a smartwatch, and additionally to evaluate the gesture recognition ability of a single sensor to identify motions that would be useful from a HCI perspective. The NUI hand gestures are detail in Table 1, and were chosen for their use in different products that currently use an NUI approach to interfacing with a computer. The Samsung Gear S2®band (Figure 1c) was used in the NUI evaluation. The results are displayed in Figure 2where relative level of different gestures can be used to distinguish one from another in certain cases.

Table 1. NUI gesture evaluation test description.

Gesture	Hold Time	Inspiration / Intent	Used in Product
Double-click	-	Historical interaction pattern	HoloLens
Wave right	-	Natural forward movement	Myo / Leap
Wave left	-	Natural backward movement	Myo / Leap
Pinch	5 sec	Learned 2D surface pattern	Smartphones
Zoom	5 sec	Learned 2D surface pattern	Smartphones
Closed fist	5 sec	Hold a chosen virtual object	Myo
Flat Hand	5 sec	Logical starting/recovery position	
Spread-open hand	5 sec	Release an object or expand a workspace	Myo
Wrist Flexion	5 sec	Move an object up	Myo
Wrist Extension	5 sec	Move an object down	Myo

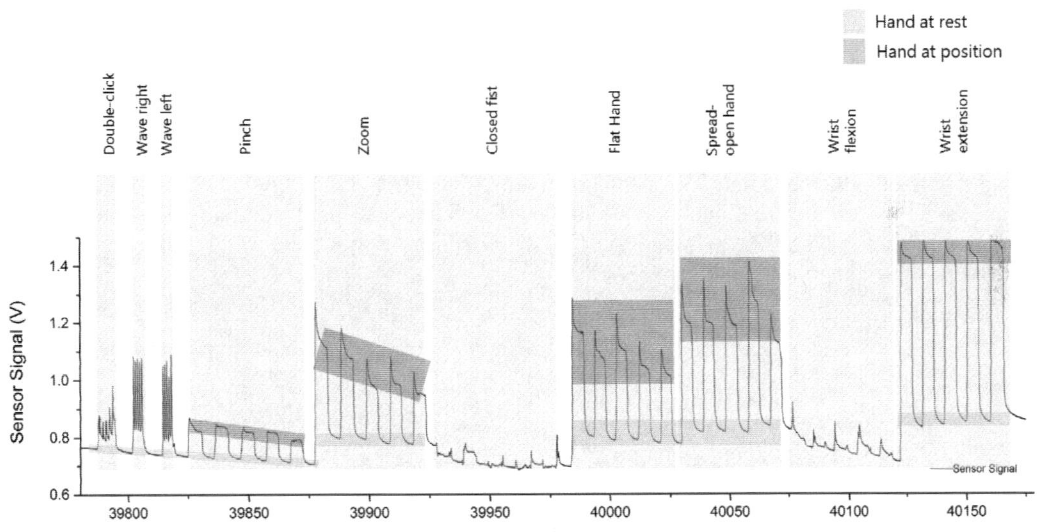

Figure 2. Sensor performance for the NUI gesture test sequence

5. Evaluation for the use of Pulse-Wave Monitoring

The pulse wave of the blood was detected at the three positions around the wrist. During the test, the subject sat in a chair with his hand on a table in front of him, the angle of his arm approximately 90° and in a relaxed state. The band sensor was secured around the wrist of the test subject so that it was tensioned but comfortable to wear. Data was acquired from the sensor for5 min at each of the three sensor positions (P1, P2, and P3). Given the position of the sensor with respect to the location of the ulnar and radial artery, it was expected that the posterior position (P3) would have the highest pulse wave signal. Figure 3 displays the raw data, which depicts the pulse wave of the test subject from

Position 3, which produced the best monitoring result where the peaks were discernable. In analyzing the waveform, the peaks and troughs were visually chosen, and then the Beats-Per-Minute (BPM) were calculated for the peak-to-peak and trough-to-trough data pairs. The average was taken for peak-to-peak and trough-to-trough data sets, and then the average of those final values was taken, resulting in a BPM of 60. This value was consistent for that test subject. Previous results of pulse-wave detection were done with the textile band sensor in previous work [2].

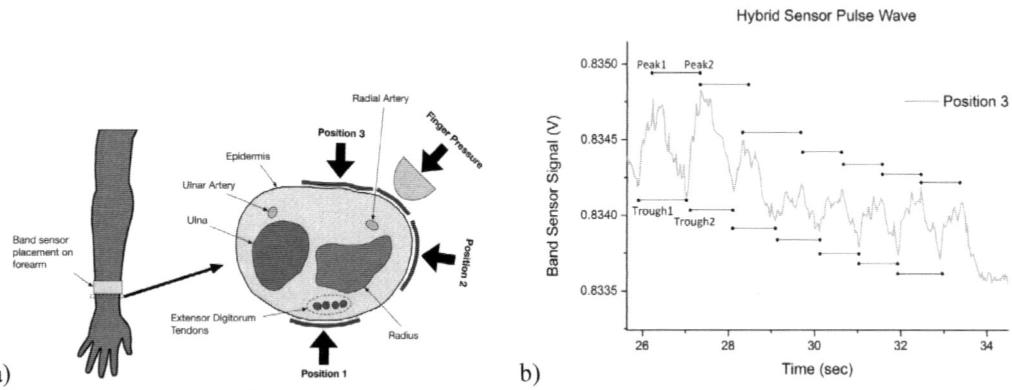

Figure 3. Positioning of the sensor on the wrist (a) and the pulse-wave raw data from Position 3 (b)

6. Evaluation for the use of Breathing Monitoring

As a first validation of the band sensor to the case of breath monitoring, a prototype chest band was developed using the textile band sensor configuration, with the SCMS material glued to an elastic textile substrate [2]. The textile sensor band was placed the band sensor near the abdomen of a patient, and then different breathing patterns were performed by the test subject, and the data analyzed. The data was collected through a smartwatch prototype, which included an accelerometer as well as two pressure sensors, which monitored the contact pressure. Therefore, the difference between the relative sensor values of the band sensor and the on-board accelerometer and pressure sensors could be compared against one another to gauge the difference in sensitivity of the different sensor types to measure breathing patterns.

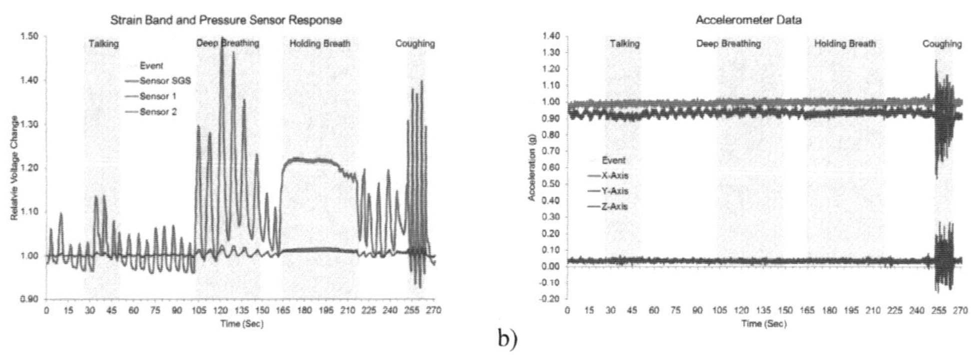

Figure 4. Sensor response for the band sensor (SGS) and pressure sensors (a), associated acceleration data (b)

The band sensor showed the highest relative signal change of the different sensors as shown in a, with the pressure sensors (a) providing better sensitivity than the accelerometer (b). Differentiation can be easily seen between normal breathing, deep breathing, coughing, and breath holding. Differentiation since the SCMS enables continuous monitoring of the strain state, it would be feasible to discern between breathing states based on the amplitude along with frequency, in order to determine the relative change in frequency and breath deepness or coughing intensity and classify the different types.

7. Conclusions

The integration of a CB/TPE sensor in to a hybrid structure of 3D printed TPE and casted silicone showed the ability to fulfill different body-mounted sensor application concepts, including gesture recognition, pulse-wave monitoring, and breathing monitoring.Although the SCMS showed good performance in a controlled state, future applications would require evaluation during normal movement of a person, where motion artifacts may corrupt the desired vital monitoring signal (motion, breathing, hear rate).

References

[1] Melnykowycz M, Koll B, Scharf D, Clemens F. Comparison of Piezoresistive Monofilament Polymer Sensors. Sensors 2014, 14, 1278-1294.
[2] Melnykowycz M, Tschudin M, Clemens F. Piezoresistive Soft Condensed Matter Sensor for Body-Mounted Vital Function Applications. Sensors. 2016, 16, 326.
[3] Almgren A F. An isotropic three-dimensional structure with Poisson's ratio =−1. J. Elast.1985, 15, 427–30.

IOP Conf. Series: Materials Science and Engineering **175** (2017) 012007 doi:10.1088/1757-899X/175/1/012007

Analysis of structure and dynamics of superfine polyhydroxybutyrate fibers for targeted drug delivery

A Olkhov[1,3], E Kucherenko[3], P Pantyukhov[1,2], A Zykova[1,2], S Karpova[2], A Iordanskii[3]

[1]Plekhanov Russian University of Economics, 113054 Moscow, Russia
[2]Emanuel Institute of Biochemical Physics of Russian Academy of Sciences, 119334Moscow, Russia
[3]Semenov Institute of Chemical Physics of Russian Academy of Sciences, 119334Moscow, Russia

E-mail: zykovaak@yandex.ru

Abstract. Creation of polymer matrix systems for targeted drug delivery into a living organism is a challenging problem of modern treatment of various diseases and injuries. Poly-3- hydroxybutyrate (PHB) is commonly used for development of therapeutic systems. The aim of this article is to examine the changes in structure and morphology of fibers in presence of dipyridamole (DPD) as model drug for controlled release. It was found that addition of dipyridamole led to disappearance of spindle-shaped nodules on fibers of PHB in comparison with pure PHB. The research of thermophysical parameters showed that specific melting enthalpy (and the degree of crystallinity) of PHB fibers increased with the addition of DPD. With the increasing of DPD content in PHB fibers, more perfect and equilibrium crystal structure was formed. According to analysis of intercrystalline regions of PHB fibers, it was found that as the crystallinity of PHB in intergranular regions rose, the corresponding decrease of radical rotation speed was observed. It was concluded that fibers of PHB can be used for creating therapeutic systems for targeted and prolonged drug delivery.

1. Introduction

Creation of polymer matrix systems for targeted drug delivery into a living organism is an actual problem of modern treatment of various diseases and injuries [1]. Poly-3-hydroxybutyrate (PHB) is used for creating these therapeutic systems in most cases. High biocompatibility, optimum time of biodegradation and the mission-related physical and mechanical properties of films and PHB fibers allow to consider this polymer as one of the most promising medical polymers [2].

2. Experimental part

2.1. Materials and Preparation of Samples

In this work, polyhydroxybutyrate of the 16F series ((BIOMER®, Germany) with a molecular weight Mw = 2.06x105 g/mol was used to produce investigated fibers. As a model drug for controlled release was used dipyridamole (2,2',2'',2'''-(4,8-di(piperidin-1-yl)pyrimido[5,4-d]pyrimidine-2,6-diyl)bis(azanetriyl)tetraethanol). Pharmacological class of dipyridamole (DPD): antiplatelet, angioprotectors and proofreaders of microcirculation, adenosynergic tools. Superfine PHB fibers were obtained by electrospinning. Spin dopes including PHB and PHB with 1%, 3% and 5% DPD in chloroform.

IOP Conf. Series: Materials Science and Engineering **175** (2017) 012007 doi:10.1088/1757-899X/175/1/012007

2.2. Measurements

The electron paramagnetic resonance spectra (X-band) were recorded on an EPR-V automated spectrometer (Semenov Institute of Chemical Physics, RAS, Moscow) at microwave power of 7 mW and modulation amplitude of 0.5 gauss. As a probe, the stable nitroxyl radical 2,2,6,6-tetramethylpiperidin-1-oxyl (TEMPO) was used. The radical was incorporated into fibers from vapor at a temperature of 50°C to a concentration not exceeding 10-3 mol/l. The DSC measurements were performed on a Netzsch DSC 204 F1 calorimeter in argon atmosphere at a heating rate of 10 K/min. The morphology of the fibers was studied via electron microscopy on Hitachi TM-1000 scanning electron microscope (Japan).

3. Results and discussion

As can be seen from Fig. 1, in the PHB fibers there is a significant number of spindle-shaped nodules of 5-7 μm in diameter and 15-20 μm in length.

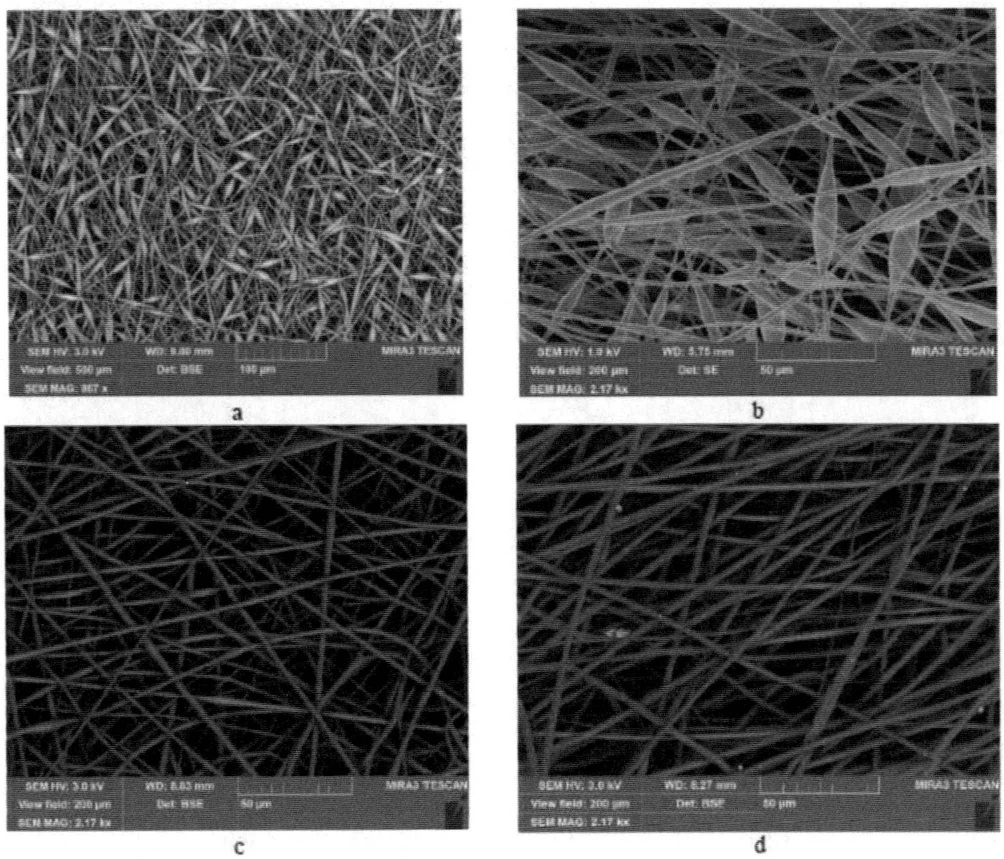

Figure 1. (a) Electron micrographs of the PHB electrospun fibers, **(b)** Electron micrographs of the PHB + 1% DPD electrospun fibers, **(c)** Electron micrographs of the PHB + 3% DPD electrospun fibers, **(d)** Electron micrographs of the PHB + 5% DPD electrospun fibers

Number of nodules decreases with increasing of concentration of drug substance (DS) and absolutely is not observed for PHB fibers with the highest concentration (5%) of DPD. The presense of nodules may be associated with insufficient conductivity of a spinning solution or with its sub-optimal viscosity.

Thermophysical and structural characteristics have been investigated by DSC method. The values

IOP Conf. Series: Materials Science and Engineering **175** (2017) 012007 doi:10.1088/1757-899X/175/1/012007

of melting enthalpy (AH) and the melting temperature (Tm) of superfine PHB fibers with DPD are presented in Table 1.

Table 1. Thermophysical Parameters of Superfine PHB Fibers with DPD Investigated by DSC

Thermophysical Parameters	PHB	PHB+1% DPD	PHB +3% DPD	PHB +5% DPD
AH, J/g	42,9	53,7	72,0	79,7
Tm., ^{0}C	168,70	168,50	166,70	169,30

It follows from the Table 1 that the specific melting enthalpy (and the degree of crystallinity) of PHB fibers increases with the addition of DPD. With the increasing of DPD content in PHB fibers, more perfect and equilibrium crystal structure was formed.

The structure and molecular dynamics of intercrystalline regions were investigated by the electron-spin resonance (ESR) method. It was found that as the crystallinity of PHB in intergranular regions rose, the corresponding decrease of radical rotation speed was observed. That caused the slowdown of molecular mobility of segments. With the increasing content of DPD the correlation time increases in the sequence: $6{,}6\text{x}10^{-9}$ (from 0%) > $7{,}1\text{x}10^{-9}$ (1%) > $8{,}8\text{x}10^{-9}$ (3%) > $9{,}0\text{x}10^{-9}$ (5%). This effect may be explained by the increase of crystallinity of PHB fibers.

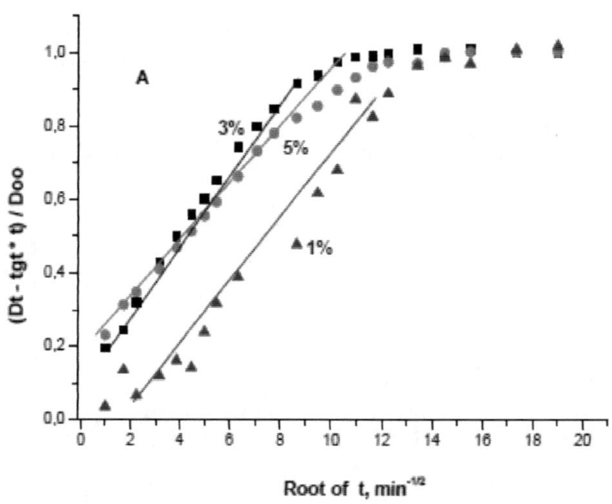

Figure 2. Ratio of absorbances excluding the destructive component (A), linear areas of kinetic curves in diffusion coordinates (B)

The kinetic curves of drug release are shown in Fig. 2, characterized by an initial non-linear time area and the final area, where the concentration of desorbed DPD increases linearly. A superposition of diffusion and kinetic processes was observed for all kinetic curves of drug release from the PHB fibers with DPD.

Nonlinear area of DS release kinetic curve is defined by the diffusion process. The abrupt change of diffusion coefficient corresponds to structural transition from spindle-shaped particles to cylindrical ones. The rate of destruction process exponentially decreases with increasing DS, what has a good agreement with the previous data. The presence of two release processes (diffusion and kinetic) suggests that DPD exists in PHB fibers in two forms. First form is the DS in "free form", and just this form is able to desorb DS from PHB fibers in accordance with the diffusion mechanism. Second form of DS exists in the immobilized state in the polymer and is characterized by an extremely low diffusion mobility, but its release from the polymer is supported by the partial degradation of PHB (weight loss with the inclusion of immobilized DS form) by zero-order equation.

4. Conclusion

In the absence of DPD or its low levels (0% and 1%) fibers have a complex geometry with the alternation of cylindrical fragments about 3.1 microns in diameter, typical for conventional fibers and spindle-shaped nodule fragments with an average diameter of 5.7 microns.

With the increasing of DPD content in PHB fibers up to 5 % these fragments disappear, and the fiber geometry becomes cylindrical only.

It is shown that the addition of DPD to PHB polymer matrix leads to a significant increase in crystallinity and to a slowdown of molecular mobility in the amorphous regions of superfine fibers.

All results, especially the impact of DS concentration on fiber shape and its dynamic characteristics are in good agreement with the thermophysical parameters and should be used for creating therapeutic systems for targeted and prolonged drug delivery.

References

[1] N. Martinho, C. Damgé, C.P. Reis, J. Biomat. Nanobiotech. **2**, 510-526 (2011).
[2] S.G. Karpova, A.A. Ol'khov, A.L. Iordanskii, S.M. Lomakin, N.S. Shilkina, A.A. Popov, K.Z. Gumargalieva, A.A. Berlin, Polym. Sci. A **1**, 76-86 (2016).
[3] Yu.N. Filatov in *Electrospinning of fibrous materials,* edited by V.N. Kirichenko (Neft i gaz, Moscow, 1997), pp.167-171.

IOP Conf. Series: Materials Science and Engineering **175** (2017) 012008 doi:10.1088/1757-899X/175/1/012008

Fibrous materials on polyhydroxybutyrate and ferric iron (III)-based porphyrins basis: physical-chemical and antibacterial properties

A Olkhov [1,3], A Lobanov[3], O Staroverova [3], P Tyubaeva [1,2], A Zykova [1,2], P Pantyukhov[1,2], A Popov [1,2], A Iordanskii[3]

[1]Plekhanov Russian University of Economics, 113054 Moscow, Russia
[2]Emanuel Institute of Biochemical Physics of Russian Academy of Sciences, 119334 Moscow, Russia
[3]Semenov Institute of Chemical Physics of Russian Academy of Sciences, 119334 Moscow, Russia

E-mail: zykovaak@yandex.ru

Abstract. Ferric iron (III)-based complexes with porphyrins are the homogenous catalysts of auto-oxidation of several biogenic substances. The most perspective carrier for functional low-molecular substances is the polymer fibers with nano-dimensional parameters. Application of natural polymers, poly-(3-hydroxybutyrate) or polylactic acid for instance, makes possible to develop fiber and matrice systems to solve ecological problem in biomedicine The aim of the article is to obtain fibrous material on poly-(3-hydroxybutyrate) and ferric iron (III)-based porphyrins basis and to examine its physical-chemical and antibacterial properties. The work is focused on possibility to apply such material to biomedical purposes. Microphotographs of obtained material showed that addition of 1% wt. ferric iron (III)-based porphyrins to PHB led to increased average diameter and disappeared spindly structures in comparison with initial PHB. Biological tests of nonwoven fabrics showed that fibers, containing ferric iron (III)-based tetraphenylporphyrins, were active in relation to bacterial test-cultures. It was found that materials on polymer and metal complexes with porphyrins basis can be applied to production of decontamination equipment in relation to pathogenic and opportunistic microorganisms.

1. Introduction

Development of polymer modification of biologically active substances is the one of the most promising field of engineering chemistry. It is focused on synthesis of polymer materials with significant biomedical properties. Ferric iron (III)-based complexes with porphyrins are the homogenous catalysts of auto-oxidation of several biogenic substances. In this process the formation of active forms of oxygen as intermediate products of reaction occurs, such as superoxide anion radicals, peroxy and hydroperoxy radicals, hydrogen peroxide. The cytostatic activity of these substances is well known. These radical and ion-radical particles cause oxidative and destructive reactions in cells, which define its antibacterial effect [1, 2]. The most perspective carrier for functional low-molecular substances is the polymer fibers with nano-dimensional parameters. The electrostatic spinning or electrospinning of polymer solution is one of the most updated technologies of obtaining such fibers.

Low cost of equipment, simplicity of instrumentation, variety of technological conditions, and wide range of obtained fibers and materials on their basis are among the advantages of electrospinning [3].

IOP Conf. Series: Materials Science and Engineering **175** (2017) 012008 doi:10.1088/1757-899X/175/1/012008

Application of natural polymers, poly-(3-hydroxybutyrate) or polylactic acid for instance, makes possible to develop fiber and matrice systems to solve ecological problem in biomedicine. These materials are biocompatible and have controlled biodegradability without formation of toxic substances [4, 5].

2. Experimental part

2.1. *Materials and Preparation of Samples*

In current work the new material on polyhydroxybutyrate (PHB) fibers and ferric iron (III) complexes with tetraphenylporphyrins basis was obtained. The structural formula of ferric iron (III)-based porphyrins is presented in Figure 1

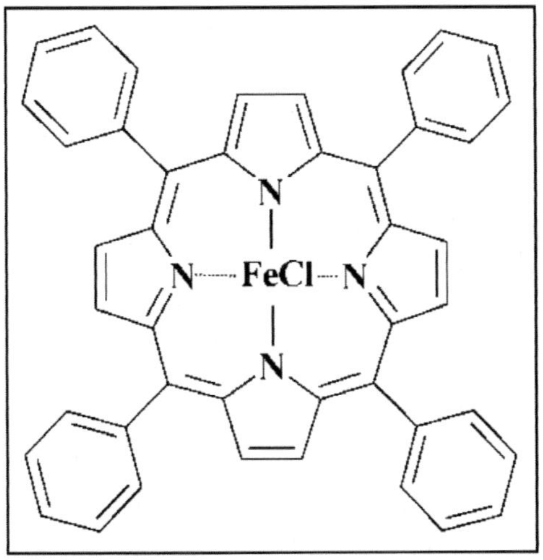

Figure 1. Ferric iron (III) complex with tetraphenylporphyrin

Ultrathin fibers of PHB fibers with addition of 1,3 and 5 % wt. of ferric iron (III)-based porphyrins (in relation to content of PHB in solution) were obtained by electrospinning method in chlorophorm solution. The single-capillary installation EFW-1 (Russia) was used with voltage 11kV. The distance between electrodes was 15 cm. The concentration of polymer in spinning solution was 7% wt. The dynamic viscosity of solution was 0,9 Pa*s and volumetric flow rate of spinning solution was (10-12)*10^5 g/s.

2.2. *Measurements*

The morphology of fibrous materials was examined by electronic microscopy method with using of electronic microscope Hitachi TM-1000 (Japan). Biocidal activity was determined by diffusion into agar medium, containing test-cultures. The parameter was defined by diameter of bacteriostasis zones. Test-cultures of *Staphylococcus aureus* P209 and *Escherichia coli* 1257 with different dilution were used in current work. Test- cultures of microorganisms were cultivated in beef-extract agar at temperature 37°C during 24 hours, then wipe samples from fibrous material were made. These samples were placed in test pit with the diameter 3 mm. Inoculations were incubated within 24 hours at temperature 37°C. Then the measurements of bacteriostasis zones were made.

3. Results and discussion

In dependence to characteristics of spinning solution and technological parameters of electrospinning, the fibers with different diameter, density, geometry can be obtained [3, 5]. In Fig. 2a and 2b the microphotographs of nonwoven fabric on PHB with addition of 1% wt. ferric iron (III)-based porphyrins are shown. As it is seen, for initial PHB significant number of spindly thickenings was observed. The most common reason of thickenings is disturbance of steady state of spinning solution outflow cause of insufficient conductivity of spinning solution or its suboptional viscosity [3]. As it is seen in Fig. 2a for single fiber the sequence of cylindrical fibers with the diameter of 1-3 μm and spindly elements with maximum diameter ~10 μm and length 20-30 μm was observed. With the addition of 1% wt. ferric iron (III)-based porphyrins average diameter of such fibers increased insignificantly 2-4 μm, but spindly structures practically disappeared. Only single local thickenings on fibers were observed. Possibly, this effect is explained by increased conductivity of spinning solution in presence of polar molecules of ferric iron (III)-based porphyrins. Following increase of concentration of ferric iron (III)- based porphyrins in spinning solution caused disappearance of thickenings on fibers. Thus, it was concluded that ferric iron (III) complex with tetraphenylporphyrin can be used as technological additive, which increases conductivity of spinning solution of PHB in chlorophorm.

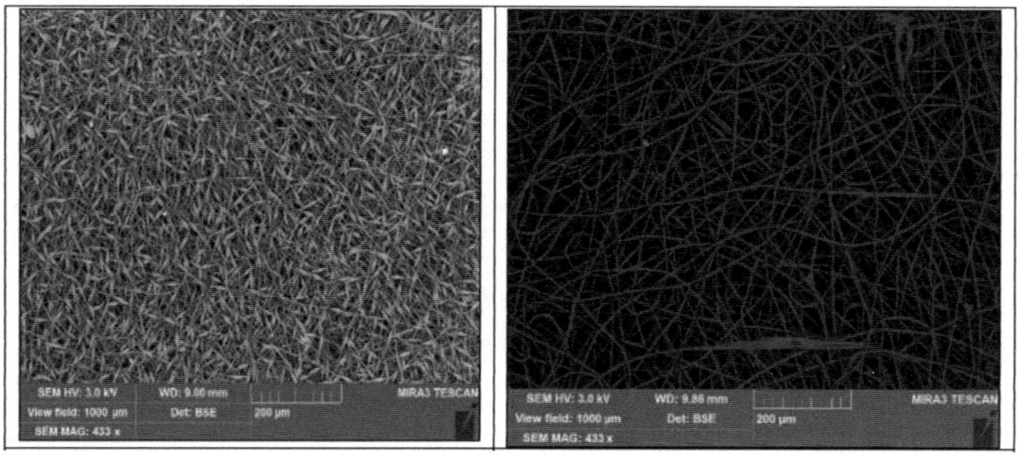

(a) (b)

Figure 2 (a). Electron microphotographs of fibrous material on PHB basis (b). Microphotograph of fibrous material on PHB and ferric iron (III)-based porphyrins basis

Biological tests of nonwoven fabrics showed that fibers, containing ferric iron (III)-based tetraphenylporphyrins, were active in relation to bacterial test-cultures. It was found that materials on polymer and metal complexes with porphyrins basis can be applied to production of decontamination equipment in relation to pathogenic and opportunistic microorganisms.

4. Acknowledgments

Authors would like to express their gratitude to A.L. Buchachenko Leading School (project NS-5226.2014.3) and to Russian Fond of Fundamental Research (project N° 15-03-03591) for financial support.

References

[1] B.D. Berezin, *Coordination compounds of porphyrins and phtalocyanins* (Science, Moscow, 1978),- p. 280.

[2] A.V. Lobanov, O.V. Nevrova, V.A. Ilatovskii,G. V. Sin'k, G.G. Komissarov,

Macroheterocycles 2, 132134 (2011).
[3] Yu.N. Filatov in *Electroforming of fibrous materials,* edited by V.N. Kirichenko (Neft and Gas, Moscow, 1997), - p. 298.
[4] M. Unverdorben, A. Spielberger, M. Schywalsky, D. Labahn, S. Hartwig, M. Schneider, D. Lootz, D. Behrend, K. Schmitz, R. Degenhardt, M. Schaldach, C. Vallbracht, Cardio-vasc. Intervent. Radiol. 2, 127 (2002).
[5] S.G. Karpova, A.A. Olkhov, A.L. Iordanskii, S.M. Lomakin, N.S. Shilkina, A.A. Popov, K.Z. Gumargalieva, A.A. Berlin, Polym. Sci. Series A, 1, 61-72 (2016)

IOP Conf. Series: Materials Science and Engineering **175** (2017) 012009 doi:10.1088/1757-899X/175/1/012009

Surface treatment and corrosion behaviour of austenitic stainless steel biomaterial

M Oravcová[1], P Palček[1], V Zatkalíková[1], T Tański[2], M Król[2]

[1]University of Žilina, Department of Materials Engineering, 010 26 Žilina, Slovak Republic

[2]Institute of Engineering Materials and Biomaterials, Faculty of Mechanical Engineering, 44 – 100 Gliwice, Poland

E-mail: monika.oravcova@fstroj.uniza.sk

Abstract. In this article results from corrosion behaviour of austenitic stainless steel AISI 316L after different surface treatments are published. "As received" surface and surface after grinding resulted in lower resistance to pitting corrosion in physiological solution than electrochemically polished in H_3PO_4+H_2SO_4+H_2O. Electropolishing also improved the surface roughness in comparison with the "as received" surface. Deposition of Al_2O_3 nanometric ALD coating improves the corrosion resistance of stainless steel in chloride-containing environment by shifting the breakdown potential toward more positive values. This oxide coating not only improves the corrosion resistance but it also affects the wettability of the surface, resulting in hydrophobic surface.

1. Introduction

Austenitic stainless steels are still widely used as biomaterials to replace structural components in the human body. They possess more superior tensile strength, fatigue strength, and fracture toughness compared to polymeric and ceramic materials. Such, austenitic biomaterials are used in medical devices such as artificial joints, bone plates, screws, intramedullary nails, spinal fixations, external fixators, stents, and dental implants. These steels have very good mechanical and technological properties combined with good corrosion resistance. The corrosion resistance of the steel increases with increase of chromium content. Their excellent resistance to uniform corrosion depends on formation of passive oxide layer on the surface. Metallic biomaterials have a tendency to spontaneously form stable oxide layer in the reaction with atmospheric oxygen or most aqueous environments and this passive layer is just few nanometers thick. The biocompatibility of the metallic biomaterial mainly depends on the properties of their surface quality and quality of native oxides that forms on the surface. Aggressive ions that are present in human body environment (chlorides and bromides) can cause local breakdown of this protective film, and enable a progress of the local pitting corrosion. This stainless orthopedic implant can corrode in body environment and release iron, chromium, and nickel ions [1-4].

Resistance of stainless steels to pitting corrosion is affected by a number of factors such as chemical composition, structure, quality of surface, concentration of aggressive ions, temperature and pH. The quality of the surface passive film is one of the most important factors of corrosion behavior of stainless steels. The passive film properties are strongly influenced by mechanical and chemical finishing processes. The mechanical treatment (grinding, polishing) affects the electrochemical and mechanical stability of the passive film and near surface layers by changing the surface reactivity. By mechanical, wearing, grinding or mechanical polishing an amorphous layer forms on the surface of metals called the "Beilby" layer. To avoid creating this layer electrochemical polishing can be used.

During the process of electrochemical polishing a high quality passive oxide film is created on the surface. This process also improves the resistance against contamination and bacteria. This treatment is very useful in medicine whereas surgical instruments and devices, joint replacements, stents and other implantable materials are electrochemically polished. Also several coating methods have been used to deposit the material's surface and to increase the corrosion resistance.One of the deposition methods can be oxide coatings realizedby the method called atomic layer deposition. By this method a few nanometers high protective oxide layer is applied on the metal's surface. The thickness of this oxide layer affects the homogeneity, the roughness, the surface energy and the wettability. The surface wettability is important for the adsorption of proteins and cell adhesion and can dictate the biocompatibility of the implanted material. Depending on the material's applicationthere is a need to achieve hydrophobic or hydrophilic surfaces in the case of biomaterial's wettability. For instance stents usedin cardiology, theyneed to have hydrophobic surface.The higher the hydrophobic surfaces it more suppresses protein and platelet adsorption and ensures good blood compatibility[3-7].

2. Atomic layer deposition

Atomic layer deposition (ALD) is a modification of chemical vapor deposition (CVD) method used for depositing surface controlled thin films one atomic layer at a time. During the coating, two or more chemical vapors or gaseous precursors react sequentially on the substrate surface and produces a solid thin film. This technology provides the ability to obtain very thin, almost defect-free nanometric coatings on the surface of different materials. In order to increase the corrosion resistance of austenitic stainless steel biomaterials, thin layersof Al_2O_3 or TiO_2 can be applied on the materials surface by ALD technology [8-9].

3. Experimental material and procedures

Commercially available low-carbon austenitic stainless steel AISI 316L was used as an experimental material with the chemical composition listed in Table 1. The specimens for ALD process were cut from rectangular bars with dimension of 10x10x5 mm and the other specimens for electrolytic polishing were purchased in sheets of 1.5 mm thickness. Microstructure of the material in initial state is presented in Figure 1 - 2. Microstructure of the experimental material is formed of polyedric austenitic grains with annealing twins and parallel lines arose by rolling.

Table 1. Chemical composition of austenitic stainless steels.

	Chemical composition (wt. %)								
	C	Cr	Ni	Mo	Mn	Si	S	P	Fe
AISI 316L	0.04	17.41	9.91	2.38	1.34	0.52	0.03	0.05	bal.
AISI 316L (sheet)	0.02	16.79	10.14	2.03	0.82	0.31	0.001	0.03	bal.

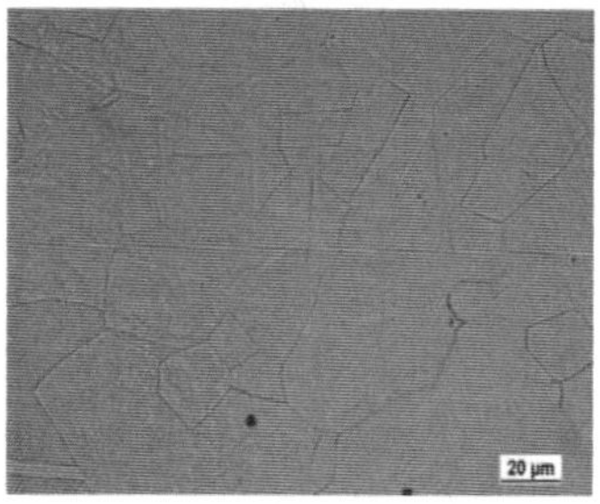

Figure 1. Microstructure of AISI 316L, cross section, initial state (Kalling´s etchant)

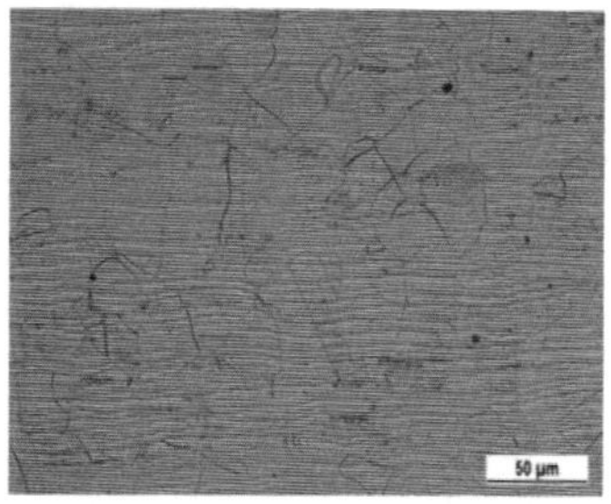

Figure 2. Microstructure of AISI 316L, longitudinal section, initial state (Kalling´s etchant)

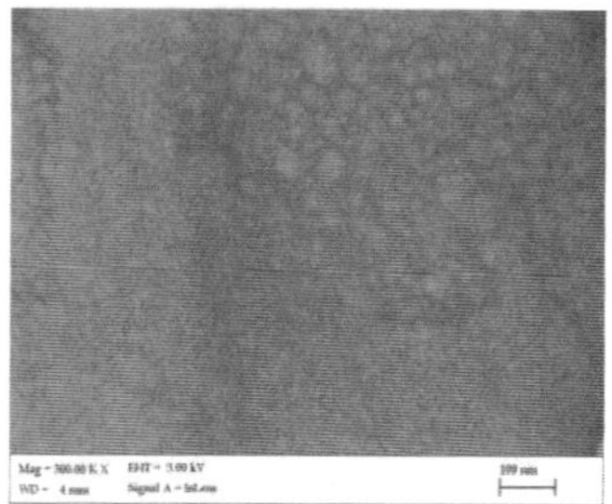

Figure 3.SEM image of ALD coating of Al_2O_3

3.1. Surface preparation

The ALD coating (Figure 3.) was deposited using PICOSUN^{TM} R-200 at the Silesian University of Technology, Gliwice. Al_2O_3 layer was obtained using trimethylaluminum and water precursors. During the Al_2O_3 ALD process, trimethylaluminum (TMA) reacts with hydroxyl groups on the surface. This reaction proceeds until the surface reaction is completed. Subsequently, TMA is carried or pumped away with the inert gas (Nitrogen). The same process is then performed with H_2O. Water reacts with methyl groups on the surface until this surface reaction is completed. The sequential exposure to TMA and H_2O represents one AB cycle. These AB cycles are repeated to achieve the desired film thickness [8-9].The number of precursor cycles of deposition was 500 cycles.The deposition was performed at a temperature of 200°C.

Other specimens were tested after electrochemical polishing, mechanical grinding, "as received" surface and electrochemically polished surface after sensitization.

Electrochemical polishing was performed in $H_3PO_4 + H_2SO_4 + H_2O$ electrolyte under the conditions listed in Table 2. The polished specimen was immersed into electrolyte and connected to the positive pole of the power source as an anode. A cathode was of the same material as the anode.

Table 2. Conditions of electrochemical polishing

Component	Content [wt.%]	Temperature [°C]	Current density [A.cm^{-2}]	Time [s]
H_3PO_4	64			
H_2SO_4	13	40	0.8	420
H_2O	23			

3.2. Potentiodynamic polarization tests

Potentiodynamic polarization tests were carried out in a three electrode cell of corrosion measuring system (VoltaLab 10 with VSP unit). Potentiodynamic polarization curves were obtained by the EC-LAB SOFT software. The potential between the sample and the electrolyte had settled for 10 minutes before the polarization. The scan range was -0.3V – 2.0 V vs the open circuit potential and the scan rate was 1mV/s. The surface of AISI 316L working electrode of 1 cm^2 area was rinsed with ethanol before measurement. The saturated calomel electrode (SCE) was applied as the reference electrode and platinum foil as a counter electrode. All experiments were performed in physiological solution represented by 0.9 % NaCl solution (i.e. 0.15 M NaCl).

3.3. Surface wettability

The surface wettability, the wetting angle,was evaluated on the samples in initial state after polishing, then samples with ALD coating and some of the samples underwent heat treatment before the measurement and polished also. Measurement of the contact angle (θ) was performed using distilled water. The measurement of a drop of liquid applied to the sample's surface was performed at room temperature (T = 23 °C) at the test stand incorporating SURFTENS UNIVERSAL goniometer by OEG and a computer software Surftens 4.5 to analyze the recorded image of the drop. Five drops of distilled water, each 1.5 μl volume, was applied on the surface of each sample. The measurement started after 15 s after application of drops. The duration of one measurement was 60 s at the sampling frequency 1 Hz.

4. Results

Potentiodynamic polarization curves express dependence of the current density i [mA.cm^{-2}] on the potential E [V]. They enable determination of the passivity region and the pitting potential E_p, which is an important electrochemical characteristic of the resistance to pitting. E_p can be located as the potential of the strong increase in current density on the polarization curve of direct measurement. When the potential reaches this critical value, current density suddenly increases, denoting the breakdown of the passive film and the beginning of a stable pit growth. Polarization curves of AISI 316L stainless steel of variously treated surfaces in physiological solution are shown in Figure 4.

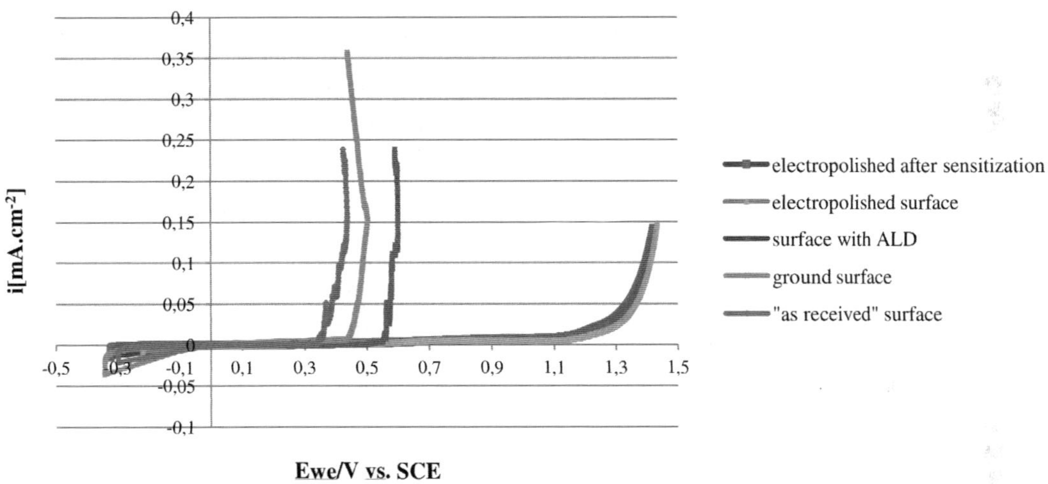

Figure 4. Polarization curves of AISI 316L

In the case of curves of"as received" surface, ground surface and the surface with $Al_2O_3$the passivity region and strong increase of current density occurred at pitting potential 0.343 V for the sample with "as received" surface; 0.439 V for the sample after mechanical grinding; 0.561 V for the sample with ALD coating. Passivity region of electropolished sample is very broad, no breakdown of passivity, which indicates initiation of pitting corrosion, was recorded. Current density suddenly increases at the potential approximately 1.2 V, which can be considered as transpassivation potential E_t. If potential is equal or higher than E_t, the surface passive film is dissolved, and active corrosion of material starts. The rise of current density at the transpassivation potential E_t points to the beginning of Cr_2O_3 dissolution and formation of soluble chromates or other compounds arose from the electrolyte which do not create a protective barrier between the metal and the electrolyte or prevents from creating the passive film on the surface. However, it should be noted that an active corrosion of tested biomaterial in transpassive region in internal human's body environment is not an option because of high E_t value, which cannot be reached spontaneously under given conditions. Electropolishing did not reveal sensitization of austenitic stainless steel which could be observed by almost the same course of polarization curve as the curve of electropolished sample without sensitization (Figure 4).

After 7 min of electrochemical polishing in electrolyte with the applied current density 0.8 A.cm^{-2} around 0.08 - 0.09 mm layer was removed from the sample's surface. This amount of removed material corresponds with the time and current density in comparison with the layer (0.02 mm) removed from the sample after electrolytic polishing published in [10].

The surface roughness of electropolished and "as received" surface of the sample was measured. The surface roughness is an arithmetic mean average stating the surface roughness parameters (R_a - roughness average, R_q - root mean square roughness, R_z - mean roughness depth) listed in the Table 3. For surface roughness measurement portable Mitutoyo SJ-210 tester was used with the standard measurement conditions: 4 x 10^{-3} N measuring force, VDA= 0.5 mm/s and λ_c= 0.8 mm. We can conclude from the results below that there is an evident decrease in surface roughness average value in the case of electropolished sample in comparison with the "as received" sample.

Table 3. Average surface roughness parameters

	R_a [µm]	R_q[µm]	R_z [µm]
Electropolished surface	0.078	0.096	0.474
"As received" surface	0.256	0.363	2.290

The results of wettability and examples of drops on the sample's surface in initial state and samples with Al_2O_3 are presented in Figure 5. and Figure 6. The ALD deposition resulted in an increase of the wetting angle above 90°. The polished surfaces in initial state are hydrophilic and the surfaces have a very low surface wettability (θ = 82.87°) and applying the ALD coating on the sample's surface resulted in a hydrophobic surface. Steel with the deposited Al_2O_3 layer showed higher wetting angle (θ = 94.39°) as compared to the values obtained in surfaces in initial polished state.

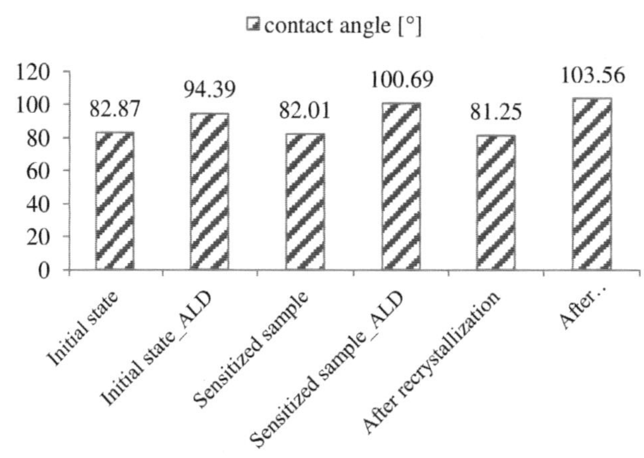

Figure 5. Results of the contact angle measurements of AISI 316L

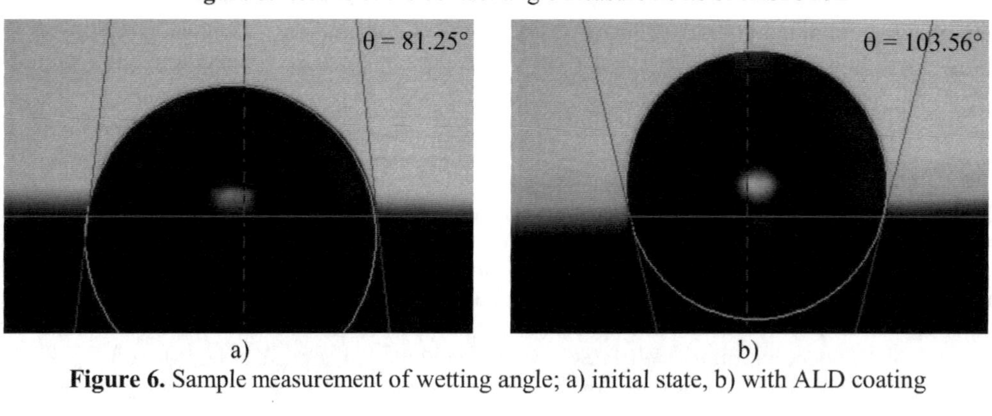

Figure 6. Sample measurement of wetting angle; a) initial state, b) with ALD coating

5. Conclusion

The results show following conclusions:

- Electrochemical polishing polishing is a conventional chemical surface finish which improves the corrosion resistance, results in higher surface quality with high surface brightness and decreases the surface roughness. Surface roughness is decreased due to the removal of particular layer from the metal's surface.
- Electrochemical polishing carried out in H_3PO_4 + H_2SO_4 + H_2O electrolyte according to recommended conditions is an efficient chemical surface treatment to increase AISI 316L resistance to pitting corrosion in physiological solution.
- "As received" surface and surface after grinding showed lower resistance to pitting corrosion than electrochemically polished ones. Therefore, such treated surfaces are not recommended for the internal human's environment.
- Deposition of Al_2O_3 nanometric ALD coating improves the corrosion resistance of stainless steel in chloride-containing environment by shifting breakdown potential towards more positive values.
- Sensitization induced by heat treatment (700 °C/10 hours) does not decrease resistance to pitting corrosion.
- Electropolishing of specimen improved the surface roughness because the "as received" surface had the average value of surface roughness R_a = 0.256 μm and the electropolished surface reached the surface roughness R_a = 0.078 μm.
- The Al_2O_3layer improved the wettability because the sample's surface resulted in a hydrophobic character.

Acknowledgement

This work has been supported by Scientific Grant Agency of Ministry of Education of Slovak republic VEGA 1/0683/15.

References

[1] M NAGASE 1995 Host Reactions to Particulate Biomaterials. Encyclopedic handbook of biomaterials and bioengineering, Vol. **1**, Part **A**, New York

[2] B D RATNER, A S HOFFMANN, F J SCHOEN, J E LEMONS 2004 *Biomaterials Science*. Elsevier, San Diego, USA

[3] Z SZKLARSKA-SMIALOWSKA 2005 *Pitting and crevice corrosion.* Houston, Nace, 590 p

[4] T LIPTÁKOVÁ 2009 Bodová korózia nehrdzavejúcich ocelí. (Pitting corrosion of stainless steels.) Žilina, EDIS – Žilinská Univerzita, 67 p. (in Czech)

[5] D VAŇA 2014 Porovnanie technologických možností plazmového a elektrochemického leštenia antikoróznych ocelí v elektrolyte. Materiálovo technologická fakulta STU, Trnava, Alumni Press, 101 p (In Czech)

[6] Y HAN, J MEI, Q PENG, E H HAN, W KE 2015 Corros. Sci. **98**, 72

[7] A NOURI 2011 Introduction to surface coating and modification for metallic biomaterials. *Surface Coating and Modification of Metallic Biomaterials*, pp 3-60.

[8] E MARIN, A LANZUTTI, L GUZMAN, L FEDRIZZI 2011 *Journal of Coatings Technology and Research* **8** (**5**) 655

[9] E MARIN, A LANZUTTI, L PAUSSA, L GUZMAN, L FEDRIZZI 2015 *Materials and Corrosion* **66** (**9**) 907

[10] S E ZIEMNIAK, M HANSON, P C SANDER 2008 *Corros. Sci.* **50** 2465

IOP Conf. Series: Materials Science and Engineering **175** (2017) 012010 doi:10.1088/1757-899X/175/1/012010

Development and application of nano-and monocrystalline materials for medical miniatures X-ray tubes

A A Pavlov[1], M L Taubin[1,2], D A Chesnokov[1] and A A Yaskolko[1,2]

[1]Federal State Unitary Enterprise "Science Research Institute "LUCH", Zheleznodorozhnaya st, Podolsk, Moscow region, 142100, Russia
[2]National Research Nuclear University MEPhI (Moscow Engineering Physics Institute), Kashirskoe highway 31, Moscow, 115409, Russia

E-mail: taubin@sialuch.ru

Abstract The concept of increase of characteristics x-ray tubes by use nano-and monocrystal materials is offered in the report.

1. Introduction

Results of own works of the authors and works of the Russian, Japanese, Korean and American experts in refractory metal nano composites with increased strength characteristics for cathodes and anodes are analyzed in this report. It is shown, that though monocrystals have shown the better characteristics in comparison with traditional polycrystals both as anodes and as cathodes, more radical way of increase of operational characteristics of the x-ray tubes is application refractory nanostructural materials.

The study of emission characteristics shows that single-crystal tungsten with crystallographic orientation of emitting plane (111) has higher emission characteristics (work function equals 4.43 eV) than polycrystalline tungsten.

In this work results of research of evaporation of tungsten are presented. It is shown that the monocrystal has essential advantages on this parameter. It is important for materials working at temperatures more 2000°C.

An increase in the emission characteristics of a cathode and the strength characteristics of an anode at the same time can be achieved by modifying standard tungsten powders (for cathodes), tungsten and molybdenum powders (for anodes) by nanoscale powders of refractory materials (grain size is10-100 nm). W, Mo, Nb, Re nanopowders are proposed to be used for producing anodes based on tungsten and molybdenum. It must increase allowable unit loads of an X-ray tube by reducing porosity, increasing strength and plasticity of anode material. Prerequisites for these proposals are some results of work where an increase in density, strength by 10-30%, and plasticity by a factor of 1.5 - 2 was achieved due to modification of steels by ultra-disperse powders of refractory compounds.

The authors carried out some experiments to determine the creep of tungsten alloyed by potassium on samples as foils 200 μm thick, used when manufacturing X-ray tube cathodes.

The results of these experiments showed that high-temperature sample annealing resulted in a sharp increase (by a factor of more than 3) in material creep resistance at a temperature of 2200°C.The authors think that this effect results from forming nanoscale little bubbles filled with potassium at the grain boundaries.

Results of this work can be used when development x-ray tubes with the increased operational characteristics.

2. Emission properties of tungsten

The problem of X-ray tubes at the present stage of developing medical equipment consists in the necessity of producing X-radiation of high intensity when providing small-size focal spots.

It is of particular importance for X-ray tubes of digital scanning X-ray units and mammography. Cathode problems are key problems for providing required roentgen optical characteristics of X-ray tubes.

The main task for having a focal spot of necessary size is electron focusing. Today, the most promising way of producing small-size focal spots is applying the design of cathodes with flat emitting surface.

A flat-shaped emitter is more efficient as compared with a spiral as it has more uniform electric field distribution and, also, large area which is important for producing X-radiation of high intensity [1].

A decrease in work function for increasing emission current is only possible when using new materials.

The results of many-year investigations of materials conformably to their application as emitters of thermionic energy converters[2,3] showed that tungsten with single-crystal structure could also be promising material for X-ray tube emitters. The results of these investigations showed higher emission characteristics, lower and uniform (as compared with polycrystalline) tungsten evaporation from surface and higher dimensional stability at high operating temperatures due to absence of recrystallization (a considerable disadvantage of emitters made of polycrystalline) up to melting point as well. Proceeding from the above-mentioned, we think that producing a cathode with a flat emitter made of single-crystal tungsten is one of the promising trends in the area under consideration.

The use of single-crystal tungsten as emitter material allows, firstly, reducing the emitter working temperature by 70°C as the work function of the crystallographic face [111] of single-crystal tungsten is (4.4±0.05) eV, being lower by 0.15 eV than the work function of polycrystalline tungsten (4.55±0.05) eV, secondly, making easier electron focusing due to lower emission contrast range of a single crystal.

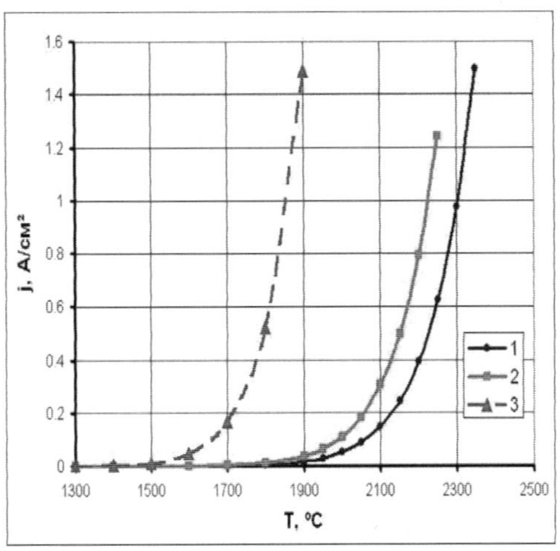

Figure 1. Emission current density of temperature dependence
1 - polycrystalline tungsten;
2 - single-crystal tungsten from plane[111];
3 - nanoscale W

The results of studying the emission characteristics (Figure 1) show that single-crystal tungsten with crystallographic orientation of emitting plane (111) has higher emission characteristics (work function equals 4.43 eV) than polycrystalline tungsten. But best results has nanoscale W (Figure 1).

3. Compare of damageability of poly- and monocrystalline tungsten

The second task for increasing the output characteristics of X-ray tubes is anode problems, i.e. a decrease of its failure.

Doping of tungsten by rhenium is not always optimum, as to mammography it is not acceptable. In this connection, the authors have offered manufactured and tested single-crystal molybdenum (for mammography) and single-crystal molybdenum coated with single-crystal tungsten (for other applications) as anode material.

It is shown that the dependence of reducing in X-ray tube dose rate for a single-crystal anode (Figure 2) differs slightly from the typical dependence for a polycrystalline molybdenum anode. After reducing the dose power by 10% during the first 1500 switches, stabilization takes place, providing large service-life of an X-ray tube.

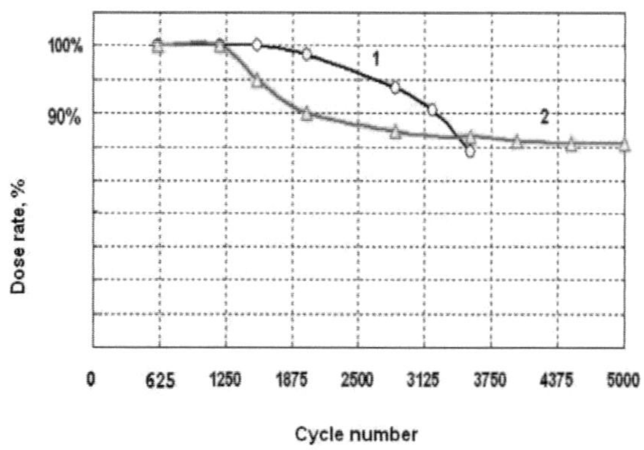

Figure 2. Dose power dependence on the number of switches 1 - polycrystal tungsten doped by rhenium; 2 - single-crystal tungsten

In work results of research of evaporation of tungsten also are presented. It is shown that the monocrystal has essential advantages on this parameter (Figure 3). It is important for materials working at temperatures more 2000°C

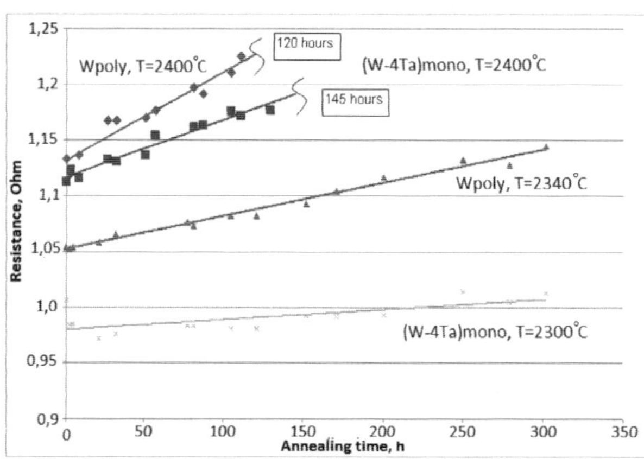

Figure 3. Evaporation of poly- and monocrystalline tungsten

4. High temperature deformation of tungsten

We obtained the effect of an increase in tungsten strength characteristics at high temperature annealings of samples as 100 μm thick foils of BA type tungsten (tungsten with silicon-aluminum and alkali additives) (Figure 4).

Investigations in the structure of this material samples after annealings (Figure 5). Allowed finding out nanodimensional (≤ 200 nm) inclusions along the grain boundaries of tungsten subjected preliminarily to heat treatment, resulting in a sharp increase (by a factor of more than 3) in material creep resistance at a temperature of 2200°C.

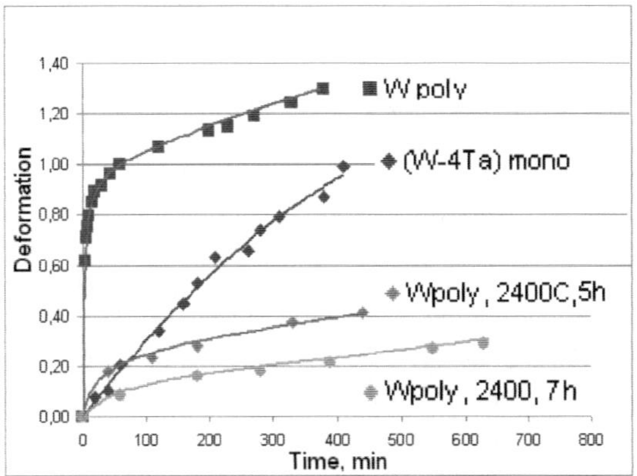

Figure 4. Deformation dependence on time at creep tests. 1 - polycrystalline tungsten without heat treatment; 2 - polycrystalline tungsten preannealed at a temperature of 2500°C for 5 hours. 3 - polycrystalline tungsten preannealed at a temperature of 2500°C for 7 hours. 4 - monocrystalline tungsten.

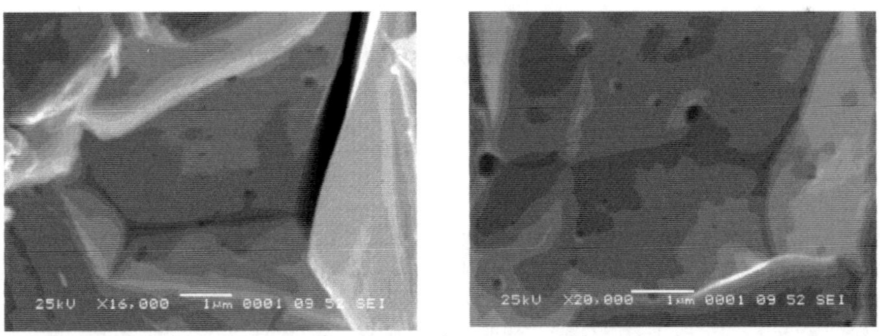

Figure 5. Microstructure of polycrystal tungsten subjected to heat treatment at a temperature of 2500°C for 5 hours

As a whole, a qualitative increase in the operating characteristics of X-ray tubes with the use of nanomaterials and monocrystal materials can be achieved by a simultaneous increase in the density of electron current (a flat cathode of nanotungsten or tungsten nano-composite) and by an increase in load-carrying capacity of an X-ray tube (anode made of nanocomposites of molybdenum with niobium or tungsten with rhenium) by providing a great number of interactions between electrons of increased density and a great number of atoms of anode target materials.

For an establishment of influence of duration of annealing on thermal stability of structure of a material on samples, of annealing within 30 minutes, the diamond pyramid puts marks. Then, these samples were annealing at temperature 2400°C within 1,5 hours and 3 hours.

Thus, temperature stability of this effect as shown is at high (up to 2400°C) temperatures. The similar effect the formation of nanosizes pores in wires of modified tungsten is described in a number of works [4-6].

5. Advantage of nanomaterials for using in X-ray units

Though single crystals showed better characteristics both as anodes and as cathodes, the application of nanomaterials may be a more radical way of increasing the operating characteristics of X-ray tubes.

For instance, the application of nanopowders (Mo, Nb) for producing anodes based on molybdenum or molybdenum alloy with niobium must exclude the influence of the technological disadvantages above-described and increase permissible unit loads to the focal spot due to a decrease in porosity, an increase in strength and plasticity of material.

Some investigation results are known [4] where the work function of nanocrystal tungsten was lower by 0.8 eV than that of coarse tungsten. This phenomenon can be explained by the fact that when forming nanomaterial, current tubes with lower work function appear in the range 10 nm wide near grain boundaries. This effect can considerably increase X-radiation intensity at the same cathode temperatures or decrease the cathode temperature at the same X-radiation intensity by 400°C. Estimations show that at the temperature at which the effect of an increase in the emission characteristics for this material is still available (T ≤1500°C), one may rely on obtaining the density of emission current equal to 10 mA/cm^2.

A number of publications have appeared lately where some results of developing X-ray tubes with the use of carbon nanotubes (CNT) as cathode materials are described.

The development engineers from the University of South Caroline together with the other researchers presented materials on developing a scanning source of X-radiation with a cathode made of CNT. The anode voltage was 40-60 kW, electric field strength - 2.3-3.1 V/cm, emission current density reached 1 A/cm^2.

In parallel with the improvement of the characteristics of cathode made of CNT, the application of nanodimensional dopes may also increase strength properties of tungsten used conventionally.

6. Conclusions

For the first time it is received nanocomposite of tungsten in the form of a foil. Structure of this material is stable up to temperatures 2200 °C.The study of emission characteristics shows that single-crystal tungsten with crystallographic orientation of emitting plane (111) has higher emission characteristics (work function equals 4.43 eV) than polycrystalline tungsten..

Results of research of evaporation of tungsten are presented. It is shown that the monocrystal has essential advantages on this parameter. It is important for materials working at temperatures more 2000°Ct.

The effect of an increase in tungsten strength characteristics at high temperature annealing of samples as 200 μm thick foils of BA type tungsten (tungsten with silicon-aluminum and alkali additives) obtained.

From experimental data it is visible, that deformation of a sample past preliminary thermal processing is lower, than for sample without preliminary thermal annealing. The similar behaviour of a material is connected to distinctions in structure of the tested samples.

The results presented are a cause for developing a new generation of X-radiation sources with increased diagnostic opportunities.

References

[1] Alekseev S, Taubin M and Yaskolko A 2009 *20th Workshop ISTC in Korea Nanomaterals and nanotechnology* (Seul) pp 46-51

[2] Alekseev S V, Taubin M L and Yaskolko A A 2008 *Proc. 2 All-Russia national congresses on radiology*

[3] Alekseev S V, Taubin M L and Yaskolko A A 2008 *Medical physics* **2**

[4] Wright P K. 1978 *Metallurgical Transactions* A **9A** 955
[5] Yamazaki S, Ogura S, Fukazawa Yand Hatae N. 1978 *High Temperature - High Pressure* **10** 329
[6] Briant C L Potassium 1993 *Metallurgical Transactions* **A2** 4A. 1073
[7] Klinkov A E, Nikolaev Yu V and Vybyvanetz V I 1991 *Proc. The second Inter.Confer. on Nuclear Power Enginering in Space. Physics of Thermionic Energy Converters* (Sukhumi) 37

IOP Conf. Series: Materials Science and Engineering **175** (2017) 012011 doi:10.1088/1757-899X/175/1/012011

Silver metal nanoparticles study for biomedical and green house applications

E Rauwel[1], L Simón-Gracia[2], M Guha[3], P Rauwel[1], S Kuunal[1], D Wragg[4]

[1]Tartu College, Tallinn Univ. of Technol., 51008 Tartu, Estonia
[2]Laboratory of Cancer Biology, Institute of Biomedicine, Centre of Excellence for Translational Medicine, University of Tartu, Ravila 14b, 50411 Tartu, Estonia.
[3]Fac. of Medicine, Dept. of Gen. and Mol. Pathology, Univ. of Tartu, Tartu, Estonia
[4]Univ. of Oslo, Dpt. of Chemistry and SMN, 0316 Oslo, Norway

Email: erwan.rauwel@ttu.ee

Abstract. Metallic nanoparticles (MNP) with diameters ranging from 2 to 100nm have received extensive attention during the past decades due to their many potential applications. This paper presents a structural and cytotoxicity study of silver metal nanoparticles targeted towards biomedical applications. Spherical Ag MNPs of diameter from 20 to 50 nm have been synthesized. The encapsulation of Ag MNPs inside pH-sensitive polymersomes has been also studied for the development of biomedical applications. A cytotoxicity study of the Ag MNPs against primary prostatic cancer cell line (PPC-1) has demonstrated a high mortality rate for concentrations ranging from 100 to 200mg/L. The paper will discuss the potential for therapeutic treatments of these Ag MNPs.

1. Introduction

Noble metallic nanoparticles (MNPs) with diameter ranging from 2 to 100 nm have received extensive attention during the past decades due to their many potential applications in fields such as catalysis, biomedicine and cancer therapy. The biocidal properties of Ag MNPs have been known for eons, therefore making the study of Ag MNPs even more attractive for biomedical applications[1]. In fact, among noble metals, silver nanoparticles possess the most efficient antibacterial properties with a very broad bactericidal and fungicidal activity spectrum and have been used for the development of new generation of antibacterial burn wound bandage, water filter, antibacterial clothing, etc. In addition, Ag MNPs have a combination of unique physico-chemical properties like chemical stability, catalytic activity, high thermal and electrical conductivity and non-linear optical properties which also make them very appealing for the development of ink and microelectronic applications. The encapsulation of MNPs into biologically friendly water soluble polymer with an extremely low cytotoxicity is a prerequisite for the development of biomedical applications in nanomedicine. In fact, the development of functionalized, nontoxic and biocompatible nanoparticles has focused a lot of interest for applications in cancer diagnostic, drug delivery and anticancer drugs [2] during the last two decades.

The biocidal properties of these Ag NPs have recently been studied against fungi and bacteria that develop on straw in order to protect straw bales that are used in green housing construction [3, 4]. This study has demonstrated the toxicity of these Ag NPs against bacteria, but fungi appear to be a more serious threat, and these Ag NPs need to be combined with another metal or chemical group to enhance their efficiency. In this paper we have studied the toxicity of these Ag MNPs against human cells and cancer cells via MTT assay. The encapsulation of MNPs into polymer with an extremely low cytotoxicity is a prerequisite for the development of biomedical application in nanomedicine. pH-sensitive polymersomes are specially convenient for intracellular delivery, because they are stable vesicles at physiological pH and release the cargo at acidic pH inside the endosomes, releasing the cargo in the cytosol [5]. For these reasons, the encapsulation of these Ag MNPs inside pH-sensitive polymersomes has also been studied for the development of biomedical applications.

2. Experimental

2.1. Synthesis
The synthesis of Ag NPs was carried out under air and is based on non-hydrolytic sol-gel method developed elsewhere [6, 7] and has been reported elsewhere [3]. The Ag NPs were encapsulated inside pH-sensitive polymersomes composed by the co-polymer $POEGMA_{20}$-$PDPA_{90}$ (poly oligo(ethylene glycol) methacrylate co-poly(2-(diisopropylamino)ethyl methacrylate) [8] following the protocol described by Chen et al. [9]. Briefly, the co-polymer dissolved in DMF (dimethylformamide) was mixed with the Ag NPs dissolved in methanol and dodecanethiol. Then water was added slowly to form the polymeric vesicles. The organic solvent was removed by dialysis and the sample was purified by centrifugation.

2.2. Characterization
XRD data were collected using a Bruker D8 Discover instrument equipped with a LynxEye detector. Cu (ka1=1.54056) radiation selected by a Ge (1 1 1) monochromator was used. Crystallite size analysis from the XRD data was carried out using full profile Scherrer methods in TOPAS, with a fundamental parameters peak shape. Transmission electron microscopy (TEM) studies were carried out on a probe corrected Titan G2 80-200 kV operating at 200 kV in STEM mode and a Tecnai 10, Philips (Netherlands) to assess the size, surface topology and morphology of Ag NPs and assembled Ag-polymeric nanoparticles, respectively. X-ray photoelectron spectroscopy (XPS) analysis was carried out on a Kratos Analytical Axis UltraDLD photoelectron spectrometer equipped with Al Ka X-ray source.

2.3. Cytotoxicity assays
Different concentrations of nanoparticles were studied via MTT assay (M). Primary prostatic cancer cell line (PPC-1) were seeded on day 0 at a density of 1000 per well in 96-well microtiter plates. On day 1, silver nanoparticles at different concentrations were added and incubation was continued for 48 hours. After 48 hours, 3-(4,5-dimethylthiazol-2-yl)-2,5-diphenyltetrazolium bromide (MTT) was added to each well (0.5 mg/ml; Sigma Aldrich), and plates were maintained at 37°C for 2 h. The medium was then discarded, and DMSO was added to each well to lyse the cells. Absorbance was measured using a multiwall spectrophotometer (Tecan, microplate reader). All MTT assay were repeated twice.

3. Results and Discussion

3.1. Structural characterizations
The XRD pattern of the prepared Ag MNPs (figure 1a) shows that silver nanoparticles are highly crystalline with the face-centred cubic structure of silver metal (JCPDS File No 87-0720) and no secondary phase of silver oxide structure is visible. From the XRD pattern an average crystallite size of 58 nm was calculated using the Scherrer method.

The morphology and size of Ag MNPs were also studied by (S)TEM. The HAADF-STEM electron micrograph of spherical silver nanoparticles is shown in figure 1b and gives an overview of the nanoparticles. Before dispersing it on a carbon coated copper grid for TEM observations, the solution had to be sonicated at room temperature for 2 min. STEM images revealed that particles are easily dispersed in ethanol and surfactant-free. The silver nanoparticle diameter ranges from 20 to 50 nm with a few bigger particles of around 100nm in size suggesting that the average crystallite size suggested by XRD is not misleading.

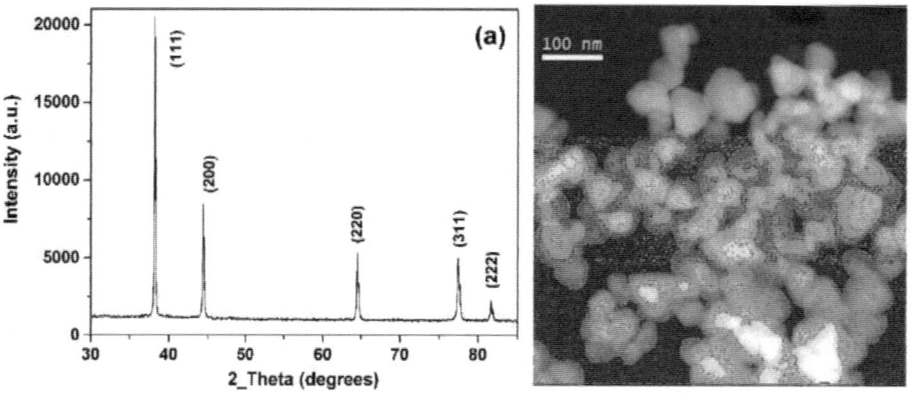

Figure 1. (a) XRD pattern of Ag metal nanoparticles, (b) STEM image overview of Ag MNP

The biocidal properties of these Ag MNPs have recently been reported [10], however, it has also been observed that the culture medium consisting of many amino acids that contain Sulphur groups shielded these biocidal properties due to the attachment of these molecules on the surface of the metal Ag NPs [4]. The metallic character of the surface promotes the bonding of Sulphur groups on the surface of Ag MNPs. To prevent such an attachment and shielding, the possibility of encapsulating these Ag MNPs with diameters around 10-50 nm inside pH-sensitive polymersomes composed by the co-polymer POEGMA-PDPA (poly oligo (ethyleneglycol) methacrylate co-poly(2-(diisopropylamino)ethyl methacrylate) [8] was studied. The formation of the hybrid metallic/polymeric nanoparticles was then investigated using TEM.

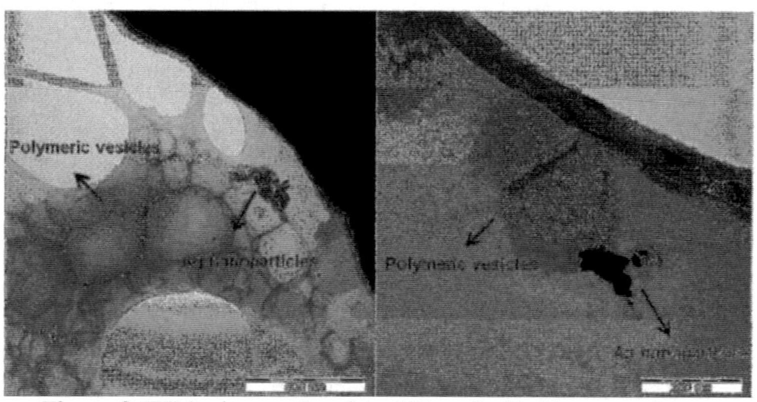

Figure 2. TEM micrographs of Ag NPs embedded into co-polymer POEGMA-PDPA vesicles. The Ag NPs are attached on the surface of the polymeric vesicles

TEM study shows that Ag MNPs are mainly localized outside of the polymer shell (Fig 9) and the Ag MNPs agglomerate on the surface of the vesicle. This study shows that the Ag MNPs need either to be functionalized to promote their encapsulation inside the polymer shell or a specific polymer containing sulfur or thiol groups should be used. We plan to test a surfactant that contains thiol groups that can attach on the metallic surface of the Ag MNPs. The encapsulation of the metal nanoparticles appears to be very important to enhance the efficiency of the metal nanoparticles for cancer therapy. The encapsulation inside other types of polymersome or inside gelatin shell will also be tested.

In order to confirm the nature of the surface of the Ag NPS, XPS measurements have been performed [3]. This study was performed 6 months after the synthesis and only binding energy peak from Ag metal were detected. Figure 3 provides the XPS spectra obtained for Ag 3d5/2, the photoelectron peak corresponds to metal Ag and confirms the metallic nature of the surface of the Ag NPs. No photoelectron peak corresponding to silver oxide was detected. Due to the high surface to volume ration present in nanoparticles, any oxidation of the Ag NPs surface would have been detected by XPS.

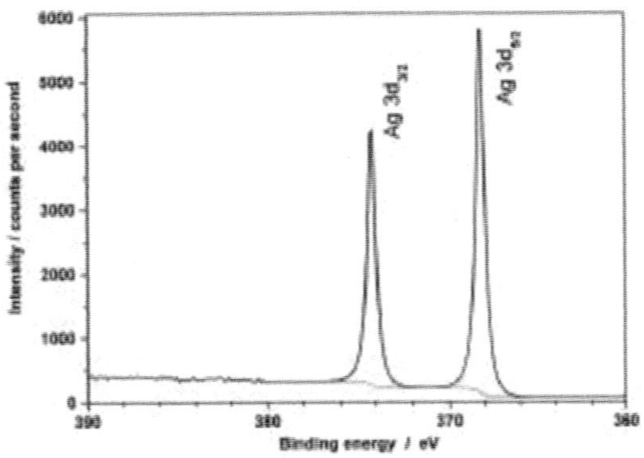

Figure 3. XPS spectra obtain on Ag NPs

3.2. Toxicity study

The toxicity of these Ag MNPs on Human Embryonic kidney (HEK 293) cells[11] as already been tested showing a moderate toxicity [3]. HEK cells are very commonly used for testing the toxicity of metal nanoparticles and the toxicity study has shown that mortality rate is over 50 % only for a concentration of silver nanoparticles of high concentration, which makes them a suitable material for biomedical application. Therefore, a similar cytotoxicity study has been performed on cancer cells and more particularly on the primary prostatic cancer cell line (PPC-1).

1.0-,

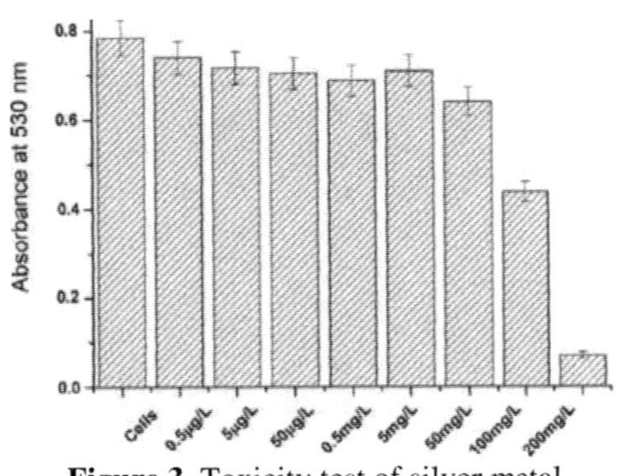

Figure 3. Toxicity test of silver metal nanoparticles on PPC-1 cells

Figure 3 shows MTT assays performed on Ag MNPs solutions of concentration ranging from 0.5 mg/L to 200 mg/L. Figure 3 shows the mean ± SEM of duplicate measurements of a representative sample of three independent experiments. This toxicity study on PPC-1 cells shows that mortality rate is over 44 % for a concentration of silver nanoparticles of 100 mg/L and shows a higher mortality rate (over 91 %) for a concentration of 200 mg/L. These results demonstrate that these Ag MNPs are potential candidates in the treatment of cancer, but their functionalization is needed.

4. Conclusions

In summary, we have reported on the synthesis of spherical silver nanoparticles with average diameters ranging from 20 to 50 nm for biomedical applications. The biocidal properties of these Ag MNPs have recently been demonstrated and the study of their encapsulation into pH-sensitive polymersomes was then performed for their potential use in cancer therapy. It has been shown that stabilisation of the metallic surface of the Ag MNPs requires the use of specific polymersomes that contain Sulphur or thiol groups. The cytotoxicity study shows that at a concentration of 100mg/L the mortality rate is over 44 % and increased to 91 % for a concentration of 200mg/L (200mg/mL). The synthesized Ag MNPs exhibit noteworthy cytotoxicity and show highly effective apoptotic activity against primary prostatic cancer cell line (PPC-1). These Ag MNPs therefore have high potential in therapeutic treatments and may be helpful for the development of anticancer therapeutics via their encapsulation into the appropriate polymersome.

Acknowledgement

Financial support from through the Estonian Research Council (grant number PUT431), Marie Curie (PERG05-GA-2009-249243), and the Estonian Road Map infrastructure and the NAMUR projects for the use of the Titan G2 80-200 Transmission electron microscope is acknowledged. The research has been supported by the European Regional Development Fund project TK134 (TAR19016). The authors acknowledge Dr. Tambet Teesalu from Laboratory of Cancer Biology of the University of Tartu for fruitful discussion and Prof. Giuseppe Battaglia from University College London for the synthesis of the polymer POEGMA-PDPA. We express our thanks to Mr. M. F. Sunding for XPS

References

[1] L. Ge, *et al.* 2014 *International Journal of Nanomedicine* **9** 2399-407
[2] A. Goldman, *et al.* 2015 *Nat Commun* **6**
[3] S. Küünal, *et al.* 2016 *International Nano Letters* **6** 191-7
[4] S. Küünal, *et al.* 2016 *Key Eng. Mater.* **674** 133-8
[5] L. Simón-Gracia, *et al.* 2016 *Molecular Cancer Therapeutics* **15** 670
[6] E. Rauwel, *et al.* 2012 *Adv. Func. Mater.* **22** 1174-9
[7] E. Rauwel, *et al.* 2011 *J. Phys. Chem. C* **115** 25227-33
[8] J. Du, *et al.* 2005 *Journal of the American Chemical Society* **127** 17982-3
[9] H. Y. Chen, *et al.* 2008 *ChemPhysChem* **9** 388-92
[10] P. Rauwel, *et al.* 2016 *Beilstein Journal of Nanotechnology* **7** 1075-85
[11] K. Sooklert, *et al.* 2016 *Int. J. Nanomedicine* **11** 597-605

IOP Conf. Series: Materials Science and Engineering **175** (2017) 012012 doi:10.1088/1757-899X/175/1/012012

The facts that the physical-chemical properties of modern tablets distinguish them from natural food lumps

A Urakov, N Urakova, A Reshetnikov, M Kopylov, A Kasatkin, D Baymurzin, R Gabdrafikov

Department of Thermal Deformation Processes, Institute of Mechanics, Ural branch of RAS, Izhevsk, Russia

E-mail: urakoval@live.ru

Abstract. It was found that pharmaceutical companies produce drugs in tablet form, physical or physical-chemical properties that are radically different from those of the properties of natural food lumps, in that adult converts food in our mouth before swallowing. It was shown that the conventional shape, color, size, volume, specific gravity, hardness, osmotic and acid activity of modern tablets impair the physical and physicochemical properties of the liquid contents of the stomach is much stronger than such "building" materials, such as chalk, clay, sand, river pebbles and gravel.The results showed, that the value of the specific hardness, deforming tablets, can distinguish modern tablets from each other by more than 5000 times. Therefore, introduction tablets inside without information of ability injuring their action leads to the fact that soft and "unsalted" tablets almost nothing damage, and too "salty" and solid tablets damage the gums, lips, tongue, teeth and dental structures. To reduce the traumatic action tablets offered standardize osmoticity, corrosion and hardness within the range of safe values for soft and hard tissues of the oral cavity and improve standard introduction of tablets in the mouth.

1. Introduction

The modern tablets are manufactured by compressing powders, as they were a hundred years ago. Therefore, they are hard and resemble natural stones. Typically, these artificially made stones look like white or gray circular shaped disks [1, 2]. Before it was found that physical and chemical properties and aggregate state of modern tablets are very similar to comparable pieces of chalk, clay and/or compressed salt [3,4]. But human stomach is not adapted for stones of any size and shape, including silica sand, crushed stone and river pebbles. Similarly, human stomach is not adapted for medical stones, or tablets. However, a man will not stay healthy after swallowing comparable number of most modern tablets [5, 6, 7].

It is paradoxical, but the fact is that there is evidence that tablet-form medications are as hard as rocks, so when a person chew the tablets, he may break his jaw, prosthesis, tooth, crowns, implants, braces or fillings and injure his gums, tongue and palate [7,8,9]. When swallowing the tablets without chewing, one can damage his esophagus and stomach, and even get ulcers [10, 11].

Shown, that tablets quality standard does not consider their hardness [12, 13]. Therefore, every manufacturer has the right to make tablets of any hardness, moreover, he is entitled not to define it or inform the consumer about it. Worse yet, today, the drugs manufacturers do not control the osmotic and acidic impact of tablets and their local irritant effect on the mucous membranes of the mouth and stomach [6, 10, 14, 15, 16]. Therefore, almost all modern tablets demonstrate high physical-chemical aggressiveness and therefore have acute local irritant and burning effects. All this contributes to the development of drug-caused iatrogenesis, suchas gingivitis, stomatitis, gastritis, gastric ulcers and tooth decay [10].

2. Materials and methods

In the laboratory settings, we studied physical and physical-chemical properties of more than 50 tablets of various medications produced by different pharmaceutical companies worldwide. We analyzed their shape, color, weight, diameter, height, volume, acidity and osmolarity. Conventional methods and equipment for drugs quality control were used for this purpose. Additionally, we measured the tablets hardness by Rockwell hardness test method. We also measured hardness at specific load in Brinnel scale (in HB units) [7].

At the same time, we monitored the dynamics of food and tablets movement inside the body, and analyzed viscosity (hardness), temperature, acidity and osmolarity of gastric contents. The studies were conducted using a plastic model of the stomach. For this purpose, we used a clear, colorless 1000 ml plastic container. To imitate food, we introduced 180 g of oatmeal, and 150 ml of milk and/or water from the tap into the container. After putting food and tablets, we added 150 ml of natural gastric juice with pH 0.8-1.2. All substances were introduced into the vessel at a temperature of +37°C. In a parallel series of experiments in the stomach cavity of the input pieces of chalk, clay, gravel and river pebbles of similar size [6].

3. Results

It is shown that all the tablets, food and water go down to the bottom of artificial gastric reservoir and move inside its cavity because of gravity, as well as river sand, pieces of chalk, clay, gravel and pebbles. We found that corrected specific gravity of all modern tablets is greater than 1 g/cm^3 and therefore all the tablets sink in gastric juice, water and milk. We also found that if the container is placed vertically, all the tablets fall in one place, and then lie still in the bottom of the cavity like river pebbles in a glass of water, despite of the liquid being added.

However, unlike the pebbles and crushed stones, tablets have aggressive impact on gastric mucosa, they corrode the stomach wall, and may lead to ulcers. The tablets that are most blamed for causing ulcers in the stomach wall are aspirin and its analogs which are known as "non-steroidal anti-inflammatory agents". Our results indicate that in the medication name present wickedness ...

We were the first to see the truth about physiological form of modern tablets !

With our theoretical, laboratory and clinical research, we were able to get rid of this pharmaceutical delusion!

Today we are sure that manufacturing tablets in a circular disk form is a mistake. This form of tablets is not compatible with the specifics of human digestive system.

Disk form of tablets is useful only for manufacturers and sellers. However, we do not have standard even for this form of tablets ! Surprisingly, today there is no standard not only for the form of tablets, but also for their sizes. Therefore, different manufacturers produce tablets in different forms and size.

Our results show that modern tablets differ in diameter and height by up to 3 times, and by up to 10 times in volume.

We decided to analyze the shape and size of the object, which people usually swallow after thorough chewing of food in their mouth. In science, this object is known as "food bolus". For this purpose, we studied the shapes and sizes of food lumps formed in the oral cavity in adult healthy people when chewing fresh bread. We found that natural bolus has a form of an olive with its largest diameter up to 1 cm, and maximum length of 2 cm [6,10]. This olive is dark, has close to zero hardness and medium elasticity, it is porous and has a specific gravity of less than 1 g/cm^3, besides, it lacks osmotic aggressiveness towards contents of the mouth and stomach.

So, it follows that God created man to swallow soft and elastic olives, but not hard disks !

Besides, we found that there is no standard for tablets chewing resistance and disintegration in gastric juice, so they are manufactured with varying chewing resistance and disintegration characteristics. All this increases the range of nonspecific physical-chemical effects of tablets inside the mouth and stomach.

In particular, our results showed that modern tablets differ in chewing resistance by up to 5000 times !(Figure 1).

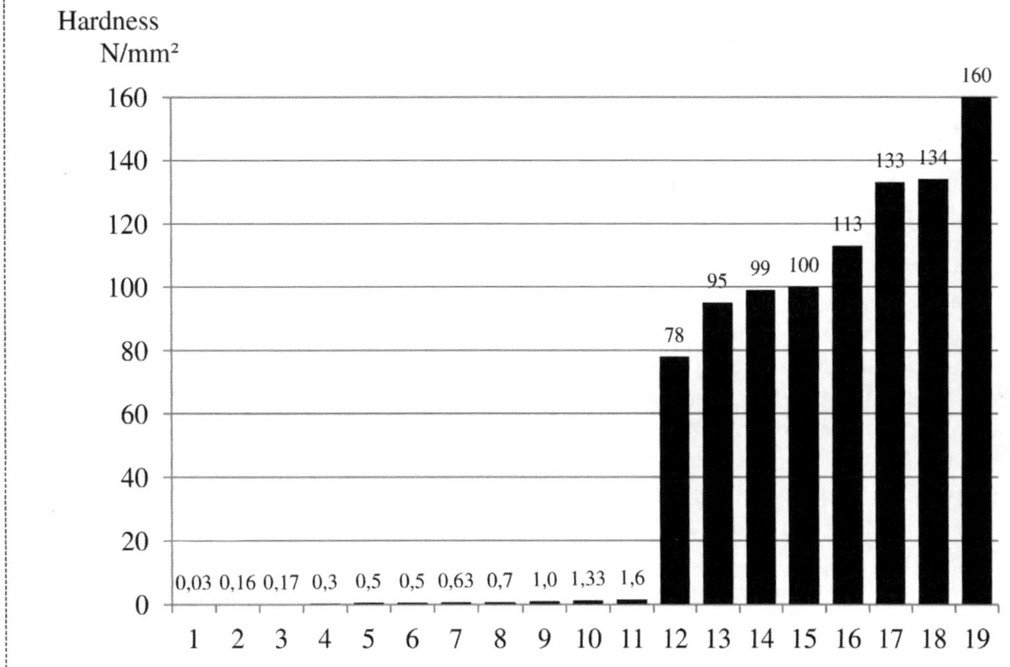

Figure 1. The values of specific pressure, deforming the tablet medicines: Ksefokam 1, 2 - Aceclofenac, 3 - Fenigidin, 4 - Dexamethasone, 5 - Acetylsalicylic acid, 6 - No-shpalgin, 7 - Elefloks, 8 – Mirloks, 9 - Analgin, 10 - Ofloxacin, 11 - Nurofen, 12 - Diclofenac sodium, 13 - Nimesulide, 14 - Pektusin 15 – Sulfosalazin, 16 – Fromilid, 17 - Prednisolone, 18 – Ampisid, 19 - Ketorol.

These data show that the specific values of the pressure deforming the tablets occupy today range from 0.03 ± 0.0001 N/mm^2 (Xefocam tablets (Lornoxicam4 mg,Nycomed) to 160 ± 0.3 N/mm^2 (Ketorol tablets, Dr. Reddy's, India).

The group, which includes a relatively soft tablets, it was presented Ksefokam, Aceclofenac, Fenigidin, Dexamethasone, Acetylsalicylic acid, No-shpalgin, Elefloks, Mirloks, Analgin, Ofloxacin and Nurofen. At the same time the most "soft" tablets (with values of hardness less than 2 N/mm^2) were tablets of Acetylsalicylic acid, Analgin, Aertal, Nurofen and No-shpalgin.

In the group with respect to solid tablets includes Sodium diclofenac, Nimesulide, Pektusin, Sulfosalazin, Fromilid, Prednisolone, Ampisid and Ketorol. At the same time the most "hard" pills (with the values of the hardness of more than 70 N/mm^2) were tablets of Sodium diclofenac, Nimesulid, Pektusin, Sulfosalazin, Fromilid, Prednisolone, Ampisid and Ketorol.

So today, no one knows the true hardness of each tablet. Therefore, today no one knows what will crumble first while chewing, a tablet or a tooth !

The results of our experiments showed that under the simulated conditions, the mixture of oatmeal porridge, gastric juice, and water had the following physical-chemical characteristics:

- Viscosity - in the range of 200 - 500 centipoise,

- Acidity - in the range of pH 4.5 - 8.0,
- Osmolarity - in the range of 240 - 340 mOsmol / l of water.

Then, we introduced into the mixture 20 small river pebbles in the form of circular disks with the diameter of about 6-20 mm, height of 2-6 mm and volume of 0.1-1.0 cm^3, and during one hour after that the physicochemical properties of the contents remained virtually unchanged.

Different results were obtained after we introduced 20 tablets of similar shapes and sizes into a similar mixture of cereal, water and gastric juice. In 30 minutes after putting them in this mixture, the contents had the following physicochemical characteristics:

- Viscosity - 100 - 300 centipoise;
- Acidity - pH 6.0 - 7.1;
- Osmolarity - 240 - 340 mOsmol / l of water.

In 30 minutes after adding 20 tablets to the "empty" plastic container (modelling taking tablets in the fasting state), tablets disintigrated halfway. The container contents appeared as suspension with non-decomposed pieces of tablets and had the following physicochemical properties:

- Viscosity value 0 - 10 centipoise;
- Acidity - pH 2.0 - 3.3;
- Osmolarity - 340 - 600 mOsmol / l of water.

Thus, we found that pharmaceutical definition ranges of mass values, diameter, height, volume, and the specific pressure, deforming tableted drugs, indicates the absence of a uniform standard in modern tablets. In particular, there is currently no acceptable limit values of their hardness. Under these conditions, the pharmaceutical companies are producing tablets with a size and hardness selected at its discretion, by accident or out of ignorance, not quality standard tablets.This has led to the fact that different drugs tablets today are of different sizes, hardness, and strength when chewed. The maximal great range of differences of tablets we found in their hardness (crushing strength tablets). Uncertain tablet hardness impairs their quality.

We believe that it is time to standardize hardness of the tablets and make them within the minimum hardness. To this end, we propose to introduce a new indicator of the quality of drugs - namely, the value of the minimum specific pressure that deforms the tablet. This indicator tablet quality can be determined with the help of the device, which is known in the mechanics of solids, called hardness Tester.The value of this indicator is invited to indicate in Brinell units, which are determined by the Rockwell method.In addition, the tablets should not sink in the gastric juice as stones. Thetabletsshouldfloatontheliquidsurface[17].

4. Conclusions

Thus, pharmaceutical companies manufacture tablet-form medications with physical or physical-chemical properties, which are entirely different from those of natural food bolus, which we have in our mouth before swallowing. It was shown that conventional shape, color, size, volume, specific gravity, hardness, osmotic and acidic properties of tablets affect the physical and physical-chemical properties of oral and gastric cavities contents. It is proved that improper physical and physical-chemical characteristics of modern tablets result in reduced medication safety for the digestive system, and increase their non-specific physical and physical-chemical aggressiveness when ingesting.

Therefore, the human digestive system happily accept pills only if their physical and physicochemical properties will not differ corresponding properties of natural food lumps produced in the oral cavity of the quality of food.

References

[1] Nikitiuk D B, Reshetnikov A P and Nasirov M R 2016 *Modern problems of science and education* **2**. URL: http://science-education.ru/ru/article/view?id=24185

[2] Urakov A L, Strelkov N S , Lipanov A M, Dementiev V B and Urakova N A 2007 *Chemical*

Physics and Mesoscopics **9 (3)** 231

[3] Strelkov N S, Urakov A L, Korovyakov A P, Muravceva O V, Korepanov M V, Urakova N A and Feigin V P 2002 *Morphological newsletter* **3** 95

[4] Urakov A L, Strelkov N S, Urakova N A, Mihailova N A, Karlova T B, Kamenschikov Y G and Feigin E P 2008 *Eksperimentalnaia i klinicheskaia gastroenterologiya* **2** 27 (in Russian)

[5] Urakov A L, Strelkov N S, Lipanov A M, Gavrilova T V, Dementyev V B, Urakova N A and Reshetnikov A P 2007 Newton's Binomial as a "formula" development of medical pharmacology (Izhevsk: Institute of Applied Mechanics Ural branch of the Russian Academy of Sciences) 192 p

[6] Urakov A L, Reshetnikov A P, Kopylov M V, Kasatkin A A and Baimurzin D Y 2016 Pills violate the phisico-chemacal state of the stomach contents, *In L. A. Gömze (Editor) 3rd International Conference on Competitive Materials and Technology Processes* (Miskolc-Lillafüred, Hungary) 40 http://dx.doi.org/10.1088/1757-899X/123/1/012007

[7] Urakov A L and Reshetnikov A P 2014 *Success of Modern Natural Science,* **9 (2)** 33

[8] Urakov A L, Urakova N A, Mihailova N A and Reshetnikov A P 2007*Meditsinskaia pomosch,* **5** 49 (in Russian)

[9] Reshetnikov A, Kopylov M, Urakov A and Urakova T 2016 Infrared diagnostics of the calluses disease from braces. *Quantitative Infrared Thermography 2016. ABSTRACTS. 13TH Quantitative Infrared Thermography*. (Gdansk: Gdansk University of Tecnology) pp 79–80

[10] Urakov A L, Urakova N A, Reshetnikov A P, et al 2008 *Med. Almanac* **2** 45

[11] Urakov A L, Urakova N A and Kozlova T S 2011 *Bullitin of Ural Academy Medical Sciences* **1 (33)** 105

[12] Urakov A L, Urakova N A, Reshetnikov A P, Abolmasov N N, Pozhilova Y V, Novikov V E, Lipanov A M, Zabokritskii N A and Bakurinskih A A 2014 Artificial food bolus and method forinstant assessment of dento-facial health with using artificial food bolus. Invention RU Patent 2533840 C2

[13] Urakov A L 2014 Development of new materials and structures based on managed physico-chemical factors of local interaction. *In L.A. Gömze (Editor) 3rd International Conference on Competitive Materials and Technology Processes, Book of Abstracts* (**ISBN 978-963-12-0334-9**) 9

[14] Urakov A L 2015 *Epitőanyag - Journal of Silicate Based and Composite Materials* **67(1)** 2 http://dx.doi.org/10.14382/epitoanyag-jsbcm.2015.1

[15] Korovyakov A P, Vakhrushev Ya M, Strelkov N S, Urakov A L, Urakova N A, Urakova T V, Korepanova M V, Muravtseva T M and Feigin V P 2003 Eksperimental'naya *i klinicheskaya gastroenterologiya* **1** 30 (in Russian)

[16] Urakov A L and Urakova N A 2006 *Biomeditcina* **4** 66

[17] Urakova N A, Urakov A L, Muravceva O V, Ovchinnikova E N, Scherbakova N V, Tulenkov A M and Perceva N A 2005 Floating tablet.Invention. RU Patent 2254121 C2

Development ceramic composites based on Al_2O_3, SiO_2 and IG-017 additive

E Kurovics[1], A Shmakova[2], B Kanev[2] and L A Gömze[1, 3, 4]

[1]Intsitute of Ceramics and Polymer Engineering, University of Miskolc, Hungary
[2]Institute of Geology Komi SC UB RAS, Syktyvkar, Russia
[3] IGREX Engineering Service Ltd, Igrici, Hungary
[4]Tomsk State University, Tomsk, Russia

E-mail:mse205@gmail.com, femgomze@uni-miskolc.hu

Abstract. Based on high purity alumina and quartz powders and IG-017 bio-original additives the authors have developed new ceramic composite materials for different industrial purposes. The main goal was to fine a material and morphological structures of high performance ceramic composites as frames for development complex materials for extreme consumptions in the future. For this the mixed powders of Al_2O_3, SiO_2 and **IG-017** bio-original additive were uniaxially pressed at different compaction pressures into disc shapes and were sintered in electric kiln under air (1) and nitrogrn (2) atmosphere. The grain size distributions of the raw materials were determined by laser granulometry. There thermo-physical properties were also determined by derivatography.

The prepared and sintered specimens were tested on geometrical sizes, microstructure and morphology by scanning electron microscopy, porosity and water absorption. In this work the authors present the results of their research and investigation.

1. Introduction

In our days the industry requires more and more low density and mechanically strong materials capable to work in high temperature environment. To satisfy these inquires more and more ceramics [1-7] and ceramic matrix composites [8, 9] or cermets [10, 11]. To increase the wear resistance and live times of machine parts or machine tools the ceramic coatings and ceramic based composites are used in very often [12-14] Generally in cases when the machine parts and machine tools have to work under high mechanical loads and working temperatures the ceramic materials and ceramic composites are probably the best [15-18].

Our aims with this research work to create ceramic matrix composites and develop artificial meta-kaolin and mullite particles on the basis of different mixes of Al_2O_3, SiO_2 and **IG-017** bio-original additives.

2. Materials and experiments

In our experiments we used alumina powders of **Nabalox 315**, silica flour from **Fehérvárcsurgó** and **IG-017** refined bio-original additives developed and produced by Igrex Ltd. From these **3** components **18** different mixtures were prepared as it is shown in **Table 1** and milled in laboratory planetary ball mill Retch PM 400 trough 20 minutes of each at 200 rotation/minute. During the grinding we did not take into consideration the mechano-chemical phase transitions occurring under fine comminution. Later in a further work the determination of bond index of crushing can be also interesting using the easy method described by Maja S TRUMIC and her coauthors [19].

There were prepared by **6** different mixtures in the following **3** main combinations of components: **I./** - Al_2O_3+IG-017, **II./** - SiO_2+IG-017 and **III./** - Al_2O_3+SiO_2+IG-017. From these **18** mixtures the specimens were uniaxially compacted at **150** MPa pressures into shape of cylindrical discs with diameters of **20** mm. From each mixture we compacted by **10** specimens. The filling weights of specimens were **10** grams in each case. In spite of the same compacting pressures and filling weights we have got specimens of different heights (Figure 1) depending on material compositions of the used mixtures.

Table 1. The volumetric relationship between the components

Number of mixtures		I./						II./						III./					
		1	2	3	4	5	6	1	2	3	4	5	6	1	2	3	4	5	6
The composition in **m%**	**Al_2O_3**	90	80	70	60	50	40	-	-	-	-	-	-	53	49	46	43	40	38
	SiO_2	-	-	-	-	-	-	90	80	70	60	50	40	31	29	27	25	24	22
	IG 017	10	20	30	40	50	60	10	20	30	40	50	60	16	22	27	32	36	39

Figure 1. Typical differences in heights of compacted items

After compacting half of the specimens were fired (pre-sintered) in normal (**air**) atmosphere and the other half of the specimens were fired (pre-sintered) in nitrogen (**N_2**) by the firing curves as it is shown in Figure 2. In both cases the maximum sintering temperature was **1250°C**. This temperature is not enough high for formation of cristobalite crystals from tridymite phases in free **SiO_2** components of the mixtures.

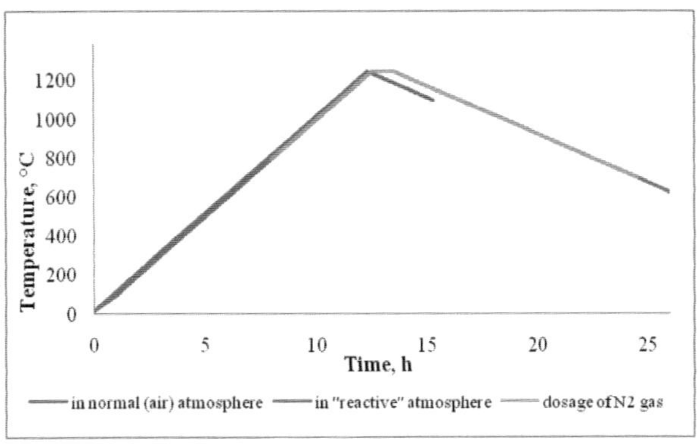

Figure 2. The sintering curves of specimens

3. Results and discussions

It is well known from practice that during sintering the **$Al_2O_3 - SiO_2$** ceramics items are shrinking and this shrinkage very strong depends on firing conditions and temperature [20-23]. Our experiments confirmed that the geometrical parameters of the pre-sintered ceramic items strong depend not only on the material compositions but on sintering conditions, including temperatures, pressures and atmospherically environments, thanking to the chemical reactions and phase transformations in solid stages and gasification or degasification of volatile component (Figure 3).

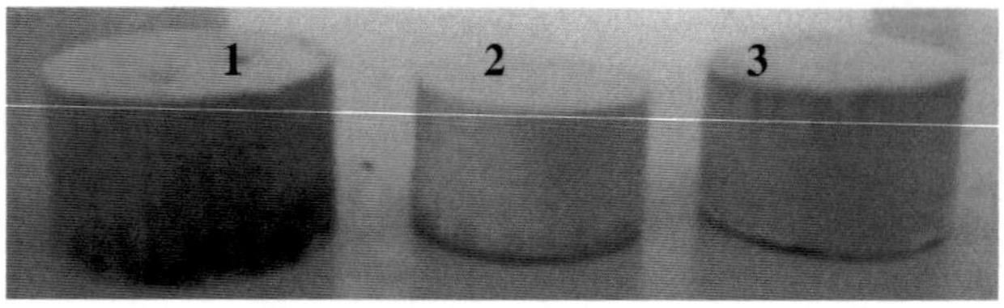

Figure 3. Changes of geometrical sizes of specimens depending on technological condition (1 - pressed green, 2 - fired in air, 3 - fired in N_2)

The results well represent that most of the bio-origin **IG-017** additives have fired out in normal (oxidation) atmosphere environment and the ceramic specimen suffered a large volume of shrinkage. This value of shrinkage is much higher than it could be expected because of the mullite formation from $\mathbf{Al_2O_3}$ and $\mathbf{SiO_2}$ components and phase transformation of the free $\mathbf{SiO_2}$ particles from β-quartz to α-tridymite crystals. At the same time in nitrogen ($\mathbf{N_2}$) atmosphere a swelling process can be observed thanking to the gasification and partly carbonization of the **IG-017** additive micro and nano particles. Depending on volume ration of the used bio-origin additive during pre-sintering (green firing) some of specimens have shrunk ("+") and some of specimens have swelled ("-") both in normal and nitrogen atmospheres as they are shown in Figure 4. In cases when the specimens were made from all three powders ($\mathbf{Al_2O_3}$, $\mathbf{SiO_2}$ and **IG-017**) we could observe only shrinkage in the normal atmosphere (1 green line) and only swelling (2 green line) in $\mathbf{N_2}$ atmosphere.

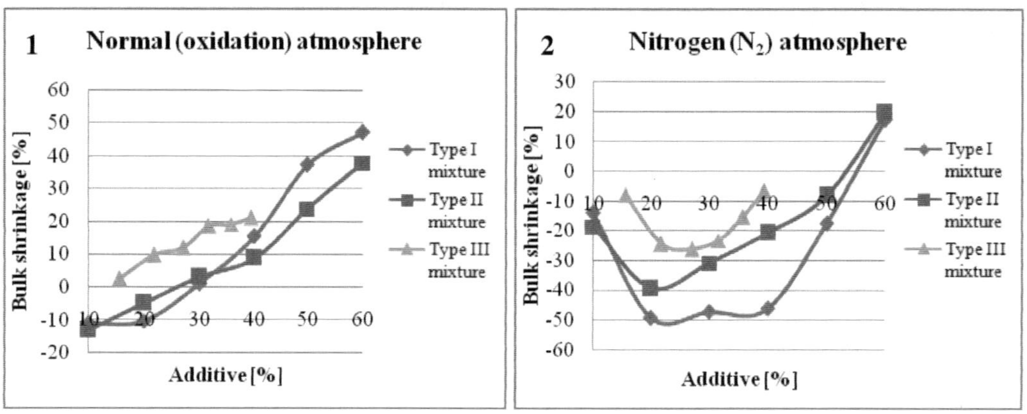

Figure 4. The change of volumes in normal (1) and in nitrogen (2) atmospheres

The burning weight losses (**I**) of specimens also were examined and determined by eq. 1.

$$I = \frac{m_1 - m_2}{m_1} \cdot 100 \quad [\%] \tag{1}$$

Where: the weight of the specimens in grams before firing ($\mathbf{m_1}$) and after firing ($\mathbf{m_2}$).
The burning weight losses of specimens are shown in Figure 5. The curves of this figure well illustrate that in cases when the portions of mixed **IG-017** additives were less than 30 m% the burning weight losses of specimens were much higher in nitrogen (2) than in normal (1) atmosphere.

The water absorption ($\mathbf{V_w}$) informs as about the open porous structure of the prepared ceramic items and can be determined as:

$$V_w = \frac{m_w - m_s}{m_w} \cdot 100 \; [\%] \tag{2}$$

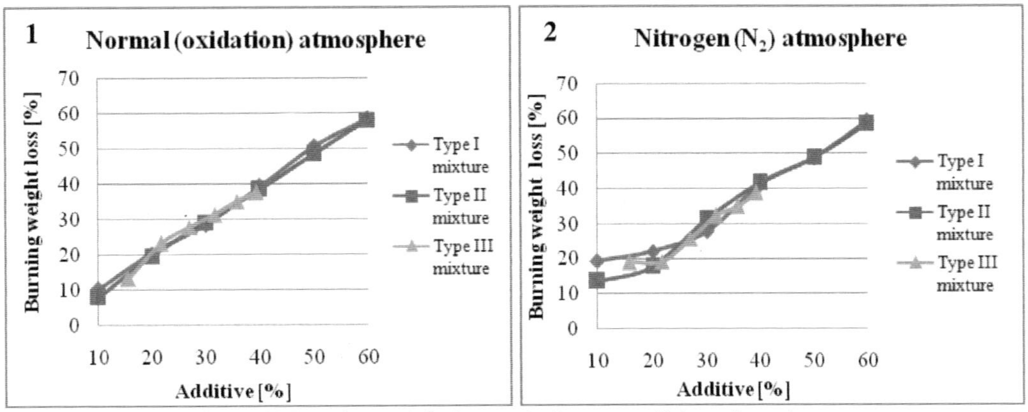

Figure 5. Burning weight losses during green firing of specimens
1 – in normal (oxidation) atmosphere, 2 – in nitrogen (N_2) atmosphere

Where: the weight of the specimens in grams before (**m**$_s$) and after boiling (**m**$_w$) in water through **2** h. The water absorptions of the green fired specimens as function of the volume of mixed IG-017 bio-origin additives are shown in Figure 6.

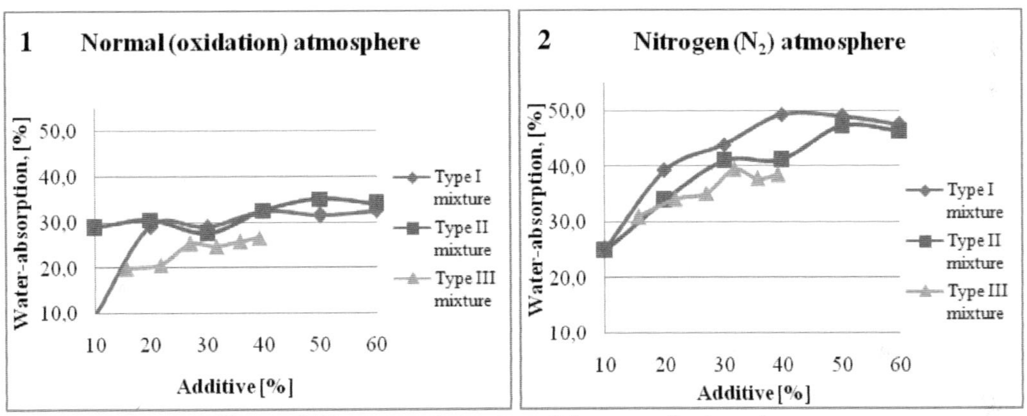

Figure 6. The water absorption of the green fired specimens
1 - fired in normal (air) atmosphere, 2 - fired in nitrogen (N_2) atmosphere

For the examination of the morphology and microstructure of the prepared specimens were used scanning electron microscopy and EDS. The typical microstructures and material compositions are shown in Figures 7-9 depending on quantity of the mixed **IG-017** additives and the firing conditions.

Element	Wt %	At %
O	30,24	42,47
Al	51,97	43,28
Si	17,8	14,24

Figure 7. Specimen prepared from III/5 mixture and sintered in normal (air) atmosphere

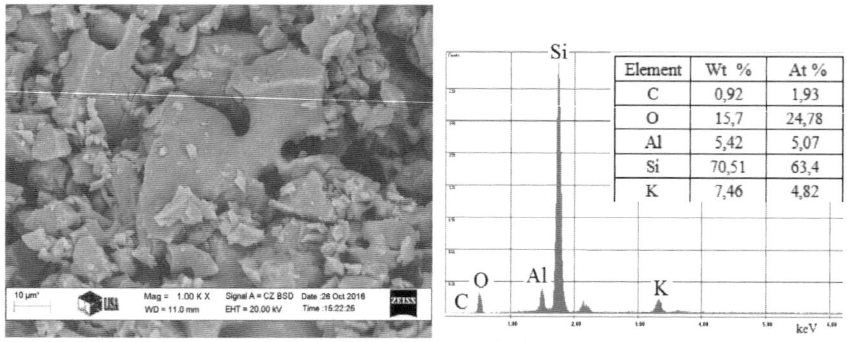

Element	Wt %	At %
C	0,92	1,93
O	15,7	24,78
Al	5,42	5,07
Si	70,51	63,4
K	7,46	4,82

Figure 8. Specimen prepared from II/3 mixture and sintered in normal (air) atmosphere

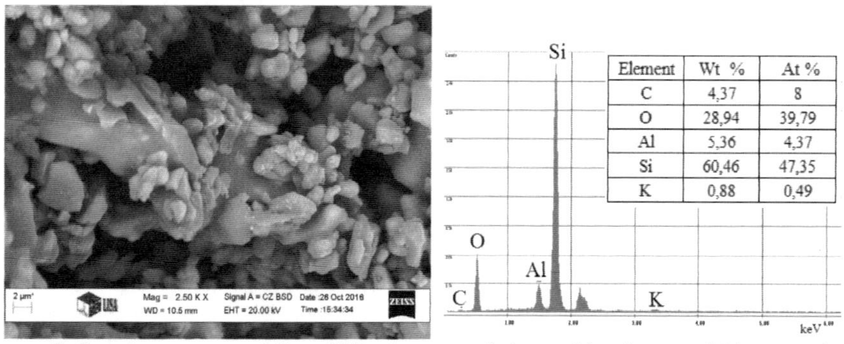

Element	Wt %	At %
C	4,37	8
O	28,94	39,79
Al	5,36	4,37
Si	60,46	47,35
K	0,88	0,49

Figure 9. Specimen prepared from II/3 mixture and sintered in nitrogen (N_2) atmosphere

The SEM pictures of II/3 specimens well illustrate that the ceramic items pre-sintered in nitrogen (**N_2**) atmosphere have much more and larger pores than the items fired in normal (oxidation) atmosphere. At the same time the items pre-sintered in nitrogen have much more small crystalline particles thanking to the reactive sintering and carbonization of part of the **IG-017** additives and forming silicon carbide particles of micron and nano sizes. This formation of carbon base particles is well illustrated in EDS pictures, where the atomic percentage of carbon (**C**) has increased from **1.93** At% several times and achieved **8** At%. Thanking to the degradation of the **IG-017** bio-origin additives, the volume of oxygen also has increased forming a new material structure of **Si-O-C** in solid phase.

4. Conclusions

Using normal (air) and nitrogen (**N_2**) atmosphere for pre-sintering were developed high porosity new ceramic specimens from alumina (**Al_2O_3**) powder, quartz flour (**SiO_2**) and **IG-017** bio-original additives. The specimens were examined on shrinkage/swelling, water absorption and morphology. The achieved pore structures can make these ceramic materials suitable for high performance, light metal alloys impregnated new composites.

The formed new material structure of **Si-O-C** is much lighter than traditional **Al_2O_3-SiO_2** ceramics and at high temperature it has an increased contact angel of wetness for most of light metal melts. Probably the developed by us new ceramic composite material composition and is pore structure can be impregnated with light metals and metal-alloys more easy and more perfect than the traditional ceramics and ceramic composites.

Acknowledgements

This research was made in framework of collaboration agreement between the University of Miskolc, Miskolc (Hungary) and Institute of Geology Komi SC UB RAS, Syktyvkar (Russia). The authors would like to say many thanks to IGREX Engineering Ltd for they support and help giving to us the IG-017 additives and Know-How to prepare mixtures and sintering the specimens.

References

[1] K Katahira, Y Watanabe, H Ohmori and T Kato 2002 *Int. Journal of Machine Tools and Manufacture* **42** 1307
[2] J Csányi and L A Gömze 2001 *Építőanyag-JSBCM* **53** 66
[3] C B Carter and M G Norton 2007 Ceramic Materials: Science and Engineering *Springer* 1-716
[4] N I Ershova and I Y Kelina 2009 *Építőanyag-JSBCM* **61** (2) 34 http://dx.doi.org/10.14382/epitoanyag-jsbcm.2009.6
[5] S P Buyakov, E S Kalatur, A S Buyakov and S S Kulkov 2016 *Építőanyag-JSBCM* **68** (3) 70 http://dx.doi.org/10.14382/epitoanyag-jsbcm.2016.12
[6] I Zhukov, S P Buvakova and S S Kulkov 2016 *Építőanyag-JSBCM* **68** (3) 74 http://dx.doi.org/10.14382/epitoanyag-jsbcm.2016.13
[7] S N Kulkov, S P Buvakova, M Chatzinikolaidou and I Kocserha 2015 *Építőanyag-JSBCM* **67** (4) 155 http://dx.doi.org/10.14382/epitoanyag-jsbcm.2015.26
[8] A Morikawa, T Suzuki, K Kikuta, A Suda and H Shinjo 2009 *Építőanyag-JSBCM* **61** (1) 2 http://dx.doi.org/10.14382/epitoanyag-jsbcm.2009.1
[9] L A Gömze and L N Gömze 2009 *Építőanyag-JSBCM,***61**(2) 38 http://dx.doi.org/10.14382/epitoanyag-jsbcm.2009.7
[10] E Rocha-Rangel, D Hernández-Silva, E Terés-Rojas, E Martines-Franco and S Diaz-De La Torre 2010 *Építőanyag-JSBCM* **62** (3) 75 http://dx.doi.org/10.14382/epitoanyag-jsbcm.2010.15
[11] S Khara, N Sharma and K Venkateswarlu 2010 *Építőanyag-JSBCM* **62** (2) 39 http://dx.doi.org/10.14382/epitoanyag-jsbcm.2010.8
[12] M M Machado López, M I Espitia Cabrera, J Faure and M E Contreras García 2016 *IOP Conf. Ser.: Mater. Sci. Eng.* **123** 012006 http://dx.doi.org/10.1088/1757-899X/123/1/012006
[13] L Kotov, V Ustyugov, V Vlasov, V Turkov, M Lasek, Yu Kalinin, A Sitnikov and E Golubev 2016 *IOP Conf. Ser.: Mater. Sci. Eng.* **123** 012009 http://dx.doi.org/10.1088/1757-899X/123/1/012009
[14] A Vereschaka, A Kutin, N Sitnikov, G Oganyan and O Sharipov 2016 *Építőanyag-JSBCM* **68** (4) 114 http://dx.doi.org/10.14382/epitoanyag-jsbcm.2016.20
[15] Z Koncsik, M B Maros, L Kuzsela and A Kovács 2016 *IOP Conf. Ser.: Mater. Sci. Eng.* **123** 012033 http://dx.doi.org/10.1088/1757-899X/123/1/012033
[16] Raether, Friedrich 2013 *Ceramic Application 1* **1** 45
[17] X L Shi, F M Xu, Z J Zhang, Y L Dong, Y Tan, L Wang, J M Yang 2010 *Materials Science and Engineering A* **527** 4646
[18] X Teng, H Liu, C Huang 2007 *Materials Science and Engineering A* **452** 545
[19] D Todorovic, M Trumic, L Andric, V Milosevic and M Trumic 2017 *Physicochem. Prob. Miner. Process* **53** (1) 321
[20] M Belmonte, M I Nieto, M I Osendi, P Miranzo 2006 *Journal of the European Ceramic Society* **26** (7) 1273
[21] E Kurovics, A Y Buzimov and L A Gömze 2016 *IOP Conf. Ser.: Mater. Sci. Eng.* **123** 012058 http://dx.doi.org/10.1088/1757-899X/123/1/012058
[22] M Rundans, I Sperberga and G Sedmale 2016 *IOP Conf. Ser.: Mater. Sci. Eng.* **123** 012042 http://dx.doi.org/10.1088/1757-899X/123/1/012042
[23] M E Abat-Javier, M Cajero-Juárez, M E Condreras Garcia 2016 *Építőanyag-JSBCM* **68** (4) 124 http://dx.doi.org/10.14382/epitoanyag-jsbcm.2016.22

IOP Conf. Series: Materials Science and Engineering **175** (2017) 012014 doi:10.1088/1757-899X/175/1/012014

Band Structure and Optical Properties of Kesterite Type Compounds: first principle calculations

S Palaz[1], H Unver[2], G Ugur[3], A M Mamedov[4,5*], E Ozbay[4]

[1]Department of Physics, Harran University, Urfa, Turkey,
[2]Department of Physics, Ankara University, Ankara, Turkey,
[3]Department of Phisics, Gazi University, Ankara, Turkey
[4]Nanotechnology Research Center-NANOTAM, Bilkent University, Ankara, 06800 Turkey,
[5]International Scientific Research Center, Baku State University, Baku, AZ-1148, Azerbaijan

E-mail: mamedov@bilkent.edu.tr

Abstract. In present work, our research is mainly focused on the electronic structures, optical and magnetic properties of Cu_2FeSnZ_4 (Z = S, Se) compounds by using ab initio calculations within the generalized gradient approximation (GGA). The calculations are performed by using the Vienna ab-initio simulation package (VASP) based on the density functional theory.The band structure of the Cu_2FeSnZ_4 (Z = S, Se) compounds for majority spin (spin-up) and minority spin (spin-down) were calculated . It is seen that for these compounds, the majority spin states cross the Fermi level and thus have the metallic character, while the minority spin states open the band gaps around the Fermi level and thus have the narrow-band semiconducting nature. For better understanding of the electronic states, the total and partial density of states were calculated, too. The real and imaginary parts of dielectric functions and hence the optical functions such as energy-loss function, the effective number of valance electrons and the effective optical dielectric constant for Cu_2FeSnZ_4 (Z = S, Se) compounds were also calculated.

Keyword: ab initio calculation; elastic constant; electronic structure; quaternary semiconductors

1. Introduction

Along the progress of growing high-quality complex compounds, the group I_2-II-IV-VI_4 quaternary semiconductors have rapidly been broad interest due to many possibilities in varying the chemical composition, and thereby optimizing the material functionality. Especially cupper based kesterite materials (Cu_2XYZ_4, X=Zn, Fe, Y=Sn, Ge, Z=S, Se, Te) have recently found increased interest as absorber layer in thin film solar cells technologies and devices [1]. The band gap energy of $Cu_2XYZ_4$are suitable for photovoltaics, and Cu_2XYZ_4 have the advantage over corresponding $CuIn(Ga)Se_2$ absorbers to involve abundant, nontoxic, and less expensive chemical elements. Cu_2XYZ_4 compounds have experimentally and theoretically [2-5] been verified to crystallize in either S_4^2 kesterite or D_{2d}^{11} stannite structure although disordered structure also has been reported [6].

Cu_2ZnSnS_4 based solar cells have recently achieved solar energy conversion efficiency of 12.6%. While Cu_2ZnSnS_4 itself has many attractive properties for solar cells, further increases in the performance will require optimisation that can be achieved through control of the materials stoichiometry and/or incorporation of extrinsic impurities, such as Fe (to replace Zn) or Se (to replace S) [7]. In particular, Fe is of interest for Si-based tandem solar cells since the lattice constant of Si lies between two end members of the alloy, Cu_2ZnSnS_4 and $Cu_2FeSnSe_4$, and an increase in the band gap could also be beneficial.

Further analysis of the material properties as well as future transport simulations of various device structures require a detailed understanding of the electronic conducting-band and valence-band structures.

In the present work, by means of DFT (Density Functional Theory) approach we examined the series of Cu_2FeSnS_4 (CFTS), and $Cu_2FeSnSe_4$ (CFTSe) assuming they crystallize in the typical S_4^2 and D_{2d}^{11} structures. For these compounds we derived structural, mechanical, optical properties. Consequently, the primary purpose of this work is to provide some additional information to the existing data on the physical properties of CFTS and CFTSe compounds by using the ab-initio total energy calculations.

2. Calculation methodology

The first principles calculation based on DFT was used with the aid of the VASP [8-10] program. The exchange and correlation potentials were Perdew-Burke-Ernzerhof method [11] based on generalized gradient approximation (GGA). The plane wave cut-off energy in the wave vector K space was 500 eV. We have performed the Brillouin–Zone integration by using 9x9x9 gamma centered special Monkhorst-Pack k-points [12]. When the total energy was stabilized within 1×10^{-7} eV, the force acting on each atom of the cell after optimization was less than 0.001 eV/Å, the residual stresses of the cell was less than 0.001 GPa, and the tolerance offset was less than 1×10^{-5} Å.

The elastic constants are calculated by the efficient stress-strain method [13] as implemented in the VASP code. The optical properties were obtained complex dielectric function $\varepsilon(\omega)=\varepsilon_1(\omega)+i\varepsilon_2(\omega)$ where the details explained in Refs. [14].

3. Results and Discussion

3.1 Mechanical Properties

The estimated independent elastic constants are tabulated in Table 1. To our knowledge, there are no experimental or theoretical data available for elastic constants. The calculated elastic constants satisfy Born criteria [15]. Therefore, it can be said that, CFTS and CFTSe compounds for considered structures are mechanically stable.

Table 1. The calculated elastic constants (C_{ij}, in GPa) , isotropic bulk modulus (B, in GPa), shear modulus (G, in GPa) and Young's modulus (E, in GPa), the longitudinal, transverse and average elastic wave velocities together with the Debye temperature for Cu_2FeSnZ_4 (Z = S, Se) compounds

Crystal	C_{11}	C_{12}	C_{13}	C_{16}	C_{33}	C_{44}	C_{66}
$Cu_2FeSnSe_4$(P4)	82.43	56.65	56.23	-4.12	84.07	40.86	43.26
Cu_2FeSnS_4(I-42m)	92.35	58.96	58.98	-	94.81	50.45	43.81
Cu_2FeSnS_4(P4)	90.68	58.24	54.24	1.47	85.81	35.30	50.69

	B	B_V	B_R	G	G_V	C_R	E	E_V	E_R
$Cu_2FeSnSe_4$(P4)	65.24	65.24	65.24	26.24	30.32	22.15	69.23	78.76	59.70
Cu_2FeSnS_4(I-42m)	70.36	70.37	70.36	31.82	35.78	27.85	82.80	91.79	73.82
Cu_2FeSnS_4(P4)	66.65	66.74	66.57	28.21	30.95	25.47	74.09	80.42	67.76

	v_l (m/s)	v_t (m/s)	v_m (m/s)	Θ_D (K)
$Cu_2FeSnSe_4$(P4)	2195	4291	2459	256
Cu_2FeSnS_4(I-42m)	2675	5037	2990	328
Cu_2FeSnS_4(P4)	2521	4847	2822	309

The isotropic bulk moduli (B_v), shear moduli (G), Young's moduli (E), Poisson's ratios, B/G ratios, elastic constants (C_{ij}) Debye temperatures, and sound velocities are calculated by using Voigt Reuss-Hill [16-18] approach. The results are tabulated in Table 1. It shows that the compressibility of these structures is relatively same after application of pressure. We note that the B/G values are all higher than 1.75. Therefore, the studied systems in its all forms can be classified as ductile materials. Our calculated values of Poisson's ratio vary from 0.31 to 0.40 for these structures. Thus indicating strong semi-metalic contribution in the intra-atomic bonding for these structures.

3.2 Electronic structure

The electronic structure plays an important role in determining of the properties of the Kesterite compounds. By using the equilibrium lattice constants the calculated band structure dispersion in k-space along high symmetry direction in the Brillouin zone, the total and partial density of states are presented in Figs. 1-3. The band structure of the CFTS and CFTSe compounds were calculated for majority spin (spin-up) and minority spin (spin-down). It is seen that for these compounds, the majority spin states cross the Fermi level and thus have the metallic and semimetallic characters, while the minority spin states open the band gaps around the Fermi level and thus have the narrow-band semiconducting nature. For these crystals the unit cell contains four atoms. In order to understand the contribution of each orbital in these atoms we also investigate the angular momentum decomposition of the atoms projected density of states (DOS). DOS show that all these orbitals are spatially delocalized. Energy fragment from -7 eV till Fermi level originated from more localized Cu (d), and delocalized Se or S(s,p), Sn(p) states with small contribution from Sn(s,d), Fe(s,p), Se or S(d), and Cu(s,p) states. Finally, the structure from conduction band minimum and above presents an admixture of Cu (s,p), Sn (p,d), Fe(s,p), Se (s,p,d) and S(p) states. One can see a strong hybridization between Cu(p), Fe(d) and Sn(d) states and also some overlap between Se or S(s) and Sn(s) states, at around -14 eV.

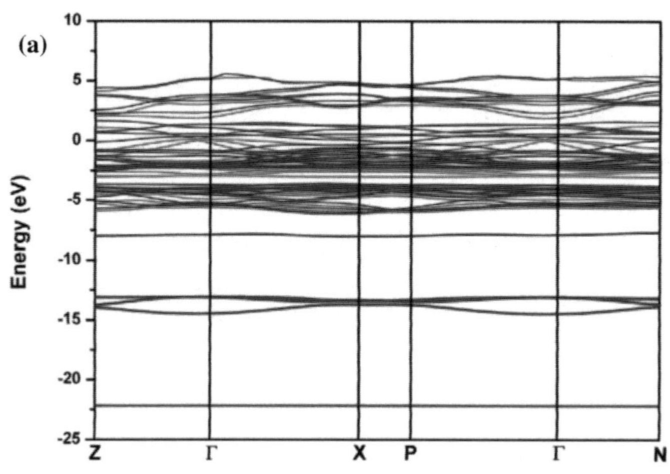

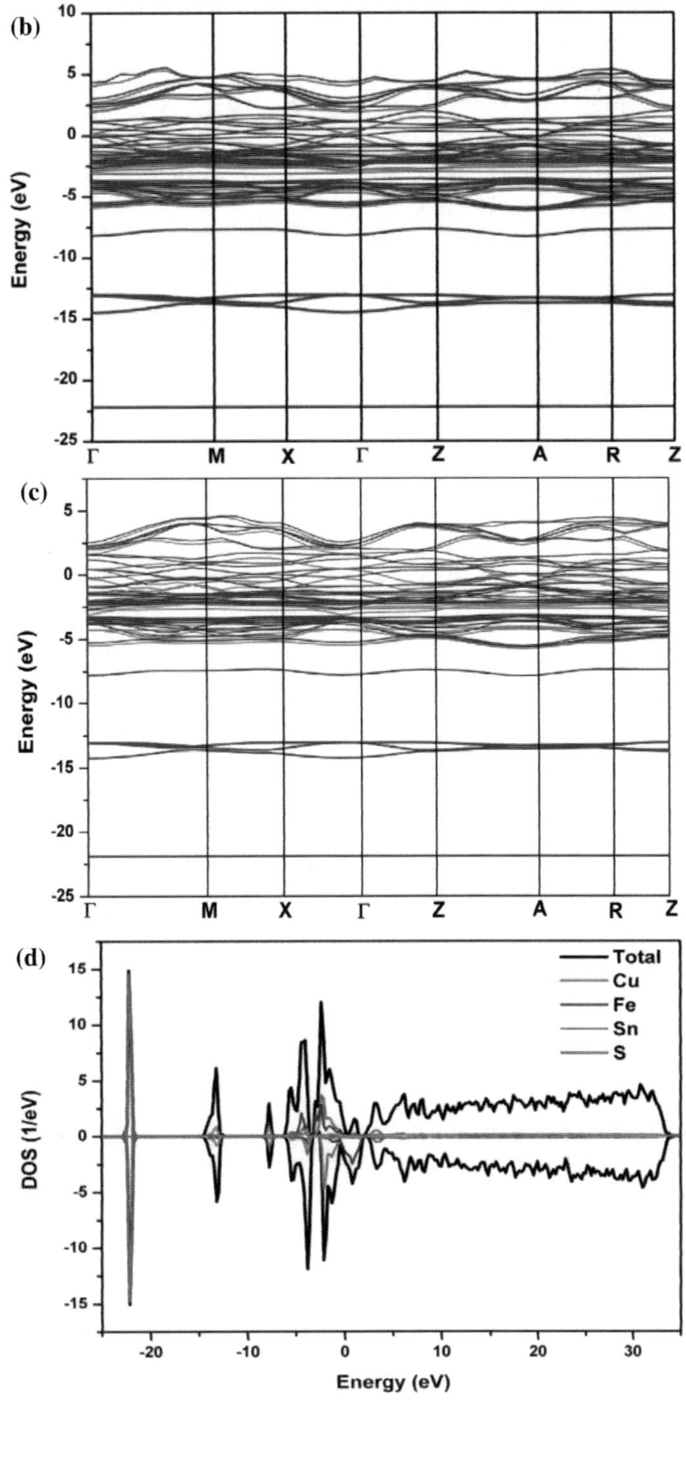
(b)
Energy (eV)
Γ M X Γ Z A R Z
(c)
Energy (eV)
Γ M X Γ Z A R Z
(d)
Total
Cu
Fe
Sn
S
DOS (1/eV)
Energy (eV)

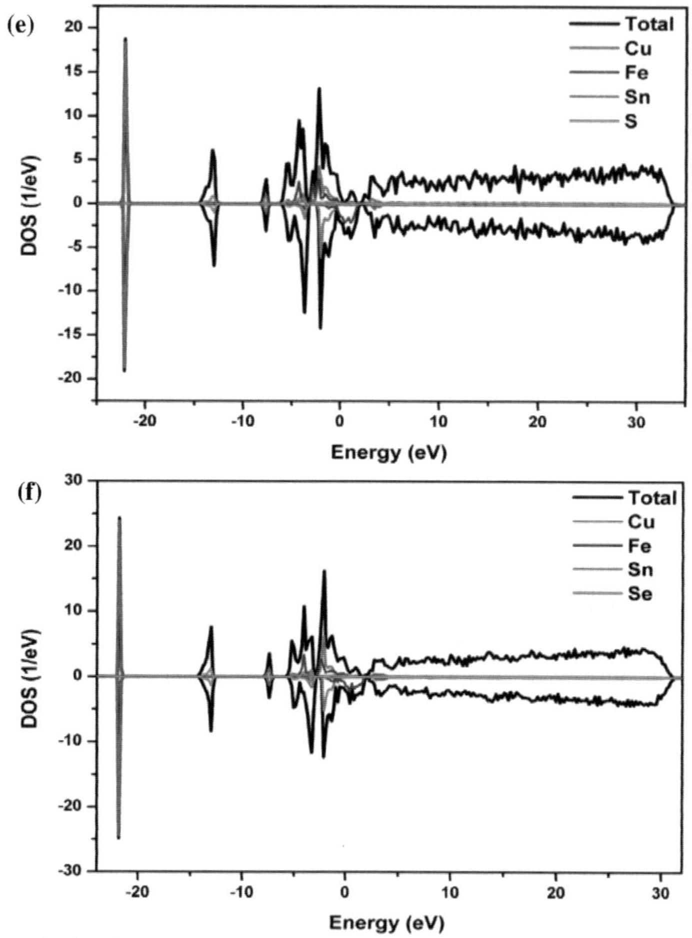

Figure 1. The calculated electronic band structure and DOS of compounds. The electronic band structure for (a) Cu_2FeSnS_4(I42M), (b) Cu_2FeSnS_4(P4) and (b) $Cu_2FeSnSe_4$(P4). DOS for (d) Cu_2FeSnS_4(I42M), (e) Cu_2FeSnS_4(P4) and (f) $Cu_2FeSnSe_4$(P4).

3.3 Optical properties

We have first calculated the real and imaginary part of $\varepsilon(\omega)=\varepsilon_1(\omega)+i\varepsilon_2(\omega)$ complex dielectric function using the Kramers-Kroning relations. The optical constant such as energy-loss function, the effective number of valance electrons and the effective optical dielectric constant have been calculated with the help of the real and imaginary part of dielectric function for these compounds. The results obtained show manner similar to our recent works [19]. The energy values of ε_1x (ε_1z) that decreasing $(d\varepsilon_1)/(dE<0)$ and increasing $(d\varepsilon_1)/(dE>0)$are zero are 7.04 eV and 16.63eV for CFTSe compound, 7.46 eV and 17.64 eV for CFTS compound. These values that the ε_1x (ε_1z) are zero are points reduced of the reflections, and show that the polarization disappears. The maximum peak values of ε_2x (ε_2z) are 0.66 eV, 4.73 eV, and 7.21 eV for CFTS compound, 0.60 eV, 4.45 eV, and 6.54 eV for CFTSe compound. These values show how much the electromagnetic wave polarizes the system, and corresponds to the electronic transitions from the valance band to the conduction band. Furthermore, energy region above 8 eV for CFTS and CFTSe compounds, respectively is the region where dispersion and transparency are low. This energy region corresponds to the region beginning of the transition between the bands. The 1.4-8.0 eV energy region for these compounds is the region where the transitions between the bands are very intense. The 1.6-8.0 eV energy region has reduced transitions between the bands. The energy region above 10 eV also corresponds to the collective vibration of valance electrons. This energy region defined as plasma oscillations is described by the energy loss function (L). The sharp maxima in the energy-loss function are associated with the existence of plasma oscillations. The optical data for CFTS and CFTSe show that in the limit of IR or visible region of the spectrum that have been studied the values of the real part of the dielectric constant lie near zero. This means that the negative contribution from the accelerating mechanism of absorption is quite small and is compensated by the positive contribution from the real and virtual interband transitions of electrons. Such a behavior of the $\varepsilon 1$function indicates the low concentration of conduction electrons in Cu_2FeSnZ_4. The presence of a Drude component in the dielectric functions of Cu_2FeSnZ_4 (D_{2d11} phase) permitted us to determine the parameters of free electrons – plasma frequency of conduction electrons E_{pl} = 19.2 eV and the effective number of free electrons - N_{eff} = 2.4x10^{22} cm^{-3}. The low values of N_{eff} indicate the formation of a pseudogap in the energy band spectrum of Cu_2FeSnZ_4. In the case of Cu_2FeSnZ_4, the zero values of ε_1 prevent making corresponding estimates for Neff. However, it is reasonable to explain the observed behavior of the dielectric properties of Cu_2FeSnZ_4 by the formation of a deeper pseudogap in the density of states. Our results concerning the low-energy interband absorption indicate a weak "growing-in" of the gap in the density of states, which was predicted theoretically for the both phases of Cu_2FeSnZ_4.

4. Conclusions

In this work, we have investigated mechanical, electronic, and optical properties of CFTS and CFTSe compounds. The results of elastic constants reveal all compounds are mechanically stable. The traditional B/G ratio indicate that considered compounds possess ductile nature. The electronic structure calculations show that the majority spin states cross the Fermi level and thus have the metallic character, while the minority spin states open the band gaps around the Fermi level and thus have the narrow-band semiconducting nature. Finally, optical properties were studied and discussed.

References

[1] Jimbo K, Kimura R, Kamimura T, Yamada S, Maw W S, Araki H, Oishi K and Katagiri H 2007Thin Solid Films.**515** 5997
[2] Katagiri H, Saitoh K, Washio T, Shinohara H, Kurumadani T and Miyajima S 2001 Sol. Energy Mater. Sol. Cells.**65** 141
[3] Moriya K, Watabe J, Tanaka K and Uchiki H 2006Phys. Status Solidi C. **3**2848
[4] Babu G S, Kumar Y B K, Bhaskar P U, and Raja V S 2008Semicond. Sci.Technol.**23** 085023
[5] Chen S, Gong X G, Walsh A and Wei S H 2009 Appl. Phys. Lett.**94**041903
[6] Schorr S 2007 Thin Solid Films.**515** 5985
[7] Persson C2010 J. Appl. Phys.**107**053710
[8] Kresse G and Furthmuller J 1996 Comp. Mater. Sci.**6** 15
[9] Blochl P E 1994 Phys. Rev. B.**50** 17953
[10] Kresse G andFurthmuller J 1996 Phys. Rev. B.**54** 11169
[11] Perdew J P, Burke K, and Ernzerhof M 1996 Phys. Rev. Lett .**77** 3865
[12] Monkhorst H J and Pack J D 1976 Phys. Rev. B.**13** 5188
[13] Kocak B, Ciftci Y O, Colakoglu K, Deligozand E, and Tatar A 2013 Materials Science and Technology**29** 925
[14] Simsek S, Koc H, Palaz S,Oltulu O, Mamedov A M and Ozbay E 2015 Ferroelectrics.**483** 26
[15] Born M 1940 Proc. Cambridge Philos. Soc.**36** 160
[16] Voight W 1928Lehrbook der kristall physic Leipsig: Teubner
[17] Reuss A and Angew Z 1929 Math Mech**9** 49
[18] Hill R 1952 Proc. Phys. Soc. London**70**65349
[19] Koc H, Simsek S, Palaz S,Oltulu O, Mamedov A M and Ozbay E 2015 Phys. Stat.Sol. (c). **12(6)** 651

IOP Conf. Series: Materials Science and Engineering **175** (2017) 012015 doi:10.1088/1757-899X/175/1/012015

Crystal chemical characteristics and physical properties of ferrous minerals as the basis for the formation of functional materials

A Shmakova[1], B Kanev[1], A L Gömze[2], O Kotova[1]

[1]Institute of Geology Komi SC UB RAS, 167982 Syktyvkar, Russia
[2] University of Miskolc, Hungary

E-mail: kotova@geo.komisc.ru, femgomze@uni-miskolc.hu

Abstract. Crystal chemical characteristics and physical properties of ferrous minerals can be criteria for search and evaluation of mineral (natural) raw for the production of functional materials. Special attention will be given to new experimental methods of transformation of minerals at different methods of influence. As a probe to identify the relationship between the actual crystalline structure of the mineral and its technological properties we used the oxidation - reduction reactions of iron ore-forming minerals. We will show that the inexpensive and affordable methods of influence at ore and technological products result in the observed Fe^{2+} — Fe^{3+} charge transfer, which result in the increase of the conversion degree of the structure and change of magnetic properties of the substance.

Introduction

Natural titanium and aluminum minerals are not only an important source of aluminum and titanium, but also raw for functional materials. Traditional deposits of aluminum and titanium ores are coming to the end, which is boosting the search for new sources and technologies to solve these tasks. This may be a non-traditional raw – poor, but affordable and common minerals, ores and rocks (or technogenic and industrial wastes), which require innovative technologies for utilization. The efficiency of the technologies depends on the mineralogical and geochemical studies of oxide ores and supergene modification products. In this article the authors focused on the problem of iron-titanium and iron-aluminum minerals. For example, the presence of non-magnetic ferrous oxides is a significant obstacle for technological processes in bauxites; and low-magnetic titanium minerals are often a cause of non-conditional concentrates [1-4]. The iron ions are involved in the redox environment of natural mineral formation. Our experiments have shown that Fe-minerals are good test objects to identify the relationship between the crystal-chemical characteristics and technological properties of minerals. During the technological processes the inexpensive and affordable methods of influence at ore and technological products result in the observed Fe^{2+} — Fe^{3+} charge transfer, which results in the increase of the conversion degree of the structure and change of magnetic properties of aluminum, titanium and iron concentrates [5-7]. A special attention should be paid to the mechanisms of natural and technogenic phase transformations at different methods of energy methods of influence (e.g. radiation-thermal) in minerals of rebellious ores and wastes to control concentrations of the valuable components in the technological processing schemes, production of new industrial products.

The aim of the work – the phase analysis of iron-titanium and iron-aluminum minerals and studies of methods of Fe^{2+} - Fe^{3+} charge transfer as the basis of innovative technologies of utilization of the observed ores.

1. Objects and Experimental procedure

The objects of study are aluminum, titanium and iron ores associated with Devonian-Carboniferous tropical and subtropical lateritic weathering crusts and their redeposition products.

The analytics was done mainly by the equipment in the Institute of Geology of Komi Science Centre UB RAS: morphological features of the ore minerals and their composition were studied by a scanning electron microscope Jeol JSM-6480 LV. Other modern methods were also used, for example, X-ray fluorescence techniques (XRF), optical-mineralogical methods (stereomicroscope MBS-10, polarizing microscope POLAM L-311), etc.

Results and discussion

Figure 1 represents a triangular diagram of TiO_2 – FeO – SiO_2 figurative points of titanium minerals from titanium concentrates of Pizhemsksoe deposit presented in [6]. According to the authors' opinion, the first stage is associated with Fe-rutile transformation into pseudorutile, the second stage is associated with Fe-rutile transformation in pseudorutile, the third stage is associated with the transformation of pseudorutile into leucoxene. All steps of the transformation of ilmenite $FeO{\cdot}TiO_2$ into leucoxene via intermediate phases: $Fe_2Ti_4O_{11}$ (Fe-rutile), $Fe_2Ti_5O_{13}$ (pseudorutile) and also leucoxene, hematite, Mn-siderite and rutile are recorded instrumentally and according to the authors' opinion they are formed within the same process described in [7]. Such a variety of iron-titanium minerals suggests that many natural processes occur with participation of iron ions. These minerals possess different physical and chemical properties, making difficult to obtain monomineral fractions of grains with similar magnetic or electrical properties. This reduces the efficiency of the separation of minerals.

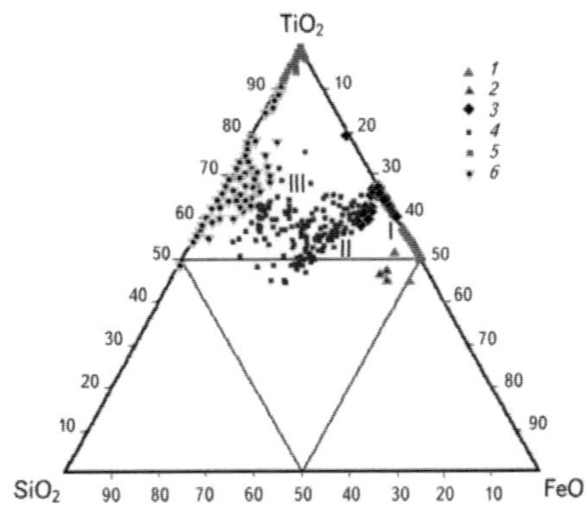

Figure 1. Triangular diagram of TiO_2 – FeO – SiO_2 figurative points of titanium mineral from titanium concentrates of Pizhemskoe deposit [6]

About 10% leucoxene contain significant amounts of iron and possess magnetic properties. The magnetic susceptability of ferrous leucoxene is variable, and its low-ferrous varieties are not removed during magnetic separation. Our experiments showed that γ-irradiation resulted in increasing magnetic susceptibility of the leucoxene at 4.5 times (Table 1). Thus, the magnetic characteristics can be easily changed for technological tasks.

Table 1. Influence of irradiation on the magnetic properties of leucoxene

Irradiation dose D, Mrad	Magnetic susceptibility, r.u.		
	High magnetic fraction	Medium magnetic fraction	Low magnetic fraction
0	1.80	1.0	0.14
1.85	1.80	1.1	0.13
2.85	3.42	1.4	0.14
3.75	7.62	1.0	0.10
4.75	17.70	4.1	0.10

For example, in [4, 8] it was shown that leucoxene sandstone was represented by non-durable grains acicular rutile and anatase, which during sintering transformed into rutile, titanium magnetite and spinel, which crystals strengthened the glass phase and interporous partitions, durability and frost resistance of the ceramic batch. This is accompanied by intensification of redox reactions between organic and ferrous components that contribute to the Fe^{2+} — Fe^{3+} transformation.

We studied the selected from Vezhayu-Vorykvinskoe deposit samples of three most high-ferrous bauxite types with different ratios of mineral forms of iron: oxide, phyllosilicate and mixed oxyde-phyllosilicate. As is known, the ferrous silicate phase in the composition of exogenous rocks is treated as chamosite, i.e. 14 Å phyllosilicate with the chlorite structure (alternating layers of T-O-T or 2:1) and composition $(Fe^{2+}, Fe^{3+}, Al)_5Al[(Si_3Al)_4O_{10}](OH)_8$, or as septechamosite (berthierine) - 7Å phyllosilicate with kaolinite structure (alternating layers of T-O or 1: 1) and composition $(Fe^{2+},Fe^{3+},Mg)_{2-3}[(Si,Al)_2O_5](OH)_6$ [2,]. The same mineral in the studied Timan bauxites, according to X-ray patterns, is 7Å phyllosilicate, and therefore can be attributed to berthierine.

According to the lithochemical properties the studied types of ferrous bauxites are aluminum standard-superferrous superhydrolizates. They are characterized by the highest content of iron within 20-35 % Fe_2O_3. By silicon module they correspond to rather high grade bauxites. The calculations of normative mineral composition show that all types of the ferrous bauxites are predominated by boehmite, which normative content is not lower than 60 mol %. The most important impurities are widely various ferrous minerals - berthierine, oxides, oxyhydroxides. It was determined that with increasing content of oxide and oxyhydroxide phases results in fast declining phyllosilicate proportion. Besides the recalculation of the chemical composition of the minerals identifies titanium oxides (this is confirmed by X-ray analysis) and X-ray amorphous siliceous substance.

Our studies showed that Timan ferrous bauxites by iron crystal chemistry can be divided into 1) oxide-oxyhydroxide with sharply dominant concentration of iron in the oxides and its distribution between the hematite, goethite and kaolinite in the ratio of about 1:0.1:0.3 (hematite-boehmite type) ; 2) oxide phyllosilicate-oxyhydroxide, in which iron is more inclined to hematite and berthierine in the ratio of 1:0.3:0.7 (hematite-berthienie-boehmite type); 3) phyllosilicate-oxyhydroxide with iron, almost entirely tied with berthierine (berthierine-boehmite type). The obtained data can be used to improve the processing technology of Timan bauxites and in particular, to minimize the aluminum loss within the structure of ferrous minerals.

An important obstacle for processing of bauxite is non-magnetic ferrous oxides - goethite-hydrogoethite and hematite-hydrohematite. Irradiation of ferrous non-magnetic materials with high-energy electrons result in the transformation of non-magnetic and low-magnetic minerals into ferromagnetic [ ]. The X-ray pattern of the sample, subjected to stepwise heating with quadruple 60 minutes exposure, lacks reflections from boehmite and goethite, but obviously amplifies reflections from hematite and adds reflections from quartz and γAl_2O_3 – spinels with defective structure. Obviously, these changes are caused by the thermal dissociation of goethite and boehmite:

1) $2FeO(OH) \rightarrow Fe_2O_3 + H_2O$;

2) $2AlO(OH) \rightarrow Al_2O_3 + H_2O$.

The analysis showed that during such heating of the bauxite sample the proportion of the magnetic fraction was consistently reduced (coefficient of pair correlation of the share with temperature r = –0.86). However, the gross value of the magnetic susceptibility was not changed [3]. This paradox can be explained by the fact that slightly larger magnetic susceptibility of hematite compared to goethite was enough to increase the yield of the magnetic fraction, but not enough for a substantial increase of gross magnetic susceptibility. It is clear that the presence of boehmite in radiation-thermal modificate of bauxite is explained by incomplete dissociation of the latter due to insufficient heating of the sample at an appropriate temperature (500–600 °C). But the appearance of two ferromagnetic phases – maghemite and magnetite – should be attributed directly to the results of irradiation. The analysis showed that the formation of such phases significantly influenced the gross magnetic properties of the modified sample – volumetric and specific magnetic susceptibility were increased in comparison to the initial sample by 6–7 times. In accordance with this during the radiation-thermal processing of the sample the proportion of the magnetic fraction increased and, which consistency is confirmed by a strong direct correlation of the magnetic susceptibility with heating temperature (r = 0.71).

Conclusion

The paper presents the mineralogical and crystal chemical studies of ferrous minerals of titanium and aluminum ores from Timan. The specifics of the Fe^{2+} and Fe^{3+} distribution in the structural positions in oxides are discussed. It was shown that the inexpensive and affordable methods of influence at ore and technological products resulted in the observed Fe^{2+} — Fe^{3+} charge transfer, which resulted in the increase of the conversion degree of the structure and change of magnetic properties of the substance. The oxidation-reduction reactions of iron ore-forming minerals are a probe to identify the relationship between the actual crystalline structure of the mineral and its technological properties. Crystal chemical characteristics and physical properties of ferrous minerals can be criteria for search and evaluation of mineral (natural) raw for functional materials.

References

[1] Kotova O B, Vakhrushev A V 2011 *Vestnik of the Institute of Geology Komi SC UB RAS* **3** 12
[2] Kotova O, Silaev V, Lutoev V, Vakhrushev A 2016 *IOP Conf Series: Materials Science and Engineering* **123** 012024 (doi: 101088 / 1757-899X / 123/1/012024)
[3] Kotova O B, Razmyslov I N, Rostovtsev V I, Silaev V I 2016 *Ore Enrichment* **4** 16
[4] Kotova O, Ozhogina E, Ponaryadov A, Golubeva I 2016 *IOP Conf Series: Materials Science and Engineering* **123** 012025
[5] Kotova O B, Ponaryadov A V, Gömze L A 2016 *Vestnik of the Institute of Geology, Komi SC UB RAS* **1** 34
[6] Makeev A B, Lyutoev V P 2015 *Ore Enrichment* **4** 16
[7] Watson E B, Wark D A, Thomas J B 2006 *Contrib. Mineral Petrol* **151** 413

Change of fine gold properties for industrial application

S Sitkevich[1]**, L A Gömze**[2]**, T Mayorova**[1]

[1]Institute of Geology Komi SC UB RAS, 167982 Syktyvkar, Russia
[2]University of Miskolc, Hungary

E-mail: sitx0113@yandex.ru

Abstract. At present we face the problem of cost-effective and efficient method for the enrichment of fine and small fractions of gold and its industrial application. New extraction technologies are required (sometimes more than a third of valuable raw are wasted). One of the trends to solve this problem is the change of magnetic susceptibility of gold.

1. Introduction

With the development of economy and science-based industries the demand for gold is increasing. This leads to a gradual involvement of poor and rebellious ores with rebellious, small and fine gold and also old tailings of processing plants. It is known that in certain objects main gold reserves are represented by small and fine fractions (up to 85% in some places in Chukotka). This significantly affects the results of enrichment: sometimes 40% of valuable raw go to the dumps. Loss is conditioned by both small dimension and morphology of gold particles.

One of the trends to solve the problem of gold recovery of specified classes is to change its natural properties, particularly magnetic susceptibility, for subsequent enrichment by a magnetic (electromagnetic) method. Detailed researches were conducted in the area of imparting magnetic properties to gold [1], taking into account the surface properties [2]. For example, according to one of the methods, the ore is treated with gaseous Fe (Ni, Co) carbonyl without changing the magnetic susceptibility of waste rock, i.e. based on chemical processes of interaction between gold surface and magnetic material [3]. However, these methods are still quite expensive. In addition to the problem of low-dimensional gold enrichment, there are certain difficulties with the extraction of a useful component from the concentrate and recleaning products.

The aim of the paper is to study the process of change of magnetic properties of fine and small gold to improve the efficiency of its extraction at the enrichment process.

2. Subjects and Methods

The objects of study were samples (collection) of gold from the Devonian paleoplacers of Timan. Dispersed gold content is characteristic for the whole producing horizon with higher concentrations in pebble conglomerates of bedrock-adjacent part of Devonian system.

In the experiments we used gold particles of flattened shape with grain size 0.1–0.4 mm from the nonmagnetic fraction. We added magnetic material to quartz sand and subjected the sample to impact test. The extraction into the magnetic fraction was carried out by magnetic separation using SIM-1 device in a weak magnetic field (less than 0.5 Tesla) at current 0,01–0,4 Å.

We used a range of modern physical and physical-chemical methods of research: optical-mineralogical (stereomicroscope MBS-10, polarizing microscope POLAM L-311), a computerized optical microscopy in translucent and reflection mode (Polam 3-312 and OLYMPUS BX51);

analytical scanning electron microscopy (JSM-64000, Jeol); X-ray fluorescence analysis (MESA-500W, Horiba) ICP-MS (ELAN 9000, Perkin Elmer); X-ray diffraction (XRD-6000, Shimadzu); IR spectroscopy (Fourier spectrometer FT-02 Intralum, Lumex), magnetometry; ferromagnetic and nuclear magnetic resonance.

3. Results and discussion

Gold occurs in association with tantaloniobates (columbite, ilmenorutile), rare earth minerals (cularite, monazite, torite), osmirides, titaniferous minerals (leucoxene, ilmenite, rutile, anatase, brookite), sulphides (pyrite, pyrrhotite), Cr-spinels, pyrope. In the form of aggregates with gold we detected: quartz, feldspars, columbite, ilmenorutile, monazite, zircon, siderite, limonite, goethite.

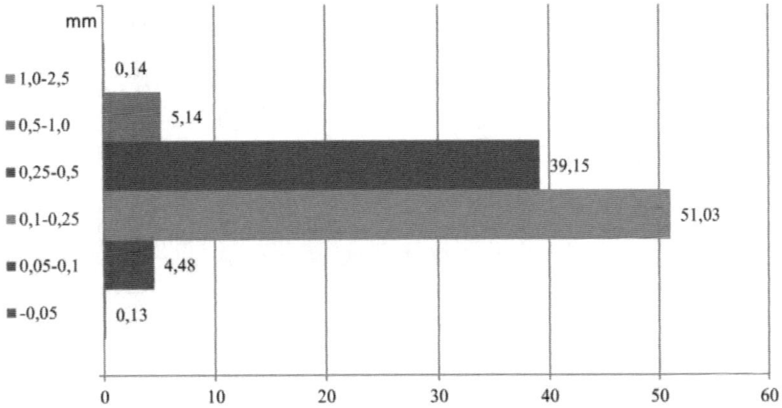

Figure 1. Histogram of the class distribution of gold

The gold is small, very small and fine (Fig. 1). It is characterized by high sorting, composition uniformity, high fineness. It is represented mostly by flattened low elongated forms. The coefficient of elongation of gold particles varies between 1.0-3.0. The modal value of grain length - 0.22 mm, width - 0.13 mm. The ratio of the modal values of gold particles of length to width defines elongation factor 1.69.

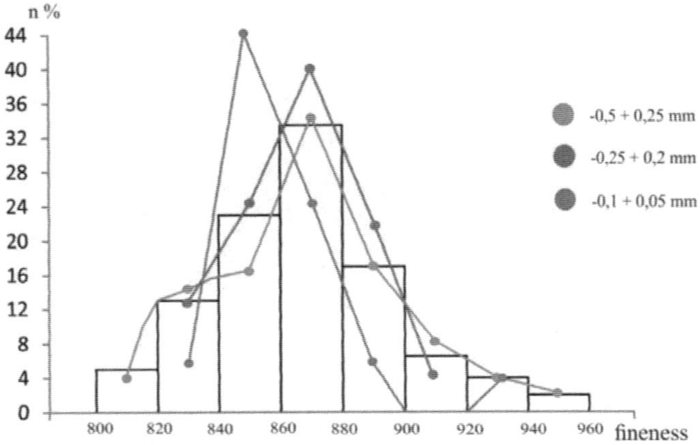

Figure 2. Histogram of gold fineness and fineness curves depending on the particle size

Figure 2 shows the curves of distribution of gold fineness into classes. Perhaps the lowest fineness values are a local feature and do not apply to all gold-bearing Devonian formations of Timan.

The permanent impurities in gold are silver, iron, copper, lead, titanium. Less common - bismuth, palladium, zirconium, mercury, arsenic, manganese, antimony, zinc (Fig. 3). The copper content ranges from 0.05 to 0.15, mercury from 0.04 to 0.15, palladium – thousandths to hundredths of a percent. Petrogenous elements: aluminum, silicon and others are always present, but their presence may be explained by inclusions.

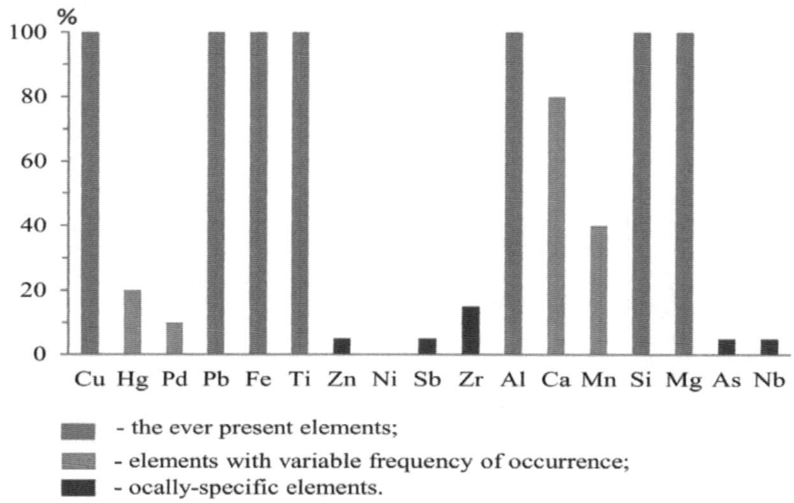

Figure 3. Frequency of occurrence of impurity elements in gold from Devonian paleoplacers of Timan

From 70 to 90% of the gold belong to subrounded (roundness factor by Powers is 40), 5-15% are subangular individuals (roundness factor – 30) and 8 to 10% are represented by angular gold with roundness factor equal to 21, and particles with size 0.2 mm or more have a better roundness [1].

The surface of the gold particles is often microscaly, low porous. However, particles of -0.5; +0.2 class are characterized by a coarse foveolar, lumpy surface. The morphology of the gold from the paleoplacer is quite diverse: from nodular and plate to spherical. After processing the gold particles are mostly lumpy (Fig.4).

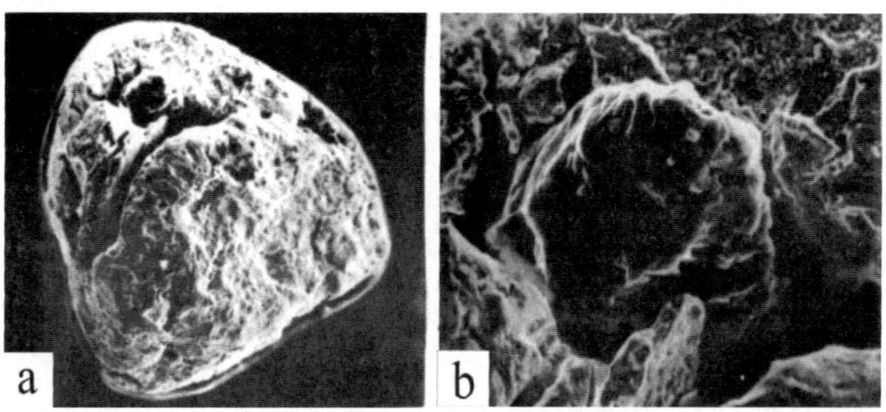

Figure 4. Gold particles after processing: a – general view of gold particle, magnified 400x; b – fragment of capture of mineral with electromagnetic properties by the gold particle, 1500x [1]

From the above characteristics it should be followed that paleoplacer gold of Timan belongs to the rebellious category, which requires application of new methods of metal extraction.

The application of high-gradient magnetic separation allows extracting up to 60% of gold to a hard fraction. Such an effect is conditioned by the following factors: 1) riveting of dispersed iron on gold particles during grinding process; 2) deposition of iron hydroxides, formed during pyrite oxidation, on gold in heaps; 3) close association of gold with magnetic minerals; 4) iron impurities in the gold. Studies of the influence of the material dimension to extraction gave the following results: 1) optimal dimension of enriched material ranges from 106 + 30 mcm; 2) separation efficiency decreases with sizes from 15 to 100 mcm, finer grains have relatively less magnetic susceptibility.

The analysis of processed gold revealed that the magnetic material got lumpy inside particles or captured by the edges of particles at the moment of deformation. Microscopic examinations with an electron microscope allow diagnosing the observed microinclusions as iron-containing minerals (Fig.5).

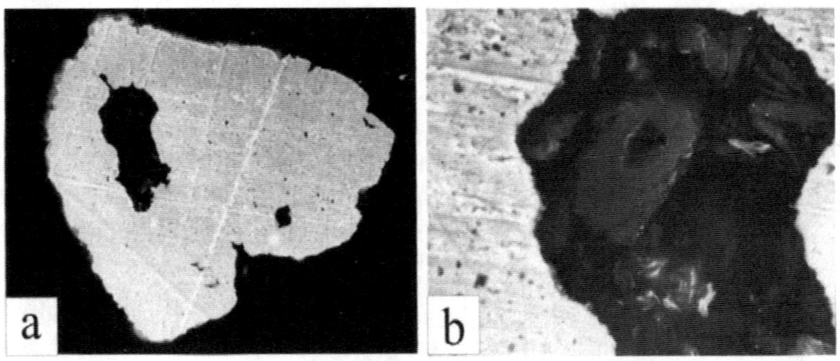

Figure 5. Entering of magnetic fraction into gold (polished section): a – general view 50x; b – fragment of the area of entering 250x

It should be noted that the iron-containing material (in this case - magnetite) should be added to the sample in the range 0.07–0.08 %. As a result, from 92–95 % to 100 % of gold can be extracted into the electromagnetic fraction from the concentrate.

4. Conclusion

Thus, changes of the magnetic properties of gold can be carried out at the stage of disintegration of rocks fast and without additional operations that will enhance the efficiency of extraction of gold of very small and fine classes. However, we must emphasize the laboratory level of researches. The solution of industrial problems requires some technical improvements of industrial installations in the technological scheme of enrichment of specific deposits of fine and small gold.

References

[1] Uskov N N, Ostashenko B A, Mayorova T P, Filippov V N 1992 *Enrichment of fine ores* 31
[2] Kotova O 2013 *IOP Conf Series: Materials Science and Engineering* **47** 170 012037 (doi: 101088 / 1757-899X/47/1/012024)
[3] Rinding I R, Turner R L, 1983 Pat. 529662 (Australia).

Damascus steel ledeburite class

D A Sukhanov[1], L B Arkhangelsky[2]and N V Plotnikova[3]

[1]ASK-MSC Company, 117246Moscow, Russia.
[2]President Union Smiths Russia, 141071Korolev, Russia
[3]Novosibirsk State Technical University, 630073 Novosibirsk, Russia

E-mail:suhanov7@mail.ru

Abstract. Discovered that some of blades Damascus steel has an unusual nature of origin of the excess cementite, which different from the redundant phases of secondary cementite, cementite of ledeburite and primary cementite in iron-carbon alloys. It is revealed that the morphological features of separate particles of cementite in Damascus steels lies in the abnormal size of excess carbides having the shape of irregular prisms. Considered three hypotheses for the formation of excess cementite in the form of faceted prismatic of excess carbides. The first hypothesis is based on thermal fission of cementite of a few isolated grains. The second hypothesis is based on the process of fragmentation cementite during deformation to the separate the pieces. The third hypothesis is based on the transformation of metastable cementite in the stable of angular eutectic carbide.It is shown that the angular carbides are formed within the original metastable colony ledeburite, so they are called "eutectic carbide".It is established that high-purity white cast iron is converted into of Damascus steel during isothermal soaking at the annealing.It was revealed that some of blades Damascus steel ledeburite class do not contain in its microstructure of crushed ledeburite.It is shown that the pattern of carbide heterogeneity of Damascus steel consists entirely of angular eutectic carbides.Believe that Damascus steel refers to non-heat-resistant steel of ledeburite class, which have similar structural characteristics with semi-heat-resistant die steel or heat-resistant high speed steel, differing from them only in the nature of excess carbide phase.

Keywords: Bulat; Damascus steel; Wootz; Indiansteel.

Introduction

In the classification of tool steel according to Prof. Geller (1968) [1] was no place for the legendary Damascus steel ledeburite class. Damascus steel removed from modern metallurgy. A relic of the past, which have not found application in the present. The legend for which no future! Why did this happen? Is it possible the triumphant return of Damascus steel on the highest step of the podium? We believe that this is possible.

Studying articles Prof. Gayev (1965) [2], Prof. Tavadze et al (1984) [3], Prof. Schastlivtsev et al (2013) [4], Sherby and Wadsworth (1985) [5], Verhoeven and Pendray (1996) [6], Prof. Barnett et al (2009) [7], in which researchers samples of ancient blades from Damascus steel. We concluded that of part of excess cementite has an unusual nature of origin differing from of the excess phases of secondary cementite, ledeburite cementite and primary cementite. The morphological features of excess cementite have the shape of an irregular octahedra and prisms (Fig. 1), which from our point of view, more similar in morphology with eutectic carbides of steels ledeburite class [8, 9].

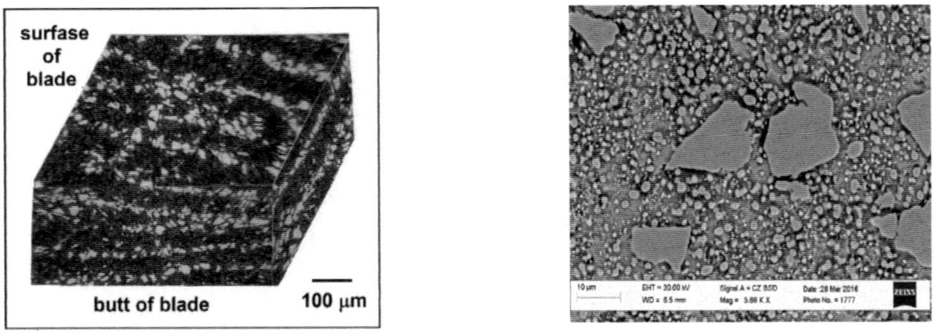

Figure 1. Morphology of carbides in Damascus steel (BU22A):
a - diagram of the structure of blade Damascus steel; b - electron microscope (eutectic carbides)

The most typical group of tool steels ledeburite class is deemed to be of die steels and high-speed steels. The structure of die steel ledeburite class is shown in Fig. 2. In these steels at melting is present of metastable ledeburite located at the grain boundaries austenite. Metastable ledeburite based on carbides of type M_6C and M_3C crushed and transforming in the stable of eutectic carbides type MC, M_2C and M_7C_3 with hexagonal structure as a result of plastic deformation (where M is alloy metal; C is carbon). Heat treatment steels ledeburite class is a complex process that includes multi-stage heating and cooling and cold processing. High-speed steel and die steels have low elasticity, which limits their use as a cutting tool subject to dynamic loads.

DIE STEEL LEDEBURITE CLASS (X12)		
MELTING	FORGING	ANNEALING
Matrix: **AUSTENITE** Excess phase: **LEDEBURITE**	Matrix: **SORBITE** Excess phase: **CARBIDES**	Matrix: **FERRITE** Excess phase: **CARBIDES**
Zonal segregation of carbon	The formation of eutectic carbides	The growth of excess carbides

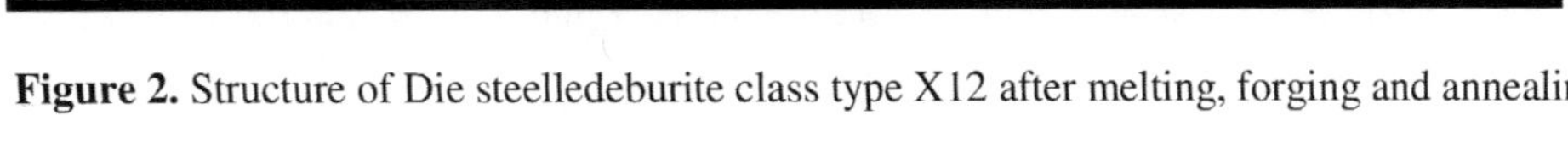

Figure 2. Structure of Die steelledeburite class type X12 after melting, forging and annealing

Extreme operating conditions of the cutting tool under dynamic loads can match Damascus steel in which the carbon content is the same as in white cast iron. The aim of this work is the study of morphological signs of excess carbide phase in Damascus steel, after the process of smelting, heat treatment and plastic deformation. The mechanism of formation of faceted carbides in Damascus steel is still not known. Question about the transformation of excess cementite in the angular carbides is one of the most interesting and important in the analyzed problem. This has not only scientific but also practical significance. Knowing the answer to this question, it is possible to control the entire range of mechanical and physical properties of Damascus steel.

Materials and methods

The object of the research was chosen the high carbon alloys after melting, which in the structure have ledeburite phase. The chemical composition of the alloys is presented in table №1. In the marking of high-speed steel P6M5 letters and numbers mean the following: P is rapid; 6 is the average tungsten weight fraction (5.8 wt.%); M is molybdenum; 5 is the average molybdenum weight fraction (4.7 wt.%). Marking die steels X12 means the following: X is chrome; 12 is the average chrome weight fraction (11.85 wt.%). In the marking of Damascus steel BU22A letters and numbers mean the following: BU is Bulat (Damascus steel) containing not more than 0.1% manganese and silicon (each individually); 22 is the average carbon weight fraction (2.25 wt.%); A is a high-quality alloy containing not more than 0.03% sulphur and phosphorus (each individually).

Table 1.

Alloys	The contents of chemical elements, %								
	C	Si	Mn	P	S	Cr	V	Mo	W
P6M5 High-speed steel	0,82	0,51	0,25	0,020	0,006	3,60	1,8	4,7	5,8
X12 Die steel	1,95	0,30	0,24	0,024	0,023	11,85	--	--	--
BU22A Damascus steel	2,25	0,065	0,024	0,002	0,004	--	--	--	--

Structural investigations were carried out using an optical microscope of a series METAM RV-21-2 in the zoom range from 50 to 1100 fold. Deeper structural investigations were carried out on scanning electron microscope CarlZeiss EV050 XVP using microanalyzer EDS X-Act. The chemical composition of the alloy controlled with the help of optical emission spectrometer ARL 3460 type. The sizes of the analyzed samples was 15x15x30 mm. Phase analysis was performed by x-ray diffractometer ARL X'tra. The diffraction patterns of samples were recorded using copper x-ray tube as an x-ray source at a voltage of 40 kV and current of 40 mA. Analysis of samples was performed in the reflection geometry without monochromatization of the incident and reflected radiation. Average recorded energy dispersive Si(Li) detector wave length of the beam was $\lambda = 0{,}15406$ nm. Diffraction patterns were recorded repeatedly in the time mode ($t = 4...9$ seconds.) with step $\Delta 2\theta = 0.02$ and 0.05 degrees.

Results and Discussion

Advanced structure of Damascus steel before forging is a high-purity of iron matrix, with excess phase of metastable ledeburite, without Widmanstatten cementite. The purity of matrix during deformation plays a decisive role. The modern level of metallurgical production does not yet allow large scale to obtain large-sized products from high-carbon alloy with a minimum amount of impurities (hundredths of a percent). Only iron and carbon! However, as noted by Prof. Taganov (2009) [10], the blacksmiths of ancient India and Persia to effectively use low productive crucible melting to obtain a Damascus steel of the blades.

In the modern sense of the purity alloys does not exclude the presence of residues of deoxidation with manganese or silicon. The presence in alloys of manganese more than 0.2%Mn reduces the growth of dendrites of austenite during solidification of the melt, changes the chemistry of carbide, perlite stabilizes at the high-temperature annealing. The silicon about 0.2%Si creates centers of graphitization. When supercooling of the melt forms a metastable cementite which capable at the heating to disintegrate with the formation of graphite and austenite. As noted by Prof. Bogachev (1952) [11], the graphitization occur by the dissolution of metastable cementite in solid solution of austenite with subsequent nucleation and growth of graphite centers.

In the book Prof. Golikov (1958) [12] indicated that the most effective way to clean of the steels from impurities is the zonal segregation. Development zonal segregation in the ingot steel contributes to the boiling during melting. Boiling steel is substantially free of non-metallic inclusions and deoxidation products (Si < 0,07%). Additive of iron oxide (FeO) in smelting improves ductility and zonal segregation of carbon, as indicated in the work of Anosov (1841) [13]. The modification of the charge iron oxide binds harmful impurities, which are removed together with the slag.

Damascus steel (BU22A) has been received melting in the crucible at temperature of 1500^{0}C with isothermal exposure for 20 minutes until complete melting of the alloy (crucible steel). At the temperature of 1200^{0}C alloy was aged for 10 hours. The result isothermal holding between liquidus and solidus is formed of large dendrites of austenite with depleted carbon content (C <0,8%). Interdendritic areas is enriched of the carbon until the eutectic content. A similar effect was noted in the work of Prof. Gayev (1965) [2]. After slow cooling with the furnace, the alloy matrix BU22A acquires a coarse structure of pearlite with interlamellar spacing of about 0.6-1.0 □m. Noticeable changes occur in the morphology of the excess carbide phase. In the plane of the thin section observed coarse conglomerates carbide formationsof about 10.0 - 30.0 □m. Metallographic signs allowed to identify them as a metastable ledeburite (Fig. 3). The Characteristic morphological feature of the metastable ledeburite is that it compared to lamellar and cell ledeburite white cast iron contains in its structure a reduced amount of micropores and has not pronounced layering. It is in composition not quite ledeburite, but still not quite the eutectic carbide.

Deformation of the cast alloy BU22A was performed using a scythe forging strikers at an angle of 45 degrees to the temperature range from 850 ^{0}C to 650 ^{0}C. Forging occurred in the intercritical interval temperature. Metastable ledeburite is under the influence of the normal stress austenitic matrix and shear stress deformation. Around metastable carbides, accumulate defects such as dislocations. When the dislocation density reaches a critical value in a metastable ledeburite occur phase change, resulting in less stable carbides ledeburite transform more stable eutectic carbides (Fig.3). The transformation of metastable ledeburite in the faceted carbides angular shape is that a new phase is formed inside the source phases. The growth carbides angular shape happens migration of interphase boundaries. In the future, the growth of excess carbides is a diffusion redistribution of the components between the carbide and solid solution of austenite. The migration of interphase boundaries in the process of transformation leads to the separation of excess carbides angular shape.

According to Prof. Taran et al (1974) [14], the angular carbides in steels ledeburite class are not eliminated by annealing and deformation. What is this angular carbide in Damascus steel? To answer this question, we attempted to determine the phase composition of carbides angular shape. Analysis of diffraction patterns showed that the main phases in BU22A are ferrite (α-Fe) and cementite (Fe_3C). Because of the interference of the overlapping lines of the secondary cementite is not possible to determine the exact composition of angular carbides. At this stage of research it is possible to speak only about what we are faced with a special morphology of cementite.

The mechanism of formation of faceted carbides angular shape in Damascus steel is still not clear. From our point of view, there are three hypotheses that explain why, under certain conditions, the cementite acquires an angular shape.

DAMASCUS STEEL LEDEBURITE CLASS (BU22A)		
MELTING	FORGING	ANNEALING
Matrix: **PEARLITE** Excess phase: **LEDEBURITE**	Matrix: **SORBITE** Excess phase: **CARBIDES**	Matrix: **FERRITE** Excess phase: **CARBIDES**
Zonal segregation of carbon	The formation of eutectic carbides	The growth of excess carbides

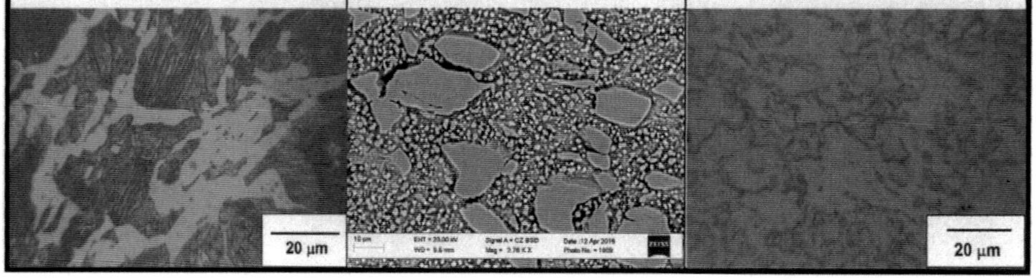

Figure 3. Structure of Damascus steelledeburite class type BU22A after melting, forging and annealing

The first hypothesis is based on the process of thermal division of the plates of cementite on the isolated single grain. Process of thermal division of cementite for steel ledeburite class has been studied in the work of Prof. Bunin et al (1969) [15]. The alloy BU22A is becomes Damascus steelduring isothermal soaking at the annealing. Prolonged annealing leads not only to thermal division of cementite plates, but also to their faceting. In the annealing process on the surface of the plate's cementite in the junctions between the grains of boundaries of austenite appear protrusions in the form of spikes. The growth of spikes of a long of grain boundaries austenite occur due to increased speed of diffusion of carbon. In the process of isothermal exposure at the annealing of cementite between of spikes of become thinner. In these places occur to the divide of cementite into parts. As a result, between the individual particles of cementite formed grains boundaries of austenite. Part of the excess carbides has an irregular trigonal-prismatic morphology. New angular carbides is formed within the original colony metastable ledeburite, so they are called "eutectic carbides", by analogy with a faceted angular carbides in alloy steels ledeburite class.

The second hypothesis is based on the process of fragmentation cementite during deformation to the separate the pieces. The traditional view on the formation of angular carbides are described in detail in book Prof. Geller (1968) [1]. It is assumed that the anomalously large angular carbides are formed because of the fragmentation of metastable ledeburite in the process of plastic deformation into separate parts (fragments). The greater the degree of deformation during forging, the stronger destroyed clusters of carbides. Crushing carbide conglomerates occur in places of a congestion of dislocations. It can be assumed that these of angular carbides are fragments of the ledeburite eutectic. However, we have convincingly shown that the structure of Damascus steel BU22A does not contain crushed ledeburite [9].

The basis of the third hypothesis laid the conclusion of Prof. Nizhnikovskaya (1982) that the plasticity of white cast irons, possible as a result to the transformation of excess carbide [16]. The essence of the hypothesis is recrystallization of metastable ledeburite in the process of plastic deformation in a stable phase of the eutectic carbide angular morphology. Non-alloy eutectic carbide is formed by the restructuring of the lattice of intermediate metastable cementite carbides in which the carbon atoms are packed in a trigonal prismatic complex, which gives of the faceted eutectic carbides.

IOP Conf. Series: Materials Science and Engineering **175** (2017) 012017 doi:10.1088/1757-899X/175/1/012017

High-temperature annealing forged Damascus steel (BU22A) at a temperature of 1000^0C and isothermal exposure for 15 minutes showed that is occurs growth of the eutectic cementite due to the secondary particles of cementite. The process of coalescence of the eutectic carbides at high-temperature annealing is consist to the diffusion transfer of carbon atoms through the solid solution of austenite. The growth of eutectic cementite is carried out by the migration of interphase boundaries. Eutectic carbides angular forms grow at the expense of smaller particles of secondary cementite. The process of coalescence of the eutectic carbides increases with long-term high temperature exposures. As a result, is occur the decarburization of the matrix due to the abnormal growth of eutectic carbides (Fig. 3). A similar process was observed in articles Prof. Gayev (1965) [2] and Prof. Schastlivtsev et al (2013) [4]. It should be noted that in the process of coalescence, eutectic carbides retain angular form whereas secondary excess carbides, usually spheroidization. The process of coalescence of the eutectic carbides is irreversible. During high-temperature annealing forged Damascus steel is terminated the process of the formation of abnormally large eutectic carbides.

Conclusion

By nature Damascus steel ledeburite class type BU22A similar to high-speed steel type P6M5 and die steel type X12, because in it is occur the same transformation at the melting and thermomechanical processing. Damascus steel containing about 1.9...2.3% carbon are the tool steels ledeburite class [17]. The cast form the structure of the Damascus steel represents a major dendrites, in the interstices of which is found of metastable ledeburite. When forging is occur transformation of metastable ledeburite in eutectic carbides angular shape. In the annealed condition the structure of Damascus steel is ferritic matrix with unevenly spaced in it large angular eutectic carbides with a size to 30.0 μm.

Pattern carbide inhomogeneity in the alloy BU22A consists entirely of angular eutectic carbides having an irregular trigonal-prismatic morphology. Similar morphology of the carbides is observed in tool steels ledeburite class (Fig. 4).

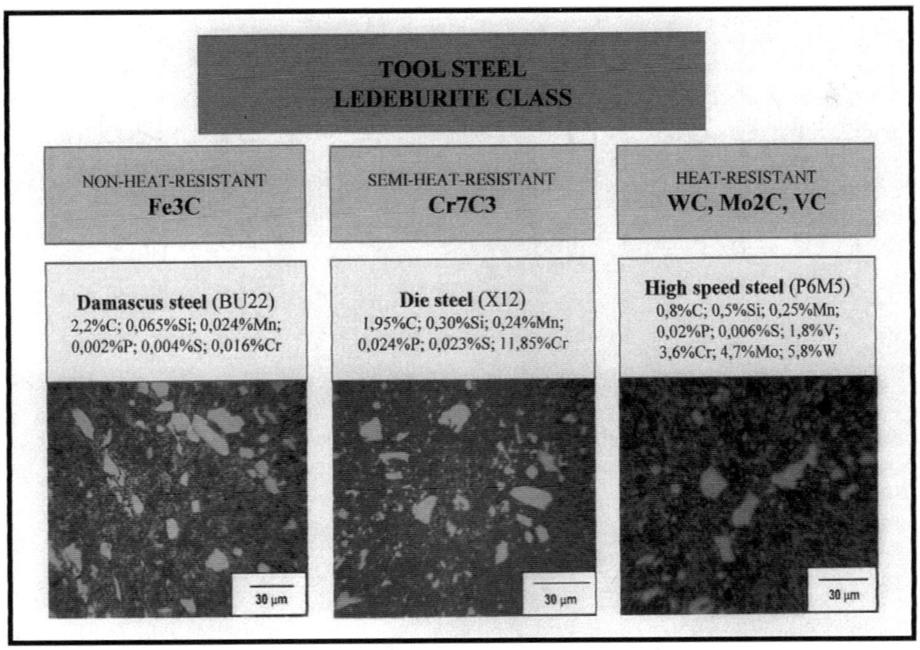

Figure 4. Classification of Tool steels ledeburite class of heat resistance of the excess carbides

Eutectic carbides in BU22A is a cementite, which has low heat resistance. Believe that Damascus steel refers to non-heat-resistant steel of ledeburite class, which have similar structural characteristics with semi-heat-resistant die steel or heat-resistant high speed steel, differing from them only in the nature of excess carbide phase.

References

[1] Geller Yu A 1968Tool steels (*Moscow: Metallurgy*) p 568
[2] Gaev I S 1965 Bulat and contemporary iron-carbon alloys *MiTOM* №9 pp 17–24
[3] Tavadze F N, Amaglobeli V G, Inanishvili G V and Eterashvili T V 1984 Electron microscopic studies of bulat steel *Soobsh. AN GSSR* (113)№3 pp 601–604
[4] Schastlivtsev V M, Urtsev V N, Shmakov A V et al 2013Structure of bulat *FMM*(114)№7 pp 650–657
[5] Sherby O D and Wadsworth J 1985 Damascus steel *Scientific American* 252 (2) pp 112-120
[6] Verhoeven J D, Pendray A H and Gibson E D1996 Wootz Damascus steel Blades *Materials Characterization*№37 pp 9-22
[7] Barnett M R, Sullivan A and Balasubramaniam R 2009 Electron backscattering diffraction analysis of an ancient wootz steel blade from central India *Materials Characterization*№60 pp 252-260
[8] Sukhanov D A and Arkhangel'ski L B 2016 Damascus steel microstructure *Metallurgist*№59 pp 818-822
[9] Sukhanov D A, Arkhangelsky L B, Plotnikova N V and Belousova N S 2016 Morphology of Excess Carbides Damascus Steel *Journal of Materials Science Research* (Vol. 5) №3 pp 59-65
[10] Taganov I N 2009 Setting of bulat legend *Kalashn. Oruzhie Boepr. Snaryazh.*№11 pp 92-97
[11] Bogachev I N 1952 Metallography of cast iron (*Moscow: Mashgiz*) p 367
[12] Golikov I N 1958 Dendritic liquation in steel (*Moscow: Metallurgizdat*) p 206
[13] Anosov P P 1841 Bulats *Russian Scientists – Metallurgists* (Moscow: GNTI Mash. Lit.) (1951) pp 38–112
[14] Taran Yu N, Nizhnikovskaya P F and Mironova E N 1974 The nature of angular carbides in high-speed steel *MiTOM* №2 pp 10–13
[15] Bunin K P, Baranov A A, Evsyukov M F and Pritomanov M I 1969 The Mechanism of division of crystals of cementite in annealing spheroidizing *MiTOM* №6 pp 19–23
[16] Nizhnikovskaya P F 1982Carbide transformations in eutectic iron-carbon alloys*Metals* №6 pp 105-110
[17] Sukhanov D A 2014 Bulat – non-alloy steel carbide class *Metallurgist*№2 pp 93–96

IOP Conf. Series: Materials Science and Engineering **175** (2017) 012018 doi:10.1088/1757-899X/175/1/012018

On the synthesis atmosphere influence in the technology of complex composite materials in the wide temperature range

A A Uspensky, S Yavshits, V Lipin, P Zhigalov, A Slobodov*

ITMO University, St.Petersburg, 197101, Russia

E-mail: aslobd@gmail.com

Abstract. Heterogeneous and homogeneous processes of synthesis for complex functional materials are studied by the methods of thermodynamic modeling and calculation, based on minimization of the Gibbs energy of the researched system. The conditions and products of gasification in the field of 0-2000 °C are determined for the synthesis conditions of a ceramic tile with complex composition (10-component system). All main products of interaction of components of ceramic mass (the system $SiO_2 - K_2O - Na_2O - Al_2O_3 - Fe_2O_3 - CaO - MgO - TiO_2 - P_2O_5 - SO_3$) and main components of air (the system $N_2 - O_2 - CO_2 - H_2O$) are found for the synthesis of material in the conditions of the air atmosphere at various temperatures. The character of physico-chemical influence of each components on the occurring processes is revealed, that determines a theoretical and calculation base of management and optimization of the synthesis engineering process for various functional materials.

1. Introduction

The synthesis atmosphere plays very important (and often decisive) role in the synthesis processes of modern functional materials. Its influence on phase and chemical materials transformations is most diverse under the synthesis of complex on composition (multicomponent, multiphase, etc.) - composites, ceramics and others.

These transformations and the systems are very complex - with multiple phases, many components, complexity and interconnection of the interactions and transitions, which take place during the synthesis, and wide range of values of state parameters, etc. Therefore, the detailed experimental research within the wide range of temperatures, compositions, and different conditions of synthesis is a very complex problem, which to date is generally unresolved [1- 3].

In rather many cases it is correct and sufficiently effective to apply methods of the thermodynamic physico-chemical modelling. These methods have principal advantages in comparison to experimental methods, letting to obtain more detailed data about the whole complex of interrelated phase and chemical transformations corresponding to functional materials and products synthesis [4-8].

Therefore, revealing phase and chemical transitions, which occur when synthesising ceramics (and many other functional materials), phase and chemical composition of the pottery, and how technological and external conditions, i.e. temperature, pressure, synthesis atmosphere, humidity, influence these compositions, all have not only purely scientific interest, but and practical application as well [6, 7].

2. Methods and objects of the research

The main objects of investigation is the raw material for production of the complex functional materials – ceramic tiles. As the basis of modelling, the raw material of a typical complex composition has been used (including all the main and auxiliary components). Its composition has been defined using analytical methods, and the main compounds have been recalculated into their oxide forms. As a result, oxide composition of the raw material corresponded to the ten-component system, which is the set of compounds-components fully and unambiguously describing both composition of the raw material, and the product. Then taking into account principal components of the synthesis atmosphere (air) the investigated multisystem can be represented as a set of the following 14 independent substances-components:

$$SiO_2 - K_2O - Na_2O - Al_2O_3 - Fe_2O_3 - CaO - MgO -$$
$$- TiO_2 - P_2O_5 - SO_3 - N_2 - O_2 - CO_2 - H_2O$$

The investigation has been undertaken with help of the methods and databases of thermodynamic modelling, which have been developed earlier, and which provide high correctness and efficiency of solution [4-8]. The equilibrium composition has been calculated under the different synthesis atmosphers and conditions, namely, in the temperature interval from low (room) temperatures to 2000°C (taking into account the possible overheating).

The considered conditions correspond to the annealing stage of the ceramic tiles production process.

In the Table below, the initial data in mass and mole expressions are given for all 14 substances-components, taking into account losses during the annealing. Using mole composition is viable from the point of view of the of chemical and phase transitions analysis.

The first ten substances are the components of the synthesized material (its amounts are given in accordance with the composition of the ceramic mass), the last four ones - the main components of air (its amounts are given in accordance with the composition of humid air atmosphere).

Table. The chemical composition of the sintered ceramic material and the synthesis atmosphere (air)

Substance / Content	SiO_2	K_2O	Na_2O	Al_2O_3	Fe_2O_3	CaO	MgO
mas.%	55,20	2,25	0,45	13,68	3,19	9,72	3,25
mol/kg	9,187	0,239	0,073	1,342	0,199	1,733	0,806
Substance / Content	TiO_2	P_2O_5	SO_3	N_2	O_2	CO_2	H_2O
mas.%	0,79	0,15	0,49	73,99	22,68	0,00045 6	1,244
mol/kg	0,099	0,011	0,117	76,53	20,53	0,0003	2,00

Let's consider synthesis in contact with the surrounding atmosphere. The composition of the atmosphere corresponds to the standard composition of the air into account of moisture (the Table). The pores volume has been taken as 10^{-4} м^3/kg, which corresponds to the medium porosity of the pottery fragment.

It is possible to nominally separate three fields within the ceramic composition, in which the conditions of synthesis and nature of the phase and chemical interactions differ significantly. Therefore, the investigation has been taken for each of these fields:

1. Bulk of the ceramic phase, which is completely separated from the surrounding atmosphere, where no contact with the atmosphere, and formation and evaporation of the gases, being a

result of the transformations taking place, are impossible;

2. Domains in the ceramic mass, which do not contact with the atmosphere, but formation and evaporation of the gases are possible. Such conditions might take place on the boundary of the material and internal closed pores, which are not filled with the air;
3. Ceramic mass, which contacts with the air, i.e. surface of the material, or the boundary of opened or closed pores filled with the surrounding atmosphere.

Separate consideration of each from these cases allows to reveal the mechanism of each constituent of the complex physico-chemical transformations (in the bulk of the material, in gaseous phase, and on the phase boundary) and the contribution of each of them.

Of course, the main ones are the first type of these interactions - in the thickness of the ceramic mass (press-powder) between its components. They determine primarily the final phase and chemical composition of the synthesized material. However, we can't ignore heterogeneous interactions of the second and the third types too - the formation of gaseous products and the influence of the synthesis atmosphere (inert gases, air, etc.).

Research as homogeneous and heterogeneous interactions is two difficult tasks, in many respects mutually independent and requires separate consideration. The main purpose of this paper is to study these heterogeneous interactions under synthesis of functional (ceramic) materials.

Next both types of the heterogeneous interactions are considered consistently. First, this type is an interaction between components of a ceramic mass with forming of gaseous products. Then - the same interactions, but with the participation of the synthesis atmosphere. Either inert gases or air (when it is admissible) are usually used as this atmosphere. The most complicated case, when air atmosphere is in contact with the ceramic composition, is considered here as a synthesis atmosphere. The most complicated case for synthesis atmosphere is considered here – when ceramic composition is in contact with air. The moisture content of the air has been taken into account too.

3. Results and discussion – gasification processes

Let's consider the case of phase and chemical transformations occurring in the bulk and pores of ceramic mass, when the surrounding atmosphere does not influence these transformations. This case is different from interactions only in bulk of the ceramic phase [7] by taking into account processes of the gaseous products yielded from reaction of the components of ceramic mass (but not only solid and liquid products).

Influence of temperature in the considered interval on the phase and chemical transformations obtained from the thermodynamic modelling of the considered system is presented in the Figure 1.

Analysis of the calculated data shows, that rising temperature form the room temperature to 1646 K, occurring in the reacting system (Figure 1) corresponds to the case of ceramic mass only for the corresponding temperature interval, when modelling has been performed for the conditions without the possibility to form gases [7]. Changes passing in the system are described by the corresponding reactions of complex oxides formation and decomposition most of all.

This shows that the processes of evaporation into vacuum become possible only at higher temperatures. Let's consider these processes.

At the temperature 1646 K calcium sulphate decomposes to yield sulphur(IV) oxide and oxygen. Release calcium oxide reacts with magnesium silicate $MgSiO_3$ and free silicon dioxide and forms calcium-magnesium silicate $CaMg(SiO_3)_2$ (dolomite):

$$2CaSO_4 + 2MgSiO_3 + 2SiO_2 \rightarrow 2CaMg(SiO_3)_2 + O_2\uparrow + 2SO_2\uparrow \quad (1)$$

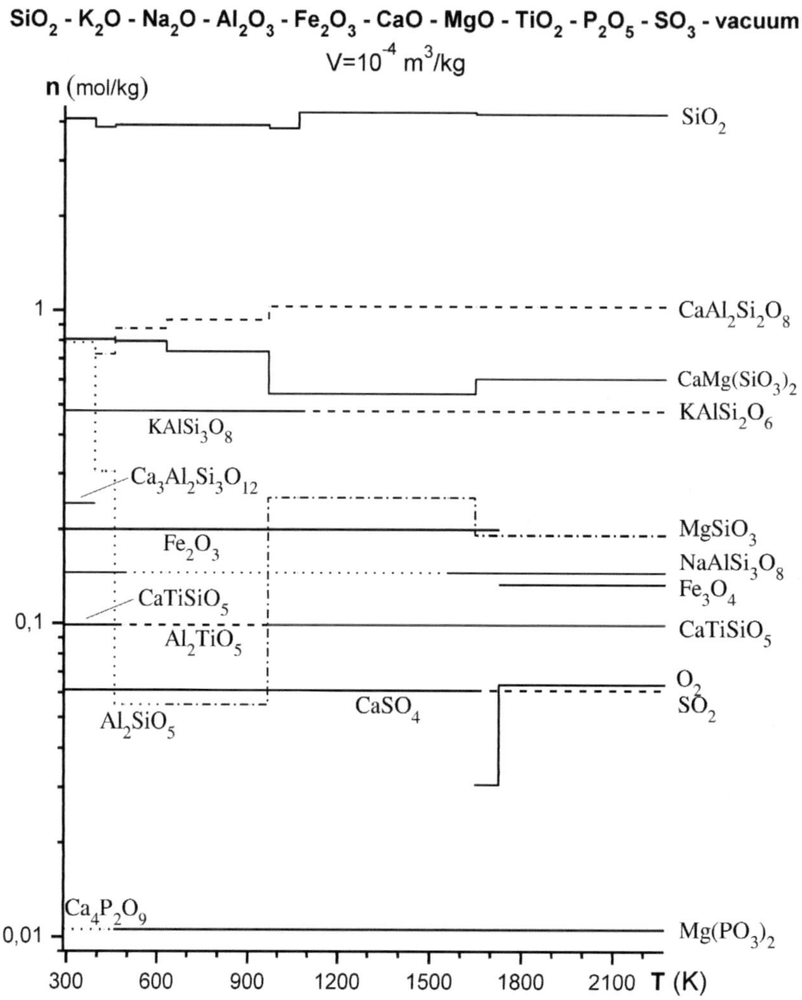

Figure 1. Influence of the temperature on the phase transformations in the airless pores of the ceramic pottery

The following heating to 1725 K results in iron(III) oxide transforming to magnetite Fe_3O_4 with evolving oxygen:

$$6\,Fe_2O_3 \rightarrow 4\,Fe_3O_4 + O_2\uparrow \tag{2}$$

It should be noted that the reaction of dolomite $CaMg(SiO_3)_2$ decomposition to silicates $CaSiO_3$ and $MgSiO_3$ under these conditions does not occur, since the calcium-magnesium silicate $CaMg(SiO_3)_2$ is stable in the absence of hematite Fe_2O_3.

Further temperature raise to 2000°C does not change chemical composition of the system.

4. Results and discussion - influence of synthesis atmosphere (air)

Now let's consider phase and chemical transitions in the ceramic mass when heated in the presence of air.

IOP Conf. Series: Materials Science and Engineering **175** (2017) 012018 doi:10.1088/1757-899X/175/1/012018

Influence of the temperature in the considered temperature range on the phase and chemical transformations, obtained as the result of thermodynamic modelling for the system under investigation, is illustrated by data in the Figure 2.

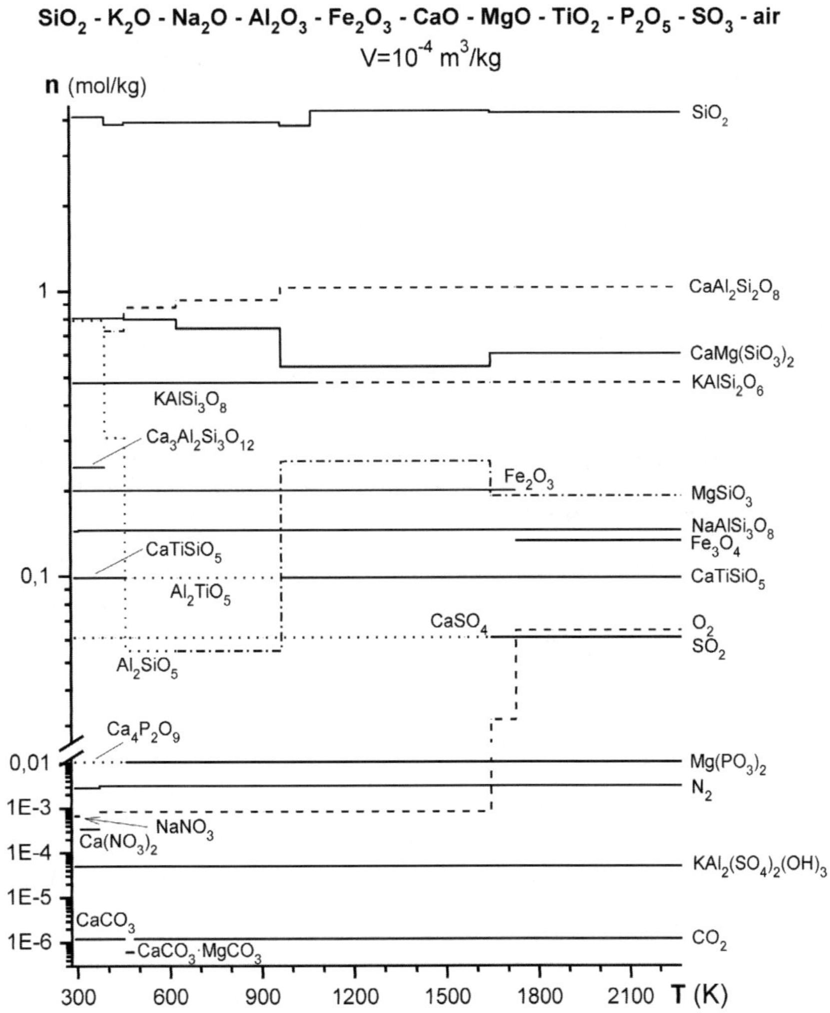

Figure 2. Temperature influence on the phase transitions in the pores of pottery with air

At room temperature in the system the following compounds form: Al_2SiO_5, $CaSO_4$, $Ca_4P_2O_9$, $CaCO_3$, $Ca_3Al_2Si_3O_{12}$, $CaTiSiO_5$, $CaMg(SiO_3)_2$, Fe_2O_3, $KAl_2(SO_4)_2(OH)_3$, $KAlSi_3O_8$, $NaNO_3$, $NaAlSi_3O_8$, SiO_2, N_2.

From this it follows that interaction of the raw ceramic components is rather similar to the cases when synthesis atmosphere doesn't influence on synthesis processes [********]. Besides this, the components of the air are considered as well. As the result of this, set of substances, which are absent in the system without air, is formed.

Oxygen, water vapours, and carbon dioxide are completely consumed when reacting with the ceramic mass.

Carbon dioxide reacts with calcium oxide to form calcium carbonate:

IOP Conf. Series: Materials Science and Engineering **175** (2017) 012018 doi:10.1088/1757-899X/175/1/012018

$$CaO + CO_2 \rightarrow CaCO_3 \quad (3)$$

Presence of water vapours in the gaseous phase results in formation of the compound $KAl_2(SO_4)_2(OH)_3$, which fully consumes them by reaction with potassium, alumina, and sulphur oxides:

$$K_2O + 2Al_2O_3 + 4SO_3 + 3H_2O \rightarrow 2KAl_2(SO_4)_2(OH)_3 \quad (4)$$

Part of the nitrogen reacts with sodium oxide and oxygen, yielding potassium nitrate:

$$2N_2 + 2Na_2O + 5O_2 \rightarrow 4NaNO_3 \quad (5)$$

This reaction goes to the complete consumption of the oxygen. Due to this, when equilibrium is reached, the gaseous phase becomes remaining nitrogen, which content in the air is approximately 77%.

In this way, one can see, that according to reactions (3)-(5), air actively reacts with initial oxides of the ceramic composition. As the result of this reaction gaseous system consists from nitrogen only at the room temperature.

Let's consider the sequence of transformations, which occur in the system upon temperature raising.

At the temperature 307 K sodium nitrate becomes unstable and decomposes reacting with calcium alumosilicate, this results in formation of calcium nitrate and sodium alumosilicate $NaAlSi_3O_8$. This reaction also involves the aluminum silicate Al_2SiO_5 and silicon dioxide:

$$2Al_2SiO_5 + Ca_3Al_2Si_3O_{12} + 6NaNO_3 + 13SiO_2 \rightarrow 3Ca(NO_3)_2 + 6NaAlSi_3O_8 \quad (6)$$

The further temperature rising to 372 K makes calcium nitrate unstable and breaks it into oxygen, nitrogen, and unstable calcium oxide. The latter reacts with aluminum silicate Al_2SiO_5 yielding calcium alumosilicate $Ca_3Al_2Si_3O_{12}$:

$$Al_2SiO_5 + 3Ca(NO_3)_2 + 2SiO_2 \rightarrow Ca_3Al_2Si_3O_{12} + 3N_2 + 7{,}5O_2 \quad (7)$$

At the temperatures 392, 424, and 446 K the solid-phase transitions of various types occur for the corresponding temperatures.

Further temperature raising results in calcium carbonate becoming unstable at 451 K and reacting with alumosilicate Al_2SiO_5 and calcium-magnesium silicate $CaMg(SiO_3)_2$ to form dolomite $CaCO_3{\cdot}MgCO_3$ and calcium alumosilicate $CaAl_2Si_2O_8$:

$$2CaCO_3 + 2Al_2SiO_5 + CaMg(SiO_3)_2 \rightarrow CaCO_3{\cdot}MgCO_3 + 2CaAl_2Si_2O_8 \quad (8)$$

At the temperature 456 K a series of interconnected phase and chemical exchange phase transformations occur (Figure 2).

At the temperature 480K dolomite becomes unstable and decomposes to carbon dioxide, and calcium and magnesium oxides, which react with silicon dioxide to form calcium-magnesium silicate $CaMg(SiO_3)_2$:

$$CaCO_3{\cdot}MgCO_3 + 2SiO_2 \rightarrow 2CO_2\uparrow + CaMg(SiO_3)_2 \quad (9)$$

Under temperature raising to the maximum calculated value of 2000 C all the interactions are the same to the ones above described for the corresponding temperatures when the absence of atmosphere, which shows that there is no interaction with the air of the ceramic material. In this case the gaseous phase has the composition corresponding to the composition of the air.

5. Conclusion

Models and quantitative data on heterogeneous interactions for conditions of synthesis of complex functional materials (10-15 components) are obtained by the methods of physico-chemical thermodynamics.

Products and conditions of gasification are determined in a wide temperatures range of ceramic materials synthesis.

The main solid-phase and gaseous products of interaction depending on temperature are determined for the conditions of materials synthesis in the air atmosphere.

Influence of the main components of air (oxygen, nitrogen, carbon dioxide, moisture, etc.) on course of phase and chemical transformations in sintered ceramic mass is revealed at temperatures from 0 to 2000 ° C.

The received results of modeling and calculations determine a basis for targeted management and optimization of the synthesis engineering process of various functional materials.

References

[1] Saito H 1987 Reliability of technical ceramics and its recovery problems *Mater. Sci. Jap.* **34** N 12 26–31

[2] Gömze A L and Gömze L N 2013 Ceramic based lightweight composites with extreme dynamic strength *Mater. Sci. Eng.* **47** 012033

[3] Uspenskaya M V, Sirotinkin N V, Gorskii V A and Goloshchapov Yu G 2006 Composites of acrylate copolymers and fullerene *Russ. J. Appl. Chem.* **79** 858-860

[4] Yokokawa H, Fujishige M, Ujiie S and Dokiya M 1988 CTC: Chemical thermodynamic computation system *J. Nat. Chem. Lab. Ind.* **83** N 11. 1–122

[5] Voronina T V and Slobodov A A 2011 Spectroscopic and thermodynamic study of heavy water Russ. J. Opt. Technol. **78** N 3 156-160

[6] Olekhnovich R O, Volkova K V, Uspenskii A A, Slobodov A A and Uspenskaia M V 2015 Synthesis of poly(acrylic acid)-co-acrylamide/bentonite polymer nanocomposite as an absorbent for removal of heavy metal ions from water *Proc. XV Intern. Multidisc. Sci. Geoconf.* **2** N 5 477-484

[7] Slobodov A, Uspenskiy Al, Yavshits S and Mischenko A 2015 Applicability of thermodynamic modelling of phase-chemical composition and rheological properties for multi-component natural and technological objects *Építőanyag – J. Silic. Based Compos. Mater.* **67**, 164–167

[8] Uspenskiy Al, Slobodov A A, Gavrilov A, Radin M and Dahno V 2016 Nuclear power plants - thermodynamic approach for ecological problems of radiation and water chemistry // *Proc. XVI Intern. Multidisc. Sci. Geoconf.* **1** 65-72

IOP Conf. Series: Materials Science and Engineering **175** (2017) 012019 doi:10.1088/1757-899X/175/1/012019

The corrosion resistance of zinc coatings in the presence of boron-doped detonation nanodiamonds (DND)

Burkat G K[1,2], Alexandrova G S[1,2], Dolmatov V Yu[1], Osmanova E D[2], Myllymäki V[3], Vehanen A[3]

[1]St. Petersburg Institute of Technology, Saint-Petersburg, Russia
[2]FSUE "SCTB "Technolog"
[3]«Carbodeon Ltd. Oy», Vantaa, Finland

Abstract: The effect of detonation nanodiamonds, doped with boron (boron-DND) in detonation synthesis on the process of zinc electrochemical deposition from zincate electrolyte is investigated. It is shown that the scattering power (coating uniformity) increases 2-4 times (depending on the concentration of DND-boron electrolyte conductivity does not change, the corrosion resistance of Zn- DND -boron coating increases 2.6 times in 3% NaCl solution (corrosion currents) and 3 times in the climatic chamber.

Keywords: detonation nanodiamonds, boron, zinc, zincate electrolytes, corrosion resistance.

Introduction

During the last years, a growing number of industries are using zincate electrolytes, which completely replace cyanide. A detailed mechanism study of the additives effect in the zincate electrolytes are still not available. The electrolyte can be operated only in the presence of the polymeric surface-active agents (surfactants) additives.

Zinc coatings on their functional properties are mainly for protection against corrosion of ferrous metals as they have more negative potential (-0,76V) than steel (-0,3V). This property of zinc coatings making it one of the most used among all the metals used in the electroplating industry. Zinc coatings are important, because 60-70% of the iron produced in the world, in fact, spent on corrosion. Zinc protects the steel from corrosion while being degraded itself.

The work [1] describes a process for the electrochemical deposition of zinc from zincate electrolyte with standard detonation nanodiamonds (DNDst) and surfactant-PEPA (polyethylene polyamine). The electrolyte composition (g / l): ZnO- 6-12, NaOH-75-120, PEPA- 1-4 DNDst - 1-10. It has been found corrosion currents decrease using DNDst of any concentration in the investigated range. The corrosion resistance of the coating is increased 1.5-2.0 times (resulting coating is not subjected to additional chromatation or phosphotation in tests).

Recommended composition (g / l): Zn - 12, NaOH - -120, Peppo - 2 DNDst - 10.

Purpose of the present work is to obtain electrochemical zinc coating high corrosion resistance.

The objective is to study the influence of the additive of boron modified DND during detonation synthesis and polymer additive A1DM [2] on the deposition of zincate zinc electrolyte.

This work presents results of coating the zincate electrolyte containing DND modified with boron at detonation synthesis and polymer additives A1DM developed by Lithuanian researchers [2]. From an environmental point of view, PEPA is biologically hard substance (does not decompose in the environment) and A1DM - biologically soft (decomposes in the environment by present the soil and water bacteria).

Selection of DND-boron is due to the positive experience of the use of boron compounds in the process of the galvanizing, its properties are described in the papers [3-5].

Experimental procedure

DND-boron obtained by detonation of charges TNT-RDX (50/50) in an aqueous solution of boric acid (H_3BO_3) 5 wt. % concentration. The ratio: the charge weight: the amount of aqueous solution of 1:10. The process was carried out in explosion chamber "Alfa-2M" volume of 2.14 m^3. The concentration of boron in the DND is 0.45 wt.%.

Zinc deposition is carried out on the steel samples. When recording cathodic polarization curves, potentiostat IPC ProMF were used, as the working electrode were used a steel sample (St20) and a steel sample of pre-coated with zinc under cathodic current density of 3 A / dm^2 and the temperature of 20°C. The cathodic polarization curves obtained in potentiodynamic mode in the potential range from E to E currentless = -2100 mV.

When recording the curves of hydrogen release a steel sample coated with galvanized zinc were used as a working electrode (the thickness of the zinc coating is not less than 10 microns). The curves were recorded in the potentiodynamic mode over currentless potential to E = -2100 mV.

The platinum electrodes were as auxiliary electrodes. The electrode potential is measured with respect to the saturated silver-chloride reference electrode with a potential equal to 0.202 V.

Obtaining hydrogen curves was carried out in a solution of 3% sodium chloride in potentiodynamic mode (1 mV / sec) relatively to silver chloride reference electrode. Anode curves were recorded in the range of -1790 mV to -1720 mV. Polarization curves were obtained on freshly zinc coating is not less than 3 times, after which the average value of corrosion currents were determined for each potential.

The dependence has been constructed according to the experimental data: i = f (-E) and Lg i = f (-E). Corrosion potential determined in dependence i = f (-E) by extrapolating the straight sections of cathodic and anodic curves.

The authors used the classical zincate electrolyte composition (g / l); NaOH-100-130, ZnO -10-14; A1DM additive ("Chimeta" company production) -8-12 ml / L, DND-boron-0.5-7.

Results and Discussion

The deposition process kinetics of the zinc-diamond coatings in the zincate electrolyte

The preliminary study the impact of supplements A1DM on the cathodic and anodic polarization of zinc in the zincate electrolyte showed its strong influence. Curve shifts towards negative potentials reaches ~ 100 mV. This additive is adsorbed at the cathode and the anode, and does not change the mechanism of the zinc discharge. The angle of Tafel curves is ~ 120 mV, which corresponds to the difficulty of attaching the first electron, and this is consistent with the literature data [6].

Against the background of A1DM was considered the impact of nanodiamond additives DND-boron on the galvanizing process kinetics in the zincate electrolyte.

It was found a good correspondence with the polarization curves of zinc electrolyte stirring or not. Tafel plots inclination angle is within the same range, about 120 mV, i.e. zinc discharge mechanism in the presence of DND-boron remains unchanged - slow stage is the attaching of the first electron. DND-boron supplement does not affect on the potentials plateau of limiting currents, they remain in the same range of potentials: the first ~ 1850mV, the second - in 1950 mV. The limiting current plateau in the curves obtained without stirring, it remains at the same current density approximately 12 A / dm^2, as in curves without DND-boron additives. The second current limit curve plateau, obtained with mixing is reduced to 2 A / dm^2, compared with no nanodiamond additive curve. This may indicate the more trapping of DND-boron supplements with stirring; its large concentration was created in the electrode region and a partial desorption A1DM additives. DND-boron adsorption reduces the current limit. With increasing the concentration of the DND additive in the stirred-boron electrolyte the second current limit decreases of 18 A / dm^2 to 16 A / dm^2 at the concentration of additive - 5 g / l. DND-boron additive has no effect on the first current limit with stirring. This suggests that the plateau of the first current limit is occupied with A1DM additive and DND-boron additive is not able to be absorbed and influence on the polarization, further, the shift of the cathode potential in the negative region leads to partial desorption of A1DM additive and there is a possibility of observed nanodiamond additives action.

Follows from the obtained data the addition of boron-DND in comparison with the one A1DM additive significantly impedes zinc ionization process, which means denser layer, which is created in the presence of DND-boron additive. Moreover, with stirring the DND-boron additive effect amplifies two times in comparison with the zinc precipitation from the electrolyte without stirring.

Investigation of zinc anodic polarization in the zincate electrolyte with additives shows good reproducibility of the experiment. Electrolyte's stirring naturally facilitates the ionization of zinc, slightly changing the angle of the Tafel dependence.

As a polarization studies result, scattering ability of zincate electrolyte was defined in the presence of additives studied by electrochemical similarity. The data are shown in Tables 1 and 2.

Table 1. The zincate electrolyte conductivity

The zincate electrolyte conductivity, [mS/cm], t = 25°C				
without DND-boron, with 2 g/l A1DM	with 0.5 g/l DND-boron + 2g/l A1DM	with 1 g/l DND-boron + 2 g/l A1DM	with 2 g/l DND-boron + 2 g/l A1DM	with 5 g/l DND-boron + 2 g/l A1DM
275	274	272	275	265,8

Table 2 - The zincate electrolyte scattering power in the presence of additives

The concentration of DND-boron, g / l	**The change of the current density, Δi mA/cm²**	**The potential change, ΔE, mV**		**The first derivative of the current potential, ΔE/Δi , Ohm**		**Electrical conductivity, æ S / cm**	**The scattering power, ΔE/Δi* æ**	
		without stirring	**stirring**	**without stirring**	**stirring**		**without stirring**	**stirring**
Only with 2 g/l A1DM	2	8,3	8,9	4,15	4,45	275	1,14	1,22
0.5	2	33.9	18.6	16.95	9.3	274	4.64	2.55
1	2	27.2	19.8	13.6	9.9	272	3.7	2.7
2	2	31.2	21.3	15.6	10.6	275	4.29	2.92
5	2	15.8	20.5	7.9	10.25	265.8	2.1	2.72

The conductivity of the zincate electrolyte in the presence of DND-boron additive is slightly reduced, except the 2 g / l concentration, at which the electrical conductivity and polarizability take the maximum value, probably at this concentration redistribution of two A1DM and DND-boron additives appears because of competitive adsorption. It is noted that the scattering power of the electrolyte increases in the presence of DND-boron additive in the electrolyte without stirring ~ 4 times as compared with the electrolyte with the A1DM additive only. The zincate electrolyte scattering power in the presence of DND-boron additive is increased by 2 - 2.5 times, even with stirring, which is a very important factor in the use of nanodiamond additives.

Addition of DND-boron in the electrolyte contributes to obtaining fine-grained structure with the creation of a composite matrix, which should improve the corrosion resistance of the coatings. Corrosion currents are calculated from the polarization curves of hydrogen (anode and cathode) recorded in 3% NaCl solution on zinc coatings obtained with the test zincate electrolyte additives, such as: A1DM and DND-boron nanodiamonds. All investigated coatings were obtained with the same current density - 3 A / dm^2 of zincate electrolyte (Figure 1, 2).

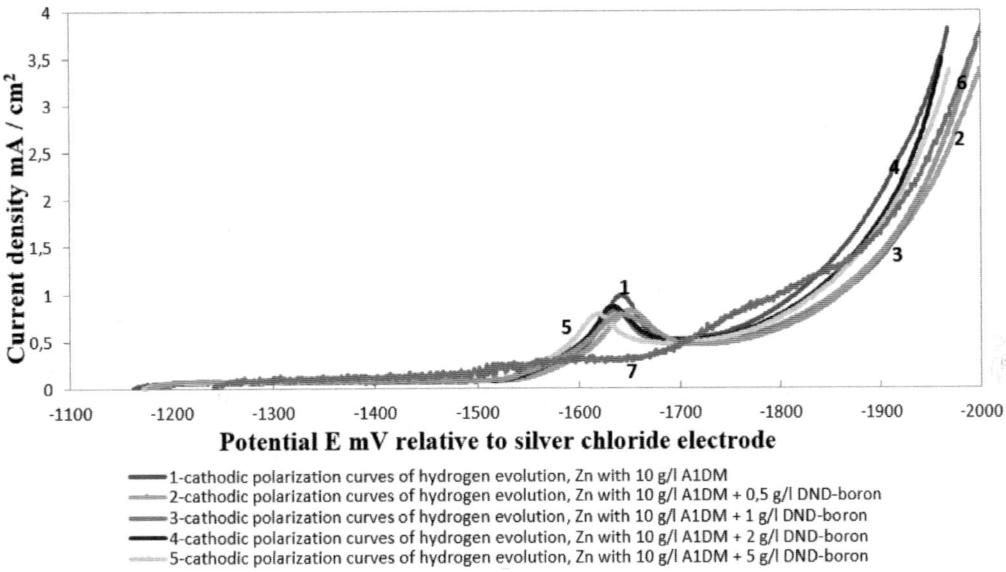

Figure 1. Cathodic polarization curves of hydrogen evolution at the zinc obtained from the electrolyte without additives (7), but with 10 g / l A1DM (1) and 0.5 (2), 1(3), 2(4), 5(5), 7 (6) / l DND-boron + 10 g / l A1DM

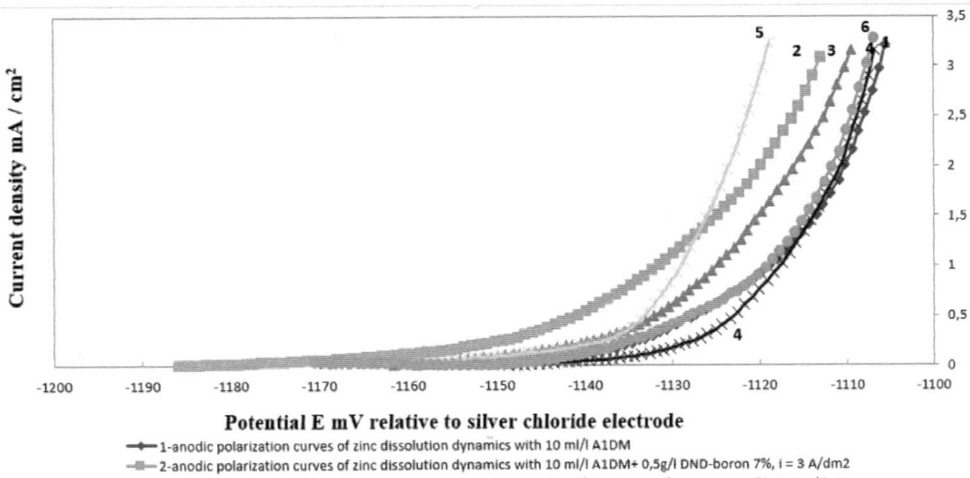

Figure 2. Anodic polarization curves of zinc dissolution obtained in the electrolyte without additives, but with 10 g / l A1DM (1) and 0.5 (2), 1(3), 2 (4), 5 (5), 7 (6) g / l DND-5-B + 10 g / l A1DM

IOP Conf. Series: Materials Science and Engineering **175** (2017) 012019 doi:10.1088/1757-899X/175/1/012019

On the cathode curves the current rise observed in 1600-1700 mV potential, possibly due to the change of the layer structure obtained in the presence of A1DM additive. Processing these curves in the semilogarithmical dependence uses to determine corrosion current (current density equal to 0.0076 mA / cm2).

Analysis of hydrogen curves data showed that in the potential range of 1600-1700 mV on the curves recorded on the coatings obtained in the presence of additives like A1DM and nanodiamonds, there is a maximum, the value of which is reduced, when added to A1DM nanodiamond additive, which may indicate the inclusion of DND-boron in the coating. This is also evidenced by zinc ionization curves in 3% NaCl solution (Fig. 2), it is observed facilitating ionization on coating obtained with nanodiamonds, when compared with coatings obtained with one A1DM additive only. All this may indicate that nanodiamond additives, including into the coating during electrodeposition of zinc, improves the structure and increases its corrosion resistance.

Data provided in Table 3, confirm the previous conclusions. Corrosion currents in 3% NaCl solution reduces practically to 3.6 times on the zinc coatings obtained in the presence of DND-boron. The fact that these additives affect the coating structure and consequently on the physico-chemical properties microhardness data presented in Table 4 shows. The addition of even small quantities of additives (0.5 g / l) increases the microhardness of the coating up to 20%, other values slightly above the level of the coatings obtained without DND-boron.

Table 3. Corrosion current value in 3% NaCl for coatings obtained at various DND-boron additive concentrations from zincate electrolyte with 10 ml A1DM (precipitation of samples was performed at 3 A / dm^2).

The content of DND-boron in the electrolyte, g / l	**Corrosion current density, $i_{corrosion\ current}$, mA /cm^2**
	$i_{electrolysis}$ = 3 A/dm^2
-	0.0076
+0.5	0.0071
+1	0.0067
+2	0.0054
+5	0.0040
+7	0.0029

Table 4. Zinc coatings microhardness obtained with different concentrations of additive DND-boron in the zincate electrolyte with 10 g of / l A1DM (at 3 A / dm^2).

The content of DND-boron in the electrolyte, g / l	**Microhardness H_μ, kgf / mm^2**
-	111
0.5	134
1	122
2	121
5	116
7	114

The data of zinc coatings corrosion testing obtained with various concentrations of DND-boron in salt fog chamber are shown in Table 5. From the analysis of the data table 5 the better results were obtained for samples deposited from the electrolyte with DND-boron additive at the concentration of 5 and 7 g / l (corrosion decreased by 3.1-3.4 times), which agrees well with corrosion currents.

Table 5. Data of zinc coatings corrosion testing obtained with the electrolyte DND-boron nanodiamond additive. (Zinc deposition temperature of 25°C, the density of zinc deposition current 3 A / dm^2, the time of the corrosion tests - 24 hours), salt fog chamber.

Number of the sample	**1**	**2**	**3**	**4**	**5**	**6**
Additive concentration, g / l	10 g/l A1DM	0,5 g/l DND-boron	1 g/l DND-boron	2 g/l DND-boron	5 g/l DND-boron	7 g/l DND-boron
Weight after coating, g	29.8074	29.9324	30.0020	30.2000	30.4026	30.1128
Weight after corrosion tests, g	29.7833	29.9096	29.9818	30.1875	30.3948	30.1057
Weight loss during corrosion tests, g	0.0241	0.0228	0.0202	0.0125	0.0078	0.0071
Appearance before the corrosion test	A smooth, uniform, shiny coating					
Appearance after corrosion tests	All samples were covered with white corrosion					

This work was carried out with full financial support of the Agreement on Subsidies provision of June 05, 2014 № 14.579.21.0001. Federal Program of the Russian Federation.

Conclusions

1. Sharing of A1DM and DND-boron additives in the zincate electrolyte composition g / l: ZnO - 10; NaOH - 100; A1DM - 6-10; DND-Boron - 5-7 allowed
   - to increase the electrolyte scattering ability (uniformity of coating) in 2,5-4,0 times;
   - corrosion resistance of Zn- DND-boron coating increased 2.5-3.0 times;
   - microhardness increased by 10%.
2. Recommended composition of the electrolyte and the deposition conditions:

   ZnO-10;
   NaOH-100;
   A1 DM-6;
   DND-boron – 5-7 g / l;
   i = 3 A/dm^2;
   t = 25°C.

Reference list

[1] *V. Yu. Dolmatov, G.K. Burkat, V. Yu. Saburbaev, A.E. Salko, M.V. Veretennikova* Preparation and properties of electrochemical composite coatings of precious and nonferrous metals with ultradispersed detonation synthesized diamonds // Journal of Superhard Materials.- 2001.- N2.- P.49-55.

[2] *G.K. Burkat, I.V. Matulaytite* Comparative studies zincate electrolytes with different surfactants, new in technologies and functional electroplating coatings // Abstracts – Leningrad. - 1990, P.57-60.

[3] *V. Yu. Dolmatov, I. I. Kulakova, V. Myllymaki, A. Vehanen, A. N. Panova, A. A. Voznyakovskii.* IR Spectra of Detonation Nanodiamonds Modified During the Synthesis/J. of Superhard Materils. -2014. – vol.36, No5. – P.344-357.

[4] *V. Yu. Dolmatov, A.N. Panova, A.A. Bochechka, M.V. Veretennikova, V. Myllymaki, S.A. Dushenok, E.V. Nikitin, G.G. Tsapyuk, E.V. Ishchenko* Influence of the synthesis conditions of detonation nanodiamonds on their thermal stability // The collection of articles XVII

International conference "Rock cutting and metalworking tools - equipment and technology of its manufacture and use," Kyiv: ISM Bakul National Academy of Sciences of Ukraine, 2014.- Vol. 17.- P.213-219.

[5] *V.Dolmatov, A.Panova, N.Lapchuk, T.Lapchuk, V.Myllymaki*. Comparative analysis of EPR and Raman scattering spectra of powders of compacted detonation nanodiamonds doped during synthesis/In the Book of Abstract of 9th International conference "New diamond and Nanocarbons" 2015, 24-29 May, 2015, Shizuoka, Japan, p.175

[6] *E.E. Starodubets* The mechanism of electrode processes involving zinc complexes (II): The expert and quantum-chemical approach, Candidate dissertation of Chemistry, -2001, Kazan. - P. 228.

IOP Conf. Series: Materials Science and Engineering **175** (2017) 012020 doi:10.1088/1757-899X/175/1/012020

Characterization of Ta-B-C nanostructured hard coatings

J Buršík[1], V Buršíková[2], P Souček[2], L Zábranský[2] and P Vašina[2]

[1]Institute of Physics of Materials, Czech Academy of Sciences, Department of Structure, CZ-61662 Brno, Czech Republic
[2]Faculty of Science, Masaryk University, Department of Physical Electronics, CZ-61137 Brno, Czech Republic

E-mail: bursik@ipm.cz

Abstract. Microstructure and mechanical properties of Ta-B-C nanocrystalline layers prepared by magnetron sputtering were studied. DC magnetron sputtering was used to prepare thin layers on rotated substrates. Various deposition parameters were tested. Microstructure of layers was studied by means of scanning and transmission electron microscopy on thin lamellar cross sections prepared using a focussed ion beam. Both undisturbed layers and the volume under relatively large indentation prints (load of 1 N) were observed. The microstructure observations were correlated with mechanical properties characterized by means of nanoindentation experiments in both the static and the dynamic loading regime. Elastic modulus, indentation hardness and fracture resistance of prepared nanostructured coatings were evaluated and discussed.

1. Introduction

Hard coatings are commonly deposited on cutting tools to protect the tool surfaces from mechanical and chemical damage and hence to improve lifetime and performance. Protective coating materials used for cutting require high stiffness and hardness to lower wear rates. On the other hand, since crack initiation and growth have to be avoided, the protective coating materials should also possess a moderate ductility. There are two criteria which are generally used for determination if the material is brittle or ductile. Pugh [1] showed that if the ratio of bulk modulus to shear modulus B/G is greater than 1.75, the material exhibits ductile metal-like behaviour. Otherwise, the material is considered brittle. Pettifor [2] showed that the value of the Cauchy pressure can also give information about ductility. Cauchy pressure is determined as the difference between elastic constants C_{12}–C_{44}. If it is negative, material is brittle. Positive Cauchy pressure, on the other hand, implies ductile behaviour.

In the last decade, a proposal for an unusually stiff and moderately ductile hard coating material Mo_2BC was proposed by Emmerlich et al [3] based on *ab initio* calculations and supported also experimentally. The calculated bulk modulus of 324 GPa was 45% larger than that of $Ti_{0.25}Al_{0.75}N$ and only 14% smaller than that of c-BN, indicating a highly stiff material. The calculated B/G ratio was 1.72 at the transition from brittle to ductile behaviour. This, in combination with a positive Cauchy pressure, suggested moderate ductility. The authors [3] showed that these properties can be understood by considering the electronic structure and particularly the extreme anisotropy (ortorhombic crystal lattice with a=0.309nm, b=1.735nm, c=0.305nm, space group *Cmcm* [4], ICSD entry no. 043318 [5]). The presence of stiff Mo-C and Mo-B layers with metallic interlayer bonding enables the intriguing and unexpected property combination. Excellent mechanical properties of Mo_2BC coatings were recently confirmed by several experimental works [6-10].

IOP Conf. Series: Materials Science and Engineering **175** (2017) 012020 doi:10.1088/1757-899X/175/1/012020

The Aachen group continued the work later by preparing Mo_2BC thin films at lower temperatures using high power pulsed magnetron sputtering [11] and by systematic theoretical study on the electronic structure and mechanical properties of broader class of similar X_2BC nanolaminated materials where X = Ti, V, Zr, Nb, Mo, Hf, Ta and W [12]. All compounds were shown to be stable with the similar orthorhombic lattice parameters with a high aspect ratio. It was shown that systems with higher valence electron concentration appear to be stiffer and ductile as inferred from both the more positive Cauchy pressure and the larger value of the *B/G* ratio. On the other hand, the calculated phase stabilities decrease as the valence electron concentration is increased. The combination of high stiffness and moderate ductility renders X_2BC compounds with X = Mo, Ta and W as the most promising candidates for protection of cutting and forming tools.

In this work we studied microstructure and mechanical properties of Ta-B-C nanocrystalline layers prepared by magnetron sputtering. Microstructure of layers was studied by means of scanning and transmission electron microscopy on thin lamellar cross sections prepared using a focussed ion beam and correlated with mechanical properties characterized by means of nanoindentation experiments.

2. Experimental

A custom built magnetron sputtering device equipped with four magnetron sputtering heads in a balanced magnetic field configuration was used for the Ta-B-C coating depositions. Samples of about 0.5 µm thin layers were deposited using magnetron sputtering of three DC driven targets: B_4C, C and Ta. The hard-metal (cemented tungsten carbide, WC-Co) substrates were ultrasonically cleaned in a degreasing agent and then placed in the chamber using a load-lock system. Prior to the deposition process the substrates were cleaned in argon plasma for 20 min. The deposition pressure was 0.3 Pa. Substrates were not heated.

Microstructure of layers was studied using a Tescan LYRA 3XMU FEG/SEM×FIB scanning electron microscope (SEM) and a Philips CM12 STEM transmission electron microscope (TEM). Thin lamellar cross sections for TEM observations were prepared using a focused ion beam (FIB) in SEM from two locations in each sample: an undisturbed layer and a central region of indentation print made with Berkovich tip with a load of 1N. TEM observations were correlated with mechanical properties measured by means of nanoindentation technique using a Hysitron dual head TI950 triboindenter equipped with diamond Berkovich tip in the load range from 50 µN to 11 mN. The hardness and reduced elastic modulus values were determined at indentation depths where the substrate influence was negligible. Moreover, microindentation using a Fischerscope H100 microindenter equipped with diamond Berkovich tip was used to study the fracture resistance of the coatings in the load range from 0.01 to 1 N.

3. Results and discussion

SEM micrographs in figure 1 show one of the large indentation prints produced with a load of 1N and two stages of thin lamella preparation. Cracks at the indentation print perimeter were often observed (see figure 1a). According to indentation loading curves the cracking started only after the maximum indentation depth was reached. The Pt protective layer for thin lamella was deposited across the central part of the indentation print (figure 1b). On the lamellar cross section (figure 1c) one can observe the large indentation depth (more than twice the layer thickness) without any visible defects at the layer/substrate interface and with the only crack at the indentation print perimeter marked by an arrow.

TEM micrographs in figure 2 depict details of the Ta-B-C layer microstructure in the indented region. The crack at the perimeter of the indentation print (figure 2a) is shown to be absorbed in the layer interior. The layer itself reveals a columnar structure of fine grains (see figure 2b and a detail in figure 2c). A thorough inspection of the lamella did not find any defects (except the crack described above) inside the layer or at the layer/substrate interface even though both the shape of the interface under indentation print and a high dislocation density in WC grains evidence heavy plastic deformation.

The depth profiles of hardness and elastic modulus in figure 3 were obtained using quasistatic nanoindentation tests with 20 partially unloading segments. The maximum load was 10 mN. The basic parameters characterizing mechanical properties of the deposited layer are hardness H=28±1 GPa and effective elastic modulus $E_{eff}=E/(1-\nu^2)$=442±2 GPa, where E is the Young modulus and ν is the Poisson ratio. Substituting the Poisson ratio 0.254 calculated *ab initio*for Ta_2BC [12] into the formula for E_{eff} the Young modulus value E=414 GPa is obtained, which is very close to the Young modulus value of 421 GPa predicted for Ta_2BC [12].

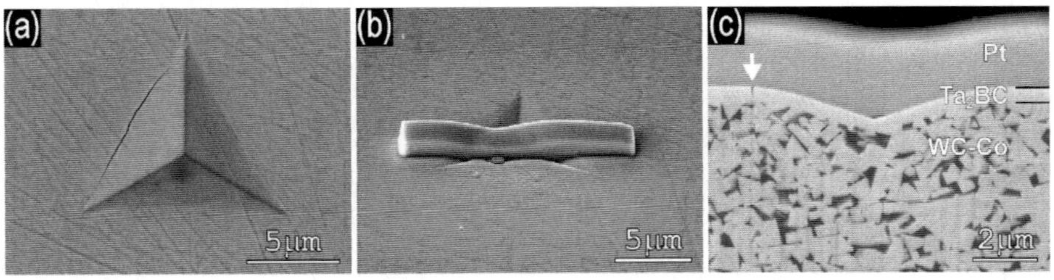

Figure 1. SEM micrographs, signal of backscattered electrons. Plan view of indentation print with a crack at the perimeter (a), a protective Pt block across the central part of indentation print defining the thin lamella position, sample surface is tilted by 55 degree (b), the lamella prior to final thinning, surface is tilted by 35 degree, a crack is marked by an arrow (c)

Figure 2. TEM micrographs: a crack at the indentation print perimeter (a), columnar structure of the layer (b) and a detail of fine grains of the layer (c)

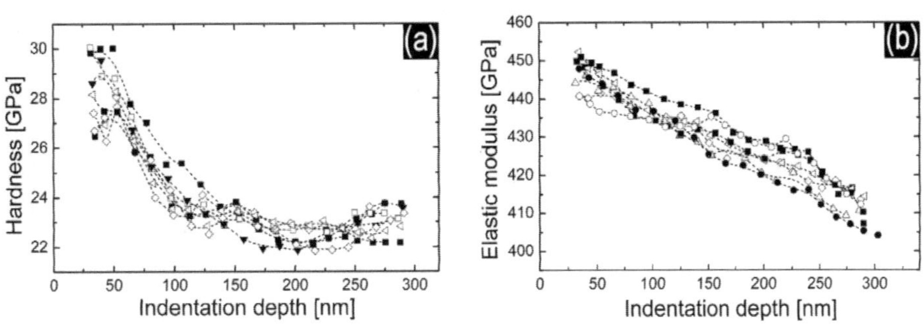

Figure 3. The depth profiles of hardness (a) and elastic modulus (b)

There exist yet other useful empirical parameters for characterization of protective coatings, such as the H/E and H^3/E^2 ratios, which are measures of the toughness and resilience of the coatings, respectively. Toughness can be defined as the ability of a material to absorb energy during deformation up to fracture. In the case of our Ta-B-C layer H/E=0.07 value indicates promising fracture toughness. The prediction of the wear resistance of the coatings can be related to the resilience of the coating, which is defined as the resistance of a material against plastic deformation. Measured value H^3/E^2=0.13GPa indicate improved wear resistance of the Ta-B-C coating.

Conclusions

Hard and ductile Ta-B-C coatings with an excellent resistance against indentation induced cracking were successfully prepared by magnetron sputtering. Complex characterisation of the local mechanical properties and microstructure was carried out showing very favorable combination of high hardness and elastic modulus together with high fracture toughness. The nanocomposite Ta-B-C coatings hardness reached 28 GPa and its effective elastic modulus was 442 GPa. TEM observations of the area under the indentation prints proved that there was no delamination at the film/substrate interface. The cracks appeared in the vicinity of the imprints only after the indentation depth substantially (more than twice) exceeded the film thickness.

Acknowledgement

This work was supported by the Czech Science Foundation (grant project No. 15-17875S).

References

[1] Pugh S F 1954 *Phil. Mag.***45** 823
[2] Pettifor D G 1992 *Mater. Sci. Technol.***8** 345
[3] Emmerlich J, Music D, Braun M, Fayek P, MunnikF and Schneider J M 2009 *J. Appl. Phys. D: Appl. Phys.***42** 185406
[4] Jeitschko W, Nowotny H N and Benesovsky F 1963 *Monatsh. Chem.* **94**565
[5] Bergerhoff G and Brown I D 1987 *Crystallographic Databases* ed F H Allen et al. (Chester, International Union of Crystallography)
[6] Djaziri S, Gleicha S, Bolvardi H, Kirchlechner C, Hans M, Scheu C, Schneider J M and Dehm G 2016 *Surf. Coat. Technol.***289** 213–218
[7] Buršík J, Kuběna I, Buršíková V, Souček P, Zábranský L and Vašina P 2016 *Def Diff Forum***368** 107
[8] Buršík J, Buršíková V, Souček P, Zábranský L and Vašina P 2016 *Rom. Reports Phys.***68** 1069
[9] Zábranský L, Buršíková V, Souček P, Vašina P and Buršík J 2016 *Eur. Phys. J. Appl. Phys.***75** 24716
[10] Buršíková V, Buršík J, Zábranský L, Souček P and Vašina P 2016 *Def Diff Forum***368**111
[11] Bolvardi H, Emmerlich J, Mraz S, Arndt M, Rudigier H and Schneider J M 2013 *Thin Solid Films***542** 5-7
[12] Bolvardi H, Emmerlich J, to Baben M, Music D, von Appen J, Dronskowski R and Schneider J M 2013 *J. Phys.: Condens. Matter***25** 045501

IOP Conf. Series: Materials Science and Engineering **175** (2017) 012021 doi:10.1088/1757-899X/175/1/012021

Relief, magnetic structure and microwave properties of composite films

L N Kotov[1], V S Vlasov[1], V A Ustyugov[1], V K Turkov[1], M Yu Dianov[1], Yu E Kalinin[2], A V Sitnikov[2], E A Golubev[3]

[1] Syktyvkar State University, Syktyvkar, Russia
[2] Voronezh State Technical University, Voronezh, Russia
[3] Komi Scientific Center of UB RAS, Syktyvkar, Russia

E-mail: kotovln@mail.ru

Abstract. Relief, magnetic structure and microwave properties of the composite (metal-dielectric) films: (CoFeZr) $_{x}$ $(Al_2O_3)_y$ (A series) and $(CoTaNb)_x(SiO_2)$ were investigated. Experimental data on the microwave magnetic characteristics were also calculated on the basis Dubowik model.

Introduction

In present time the magnetic, magnetooptical and magnetoacoustic properties of thin composite films are widely investigated. The attention in this area is caused by opportunity of new information processing devices building (spintronics), rf tuning filters. Also we have the interest to unusual behavior of the magnetization, the domain structure formation, propagation of the electromagnetic waves in these nanosized objects.

The metal-dielectric composite films describing in the work have essential nonhomogeneous structure. The parameters of the structure depend on the metal/dielectric ratio. We can reveal the regularities of the average particles size changes in the atomic force microscopy topographic images, the regularities of the particle size distribution and the films thickness changes with varying metal/dielectric ratio.

The metal phase of the films is super-paramagnetic or ferromagnetic depending on the films thickness. This fact makes difficult the analysis of the microwave properties. We develop the known model for describe the composite structures consisting of separate granules and bigger labyrinth formations. These formations appear at big metal concentrations by convergence of the small particles.

Object and investigation methods

The two series of the films $(Co_{0.3}Fe_{0.3}Zr_{0.1})_x(Al_2O_3)_y$, where $0.25<x<0.65$ (A series) and $(Co_{86}Ta_{12}Nb_2)_x(SiO_2)_y$, where $0.3<x<0.6$ (B series) were investigated. The percolation threshold for these series was about $x=0.45$. The investigated films were prepared by ion beam sputtering method on the polycrystalline glass substrate. For preparing of the films the alloyed metal target with size $270\times70\times14$ mm^3 were applied. The dielectric plates with size $70\times9\times2$ mm^3 were fixed with changing interval on the targets. It gives us in one technological process to obtain the samples with continuous changes of the metal/dielectric ratio. We used the argon in the chamber with pressure 8×10^{-5} Torr.

The films were amorphous. The ferromagnetic phase in the films is localized in the granules and doesn't form compounds with dielectric. The chemical composition and thickness of the films were determined using a scanning electron microscope JSM-6400. For investigation of the films structure the scanning of the films surface was performed by the atomic force microscope (AFM) ARIS-3500 (Burleigh Instrument Co, USA) with maximal scanning area 70×70 μm^2. The standard pyramidal shape silicon cantilevers with nose rounded radius ≈ 10 nm and stiffness 0.1 N/m were used. We made the calibration of the image scale by certificated test specimens.

Thopography and phase contrast of amorphous composite films

Let's consider topographic imaging of the films $(Co_{45}Fe_{45}Zr_{10})_x(Al_2O_3)_{1-x}$ at the different metal phase concentrations. The typical imaging shown on the fig. 1.

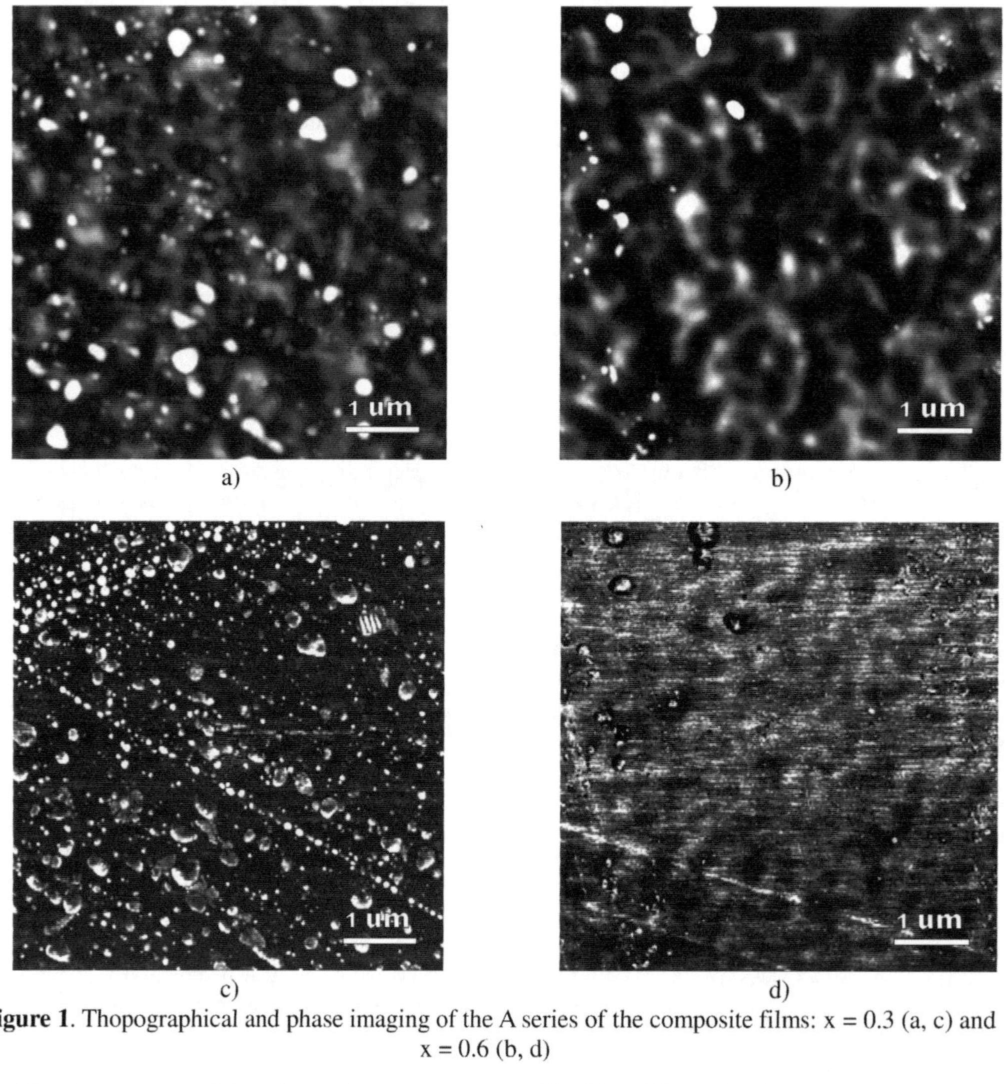

Figure 1. Thopographical and phase imaging of the A series of the composite films: x = 0.3 (a, c) and x = 0.6 (b, d)

The Fig. 1 shows essential nonhomogeneous structure of the films. We can't select separate granules in this structure. It is right for the small and big metal concentrations. But the relief of the film is less nonhomogeneous at big metal concentration and we can extract arbitrary big rolls-and-swells. This fact shows more homogeneous growth of the film after percolation threshold [3]. The bright white blobs on the figures correspond to film defects and we can exclude the blobs from consideration.

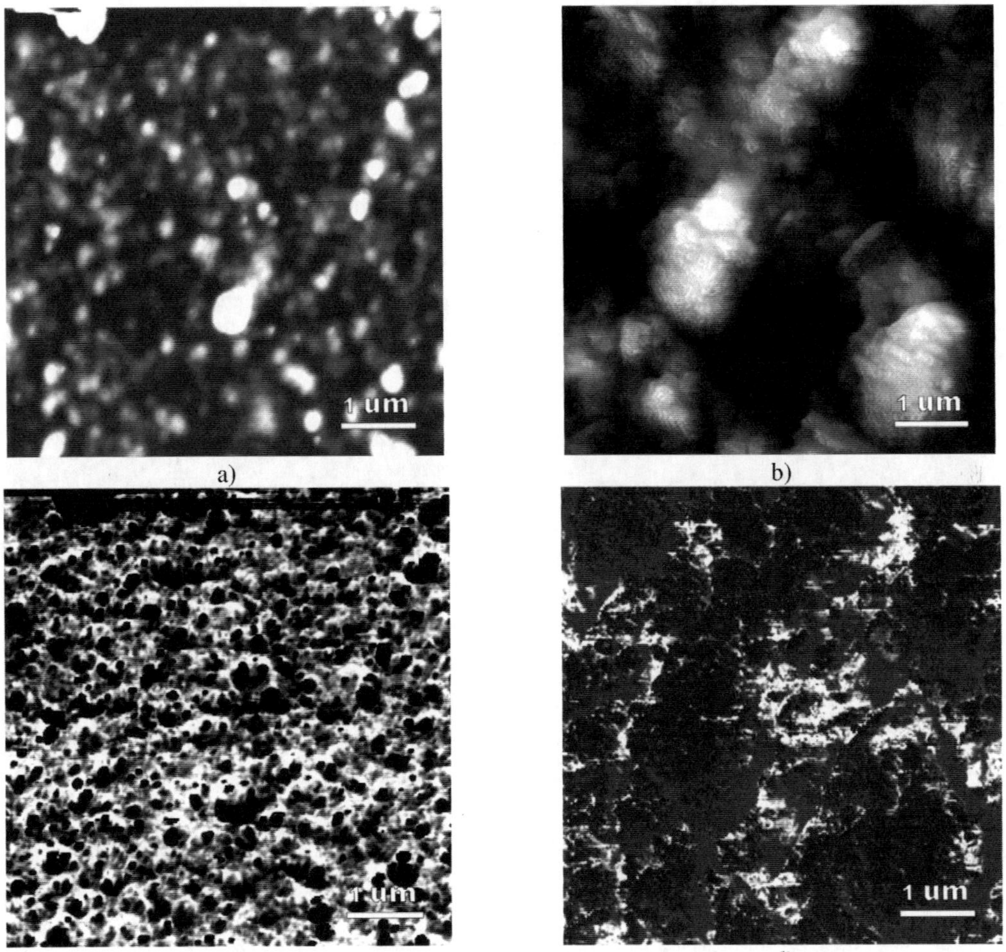

c) d)

Figure 2. Thopographical and phase imaging of the B series of the composite films: x = 0.35 (a, c) and x = 0.55 (b, d)

We can see that average particles size increases with growth of metal concentration comparing the topographical images before and after percolation threshold. This is due to increase of the common homogeneity of the chemical films compound.

We can see also that metal and dielectric don't form the granules, the components have homogeneous distribution in the film volume.

Let's consider the peculiarities of the topographical imaging for B series of the films (fig. 2). The surface of the B series of the films is similar to the surface of the A series of the films at small metal concentration and show the rolls-and-swells matrix. However, at big metal concentration we can see granular objects with elliptic shape and 1.5 nm size. We connect this fact with more homogeneous compound for the B series of the films. The basic component of the B series of the films is Co with adding of small amount of the other components which give amorphous condition.

We can find the building blocks with high and low densities on the phase image. These building blocks corresponds metal and dielectric. It is possible to see the big 3 μm granules with metal phase concentration increase. We can see that areas of the topographic elevation correspond to areas with low density if we compare phase images with topographical images.

Investigation of magnetic domains of the composite films

For investigation of the magnetic properties of the A series we used cantilever MFM10 (NT-MDT). The thickness of the magnetic covering of the AFM probe was 30–50nm, rounded radius was about 30nm. To exclude the influence of surface relief to magnetic force imaging firstly we define the surface relief in traditional regime of AFM. During the next passage on the scanning place the AFM probe move on the surface on the trajectory duplicate the relief of the specimens for excluding of the Van der Waals forces interactions with surface. In this case the AFM probe is in the field of specimens magnetic fields. We got the magnetic domain structure image as result of these forces registration.

The investigations showed that at metal phase concentration lower than 38% the magnetic structure in the films specimen it is not seen. We can see the sharp stripe or net domain structure at higher concentrations. Fig. 3 shows the topographical images of A series of the films with various metal phase concentration with superimposed images of the magnetic domains contour lines. The areas with uniform magnetization (domains) denoted by translucent filling inside the contour lines.

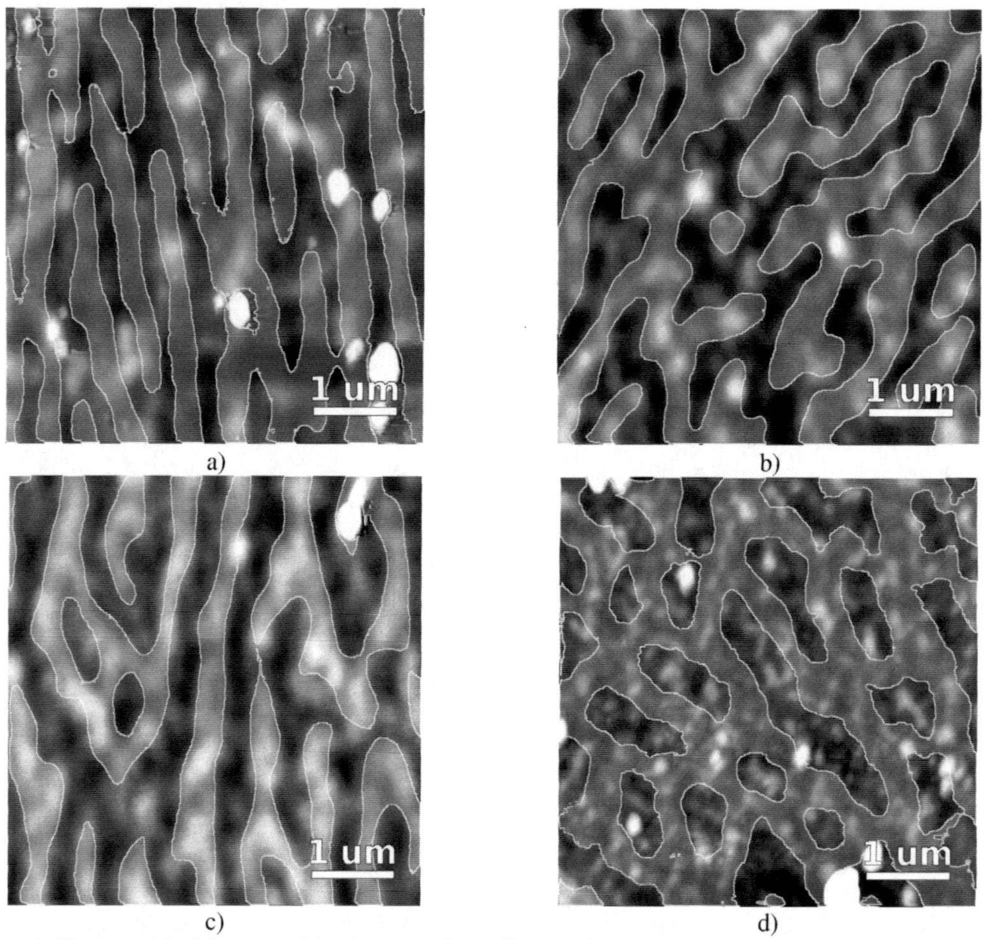

Figure 3. Topographical images of the A series of the films with superimposed images of the magnetic domains for the specimens with metal phase concentration a) 38%; b) 45%; c) 48%; d) 55%

IOP Conf. Series: Materials Science and Engineering **175** (2017) 012021 doi:10.1088/1757-899X/175/1/012021

We can see the stripe domain structure with the period about 0.4 μm at 38% metal phase concentrations. The length of the visible on the image domains is 5 μm and more. We observe the areas of uniform magnetization coincide with the areas of the topographic elevation usually but sometimes contrary coincide with throats. This occurs due to loose structure of the films before the percolation threshold. That is why there is no strict correspondence between topographic elements and compound materials.

We can see the sharp stripe domain structure associated with topographic elements reaching the percolation threshold. The magnetic domains have eccentric shape and thickness until 0.5 μm. Besides the stripes of domains can join in the more complicate formations. It seems it's connected with join of the metal granules in the film volume with increase of the metal concentration.

The domain structure has the net shape with weak connection with topographic nonuniformity at 55% metal phase concentration. This is due to the granular structure weakly appearing when a lot of materials carry the substrate and on the film surface we can observe evenness [4]. The magnetic structure in this case exact appears due to the high concentration of the ferromagnetic. The average thickness of the magnetic net wall is 0.4 μm. The average size of the cell in the net is 0.5×1.0 μm^2 and the cells have the elongate shape.

Model of the films for magnetic parameters calculations

Let the plane of the composite film coincides with the plane XY, magnetic bias field is directed along the axis Z. The particles included in the dielectric matrix is approximated by ellipsoids of rotation with the demagnetization factor

For the theoretical calculations of the microwave characteristics we present film as ensemble of particles with elongated ellipsoidal shape. According to the Dubowik model the dipole-dipole interaction energy we can write following way:

$$F_D = \frac{1}{2}\sum_{i=1}^{3} N_{ii}^{M}\langle M_i\rangle^2 V_M + \frac{1}{2}\sum_{i=1}^{3} N_{ii}^{P}\left(\langle M_i^2\rangle - \langle M_i\rangle^2\right) V_M,$$

where V_M - ensemble volume, $\hat{N}_{ii}^{M}$ - the macroscopic demagnetizing ensemble tensor, $\hat{N}_{ii}^{P}$ - demagnetizing tensor of separated granule. If the ensemble consists of rotation ellipsoids or cylinders then demagnetizing factors tensor we can present in following way:

$$\hat{N}_{ii}^{P} = 4\pi \cdot diag(\epsilon; \epsilon; 1-2\epsilon),$$

where $\epsilon = \frac{N_{\parallel}}{4\pi}$ - elliptic coefficient.

In the case of granules magnetized perpendicularly to ensemble plane the density of the anisotropy energy we can write following way [5]:

$$F_D = 2\pi M_s^2 f^2 + 2\pi M_s^2 f(f-1)N_{\perp},$$

where f - volume concentration of the particles in ensemble. For investigated in the work composite films the concentration f we can present in following way:

$$f = \frac{V_m}{V_m + V_d},$$

where V_m - the volume of metal phase, V_d - the volume of dielectric phase in composite material.

The density of the free energy of the composite we can write:

$$w = -M \cdot H + \frac{1}{2} f^2 M N_{film} M + \frac{1}{2} f(1-f) M N_{part} M,$$

where N_{film} - demagnetizing factors tensor of the film, N_{part} - demagnetizing tensor of separated particle. The demagnetizing factors of the particle we can find by Osborn's formula.

The particle sizes were obtained by AFM images.

Let's involve notation $Q = \left(f - (1 - f)(N_{\parallel} - N_{\perp})\right)$.

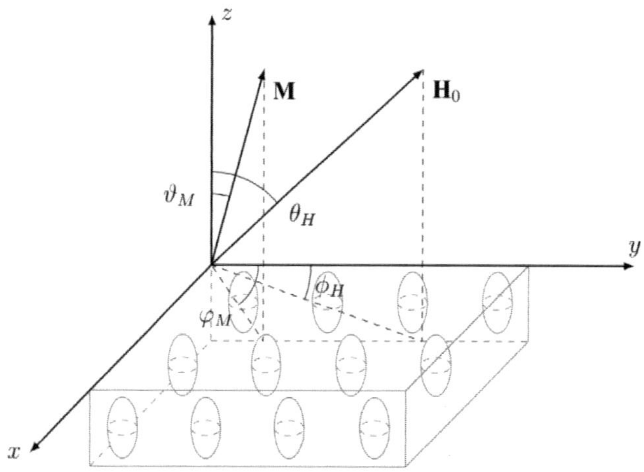

Figure 4. Geometry of the problem

For obtain the resonance condition at perpendicular relative to film plane dc field (fig. 4) we should select the axes x as the polar axes. We can write the free energy of the ensemble in the following way:

$$w = -H_0 M_s f sin\varphi sin\vartheta + \frac{1}{2} M_s^2 f^2 sin^2\varphi sin^2\vartheta +$$

$$+\frac{1}{2} M_s^2 f(1 - f)(N_{\parallel}(sin^2\vartheta cos^2\varphi + cos^2\vartheta) + N_{\perp} sin^2\varphi sin^2\vartheta)$$

The equilibrium conditions for the magnetization vector we can present in following way:

$$\varphi_0 = \frac{\pi}{2}, \qquad sin\vartheta_0 = \frac{H_0}{4\pi M_s Q}, \text{ если } H_0 < 4\pi M_s Q \text{ и}$$

$$\varphi_0 = \frac{\pi}{2}, \qquad \vartheta_0 = 0 \text{ при } H_0 \geq 4\pi M_s Q.$$

Using Smith-Beljers procedure we can obtain the resonance equation:

$$\left(\frac{\omega_{res}}{\gamma}\right)^2 = H_0^2(2f^2 + 3f + 1) - (4\pi M_s)^2 Q^2 (f^2 + f).$$

The calculation result shown in fig. 5. We can see that the model have limited range of applicability. This range doesn't include area of the high metal concentrations. The reason is the join of the metal particles and forming of the complicated volume aggregates at high metal concentrations. The calculation of demagnetizing factors of these aggregates calls for the special methods take into account the exchange interaction among the magnetic particles [6].

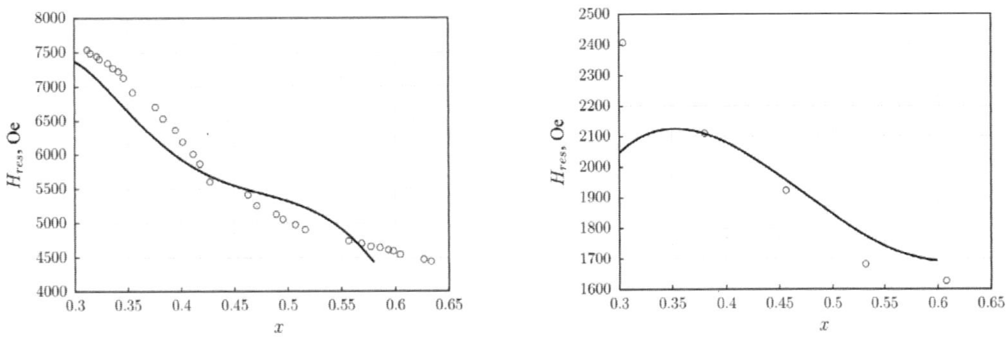

a) b)

Figure 5. The ferromagnetic resonance field for the A series of the films (a) and (b) for the B series of the films. Circles - the experimental values, solid line - the result for the model calculations.

Acknowledgements
This work was supported by RFBR (grant 17-02-01138-a).

References

[1] Kotov, L. 2008 *Advanced Materials Research*, **47**, p. 706.
[2] Kotov, L. et al. 2012 *Journal of Nanoscience and Nanotechnology*, **12**, p. 1696.
[3] Turkov, V. et al. 2013 *Bulletin of the Russian Academy of Sciences. Physics*, Vol. 77, **10**, pp. 1223–1227.
[4] Antonets, I. et al. 2016 *Technical Physics*, Vol. 61, **3**, pp. 416–423.
[5] Butera A. et al., 1999 Ferromagnetic resonance in as-deposited and annealed Fe-SiO_2 heterogeneous thin films, *Phys. Rev*. B.**60**. pp. 12270−12278.
[6] Golov A. 2015 *Solid State Phenomena* **233-234** pp. 485-489.

Pulsed Nd:YAG deposition of nanostructured FeS_{1-x} containing meta-stable phases

T Křenek[1]*, S Karatodorov[2], R Medlín[1], V Mihailov[2], J Savková[1]

[1]University of West Bohemia, Research Centre New Technologies 30614 Plzen, Czech Republic
[2]Institute of Solid State Physics - Bulgarian Academy of Sciences, 1784 Sofia, Bulgaria

E-mail: tkrenek@ntc.zcu.cz

Abstract.Pulsed near-IR laser irradiation of ferrous sulfide (FeS) in a vacuum allows a non-congruent ablation and deposition of nanostructured FeS_{1-x} thin films. Deposition has been performed on Al, Ta and Cu unheated substrate and analyzed by scanning (SEM) and high resolution transmission electron microscopy (HRTEM) and electron diffraction. Morphologically, the similar homogeneous, dark, metallic and adhesive appearanceshave been revealed for all the coats deposited on various substrates (by SEM). However, using HRTEM in agreement with electron diffraction, different phase composition on various substrates has been detected. Cubic pyrite phase (FeS_2) has been detected on Ta substrate. Cubic pyrite (FeS_2) and metastable rhomboedric smythite Fe_9S_{11} have been found in case of Al substrate. Cubic pyrite (FeS_2), metastable rhomboedric smythite Fe_9S_{11} and metastable orthorhombic marcasite (FeS_{2m}) revealed HRTEM analysis of the film on Cu substrate. In case of all deposits the detected crystalline nanograins were surrounded by amorphous matrix.

1. Introduction

Iron sulfide thin films attract attention because of its potential application for solar cell materials [eg. 1] and due to interesting electrical semiconducting [eg. 2] and magnetic [eg. 3]properties of this material. Although another solar cell materials (eg. such as cadmium, lead, indium or selenium) exhibit higher efficiencies, FeS non-toxicity, abundance, and cheap price make this candidate attractive. The iron sulfide (Fe-S) system represents a complex phase diagram including seven phases: pyrite (cubic- FeS_2), marcasite (calcium chloride structure-FeS_2), pyrrhotite-IT ($Fe_{1\_x}S$), pyrrhotite-4M (Fe_7S_8), Fe_9S_{10}, greigite (cubic spinel- Fe_3S_4), troilite- 2H (FeS) and mackinawite ($Fe_{1+x}S$) [4]. Only pyrite (FeS_{2p}) and pyrrhotite (Fe_{1-x}) are stable.

Iron sulfide thin coats have been obtained by ion beam and reactive sputtering(FeS_2) [5], vacuum thermal evaporation (FeS_2) [6], chemical spray pyrolysis (FeS_2)[7], sulfurization of iron oxides to FeS_2 [8],atmospheric-or low-pressure metal–organic chemical vapor deposition (AP or LP MOCVD; FeS_2) [eg.9], FeS_2 thin films using LPCVD of iron pentacarbonyl [$Fe(CO)_5$] hydrogensulfide, and *tert*-butyl sulfide as precursors [10] and flash evaporation(FeS_2) [11]. Iron sulfide nanoparticles were prepared using high-energy mechanical milling [12].

There have been published only few studies on laser ablative deposition of iron sulfide were interaction of reactive collisions with unheated substrate surfaces, were explored. The laser ablative deposition of pyrite (FeS_2) on aluminium and silica substrates resulted in the deposition of films including FeS constituents under the both higher and room temperature [13].

Recently we have reported on the pulsed IR laser ablative deposition of ferrous sufide FeS on unheated silica, tantalum and copper substrates [14]. Herein we continue on near-IR Nd:YAG pulsed laser deposition on unheated Al, Ta, Cu substrates.

2. Experimental

The 1064 nm near IR laser irradiation and deposition experiments are conducted in 10^{-2} Torr vacuum in a metal reactor with 343 mL volume. The metal reactor[15] has three borosilicate glass windows and it is connected to vacuum manifold and pressure transducer. The vacuum system used is Lavat AV 63.For the deposition, we use a pulsed Nd:YAG Quanta Ray GCR3 laser. The pulse duration at 1064 nm is 8 ns and the pulse energies used are up to 165 mJ. The repetition frequency is set to the maximum value of 20 Hz and the number of pulses for each sample is 2500. The radiation is focused witha 25 cm focal distance glass lens, producing up to 600 Jcm^{-2} energy fluence at the target surface. The target is a FeS pellet with diameter 8 mm and height 5 mm positioned in the centre of the reactor. The substrate is positioned vertically above the target on top of a 9,5cm diameter glass cylinder. Three different types of substrates are used: tantalum, aluminium and copperfoils. Each metal substrate is pierced in the middle to allow the passing of the laser beam.

Transmission electron microscopy (TEM) analysis (particle size and phase analysis) was carried out with a Transmission Electron Microscope JEM 2200FS (Shottky) from JEOL operated at 200kV with CCD Gatan (Digital Micrograph software), in-column Omega energy filter for EFTEM and EELS analysis, STEM mode with HAADF detector and EDS $80mm^2$ SDD (Silicon Drift Detector) X-Max detector from Oxfordon scraped samples that were subsequently dispersed in ethanol followed by the application of a drop of diluted suspension on a polymer/carbon coated Cu grid. The diffraction patterns were evaluated using the database JCPDS-2 and ProcessDiffraction software package [16]. Scanning Electron Microscope (SEM) analysis was carried out with a JEOL JSM 7600F autoemission microscope with EPMA $50mm^2$ SDD X-MAX EDS.The FeS pellet was made at 100 atm. on a hydraulic press from a commercially available iron sulfide powder (FeS, 99% Fe, Aldrich).

3. Results and discussion

The near IR laser irradiation of the FeS pellet in the vacuum leads to the pellet ablation and to the formation of visible luminescence (plasma) zone exhibits as a bluish plume which is filling whole glass cylinder [15]. The highly focused pulsed laser irradiation of FeS pellet caused creation of crater of ejected particles on adjacent of Ta, Al, Cu, where they are deposited as solid films. The coats on all selected substrates exhibit the same homogeneous, dark, metallic and adhesive appearances.

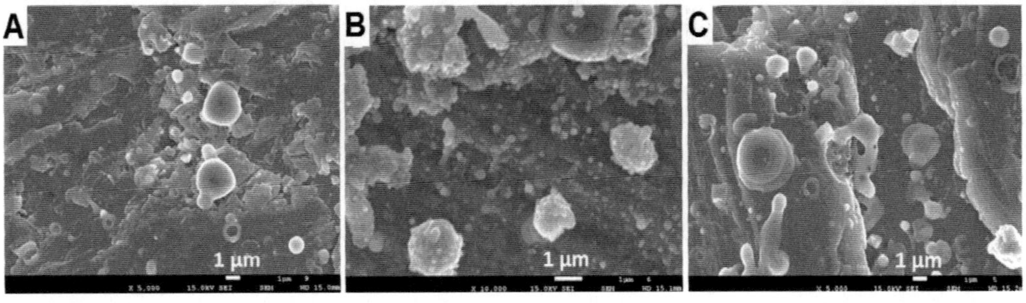

Figure 1. Morphology of FeS deposits on (A) aluminium, (B) copper, (C) tantalum substrate

The SEM images (Figure1) of the coats deposited on Al (Figure 1 A), Cu (Figure 1 B) and Ta (Figure 1 C) show round-shaped particles which dimension span from tens of nm up to units of µm on the flat discontinuous areas. EDX analyses revealed similar composition of all three samples. Round shaped particles, flat areas as well as ring-like objects exhibit heterogeneous composition with deficient of S. The S/Fe atomic percent ratio spans from 1:1 up to 1:2. The X-ray diffraction analysis of the commercial FeS sample allows the estimation of the relative amounts of crystalline troilite (~63%), pyrrhotite (~24%) and alpha-Fe (13%) it contains [14].

The phase composition of the prepared samples was studied by electron diffraction. The electron diffraction (Figure 2 A) of FeS deposit on Ta substratewhich is generally considered as an inert substrate reveals the presence of stable cubic pyrite (JCPDS file 00-001-1295) phase.The HRTEM image (Figure 2 B) shows partially crystallized structure with interlayer spacing d = 0.192 nm which corresponds with cubic pyriteinterplanar distance 220, according to JCPDS PDF 00-001-1295. From the HRTEM image is visible that the deposit has amorphous matrix with many small nanocrystals with the size about 3-5 nm.

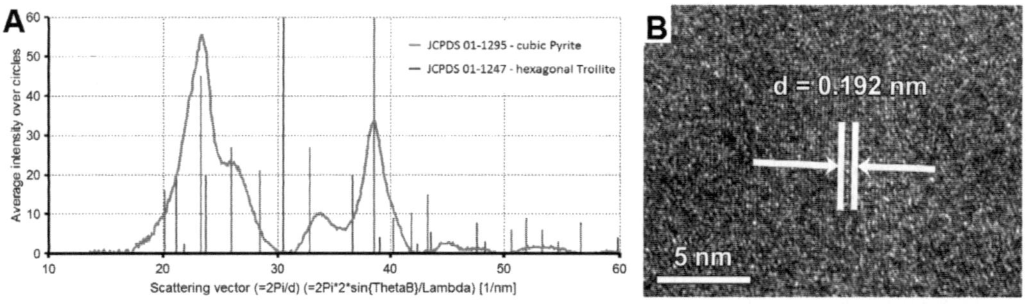

Figure 2. (A) Electron diffraction of FeS deposit on Ta substrate, (B) HRTEM image of FeS deposit on Ta substrate depicting cubic pyrite FeS_2 phase

The FeS thin film deposit on aluminium substrate exhibits the presence of two crystalline phases. In agreement with electron diffraction (Figure 3 A) the deposit consists of stable FeS_2cubic pyrite (01-1295) and interestingly of unstable Fe_9S_{11} rhombohedral smythite phase (JCPDS 10-0437). The phases show HRTEM images (figure 3B, C)where interlayer spacing d = 0.295 nm (012) and d = 0.270 nm (200)corresponds with smythite phase and pyrite phase respectively.

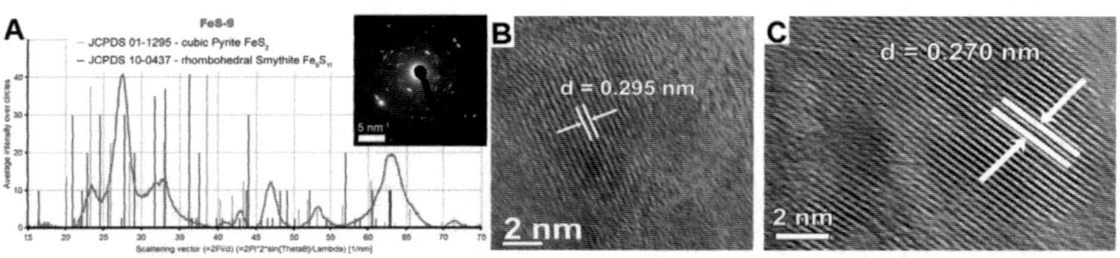

Figure 3.(A) Electron diffraction of FeS deposit on Al substrate; (B) HRTEM image of FeS deposit on Al substrate depicting unstable Fe_9S_{11} rhombohedral smythite and (C) cubic pyrite FeS_2 phase

Ablative deposition on copper substrate led to the formation of highly multiphase structure composed of FeS_2cubic pyrite, unstable Fe_9S_{11}rhombohedral smythite and FeS_2orthorhombic marcasite (JCPDS 00-003-0799). Electron diffraction assignment is given in Figure 4 A.The presence of unstable FeS phases confirm also HRTEM analyses (Figure 4 B, C) whose using allowed detection of d = 0.258 nm and d = 0.169 nm interlayer spacing which fitswith rhombohedral smythite(107) and orthorhombic marcasite phase (002) respectively.

IOP Conf. Series: Materials Science and Engineering **175** (2017) 012022 doi:10.1088/1757-899X/175/1/012022

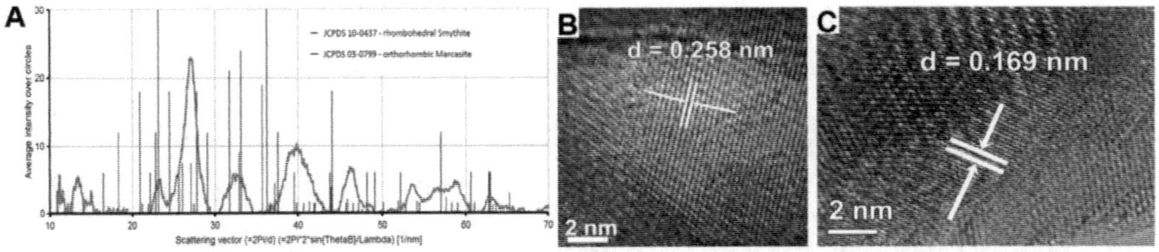

Figure 4.(A) Electron diffraction of FeS deposit on Cu substrate; HRTEM image of FeS deposit on Ta substrate depicting (B) unstable Fe_9S_{11} rhombohedral smythite, (C) FeS_2 orthoromic marcasite phase (002)

4. Conclusion

Pulsed near-IR laser deposition of FeS in a vacuum on Al, Ta and Cu substrate has been performed and the thin films were analysed by scanning (SEM) and high resolution transmission electron (HRTEM) microscopy and electron diffraction. SEM analysis revealed homogeneous, dark, metallic deposit round-shaped and ring-like particles. Via HRTEM and electron diffraction have been detected stable cubic pyrite phase (FeS_2) on Ta, Cu and Al substrate, metastable rhomboedric smythite Fe_9S_{11} on Al and Cu substrate and metastable orthorhombic marcasite (FeS_{2m}) on Cu substrate. These results represent the first examples of ablative deposition of unstable crystalline phases of ferrous sulfide.

Acknowledgements

The result was developed within the CENTEM project, reg. no. CZ.1.05/2.1.00/03.0088, cofounded by the ERDF as part of the Ministry of Education, Youth and Sports OP RDI programme and, in the follow-up sustainability stage, supported through CENTEM PLUS (LO1402) by financial means from the Ministry of Education, Youth and Sports under the National Sustainability Programme I and the project INERA (REGPOT-2012-2013-1).

References

[1] Hofmann W K, Birkholz M 1990 *Solar Energy Materials* **20** 149
[2] Lehner S W, Savage K S, Ayers J C 2006 *Journal of Crystal Growth* **286** 306
[3] Takahashi T 1973 *Solid State Communications* **13** (9) 1335
[4] Soon J M, Goh L Y, Loh K P 2007 *Applied Physics Letters* **91** 084105-1.
[5] Birkholz M, Lichtenberger D, Hoepfner C, Fiechter S 1992 *Solar Energy Materials and Solar Cells* **27** 243
[6] Rezig B, Dalma H, Kanzai M 1992 *Renewable Energy* **2** 25
[7] Smestad G, Da Silva A, Tributsch H, Fiechter S, Kunst M, Meziani N, Birkholz M 1989 *Solar Energy Materials* **18** 299
[8] Smestad G, Ennaoui E, Fiechter S, Tributsch H, Hofman W K, Birkholz M 1990 *Solar Energy Materials* **20**149
[9] Thomas B, Cibik T, Hoepfner C, Diesner D, Ehlers G, Fiechter S, Ellmer K 1998 *Journal of Materials Science* **9** 61
[10] Schleigh D M H, Chang S W 1991 *Journal of Crystal Growth* **112** 737
[11] Ferrer IJ, Sanchez C 1991 *Journal of Applied Physics* **70** 2641
[12] Chin P P, Ding J, Yi JB, Liu B H 2005 *Journal of Alloys and Compounds* **390** 255
[13] Yokoyma D, Namiki K, Yamada Y J 2006 *Radioanal. Nucl. Chem.***268** 283
[14] Pola J, Urbanová M, Pokorná D, Bezdička P, Kupčík J, Křenek T 2014 *RSC Advances* **4** 11543
[15] Karatodorov S, Mihailov V, KřenekT, Grozeva M 2016 *Journal of Physics: Conference Series* **700** 012001
[16] Lábár J L 2005 *Ultramicroscopy* **103** 237

A hybrid surface modification method on copper wire braids for enhancing thermal performance of ultra-thin heat pipes

W K Sheng, H T Lin, C H Wu, L S Kuo and P H Chen

Department of Mechanical Engineering, National Taiwan University, 10617 Taipei, Taiwan

E-mail: phchen@ntu.edu.tw

Abstract. Copper is the most widely used material in heat pipe manufacturing. Since the capability of wick structuresinside a heat pipe willdominate itsthermal performance, inthis study, we introduce a hybrid surface modification methodon the copper wire braids being inserted as wick structure into an ultra-thin heat pipe.The hybrid method is the combination of a chemical-oxidation-based method and a sol-gel method with nanoparticles being dip-coated onto thebraid. The experimentaldatashow thatbraids under hybrid treatmentperform higher water rising speed than the oxidizedbraidswhile owninghigherwater net weightthan those braids being only dip-coated with nanoparticle.

1. Introduction

Delivering heat at high rates over considerable distance with extremely small temperature drop, the heat pipe is a both smart and effective thermal management solution with simple construction and without any external pumping power. Figure 1 displays the basic working principle of the heat pipe. A heat pipe has wick structure on the inner surface and a small amount of working fluid delivering thermal energy at saturated state. It is composed of three sections: evaporator, where fluid absorbs heat and vaporizes; condenser, where vapor rejects heat and condenses; and an adiabatic section in which liquid and vapor flow in opposite direction back to evaporator and condenser, respectively.

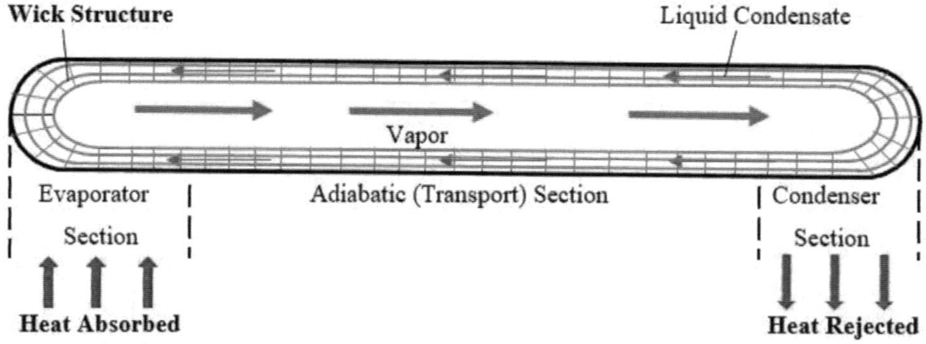

Figure 1. Basic working principle of the heat pipe

Classic examples for heat pipe application are in the market of consumer electronics [1-4]. As high performance and downsizing having been prerequisites in this marketin recent years, ultra-thinheat pipes (thickness lower than 2mm)or flat plate heat pipes are attracting more and more attention [5-8]. In this study, we introduce a hybrid surface modification method for the copper mesh being inserted as wick structure into ultra-thin heat pipes.

2. Experimental process

2.1. Preparation ofsurface modification

The hybrid treated copper wire braid in the experiment undergoes a chemical-oxidation-based process [9-11] and then being dip-coated out of a sol-gel solution containing SiO_2 nanoparticles [12, 13].

2.1.1. Chemical-oxidation-based process. Under this process, thenanostructuredCuO films are formed by immersing copper materials into a hot alkaline solution composed of $NaClO_2$, NaOH, $Na_3PO_4 \cdot 12H_2O$ and DI water [14].After 6 min, the color of the immersed copper material will become black and the solution will become apparently light blue because of the presence of $Cu(H_2O)_4^{2+}$ ions.

2.1.2. Sol-gel method. This process of surface modification is a coating method also for transforming surface wettability from superhydrophilic to superhydrophobic on copper materials [15]. The SiO_2 particles with 40 nm in diameter are mixed in the sol-gel solution.Since the dip-coating speed has a great influence on surface property after modification, two specific values of dip coating speed are set in the experiment, which are 3cm/min and 10cm/min.

2.2. Copper wire braid settlement

Following the same settlement introduced by Hsu et al. [16], figure 2 shows the definition of wetted height. The total length of a single copper wire braid is 11.0cm and all the braids arestuck at its top end onto a metal ring hanging with the same level on a plastic plate. 5 braids are weighed together at a time after immersing its bottom end into water for 75 seconds.

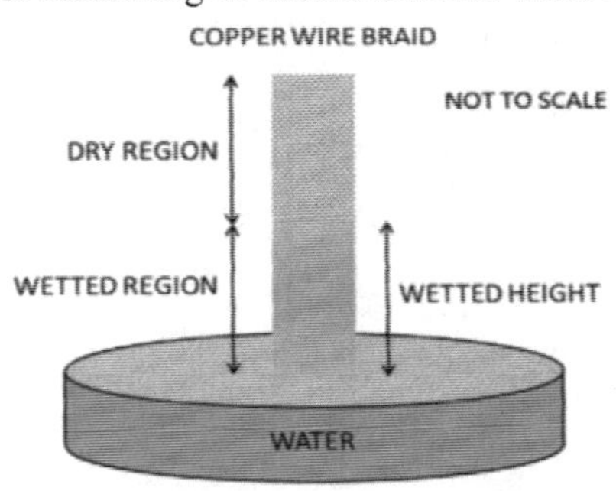

Figure 2. The scheme of the experimental setup for wetted height measurement on the copper wire braid

2.3. Wetted height observation

The black colour of the oxidized copper surface blurred the water boundary on the braid. To tackle with this issue, Tang et al. introduced an IR thermal imaging method [17].Due to the distinct difference in infrared emissivity between braid material and water, different temperature distributions were clearly displayed in IR thermal images. Figure 3 demonstrates the sequential images captured by IR camera.

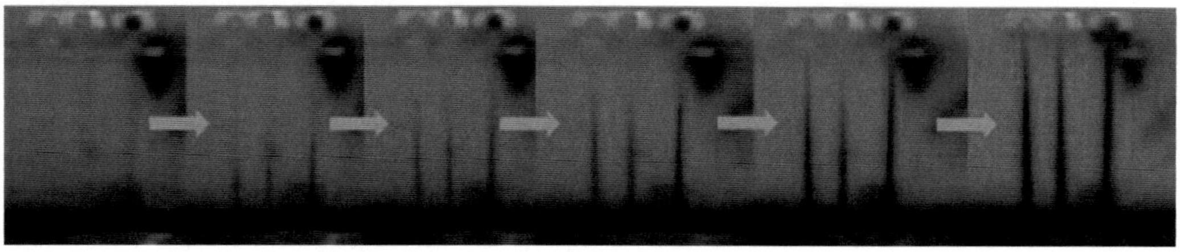

Figure 3. Sequential images captured by IRcamera

3. Results and discussion

3.1. SEM images ofsurface modification

In figure 4, the scanning electron microscopy (SEM) images are shown for the 4 different kind of copper wire braids, which are (a) plain, (b) dip-coated with SiO_2nanoparticles, (c) oxidized in hot alkaline solution, and (d) hybrid treated.The first three photos are citedfrom those shown in ref. [16]andthe fine flake-shaped structure of CuO can be seen in figure 4(c), which is consistent with results shown in ref. [14]. The hybrid structure in figure 4(d) can be found remarkably straightforward"hybrid".

However, the individual SiO_2 nanoparticles could not be easily observed in figure 4 for its scale being too small. Although the copper surfaces made by different modification method have different surface structures, copper surfaces in figure 4(b), figure 4(c) and figure 4(d) are all found to be superhydrophilic, which is experimentally proved being helpful for enhancing the wetted height. The coarser surface made by chemical-oxidation-based method implies that there are voids created in the modification process which is also found leading to an enhancement in water net weight.

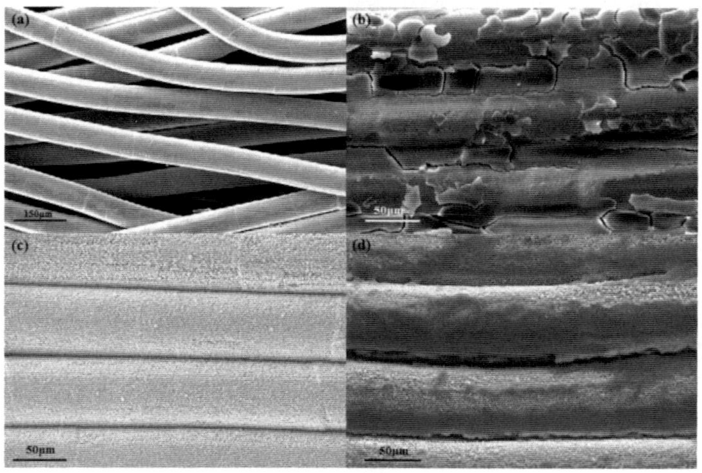

Figure 4. SEM images for copper wire braids being (a) plain, (b) dip-coated with SiO_2nanoparticles, (c) oxidizedin hot alkaline solution, (d) hybrid treated.

3.2. Water net weight

The experimental data of weight measured are shown in Table 1. For the sake of conciseness, the calculated standard deviations of those measured weight data are not shown in the table since it is almost an order smaller than the max error of the scale, which is 0.01 gram, stated by the manufacturer. Water net weight of a copper wire braid would refer to the capability of working fluid transportation by wick structure in heat pipes. Oxidized copper wire braid is found having higher porosity on the surface, which could lead to the result of higher water net weight than the dip-coated braids. After 75 seconds the bottom end being touched and immersed into water, as all the samples are supposed to be fully wetted, those braids underwent oxidation process are showing obviously higher water net weight than those braids being only dip-coated.

Table 1. Measured weight changes for copper wire braids under various surface treatment

	Dip-coated (10cm/min)	Hybrid (10cm/min)	Dip-coated (3cm/min)	Hybrid (3cm/min)	Oxidized
Dry weight (g)	26.96	26.95	26.93	27.01	26.94
Gross weight (g)	27.22	27.25	27.19	27.31	27.25
Water weight (g)	0.26	0.30	0.26	0.30	0.30

3.3. Water rising speed

The detailed wetted height measurement is as the following. First, count the vertical black-pixel-numberin each IR image with software and find the ratio between this black-pixel-number to that of a wire braid with full wetted height. Multiply the ratio obtained above by the actual full length of the copper wire braid and so comes the actual wetted height.

The water rising speed can be clearly compared among copper wire braids under different surface modification process. It is conspicuous that curves with lager slope have higher water rising speed.As shown in figure 5, dip-coated braids share higher water rising speed than others; meanwhile, the hybrid treated braids have noticeable higher value in wetted height as well as in water rising speed than theoxidized copper wire braids.

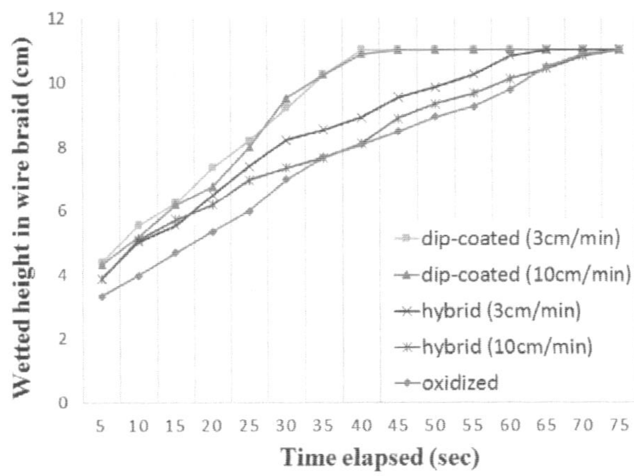

Figure 5. Wetted height against time elapsed for copper wire braids under various surface treatment. The dip-coating speed in each method is declared within parentheses

4. Conclusion

In conclusion, the feasibility of the copper wire braids applied with hybrid surface modification techniques to improve the performance of wick structure in an ultra-thin heat pipe was experimentally investigated. The hybrid surface modification method is introduced to fabricate a hybrid surface structure and from the SEM results, it is shown that the hybrid surface processed by the two methods share the two distinct surface property. The flake-shaped CuO structure may refer to a higher surface porosity and the area covered by the SiO_2nanoparticles retains a superhydrophilic surface property.

The hybrid treated copper wire braids performhigher water rising speed than the oxidized copper wire braids while owning higher water net weight than those braids being only dip-coated and this fact indicates that this hybrid surface modification method on copper wire braids could lead to a superior thermal performance of those ultra-thin heat pipes containing the braid-like wick structure inside.

Acknowledgments
The authors would like to thank the assistance for the SEM images from Prof. Chao-Sung Lin, Department of Material Science and Engineering, National Taiwan University. The financial support is from the Ministry of Science and Technology, Taiwan, ROC, under the grant numbers: 105-2221-E-002-107-MY3, 104-2218-E-002-004, 102-2221-E-002-133-MY3, and 102-2218-E-002-022.

References

[1] Moon S H, Hwang G, Yun H G, Choy T G and Kang Y L2002 Improving thermal performance of miniature heat pipe for notebook PC cooling *Microelectron. Reliab.***42**135-140
[2] Pastukhov V, Maidanik Y F, Vershinin C and Korukov M 2003 Miniature loop heat pipes for electronics cooling *Appl. Therm. Eng.***23** 1125–1135
[3] Chang J C S, Wang C H, Liu J K, Feng Y, WangY and Huang C Y 2006 A study of composite wick heat pipes for notebook cooling *8th Int. Heat Pipe Symp. (Kumamoto)* pp. 24-27
[4] Vasiliev L 2008 Micro and miniature heat pipes–electronic component coolers *Appl. Therm. Eng.***28** 266–273
[5] Liou J H, Chang C W, Chao C and Wong S C 2010 Visualization and thermal resistance measurement for the sintered mesh-wick evaporator in operating flat-plate heat pipes *Int. J. Heat Mass Transf.***53** 1498–1506
[6] Wong S C, Liou J H and Chang C W 2010 Evaporation resistance measurement with visualization for sintered copper-powder evaporator in operating flat-plate heat pipes *Int. J. Heat Mass Transf.***53** 3792–3798
[7] Lin K T and Wong S C 2013 Performance degradation of flattened heat pipes *Appl. Therm. Eng.***50** 46-54
[8] Li Y, He J, He H, Yan Y, Zeng Z andLi B 2015 Investigation of ultra-thin flattened heat pipes with sintered wick structure *Appl. Therm. Eng.***86** 106-118
[9] Enright R, Dou N, MiljkovicN, Nam Y, Wang E N 2012 Condensation on Superhydrophobic Copper Oxide Nanostructures *3rd Micro/Nanoscale Heat & Mass Transfer Int. Conf.(March 3–6, Atlanta)*
[10] NamY and JuYS 2012 A comparative study of the morphology and wetting characteristics of micro/nanostructured Cu surfaces for phase change heat transfer applications*J. Adhes. Sci. Technol.* 1–14
[11] Nam Y, Sharratt S, Byon C, Kim S J, Ju Y S 2010 Fabrication and Characterization of the Capillary Performance of Superhydrophilic Cu Micropost Arrays*J. Microelectromech. Syst.***19** 581–588
[12] Liu Z, Zhang X, Murakami T and Fujishima A 2008 Sol–gel SiO2/TiO2 bilayer films with self-cleaning and antireflection properties *Solar Energy Materials and Solar Cells***92** 1434-1438
[13] Kesmez Ö, Çamurlu H E, Burunkaya E and Arpaç E 2009 Sol–gel preparation and characterization of anti-reflective and self-cleaning SiO2–TiO2 double-layer nanometric films *Solar Energy Materials and Solar Cells***93** 1833-1839
[14] Miljkovic N, Enright R, Nam Y, Lopez K, Dou N, Sack J and Wang E N 2012 Jumping-dropletenhanced condensation on scalable superhydrophobic nanostructured surfaces *Nano Lett.***13** 179–187
[15] Hsu C C and Chen P H 2012 Surface wettability effects on critical heat flux of boiling heat transfer using nanoparticle coatings *Int. J. Heat Mass Transf.***55** 3713–3719
[16] Hsu C C, Wu C H, Sheng W K, Chen M, Kuo L S and Chen P H 2016 Improvement of water wetting capability of copper wire braids bysurface modification approaches *Int. Commun. Heat Mass***77**155-158
[17] Tang Y, Deng D, Lu L, Pan M and Wang Q 2010 Experimental investigation on capillary force of composite wick structure by IR thermal imaging camera *Exp. Therm. Fluid Sci.***34** 190-196

Thermodynamic physico-chemical modelling and calculation for the synthesis process of modern functional materials

A B Uspensky, R Ralys, D Kremnev, M Radin, A Slobodov*

ITMO University, St.Petersburg, 197101, Russia

E-mail: aslobd@gmail.com

Abstract. The method of thermodynamic modeling of phase and chemical transformations for synthesis conditions of multicomponent functional materials is presented. Calculation of the phase transformations taking place in the conditions of synthesis and operation of ceramic materials on the basis of the system $SiO_2 - K_2O - Na_2O - Al_2O_3 - Fe_2O_3 - CaO - MgO - TiO_2 - P_2O_5 - SO_3$ in the field of temperatures from 0 to 2000° C is carried out. The sequence of phase and chemical transformations between components of synthesizable material and their reaction products, depending on temperature, is founded.

1. State of the problem

It is required that modern functional materials have very different exploitation characteristics, such as strength, thermal and chemical stability, moisture absorption, rheology, etc. Nature of the phase and chemical transformations which occur during the synthesis from raw materials and composition of the product define these characteristics [1, 2]. Due to presence of multiple phases, many components, complexity and interconnection of the interactions and transitions, which take place during the synthesis, and wide range of values of state parameters, the detailed experimental research within the wide range of temperatures, compositions, and different conditions of synthesis is a very complex problem, which to date is generally unresolved [1-3].

In some cases it is correct and sufficiently effective to apply methods of the thermodynamic physico-chemical modelling. These methods have principal advantages in comparison to experimental methods, letting to obtain more detailed data about the whole complex of interrelated phase and chemical transformations corresponding to functional materials and products synthesis [4-9].

However, this method turns out to be effective, i.e. yielding reliable, quantitatively justified data, only when several conditions are met [5-7]. First of all, the modelling and computational apparatus used to describe these systems should be correct, effective, and adequate to the real system. In second, thermodynamic characteristics of the initial compounds and possible interaction products should be valid. In third, interpretation, analysis, and generalization of the results obtained from thermodynamic modelling should be correct.

2. Objects and methods of investigation

As the main object of the investigation it has been considered the process of synthesis cranial base of the ceramic tiles. At the first stage of investigation, the physico-chemical modelling of the synthesis of material, which meets certain criteria, has been undertaken. Then, based on the modelling results, it has been required to correctly interpret these results, reveal mechanism of the occurring processes, and discover principles of influence of the state parameters, such as temperature, moisture content, synthesis atmosphere, on the phase and chemical transformations during the synthesis.

As it is known, ceramics is synthesised via annealing of the raw materials till the certain temperature — the annealing temperature. As a result of the interaction of the components in the raw material, which takes place under elevated temperatures, the phase and chemical composition of the annealed product becomes different from the initial, and the product acquires such properties, as hardness, strength, etc. It is namely phase and chemical composition of the synthesised product defines its exploitation properties/

Subject to the forgoing, the primary object of investigation is the raw material of the ceramic synthesis. As the basis of modelling, the raw material of a typical complex composition has been used (including all the main and auxiliary components). Its composition has been defined using analytical methods, and the main compounds have been recalculated into their oxide forms. As a result, oxide composition of the raw material corresponded to the following ten-component system, which is the set of compounds-components fully and unambiguously describing both composition of the raw material, and the product:

$$SiO_2 - K_2O - Na_2O - Al_2O_3 - Fe_2O_3 - CaO - MgO - TiO_2 - P_2O_5 - SO_3$$

The investigation has been undertaken with help of the methods and databases of thermodynamic modelling, which have been developed earlier, and which meet the requirements described earlier [5-9]. The equilibrium composition has been calculated under the different synthesis conditions, namely, in the temperature interval from room temperature to 2000 °C (taking into account the possible overheating).

The considered conditions correspond to the annealing stage of the ceramic tiles production process.

In the table below, the initial data in mass and mole expressions are given, taking into account losses during the annealing. Using mole composition is viable from the point of view of the of chemical and phase transitions analysis.

Table. The chemical (oxide) composition of the press-powder tiles

Substance / Content	SiO_2	K_2O	Na_2O	Al_2O_3	Fe_2O_3	CaO	MgO	TiO_2	P_2O_5	SO_3	Loss on ignition
mass.%	55,20	2,25	0,45	13,68	3,19	9,72	3,25	0,79	0,15	0,49	10,83
mol/kg	9,187	0,239	0,073	1,342	0,199	1,733	0,806	0,099	0,011	0,117	–

It is possible to allocate two areas types within the ceramic materials synthesis, in which the conditions of synthesis and nature of the phase and chemical interactions differ principally. It's the homogeneous area of ceramic mass (which is completely separated from the surrounding atmosphere) and the heterogeneous area ceramic mass contact with the environment (the atmosphere of synthesis).

Certainly, the main interactions are the first type ones – in the thickness of the ceramic mass (press-powder) between its components. They determine, mainly, the final phase-chemical composition of synthesizable material. Therefore the main purpose of this work is to study these homogeneous interactions in the thickness of the ceramic mass in a wide range of temperatures corresponding to different stages of functional (ceramic) materials synthesis.

3. Modelling results analysis

Let's consider the transformations taking place in the bulk of ceramic mass. Under these conditions interaction for components takes place without atmosphere, and it is impossible for gases to evolve as a result of interactions, because the processes take place in the bulk of the solid phase.

As soon as sulphur dioxide is a gas, its presence in the system is nominal and is taken into account

IOP Conf. Series: Materials Science and Engineering 175 (2017) 012024 doi:10.1088/1757-899X/175/1/012024

as amounts of sulphur and oxygen in the system, rather than sulphur dioxide itself.

Temperature influence on the phase and chemical transformations in the temperature range considered, obtained as a result of the thermodynamic modelling of the system under investigation, is presented in the Figure.

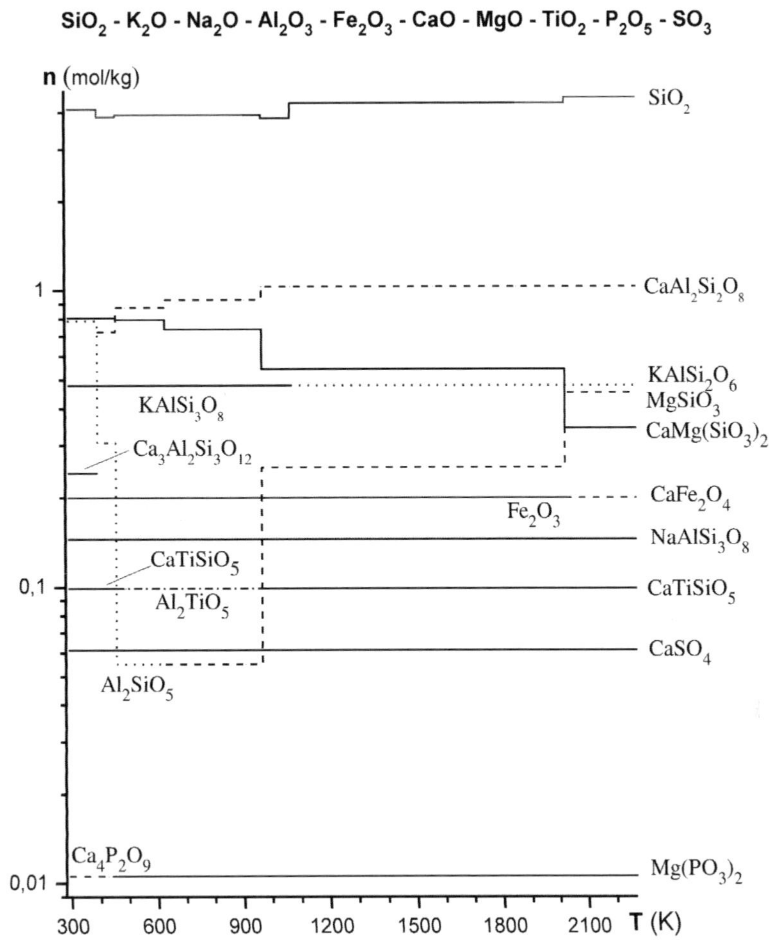

Figure. Temperature influence of the phase transitions in the bulk of the pottery fragment

Let's consider the sequence of transformations, which take place when the temperature is rising.

Under the room temperature complex compounds form from simple oxides as a rule: Al_2SiO_5, $CaSO_4$, $Ca_4P_2O_9$, $Ca_3Al_2Si_3O_{12}$, $CaTiSiO_5$, $CaMg(SiO_3)_2$, Fe_2O_3, $KAlSi_3O_8$, $NaAlSi_3O_8$, SiO_2.

Comparison of the initial composition and the data obtained allows to conclude, that the following transformations take place in the system.

Magnesium oxide MgO interacts the most actively with calcium oxide and silicon oxide. It manifests through the formation of silicate of calcium and magnesium $CaMg(SiO_3)_2$:

$$MgO + 2SiO_2 + CaO \rightarrow CaMg(SiO_3)_2 \quad (1)$$

Iron does not form complex compounds and remains in the form of iron (III) oxide — hematite.

All sulphur, being initially in the form of sulphur trioxide, reacts with calcium oxide, and yields calcium sulphate:

$$SO_3 + CaO \rightarrow CaSO_4 \quad (2)$$

Phosphorus (V) oxide reacts with calcium oxide to form complex calcium phosphate $Ca_4P_2O_9$. The process completely consumes phosphorus (V) oxide:

$$P_2O_5 + 4CaO \rightarrow Ca_4P_2O_9 \quad (3)$$

Titanium oxide reacts with calcium and silicon oxides and completely transforms titanium and calcium silicates:

$$TiO_2 + SiO_2 + CaO \rightarrow CaTiSiO_5 \quad (4)$$

Sodium present in the system in the form of sodium oxide reacts with alumina and silicon oxides, entirely binding into sodium silicoaluminate:

$$Na_2O + 6SiO_2 + Al_2O_3 \rightarrow 2NaAlSi_3O_8 \quad (5)$$

Potassium oxide fully binds into potassium silicoaluminate by reacting with alumina and silicon oxides:

$$K_2O + 6SiO_2 + Al_2O_3 \rightarrow 2KAlSi_3O_8 \quad (6)$$

As it has been shown earlier, calcium oxide yields calcium sulphate, calcium phosphate, and magnesium and calcium silicate, in accordance with reactions (1)-(4). This shows high reactivity of calcium oxide.

Remainder of calcium oxide reacts with alumina and silicon oxide, forming $Ca_3Al_2Si_3O_{12}$:

$$3CaO + 3SiO_2 + Al_2O_3 \rightarrow Ca_3Al_2Si_3O_{12} \quad (7)$$

Alumina oxide has high reactivity as well. As it has been described earlier, it reacts with sodium, potassium, calcium, and silicon oxides according to equations (5)-(7), and excess of it forms alumina silicate Al_2SiO_5 together with silicon oxide:

$$SiO_2 + Al_2O_3 \rightarrow Al_2SiO_5 \quad (8)$$

Silicon, which did not bind with the compounds mentioned above, remains in the form of silicon dioxide SiO_2.

In this way, the reactions (1)-(8) describe how initial oxides interact at the room temperature.

When the temperature rises, some of the substances undergo structural and chemical transformations to yield new compounds.

Thus, at 392 K complex calcium alumosilicate $Ca_3Al_2Si_3O_{12}$ becomes thermally unstable, and reacting with silicon dioxide and alumina silicate Al_2SiO_5 forms a simpler alumosilicate $CaAl_2Si_2O_8$. This transformation is described via the following reaction:

$$2Al_2SiO_5 + Ca_3Al_2Si_3O_{12} + SiO_2 \rightarrow 3CaAl_2Si_2O_8 \quad (9)$$

At the temperature 424 K alumina silicate Al_2SiO_5 undergoes phase transition from kyanite to andalusite.

At the temperature 446 K potassium alumosilicate $KAlSi_3O_8$ undergoes another phase transition.

The further temperature rise till 456 K results in the changes considered below. With no regard for the stoichiometric relationships these changes are described using the following transition:

$$Al_2SiO_5 + Ca_4P_2O_9 + CaTiSiO_5 + CaMg(SiO_3)_2 \rightarrow$$
$$CaAl_2Si_2O_8 + Mg(PO_3)_2 + SiO_2 + Al_2TiO_5 \quad (10)$$

As the result of this transition complex calcium phosphate $Ca_4P_2O_9$ and calcium-titanium silicate $CaTiSiO_5$ are completely consumed.

Data analysis shows that there are two independent transformations in the system. Let's consider them both separately.

The first transformation involves calcium-titanium silicate $CaTiSiO_5$. At the temperature 456 K it

IOP Conf. Series: Materials Science and Engineering **175** (2017) 012024 doi:10.1088/1757-899X/175/1/012024

becomes thermally unstable and decays reacting with Al_2SiO_5 and forms calcium alumosilicate $CaAl_2Si_2O_8$ together with titanium aluminate Al_2TiO_5. During this transformation the excessive silicon yields free dioxide:

$$2Al_2SiO_5 + CaTiSiO_5 \rightarrow CaAl_2Si_2O_8 + Al_2TiO_5 + SiO_2 \quad (11)$$

The second transformation from those mentioned above takes place because calcium phosphate becomes thermally unstable, and reacts with calcium-magnesium silicate to yield magnesium metaphospahte, and the excess of unstable calcium and silicon oxides:

$$Ca_4P_2O_9 + CaMg(SiO_3)_2 \rightarrow Mg(PO_3)_2 + 5CaO + 2SiO_2 \quad (12)$$

Calcium and silicon oxides then react with alumosilicate Al_2SiO_5 and form calcium alumosilicate $CaAl_2Si_2O_8$ according to the following reaction:

$$CaO + SiO_2 + Al_2SiO_5 \rightarrow CaAl_2Si_2O_8 \quad (13)$$

Merging (12) and (13) results in:

$$5Al_2SiO_5 + 3SiO_2 + Ca_4P_2O_9 + CaMg(SiO_3)_2 \rightarrow Mg(PO_3)_2 + 5CaAl_2Si_2O_8 \quad (14)$$

From the reaction (14) it follows that in order for all calcium oxide, excess of which forms as a result of the reaction (12), to bind into calcium alumosilicate $CaAl_2Si_2O_8$ it is necessary the presence of silicon dioxide in the system. This reaction consumes the silicon dioxide.

In this way, reactions (11) and (14) describe transformations in the system at 456 K.

Further temperature rise results in phase transition of the sodium alumosilicate $NaAlSi_3O_8$ from albite to analbite.

At the temperature 627 K calcium-magnesium silicate $CaMg(SiO_3)_2$ becomes thermodynamically unstable in the presence of alumosilicate Al_2SiO_5 and as the result of their interaction forms calcium alumosilicate $CaAl_2Si_2O_8$ and magnesium silicate $MgSiO_3$. This reaction goes to complete disappearance of alumosilicate Al_2SiO_5 from the system:

$$Al_2SiO_5 + CaMg(SiO_3)_2 \rightarrow CaAl_2Si_2O_8 + MgSiO_3 \quad (15)$$

Further temperature rise to 906 K results to phase transition of quartz to trimidite.

At the temperature 966 K calcium-magnesium silicate $CaMg(SiO_3)_2$ becomes unstable and reacts with the complex alumina and titanium oxide Al_2TiO_5 and silicon dioxide. As the result of this interaction calcium alumosilicate $CaAl_2Si_2O_8$, calcium and titanium silicate $CaTiSiO_5$ and magnesium silicate $MgSiO_3$:

$$2CaMg(SiO_3)_2 + SiO_2 + Al_2TiO_5 \rightarrow CaAl_2Si_2O_8 + CaTiSiO_5 + 2MgSiO_3 \quad (16)$$

This reaction goes until Al_2TiO_5 completely reacts.

Further temperature rise to 1065 K results to potassium alumosilicate $KAlSi_3O_8$ decaying to yield simpler alumosilicate $KAlSi_2O_6$ and silicon dioxide:

$$KAlSi_3O_8 \rightarrow KAlSi_2O_6 + SiO_2 \quad (17)$$

Further changes in the system occur only at the temperature 1559 K when the phase transition of potassium alumosilicate $NaAlSi_3O_8$ takes place.

At the temperature 2018 K calcium-magnesium silicate becomes unstable in the presence of martite. As the result of their interaction forms the double calcium-iron oxide $CaFe_2O_4$, magnesium silicate $MgSiO_3$, and free silicon dioxide:

$$CaMg(SiO_3)_2 + Fe_2O_3 \rightarrow CaFe_2O_4 + MgSiO_3 + SiO_2 \quad (18)$$

After the reaction (18) part of the calcium-magnesium silicate remains in the system, because the amount of iron oxide is insufficient for the complete disappearance of the silicate.

Further temperature rise to 2000 °C does not result in chemical changes of the chemical composition.

IOP Conf. Series: Materials Science and Engineering **175** (2017) 012024 doi:10.1088/1757-899X/175/1/012024

4. Conclusion

In this way, justified quantitative data are obtained and principle of the temperature influence on the phase and chemical transitions in the ceramic mass are obtained by methods of thermodynamic modeling of phase-chemical transformations, thus allowing to control the technological synthesis process.

Rigorous interrelation between stage, i.e. temperature, and corresponding phase and chemical transformations, and the composition of the synthesized material is established.

The temperature impact on the sequence of phase transitions of the complex oxides and silicates, forming the resulting ceramic material, is attained.

References

[1] Materials Handbook for refractories, traditional and advanced ceramics 1989 *Ceramic Ind. J.* **1** 21–124

[2] Egész Á and Gömze L A 2015 Qualification Methods of Al_2O_3 Injection Molding Raw Materials. *J. Phys.: Conf. Series* **602** 012023

[3] Smith W R Computational aspects of chemical equilibrium in complex systems 1980 *Theor. Chem. Adv. Perspect.* **5** 185–259.

[4] Cox J D, Wagman D D and Medvedev V A 1989 *CODATA key values for thermodynamics* (N.Y.: Hemisphere Pub. Corp.)

[5] Veiko V P, Slobodov A A and Odintsova G V 2013 Availability of methods of chemical thermodynamics and kinetics for the analysis of chemical transformations on metal surfaces under pulsed laser action *Laser Physics* **23** 066001.

[6] Olekhnovich R O, Glazacheva E N, Uspenskii A B, Slobodov A A, Uspenskaia M V 2015 *Proc. XV Intern. Multidisc. Sci. Geoconf.* **2** N 3 207-214

[7] Slobodov A, Uspenskiy An, Ralys R and Kremnev D 2015 Thermodynamic modelling of phase-chemical transformations as the method for study of rheological properties of substances *Építőanyag – J. Silic. Based Compos. Mater.* **67** 159–163

[8] Uspenskiy An, Slobodov A A, Kremnev D, Sochagin A and Klepova A N 2016 Phase-chemical problems of technology optimization and cleaning - thermodynamic research and modeling *Proc. XVI Intern. Multidisc. Sci. Geoconf.* **1** 769-776

[9] Ralys R V, Uspenskiy A A and Slobodov A A 2016 Deriving properties of low-volatile substances from isothermal evaporation curves *J. Non-Equilibrium Thermodynamics* **41** 3-11.

IOP Conf. Series: Materials Science and Engineering **175** (2017) 012025 doi:10.1088/1757-899X/175/1/012025

Porous Ceramic Composite ZrO_2(MgO)-MgO for Osteoimplantology

A S Buyakov[1,2], S N Kulkov[1,2]

[1]Strength and Design Department, Physical-Technical Faculty, Tomsk State University, 634050 Tomsk, Russia.
[2]Laboratory of Physics of Nanostructured Functional Materials, Institute of Strength Physics and Materials Science SB RAS, 634055 Tomsk, Russia

E-mail: Alesbuyakov@gmail.com

Abstract. Pore and crystalline structure, biocompatibility of ceramic composite ZrO_2(MgO)-MgO were studied. The main mechanical characteristics were determined and it has been shown that compression strength directly depends on microstresses obtained from X-ray data.In-vitro studies of mesenchymal stromal stem cells (MMSC), cultivated on material surface are shown that cell proliferation and differentiation of MMSC goes throw osteogenic type.

1. Introduction
Bone integrity regeneration in terms of postoperative recovery and aesthetic due to the cancer resection or injuries is one of very important issues. An application of ceramics for using as osteoreplacement material has a special interest now. The most actively developed studies in this area are investigations of zirconium ceramic (ZrO_2) included in ISO register as a material for bone replacement. Moreover, magnesium is involved in protein synthesis and stabilization of DNA molecules, which can improve regeneration processes. Using of such ceramics can provides of bone tissue integration with an implant and prevents the risk of forming adverse biological response from the surrounding tissues.

However, successful osteo-integration depends not only on biocompatible of the material, but also on the structure-phase state and mechanical parameters [1, 4]. Mechanical characteristics of osteoreplacement material should correspond to bone tissue properties [5]. Strength of the material is mostly determined by its microstructure that is why information about the fine crystalline structure parameters is very important for obtaining the necessary strength of osteoreplacement material.

The aim of this work is to study influence of composition, micro- and fine crystal structure on mechanical properties of the porous ceramic composite ZrO_2(MgO)-MgO and to study biocompatibility of the material.

2. Materials and Methods
Ceramics with different MgO concentrations with a pore volume 50% sintered at 1600°C were studied. Porosity level was achieved by adding ultra-high-molecular-weight polyethylene particles.

Material microstructures were studied by scanning electron microscope Tescan VEGA 3. Fine crystalline structure parameters were measured from X-ray diffraction data and compressive strength of were studied by using universal testing machine Devotrans DT.

Biocompatibility of ceramics was studied by in vitro definition of mesenchymal stromal stem cells (MMSC) response on the surface of ceramic samples. To determine the ability of MMSC to directed osteogenic differentiation during their cultivating on the surface of porous ceramic materials were made a detection alkaline phosphatase, which if the first marker of osteogenic differentiation, produced by cells.

3. Results and Discussion

Figure 1 represents distribution of pores inside of ceramic samples: with an average size 103 microns and 28μm.

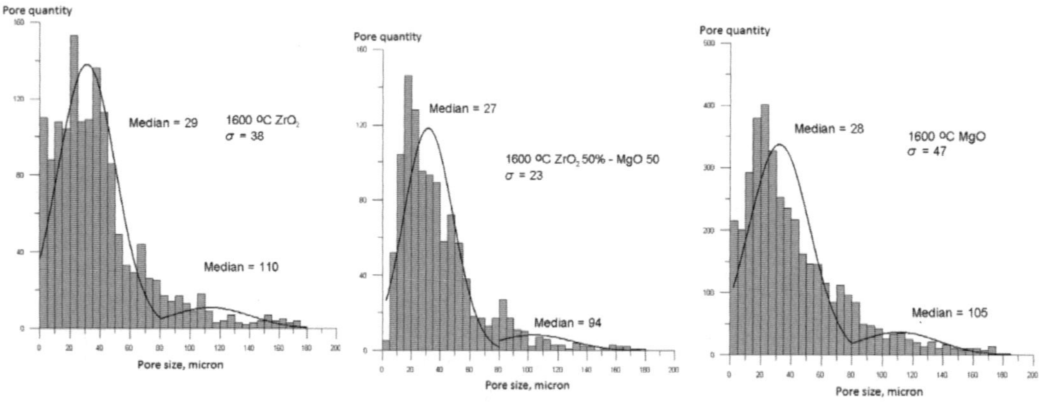

Figure 1. Pore size distribution according to the composition: **a**. $ZrO_2(MgO)$; **b**. 50% $ZrO_2(MgO)$ + 50% MgO; **c**. MgO

From X-ray data were determined the coherent diffraction domain sizes (CDD) and crystalline lattice microdistortions. This data was compared with the experimental results of ceramics compressive strength. Analyses of "stress-strain" diagrams are shows that the highest compressive strength corresponds to pure MgO ceramic – 33 MPa. ZrO_2 concentration increasing proceeds with decreasing of compressive strength to 18 MPa, which is correspond to the spongy bone ultimate strength [6].

On the Fig. 2 is shown the dependence of compressive strength vs. microstresses. As one can see strength directly depends on microstresses, this means that grain boundaries are very important in mechanical properties formation [7, 8].

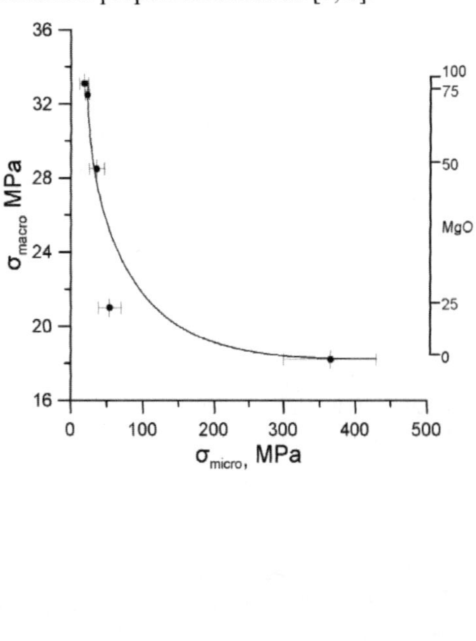

Figure 2. Compressive strength vs.microstresses

Strength, Fig. 3a, and elastic modulus, Fig. 3b, dependence on microstresses in MgO crystalline lattice can be determined as the linear. The dependence of strength vs. microstresses in ZrO_2 crystalline lattice is not obvious and was not found. It can be seen that elastic modulus and strength increasing mostly depends on microstresses in MgO crystal grains.

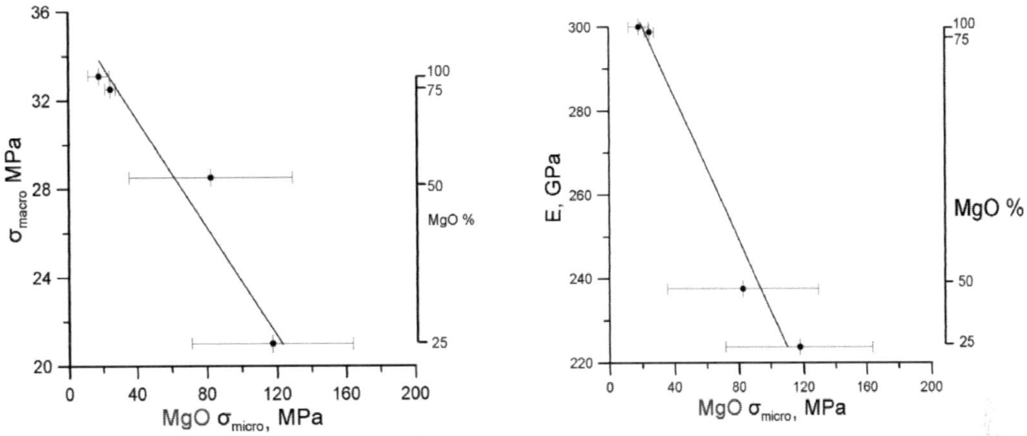

a **b**

Figure 3. Linear dependences from MgO microstresses of: **a**. ceramics compressive strength; **b**. ceramics elastic modulus

Studying of MMSC vitality and osteogenic differentiationability during their cultivating on the porous ceramic samples in osteogenic media was carried out by analysis of material cytotoxicity and alkaline phosphatase detection, which is a first marker of osteogenic differentiation.

Figure 4 a. shows undifferentiated MMSC (dark areas on the picture) in marker-media, which do not express or weakly express alkaline phosphatase and give only background staining. After 7 days of culturing various degrees of cells propagation with the highest activity on the composition with 25% of MgO concentration were observed. Figure 4 b. shows the results after 14 days of cultivation in osteoinductive media: cells differentiate into osteogenic direction, grow, agglomerate and give saturated media staining with colorimetric detection of alkaline phosphatase. The mean cell viability on the ceramic surfaces was about 93%, which is comparable to cell viability in a control sample. In addition, the presence of cell clusters in the pores should be noted, which may be described by their proliferation.

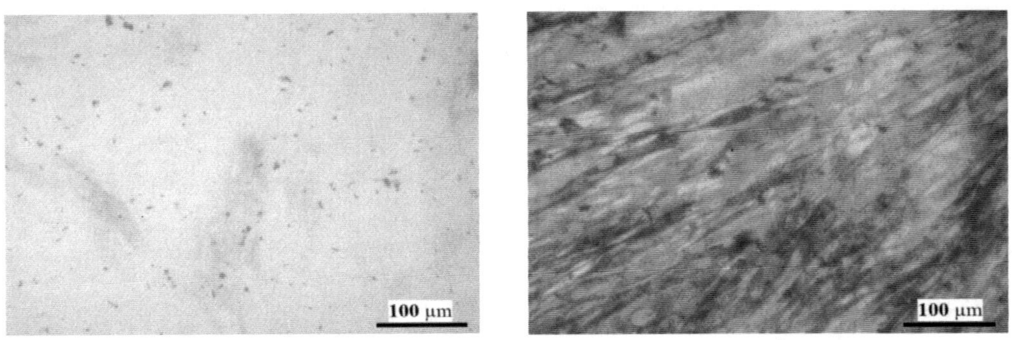

a **b**

Figure 4. Aalkaline phosphatase detection on: **a**. first day of experiment; **b**. 14th experiment day

4. Conclusions

It was shown that during sintering of porous ceramic the bimodal porosity structures with the mean size 28 and 103 μm were formed. Strength directly depends on microstresses and at high microstresses ceramic has a low strength. Ceramics strength dependence vs. microstresses in MgO can be determined as a linear. In vitro studies shown that the tested materials are not cytotoxic - cells vitality was 93%. MMS cells on the surface of ceramic samples also had osteogenic differentiation ability.

Acknowledgments
The research was carried out with the partial financial support of the Ministry of Education and Science of the Russian Federation RFMEFI60714X0069.

References

[1] Hulbert S F, Young F A, Mathews R S, Klawitter J J, Talbert C D, Stelling F H, 1970 *Journal of biomedical materials research* **4** (3) 433
[2] Kulkov S N, Buyakova S P, Gömze L A, 2014 *Építőanyag-JSBCM* **66**, (1) 2 http://dx.doi.org/10.14382/epitoanyag-jsbcm.2014.1
[3] Salerno A, Oliviero M, Di Maio E, Iannace S, Netti P A, 2007 *Journal of applied polymer science* 2007 **106** (5) 3335
[4] Gömze L A 2016 APPLIED MATERIALS SCIENCE I. Compilation of Selected Scientific Papers, *Published by IGREX Ltd,* **ISBN 978-963-12-6600-9** pp 1-189
[5] Currey J D, Butler G, 1975 *J Bone Joint Surg Am* **57** (6) 810
[6] Carter D R, Hayes W C, 1976 *Science* **194** (1) 174
[7] Kumar S, Kurtz S K, Agarwala V K, 1996 *Actamechanica* **114** (4) 203
[8] Kulkov S N, Buyakova S P, Kalatur E S, Buyakov A S, 2016 Épitőanyag–JSBCM **68** (3) 70 http://dx.doi.org/10.14382/epitoanyag-jsbcm.2016.12

IOP Conf. Series: Materials Science and Engineering **175** (2017) 012026 doi:10.1088/1757-899X/175/1/012026

Annealing and structural properties of composite films

L N Kotov[1], V A Ustyugov[1], V S Vlasov[1], V K Turkov[1], M Yu Dianov[1], I V Antonets[1], Yu E Kalinin[2], A V Sitnikov[2], E A Golubev[3]

[1]Syktyvkar State University, Syktyvkar, Russia
[2]Voronezh State Technical University, Voronezh, Russia
[3]Komi Scientific Center of UB RAS, Syktyvkar, Russia

E-mail: kotovln@mail.ru

Abstract. The composite films were investigated by AFM methods before and after annealing. Topographic and phase-contrast AFM images of the composite films at different annealing temperature were obtained. The separate metal granules and larger-scale labyrinth-like formations were described. These formations appear by the process of the film growth, also by film annealing. Strong changes of the structural properties of the films are observed after the percolation transition. The significant changes of the structural properties are connected with nanostructural transformations in the metal granules topology and presence of metal crystal phase.

Introduction

The investigation of thin composite films is a dynamically developing branch of condensed matter physics. The composite films can distinguish nanoscale systems that have significant nonlinear phenomena and quantum effects. Such effects as giant magnetoresistance, quantum Hall effect, strong magneto-optical response are characterized the films [1, 2]. Giant values of the magnetic permeability in the microwave range have been obtained for the composite films [3]. New information of processes occurring in the composite films by nanoscale systems allow to produce materials with specified or controllable magnetic properties, and makes possible to create new high-density magnetic storages and magnetic RAM. At the same time, the relationship of topology structure and the physical properties of the composite films were not studied enough. The structure of the metal-dielectric composite thin films described in article is very inhomogeneous and strongly depends on metal/dielectric concentration ratio. The analysis of the atomic force microscope (AFM) topographic images makes possible to reduce the relationship between the average size of the surface nanoparticles and the metal concentration. Areas of metal (granules) of the films can be in superparamagnetic or ferromagnetic condition depending on film thickness, which complicates the analysis of microwave magnetic properties. In the paper the separate metal granules and larger-scale labyrinth-like formations are described. These formations appear by the process of the film growth, also by film annealing and partial fusion of small granules.

Upon annealing, the structure of the films is significantly modified. Electric and magnetic properties of the composite films also changes with the structure changing [2]. In particular, this leads to a shift of the percolation threshold. The microwave magnetic characteristics (the line width of ferromagnetic resonance (FMR) and the position of the resonance) are changed [3, 5].

Object and investigation methods

This work investigates the structure of composite films, which have a composition of $(Co_{45}Fe_{45}Zr_{10})_x(Al_2O_3)_{1-x}$, where 0.25<x<0.65 (A series). The percolation threshold of these films is about x=0.45 [1]. All samples were prepared by ion-beam sputtering and vacuum deposition of the target material on the polycrystalline glass substrate [3]. The target was composed of a slab (270×70×14 mm^3) of Co-Fe-Zr alloy and was perpendicular to the longitudinal axis dielectric plates (70×9×2 mm^3). It was possible to vary the dielectric width, and consequently to form the dielectric concentration gradient on the substrate. The deposition was performed in the argon atmosphere at pressure P(Ar) = 4×10^{-2} Pa. The deposited layer thickness was determined by the deposition time. Chemical composition and thickness of the films were obtained by scanning electron microscope JSM-6400. The surface topographic and the phase-contrast structure were investigated by atomic force microscope (AFM) ARIS-3500 (Burleigh Instrument Co, USA) with scanning field 70×70 um^2. Surface scanning was carried out using NC10 (NT-MDT) cantilevers. The radius of tip curvature was about 10 nm. The investigations were performed at room temperature. In order to obtain exact metric data the scale calibration was performed using certified test samples.

The films samples were annealed at various temperatures to determine the influence of annealing on the films structure [3,5]. Temperature control was carried out using a chrome-alumel thermocouple. The temperature changed from 530 K to 830 K with the step 50 K, the error was about 2%.

Annealing influence and structure of composite films

Topographic and phase-contrast images of unannealed composite films $(Co_{0.3}Fe_{0.3}Zr_{0.1})_x(Al_2O_3)_y$, where 0.25<x<0.65 were obtained (Fig. 1). The images of films with different ratio of the metal-dielectric are shown on figure. It can be seen that the films are highly heterogeneous structure in which the individual granules cannot be selected. It is also observed for both low and high metal concentration. However, in the case of high concentration of the metal the relief film is less heterogeneous, and it is possible to separate relatively large pores and a hills. A more uniform growth of the film beyond percolation takes place. White spots in the images correspond to the surface pores of the film. Phase-contrast AFM images of the films shows that at low concentration of the metal there is strong mixing of metal and dielectric and there is no separated particles. In contrast, if the metal concentration is high, particles during film deposition growth are combined into labyrinth-like structures with multiple particles (width size and bridges between them up to 0.7 μm).

The annealing of the films in air atmosphere was carried out. Under such conditions the surface of the film is subjected to partial oxidation, but evaporating of the film components is less than by vacuum annealing. It is seen from the Fig.2 after the weak annealing on the film the surface granular structure appears. It is possible to separate the granules with the size of 0.2 μm at low concentrations. However, their positions are irregular and granules are merged into complex shape forming. The structure is more contrast at the high metal concentration. Phase density image shows that original homogeneous mixture of the components breaks down into Al_2O_3 dielectric matrix with small metal inclusions. The average granule size after the annealing in the case of high metal concentration becomes smaller than mean size of the film separate region before the annealing.

The most significant changes observed after annealing for the films at a high temperature (550°C (Fig. 3)). In this case structural phase transition occurs and AFM images. The clear granular structures for both high and low metal concentrations are shown. The regular arrangement of metal particles in a dielectric matrix is observed at low concentration of metal. Thanks to the fact that the melting of the film material leads to a rearrangement of the film structure and the formation of the ellipsoidal shape particles. It should be noted that annealing the crystallinity of the granules takes place. Before annealing the film was amorphous. After a high temperature annealing at t = 550 C metallic crystalline phase grains appear. The size of crystalline regions increases from 2 nm at x = 0.3 to 8 nm at x = 0.7 [1]. The granulated structure of annealed films is become more competitive.

If the metal concentration is high, the average difference in height of topological irregularities remains big because the individual metal particles have the same size and there is "piling up" to each other.

IOP Conf. Series: Materials Science and Engineering **175** (2017) 012026 doi:10.1088/1757-899X/175/1/012026

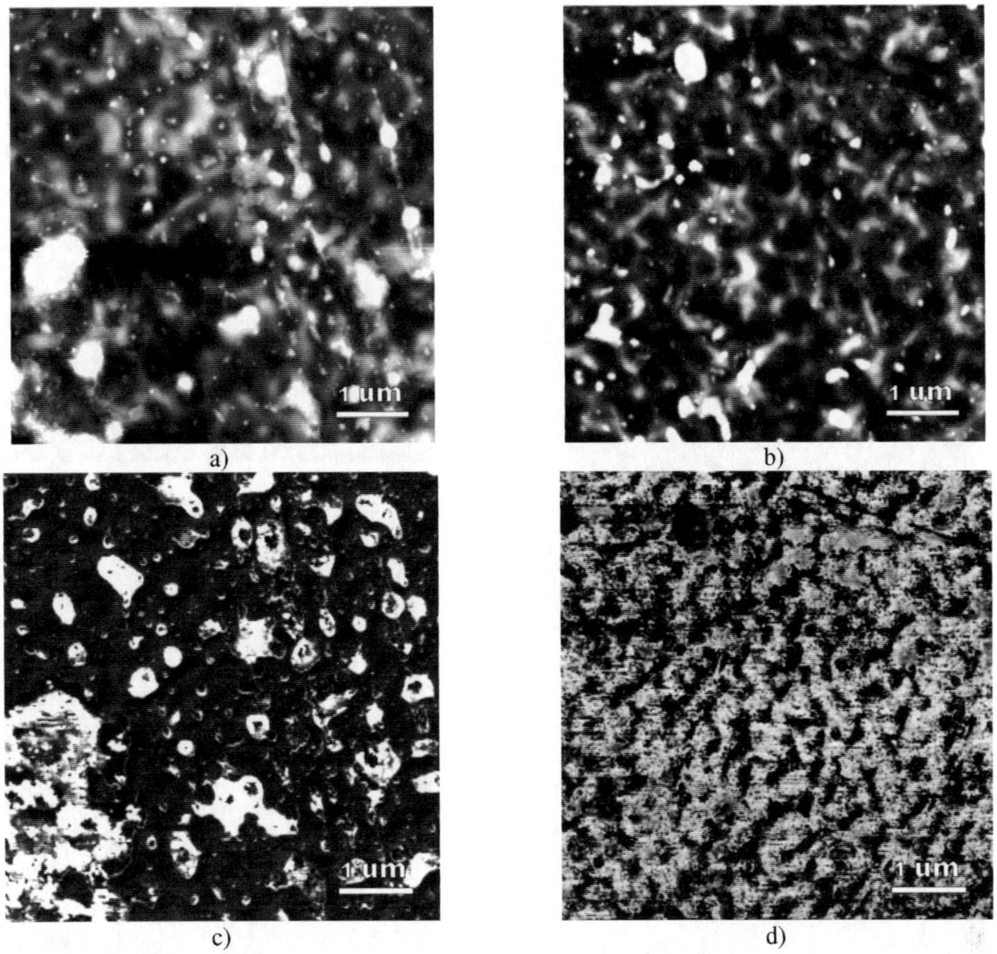

Figure 1. Topographic and phase-contrast AFM images of the unannealed films: x = 0.35 (a, c) and x = 0.65 (b, d)

IOP Conf. Series: Materials Science and Engineering **175** (2017) 012026 doi:10.1088/1757-899X/175/1/012026

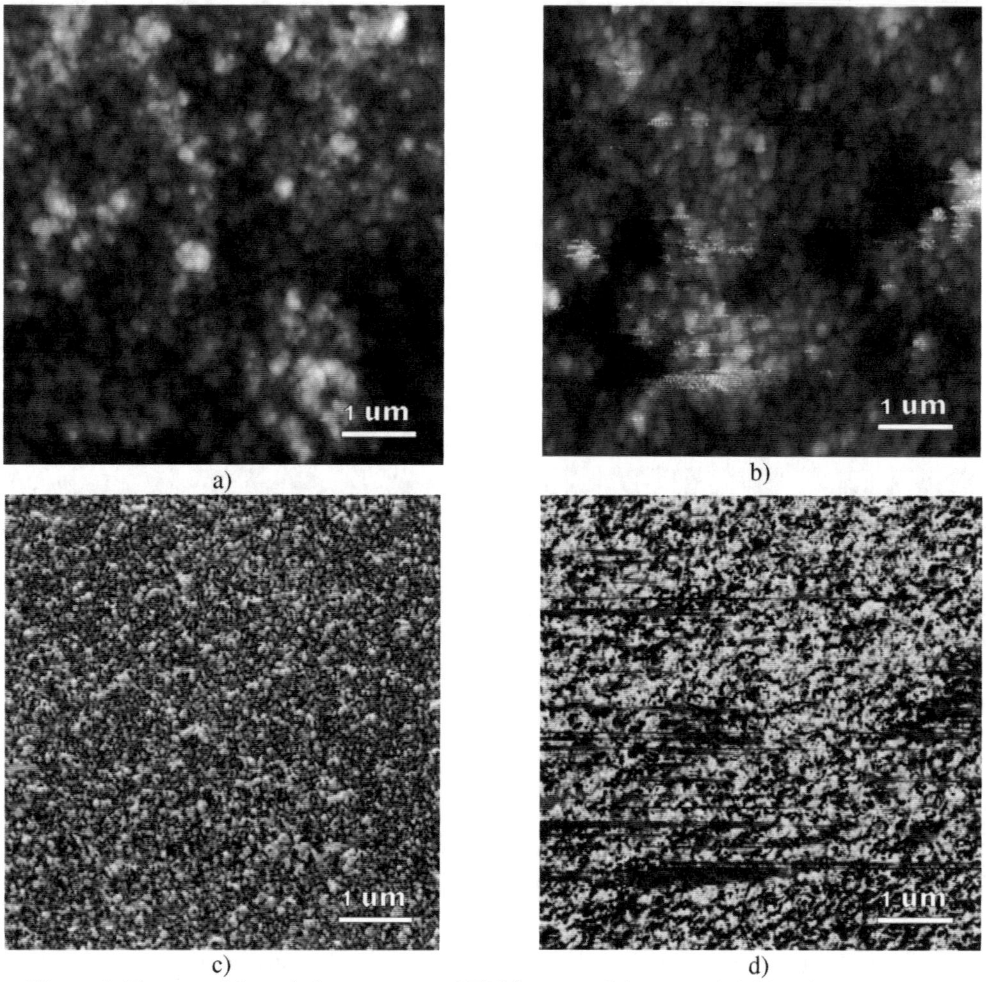

Figure 2. Topographic and phase-contrast AFM images of the annealed films by temperature at 350°C: x = 0.35 (a, c) and x = 0.65 (b, d)

IOP Conf. Series: Materials Science and Engineering **175** (2017) 012026 doi:10.1088/1757-899X/175/1/012026

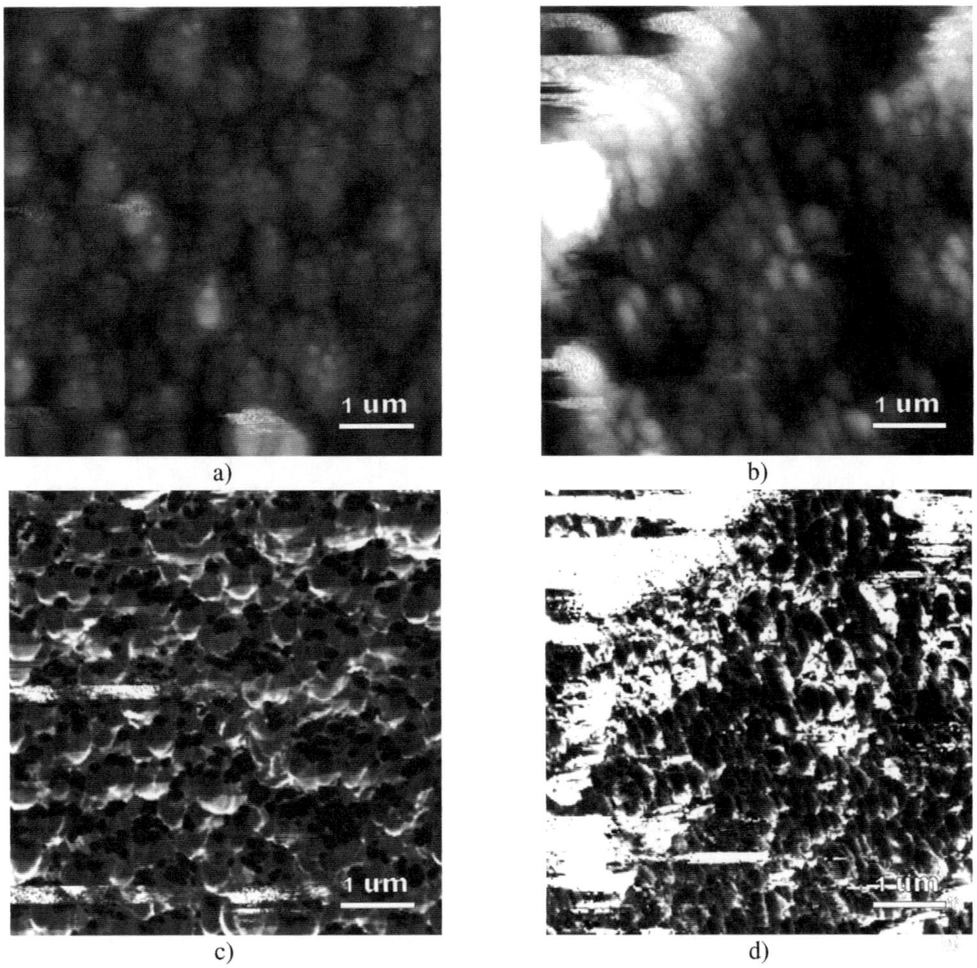

Figure 3. Topographic and phase-contrast AFM images of the film after annealing by temperature of 550°C: x = 0.35 (a, c) и x = 0.65 (b, d).

Conclusion

Thus the composite films before and after annealing were investigated by AFM methods. Topographic and phase-contrast AFM images of the annealed films at different annealing temperature were obtained. Topology and particle size distribution of composite films were investigated in relation to the annealing. The significant changes of the structural properties are connected with nanostructural transformations in the metal granules topology and presence of metal crystal phase. The mean particles radius increases by increasing of the annealing temperature. The particles anisotropy changes as the result of annealing. Changing of the particle anisotropy can occur due to the increase of the crystal field of metal particles.

This work was supported by RFBR (grant 17-02-01138-a).

References

[1] Kotov, L. et al. 2008 *Advanced Materials Research*, 47-50, p. 706.
[2] Kotov, L. et al. 2012 *Journal of Nanoscience and Nanotechnology*, 12, p. 1696.
[3] Kotov, L. et al. 2016 *IV-IC-CMTP. Book of Abstracts*, Hungary p.75.
[4] Kotov, L. et al. 2007 *Journal of Magnetism and Magnetic materials*, 316, P. e20.

[5] Turkov, V. et al. 2013 *Bulletin of the Russian Academy of Sciences. Physics*, 77, p. 1223.
[6] Kotov, L. et al. 2016 *IOP Conf. Series: Materials Science and Engineering* 123, 012009.

Interesting behavior of polymers containing multiwall carbon nanotubes

A Adamne Major

Pallasz Athene Universty,Materials Technology Department,6000 Kecskemet, Hungary

E-mail: major.andrea@gamf.kefo.hu

Abstract. Mixing is a very important polymer process. Nanocomposites were made by a new type of shear mixer, IDMX. The nanocompostes contained different amount of multiwall carbon nanotubes. Test pieces were prepared by injection moulding method. Thermal, flowing and mechanical properties were measured.

1. Introduction

Nowadays polymers have important rule in the field of technological researches. They are used instead of metals or ceramics. Polymeric materials are used to very different purposes. Sometimes we have to produce products with interesting properties and however there is no material which fulfills all of the requirements. In these cases we can produce composites. Composites consist of different types of materials, for example polymer – fiber, polymer – glass, polymer – wood, polymer - ceramics composites.

The new researches are focused on nanomaterials, nanocomposites. The most interesting features can be reached when we produce nanocomposites. Polymer nanocomposites consist of two different parts: the polymer matrix and the nanoparticle. The nanoparticle is often the carbon nanotube.The carbon nanotubes belong to the family of fullerenes. There are single wall and multiwall carbon nanotubes (Figure 1 and 2).

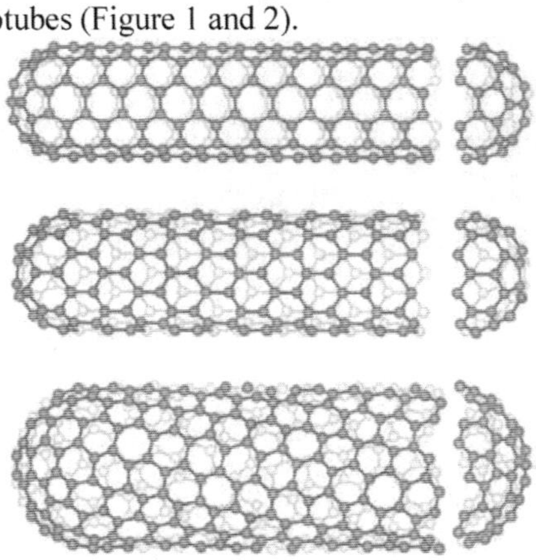

Figure 1. Three different types of single wall carbon nanotube [14]

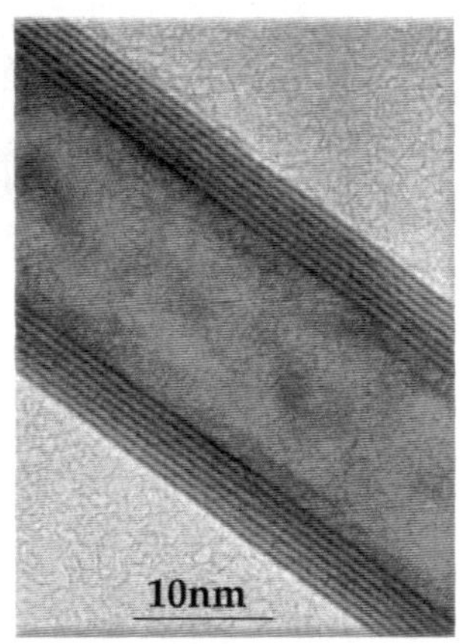

Figure 2. TEM image of multiwall carbon nanotube [15]

IOP Conf. Series: Materials Science and Engineering **175** (2017) 012027 doi:10.1088/1757-899X/175/1/012027

The diameter of the nanotubes is 1-50 nm, the length of them can be longer than 10 μm. The carbon nanotubes were discovered by Iijima in 1991 when he produced fullerenes[1]. From that time carbon nanotubes are in the focus of the researchers of every science. The carbon nanotubes have very interesting mechanical and conductive properties [2-13]. Carbon nanotubes can be used in nanoelectronics and they can be used as detector of trace elements because the conductivity of the nanotube changes when elements are adsorbed on them. They have special mechanical properties; their Young modulus is about 1 TPa which is a higher value than the modulus of steel (about 200 GPa). The carbon nanotubes are very flexible; they can be used as reinforcement in composites.

2. Experimental

During our research we used reground polycarbonate (ANJALON J100V,J&A Plastics GmbH), ABS (POLYMAN HH3, Polyman Plastics Inc.)and multiwall carbon nanotube master blend (MB-3020-01 Hyperion Catalyst, USA). The master blend contained 15% multiwall carbon nanotube and 85% polycarbonate as matrix. Concentration series were produced. The composites contained 1% and 1,5% multiwall carbon nanotubes.

The composites were produced by a new type of shear mixer, called IDMX mixer (Figure 3). Because of its specific geometry this mixer is very efficient. The mixer has got stators (Figure 4) and rotors (Figure 5) which provide dispersive and distributive mixing as well.

Test pieces were injection molded by Arburg Allrounder Injection molding machine.

The mechanical (Charpy and tensile properties) and the flowing (MVR) properties of the nanocomposites were investigated. The structures of the nanocomposites were investigated by SEM and LP-FTIR methods.

Figure 3. IDMX mixer

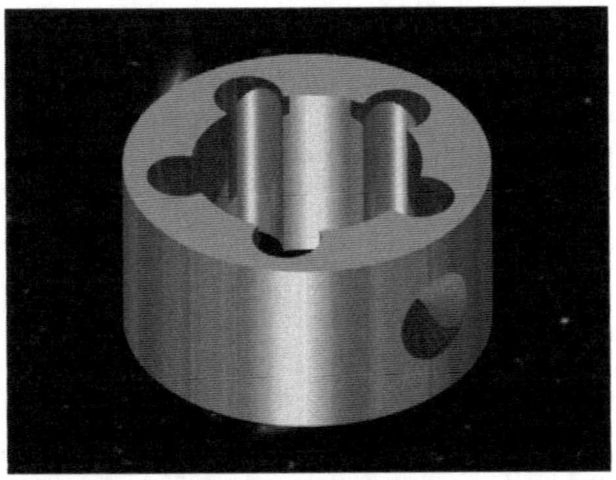

Figure 4. Stator

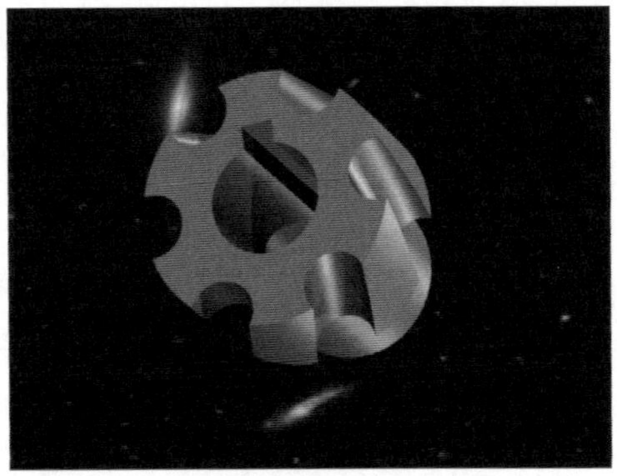

Figure 5. Rotor

3. Results and Discussion

The change of the impact strength of the nanocomposites is shown in Figure 6. It was found that the impact strength of the composites is worse than the impact strength of polycarbonate.The impact strength of the composites containing carbon nanotube was at a lower level than the value of the reference composites. The difference between the 1% carbon nanotube and the 1,5% carbon nanotube containing nanocomposites was not high. The multiwall carbon nanotube content causes lower impact strength.

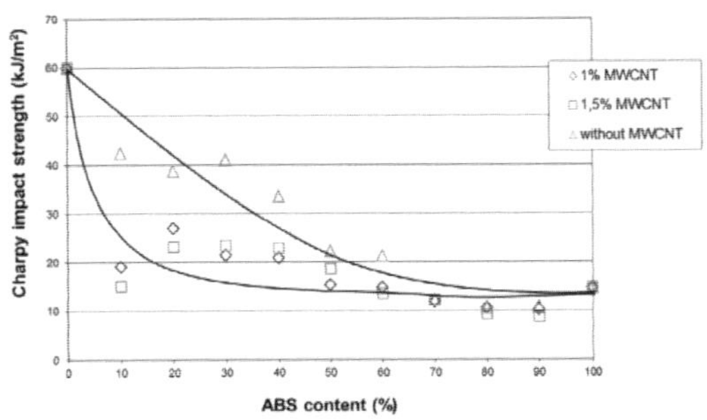

Figure 6. The impact strength of the nanocomposites.
Experiment.

In Figure 7 we can see the change of the yield strength of the nanocomposites.The yield strength of every composite is lower than the pure polycarbonate has. In our experiments, there is no significant difference between the blends having different carbon nanotube content. The change of the yield stress vs. ABS content of the blends it was found that more or less linear correlation exists.

In Figure 8 and Figure 9 the MVR value and the hardness of the different composites are shown respectively.

The flow properties of the blends change according to the composition. It was found that the carbon nanotube content slightly decreases the MVR value of the blend; however this change is not too much.

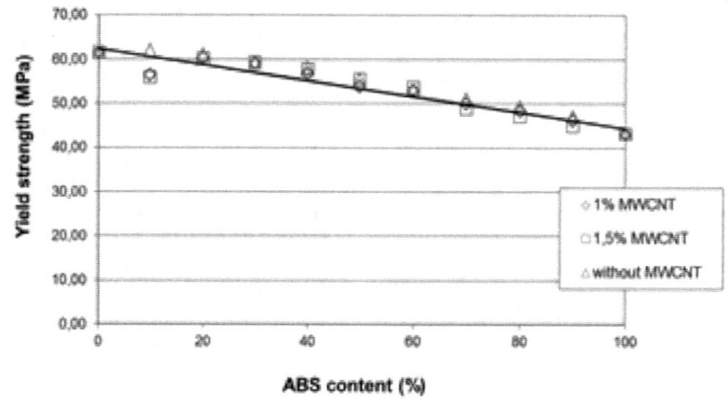

Figure 7. The yield strength of the nanocomposites. Experiment.

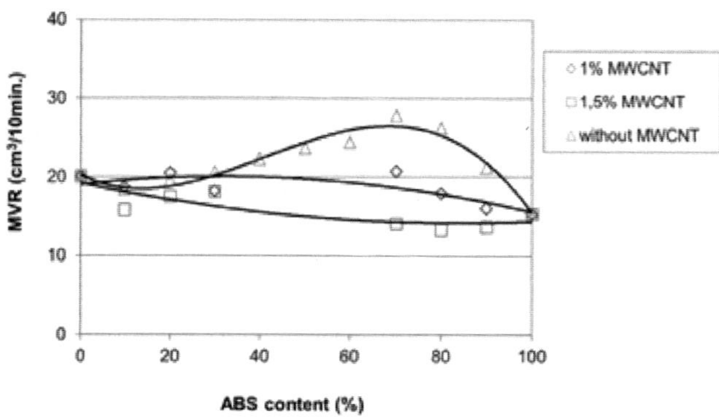

Figure 8. MVR of the nanocomposites. Experiment

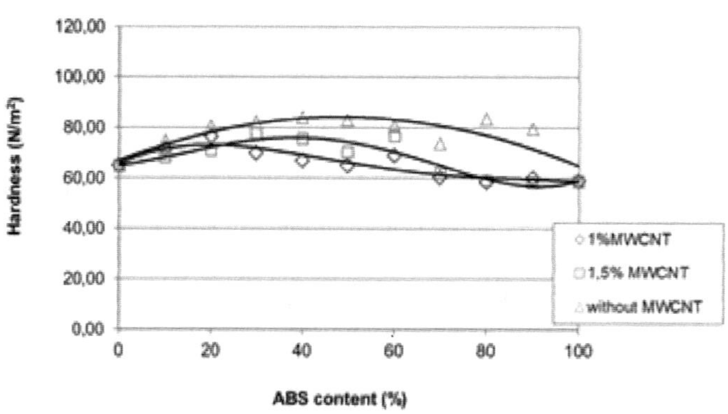

Figure 9. The hardness of the nanocomposites. Experiment.

The hardness of the blends shows only a slight change. Increasing the PC content the hardness increases. However the change in the range up to 20% PC is negligible. The measured hardness is practically the same as the hardness of the pure ABS.

In Figure 10-12 the SEM images of the raw materials and the SEM images of two nanocomposites are shown.

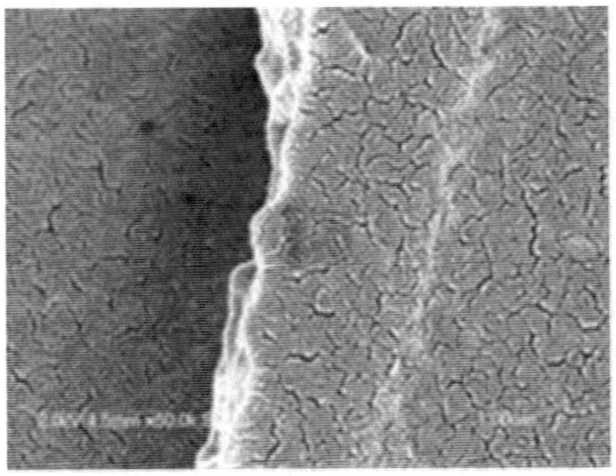

Figure 10. SEM image of polycarbonate. Experiment.

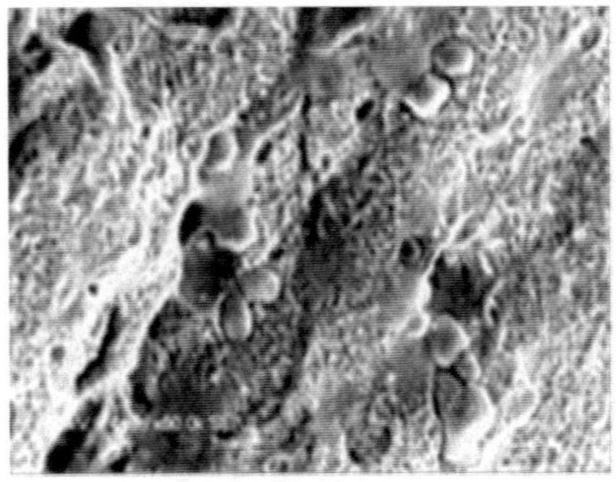

Figure 11. SEM image of ABS. Experiment.

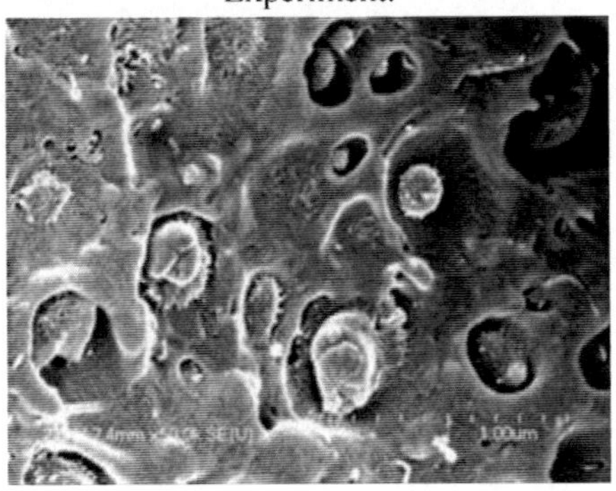

Figure 12. SEM image of nanocomposite containing 1% MWCNT, 89% PC, 10% ABS. Experiment.

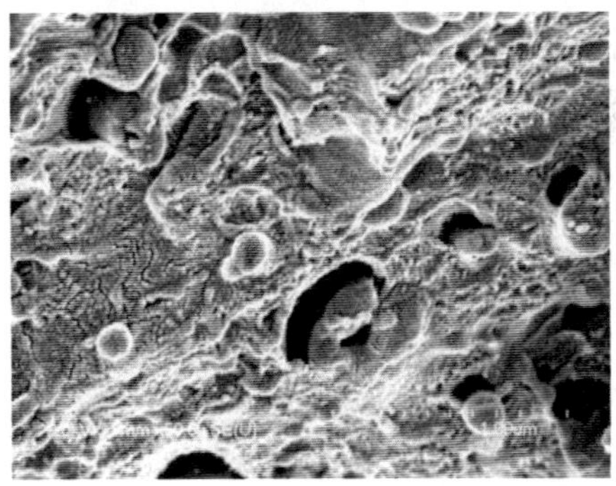

Figure 13. SEM image of nanocomposite containing 1% MWCNT, 19% PC, 80% ABS. Experiment.

It can be seen that the structure of the raw materials and the nanocomposites are very different. The carbon nanotube can be seen on the fractured surfaces. It is important to emphasize that the distribution of the nanotube in the matrix material is more or less uniform. We did not find agglomerates in the materials. Having smaller magnification, it is clear from the SEM investigation that the structure of the pure materials has changed.

We investigated the inner structure of the nanocomposites by LP-FTIR method. We used CO_2 laser and the structure of the remaining material was studied by SEM method. The SEM images of a nanocomposite can be seen in Figure 14 and 15.We found that the remaining structure is different. The structure of the nanotubes can be seen. A continuous physical network can be investigated. This network means a cohesion force in the matrix.

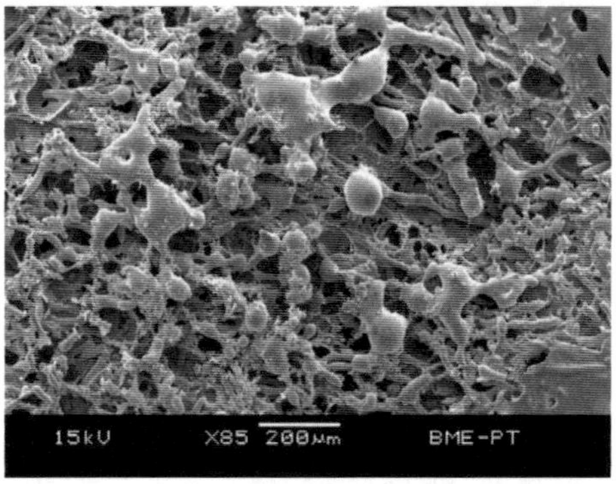

Figure 14. SEM images of nanocomposite containing 20% ABS, 79% PC, 1% MWCNT after LP-FTIR. Experiment.

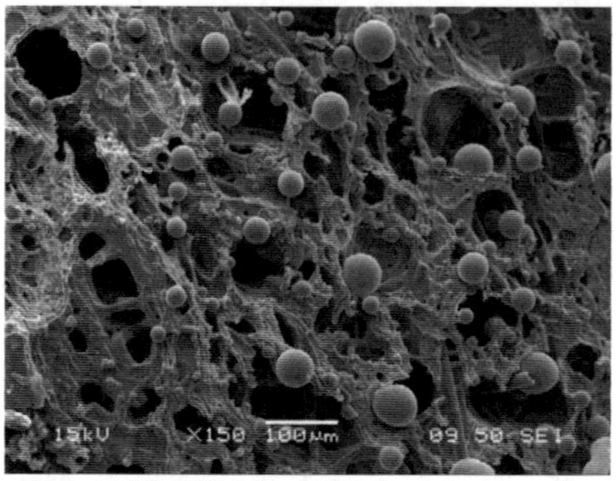

Figure 15. SEM images of nanocomposite containing 20% ABS, 79% PC,1% MWCNT after LP-FTIR. Experiment.

4. Conclusion

Polycarbonate – ABS – multiwall carbon nanotube nanocomposites were prepared by IDMX mixer. Test pieces were injection moulded and investigated. Multiwall carbon nanotubes were mixed in two-phase system, the mixing was efficient. Mechanical and flowing properties were determined.It was found that the carbon nanotube increases slightly the surface hardness of the materials. Using SEM technique we found that the prepared carbon nanotube compositesshow more or less uniform distribution of thecomponents. This structure was supported by the change the yield strength of the nanocomposites. Using LP-FTIR method it was found that the carbon nanotubes continuous physical network.

Acknowledgements

Many thanks to DesiCsongor for the support of the IDMX®mixer.
In memoriam Victor Csongor.

References

[1] S. Iijima: Nature **354**: 56-58, 1991.
[2] M. S. P. Shaffer, A. H. Windle: Adv. Mater, **11**(11): 937-41, 1999.
[3] R. Blake, Y. K. Gun'ko, J. Coleman, M. Cadek, A. Fonseca, J. B. Nagy, et. al.: Am. Chem. Soc., **126**:10226-7, 2004.
[4] L. Xie, F. Xu, F. Qui, H. Lu, Y. Yang: Macromol.**40**:3296-305, 2007.
[5] S. J. Park, M. S. Cho, S. T. Lim, H. J. Choi, M. S. Jhon: Macromol. Rapid Commun., **24**:1070-3, 2003.
[6] D. Qian, E. C. Dickey, R. Andrews, T. Rantell: Appl. Phys. Lett. **76** (20): 2868-70, 2000.
[7] B. Safadi, R. Andrews, E. A Grulke: Appl. Polym. Sci., **84**(14): 2660-9, 2002.
[8] Z. Jin, K. Pramoda, G. Xu, S. H. Goh: Chem. Phys. Lett., **337**(1-3): 43-7, 2001.
[9] R. Andrews, D. Jacques, D. L. Qian, T. Rantell: Acc. Chem. Res. **35**(12): 1008-17, 2002.
[10] P. Pötschke, A. R. Bhattacharyya, A. Janke, H. Goering: Compos. Interf. **10**(4-5): 389-404, 2003.
[11] R. E. Gorga, R. E. Cohen: Polym. Sci. Part B: Polym. Phys. **42**(14): 2690-702, 2004.
[12] O. Meincke, D. Kaempfer, H. Weickmann, C. Friedrich, M. Vathauer, H. Warth: Polymer **45**(3): 739-48, 2004.
[13] T. X. Liu, I. Y. Phang, L. Shen, S. Y. Chow, W. D. Zhang: Macromol. **37**(19): 7214-22, 2004.W. D. Zhang, L. Shen, I. Y. Phang, T. X. Liu: Macromol. **37**(2): 256-9, 2004.
[14] http://www6.ufrgs.br/lacer/gmn/imagens/nano1.jpg
[15] www.nanopicoftheday.org/2003Pics/WS2%20NanotubeTEM.htm

IOP Conf. Series: Materials Science and Engineering **175** (2017) 012028 doi:10.1088/1757-899X/175/1/012028

Combined Mechanical and Electrical Study of Polymers of Biological Origin

G Zsoldos, K Szoda, K Marossy

University of Miskolc Institute of Ceramics and Polymer Engineering, H-3515 Miskolc-Egyetemváros

E-mail: femzsgab@uni-miskolc.hu

Abstract. Thermally Simulated Depolarization Current measurement is an excellent but not widely used method for identifying relaxation processes in polymers. The DMA method is used here to analyze the mechanical changes depend on temperature in biopolymers. The two techniques take advantage of the energy changes involved in the various phase transitions of certain polymer molecules. This allows for several properties of the material to be ascertained; melting points, enthalpies of melting, crystallization temperatures, glass transition temperatures and degradation temperatures. The examined biopolymer films are made from biological materials such as proteins and polysaccharides. These materials have gained wide usage in pharmaceutical, medical and food areas. The uses of biopolymer films depend on their structure and mechanical properties. This work is based on pectin and gelatin films. The films were prepared by casting. The casting technique used aqueous solutions in each case of sample preparation. The manufacturing process of the pectin and gelatin films was a single stage solving process.

1. Introduction

The past few years have witnessed rapidly expanding interest in food additives as sources of biomolecules with potential to replace synthetic polymers in medical applications with biocompatibility, bioactivity, biodegradability for unique applications.

Due to biomolecules having become more available, ever increasing demand for high-performance "natural" matrices for biomedical and pharmaceutical applications such as organ regeneration and tissue engineering, gel like bandages, wound dressing, medical structure line, artificial limb, controlled drug delivery systems, films, contact lenses and capsules for oral ingestion.[1]

Biopolymer films and biofoils are formed from natural polymers, of animal or plant origin, such as polysaccharides, lipids and proteins. The biopolymers are neutral and always renewable, because they are made from plant materials which can be grown indefinitely.

Pectin is s a structural heteropolysaccharide contained in the primary cell walls of terrestrial plants. it is mainly extracted from citrus fruits, and is used in food as a gelling agent, particularly in jams and jellies. It is also used in dessert fillings, medicines, sweets, as a stabilizer in fruit juices and milk drinks, and as a source of dietary fiber [2].

Gelatin is a mixture of peptides and proteins produced by partial hydrolysis of collagen extracted from the bovine, pork fish or poultry skin, bones, and connective tissues of animals. Gelatin is a typical material used in the medical or pharmaceutical fields [3].

2. Experimental section

2.1. Materials

The biopolymer films were produced from pectin and gelatin (food additive) with different percentage of glycerol. For using casting techniques the following describes the method for each material. In the case of pectin and gelatin 2 m/m % aqueous solutions were made at room temperature, and drying in Petri dish in 35°C and 45% humidity. The films were then stored in desiccators at a relative humidity of 25%, for a period of three days to reach the equilibrium moisture in this environment before characterization. After sample preparation examine the movement of the molecules with DMA and TSD method.

2.2. Dynamic Mechanical Analysis

Dynamic mechanical analysis (abbreviated DMA) is a technique used to study and characterize materials. Dynamic Mechanical Test was carried out on strips cut from plates of about 1 mm thickness with Rheometrics Scientific MK-III instrument in dual cantilever mode at 1Hz frequency and 2°C/min heating rate. Glass transition temperatures were determined as the maximum temperature of mechanical loss factor (tg δ) maximum.

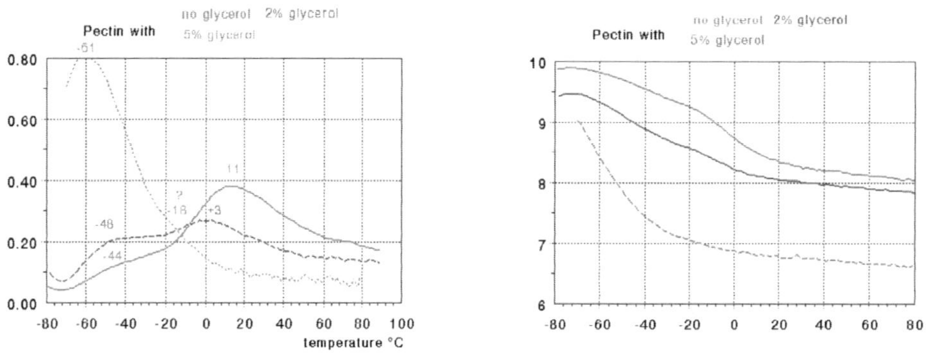

Figure 1. DMA curves (tgδ left, E' right) of pectin

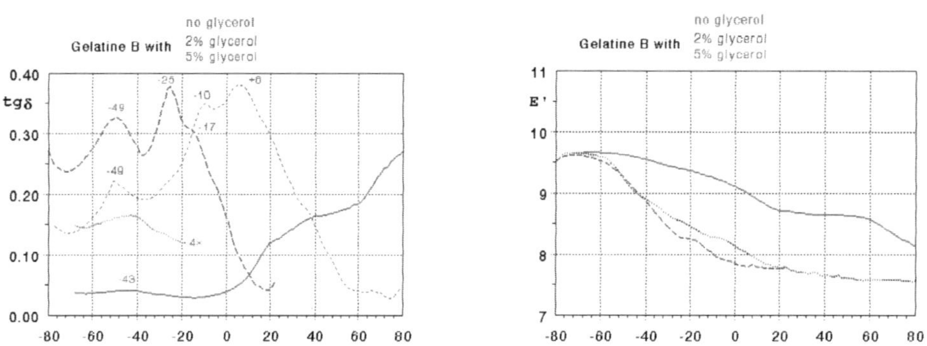

Figure 2. DMA curves (tgδ left, E' right) of gelatin

Both pectin and gelatin (Fig. 1 and Fig.2) has multiple transitions in the investigated temperature range. Interestingly, the transition of lower temperature is not sensitive to presence of glycerol; the higher temperature ones are strongly dependent. The plasticizing effect is very strong for it.

In case of pectin the transition temperature goes below the transition temperature of low peak with 5% glycerol, the transitions are fused into each other.

In case of gelatin at least three transitions are present in the material. Similarly to the pectin the transition of lowest temperature is also almost independent of glycerol content. The high temperature transition is a doublet and the glycerol acts for both process. The peaks are unified without glycerol but split transitions are detected with even low, 2% glycerol concentration.

The nature of low temperature transition is still unknown.

2.3. Thermally Simulated Depolarization Current measurements (TSD)

TSD experiments were carried out on a SETARAM TSC II instrument using helium gas as heat transfer medium.

- The sensitivity of the instrument is 10^{-15}A.
- Disk samples were used with sputter-coated gold electrodes
- Polarizing temperature: 125°C
- Polarization time: 10 minutes
- Polarizing field: 300 V/mm
- Cooling: 5K/min
- Start temperature: -150°C
- Heating rate: 5°C/min (after 10 minutes isothermal at -150°C)
- Final temperature: 125°C

Although the basic principle of the TSDC is simple, there are two difficulties. The first one is technical, measuring extremely low currents, in case of polar polymers through even 5 decades. The second one is the evaluation and interpretation of the depolarization curves. For the analysis of the depolarization processes the method is described in [4], although this paper deals with PVC, i.e. a strongly polar polymer.

Supposing that the relaxation processes follow the Arrhenius law, the relaxation time of dipoles can be calculated as follows

$$\tau_{(T)} = \tau_0 e^{\frac{A}{RT}} \tag{1}$$

where; A is the activation energy of the process.

P_0 polarization is frozen in the polymer during the cooling in electric field. Heating the sample in zero field (short circuited) the decaying polarization at t time is:

$$P_{(t)} = P_0 e^{\left(-\int_0^t \frac{1}{\tau} dt\right)} \tag{2}$$

the liberating charges produce current. The current density is

$$j_{(t)} = -\frac{P_0}{\tau} \exp\left(-\int_0^t \frac{1}{\tau} dt\right) \tag{3}$$

Using a constant b heating rate the temperature of the sample is

$$T = T_0 + bt \tag{4}$$

and the depolarization current density is

$$j_{(T)} = -\frac{P_0}{\tau} \exp\left(-\frac{1}{b}\int_{T_0}^{T} \frac{1}{\tau} dT\right) \tag{5}$$

Replacing the $\tau(T)$ from eq. (1)

$$j_{(T)} = -\left(\frac{P_0}{\tau_0}\right)\exp\left[-\frac{A}{RT} - \frac{1}{b\tau_0}\int_{T_0}^{T}\exp\left(-\frac{A}{RT}\right)dT\right] \tag{6}$$

The equation can be used for I(T) current, too, if the d thickness and A area of the sample is known. Let $C = AP_0/\tau_0$ and $B = A/Rb\tau_0$, replace $A/RT \equiv s$

$$I_{(s)} = C\exp\left\{s - B\left[e^{-s}(s^{-2} - 2s^{-3} + 6s^{-4}...\right]_{s_0}^{s}\right\} \tag{7}$$

This equation describes an asymmetric peak as a function of temperature and the A activation energy can be calculated from the half width of the peak as:

$$A = 2.406R\frac{1}{\frac{1}{T_l} - \frac{1}{T_u}} \tag{8}$$

or both from the maximum and lower half

$$A = 1.443R\frac{1}{\frac{1}{T_l} - \frac{1}{T_m}} \tag{9}$$

and the maximum and upper half

$$A = 0.962R\frac{1}{\frac{1}{T_m} - \frac{1}{T_u}} \tag{10}$$

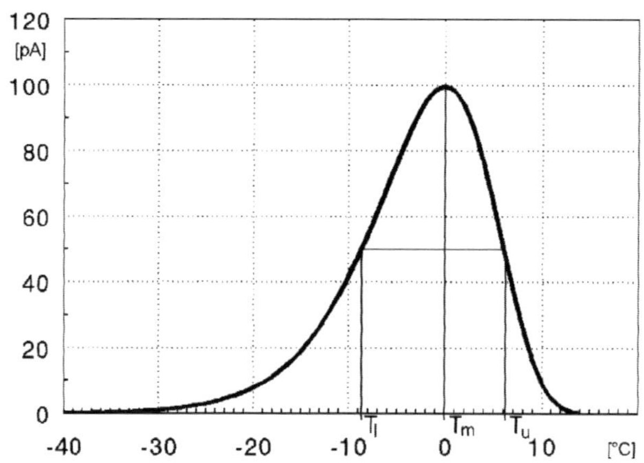

Figure 3. The shape of an ideal depolarization peak calculated by eq. (7) and the temperatures for determination of activation energy

The σ surface charge density can be obtained from the full charge Q and the A area of the sample.

$$\sigma = Q/A \tag{11}$$

The surface charge density can be used for determination of permittivity change of transition called oscillator strength (ε_0-ε_∞)

$$\Delta\varepsilon = \sigma/E\varepsilon_0 \tag{12}$$

This is the most characteristic to the transition, because does not contain dimensions and electric field. For evaluation of depolarization current spectra our previously developed software was used. This program calculates the currents for 1 $kVmm^{-1}$ field and $30cm^2$ electrode area and all current data are picoampers.

In case of zero external field, i.e. in short circuited state the measured depolarization current is the sum of currents of individual processes.

$$I_{d(T)} = \sum_{i=1}^{n} I_{d(T)n} \tag{13}$$

The eq. (8) with (9) or (10) can be used for resolving partly overlapping peaks.

The thermally stimulated discharge current method has effective frequency dependence on the activation energy and temperature range of transition. Although the determination of effective frequency has about 20% uncertainty, the effective frequency of these measurements are in the 10^{-3} Hz range.

$$\nu_{eff} = 0.113\frac{bA}{RT_m^2} \tag{14}$$

Biopolymers are not hard to investigate by depolarization method because of their polar nature. The polarization processes are quick, independent of the mobility of the polymer backbone. However, oxidative degradation and humidity can cause enough polar –C=O groups. Even the glass transition does not produce detectable depolarization current signal in the -110 – 125°C range. It must be mentioned that the depolarization currents and the relaxation strength values are at least two orders of magnitude smaller than in case of polar polymers, e.g. PVC [4].

Pectin has two extremely intensive peaks in the 0-50°C range (Fig. 4 and 5). These peaks might origin from segmental movements but their intensity (I_{max}) is amplified by ionic contamination. Because of the acidic nature of pectin, it contains free charge carriers. Low temperature peaks are also multiple ones showing independent molecular motions in the material.

According to the TSD curve the gelatin also contains more polar components (Fig. 6). There are no extremely high current peaks; the gelatin do not traps ions. The high temperature peak is undoubtedly twin process supporting the DMA results.

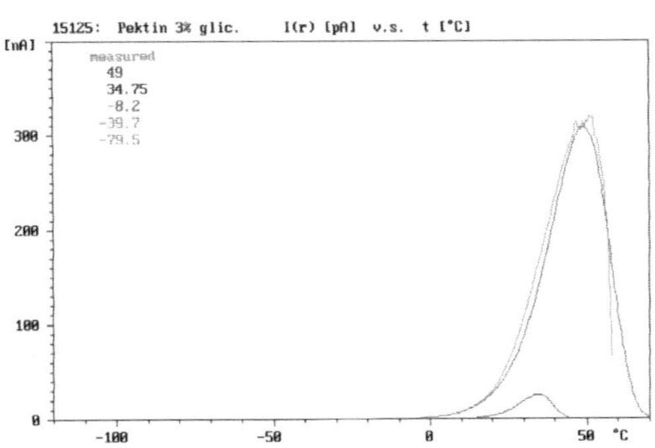

Figure 4. Pectin TSD curve before and after resolution

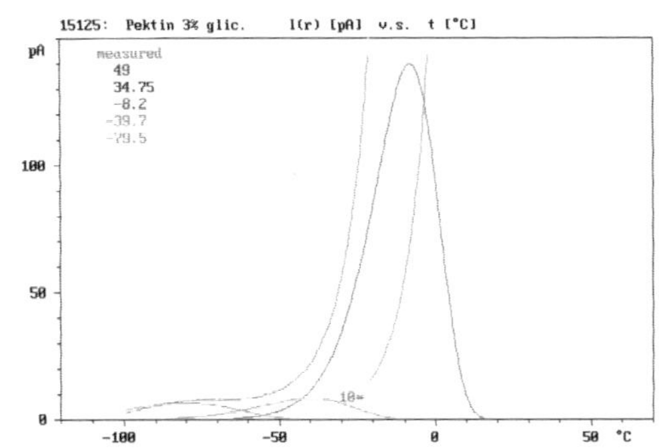

Figure 5. Pectin TSD curve before and after resolution (Low temperature peaks magnified)

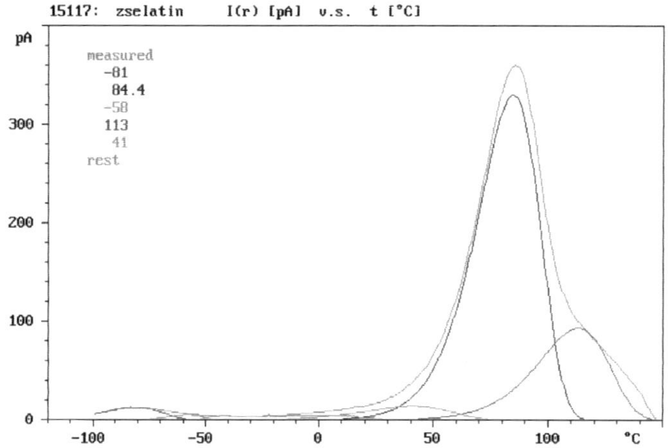

Figure 6. Gelatin TSD curve before and after resolution

TSD spectra were resolved by our own software. Data of resolved transitions are summarized in Table 1.

Table 1. TSD results of pectin and gelatin biofilms

Examined material; peaks	Peak location; ϑmax [°C]	Peak height; I_{max} [pA]	Activation energy A_e [kJ/mol]
Pectin			
1	49	308000	72
2	34.75	25700	150
4	-8.2	140	55
5	-39.7	6.7	30
6	-79.5	6.7	16.5
Gelatin			
1	-81	12.7	23
2	84.4	330	76
4	113	93	78
5	41	14	42

3. Summary

The aim of this work was to perform DMA and TSD measurements pectin and gelatin biofilms. Both techniques revealed that these bio-materials have multiple transitions in the -120 – 80°C temperature range. These transitions are completely different in terms of compatibility with very polar plasticizer, namely the glycerol.

The difficulty of DMA method is that we work at the lowest force limit of the instrument because thin films can be produced only.

The TSD requests thin film specimens, but the method is sensitive to ionic contaminations. The resolving power of TSD is excellent. Using plasticizing additives of different polarity, we intend to

IOP Conf. Series: Materials Science and Engineering **175** (2017) 012028 doi:10.1088/1757-899X/175/1/012028

identify the nature of the transitions found. Both methods are suitable for these investigations and well complement each other.

References

[1] *Handbook of Materials for Medical Devices*, ASM International 2003

[2] Braconnot, Henri: "Recherches sur un nouvel acide universellement répandu dans tous les vegetaux" (Investigations into a new acid spread throughout all plants) 1825*Annales de chimie et de physique*, series 2, **28** 173-178.

[3] *Gelatin Manufacturers* Institute of America: Gelatin Handbook 2012

[4] K. Marossy, “Assessment of activation energies of transitions by thermally stimulated discharge current (TSD) experiments“ 1989 Polymer Bulletin, vol. 22, pp. 213-220

IOP Conf. Series: Materials Science and Engineering **175** (2017) 012029 doi:10.1088/1757-899X/175/1/012029

Numerical simulation of mechanical behaviour and prediction of effective properties of metal matrix composites with consideration for structural evolution under shock wave loading

V V Karakulov[1], I Yu Smolin[1,2] and S N Kulkov[1,2]

[1]National ResearchTomsk State University, 634050 Tomsk, Russia
[2]Institute of Strength Physics and Materials Science SB RAS, 634055 Tomsk, Russia

E-mail: valery@ftf.tsu.ru

Abstract. Mechanical behaviour of stochastic metal-ceramic composite materials under shock wave loading was numerically simulated on mesoscopic scale level. Deformation of mesoscopic volumes of composites whose structure consisted of a metal matrix and randomly distributed ceramic inclusions was simulated. The results of numerical simulation were used for numerical evaluation of effective elastic and strength properties of metal-ceramic materials with different values of volume concentration of ceramic inclusions. The values of the effective mechanical characteristics of investigated materials were obtained, and the character of the dependence of the effective elastic and strength properties on the structure of composites was determined. It is shown that the dependence of the values of the effective elastic moduli on the volume concentration of ceramic inclusions is nonlinear and monotonically increasing. The values of the effective elastic limits increase with increasing concentration of the inclusions, however, for the considered composites, this dependence is not monotonic.

1. Introduction

Now metal-ceramic composite materials are widely used in various fields of industry and are often used under extreme operating conditions [1, 2]. In this connection, it is necessary to adequately numerically evaluate the elastic and strength properties and predict the mechanical behaviour of these materials under intensive energy impacts.

In the simulation of deformation of composite materials, composites are often taken as homogeneous or quasi-homogeneous. However, these materials present a complex of components with different physical and mechanical properties. The components form the structure of composites. The structure and its evolution during deformation can have a significant influence on the mechanical properties and mechanical behaviour of composite materials [3, 4]. However, the degree and nature of this influence is still incompletely understood. Therefore, the problem of studying mechanical properties and predicting mechanical behaviour of composite materials under intensive dynamic loading conditions remains relevant.

In this connection, the present paper aims at the numerical simulation of mechanical behaviour under loading by plane shock waves and study of effective elastic and strength properties of metal-ceramic composites with consideration for structural evolution during deformation of composites.

2. Simulation of mechanical behaviour of the metal-ceramic composite material

The paper employs a plate of the composite material consisting of a matrix and reinforcing inclusions to simulate loading by a plane shock wave. The simulation is performed on a rectangular fragment of the plane section of the plate along the direction of the shock wave front.

The mechanical behaviour of the composite under the considered loading conditions is described using the physical and mathematical model of the two-phase condensed heterogeneous medium with an explicit description of its structure [5-7]. The Johnson–Holmquist model of the damaged elastic-brittle medium is used for the mechanical behaviour of ceramic inclusions and the Johnson–Cook model of the elastic-viscoplastic medium for the aluminium matrix.

The used physical and mathematical model assumes the heterogeneous medium as a set of interrelated structural elements, namely, a matrix and inclusions. Inclusions are arbitrary in shape and randomly distributed in the matrix. Mechanical interaction between structural elements occurs along internal contact surfaces, i.e., matrix - inclusion interfaces. Within interfaces of each structural element the medium is taken as homogeneous and isotropic while in transition through the interface physical and mechanical properties of the medium change abruptly. Geometry of the simulated area and the number of structural elements are so chosen as to determine effective values of parameters of the mechanical state of the medium by averaging of calculated local values. The simulated areas of the two-phase heterogeneous medium with the model structures are shown in figure 1.

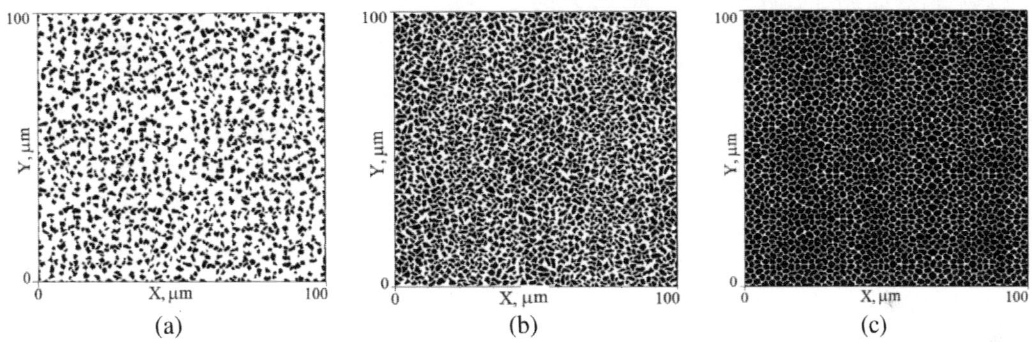

Figure1. Simulated areas of the two-phase heterogeneous medium with the model structure of the composite composed of the matrix (light region) and arbitrary-shaped inclusions (dark regions). The characteristic size of inclusions is 5 μm, the volume concentration is: (a) 25%, (b) 50%, (c) 75%

Effective parameters of the mechanical state of the heterogeneous medium loaded by a plane shock wave are determined by volume averaging of local values of the state parameters in thin flat layers perpendicular to the shock front direction. A method for determining effective parameters of the mechanical state is described in [8, 9].

3. Investigation of the effective elastic and strength properties of metal-ceramic composites

The model of mechanical behaviour of the composite under shock-wave loading and method for determining effective parameters of the state are employed to predict effective mechanical properties of metal-ceramic composites Al-B_4C and Al-SiC. Composites contain 25%, 50% and 75% ceramic inclusions by volume.

The simulation results for propagation of plane shock waves in metal-ceramic composites are drawn on to plot dependencies of effective shock wave velocity $\langle D \rangle$ on effective mass velocity $\langle Uy \rangle$ (figure 2). The plotted dependences in the considered range of shock compression are linear $\langle D \rangle = C_B + \lambda \langle Uy \rangle$, which agrees well with the experimental and theoretical data. The results prove that concentration of ceramic inclusions affects bulk sound velocity C_B and proportionality factor λ.

IOP Conf. Series: Materials Science and Engineering **175** (2017) 012029 doi:10.1088/1757-899X/175/1/012029

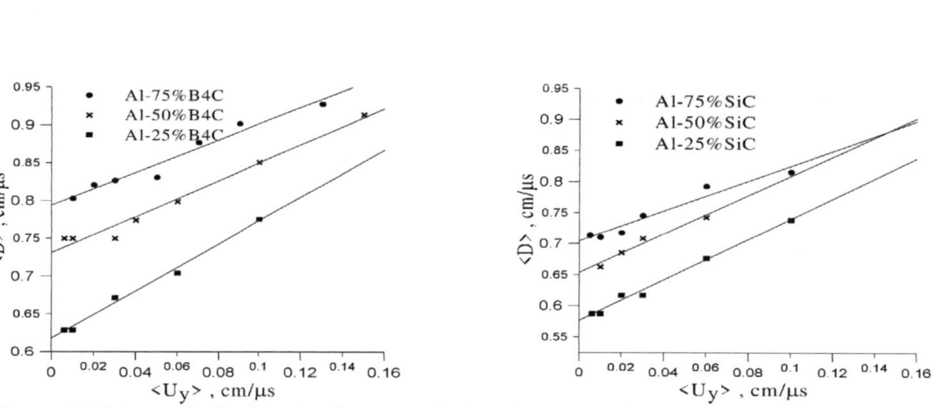

Figure 2. Values of effective shock wave velocity<D>versus effective mass velocity<Uy>in composite materials. Solid lines exhibit linear approximation of the simulation results

The numerical simulation of the loading by a plane shock wave of composites makes it possible to determine values of effective bulk, longitudinal and shear sound velocities of investigated materials. The simulation data on effective sound velocities are used for determining values of effective elastic modules of composites under normal conditions.The dependences of values of the effective bulk modulus on volume concentration of ceramic inclusions are presented in figure 3.

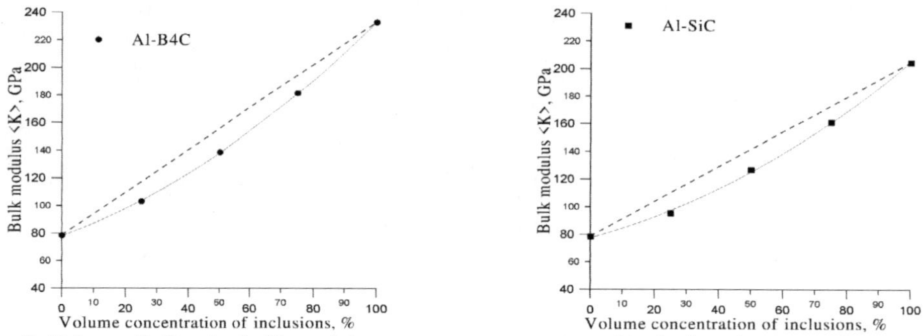

Figure3. Values of effective bulk modulus versus concentration of inclusions in composite materials. Solid lines exhibit the square approximation of the simulation results, dotted lines exhibit linear approximation of component's properties of investigated composites

Values of the effective shear modulus in dependence on volume concentration of ceramic inclusions are presented in figure 4.

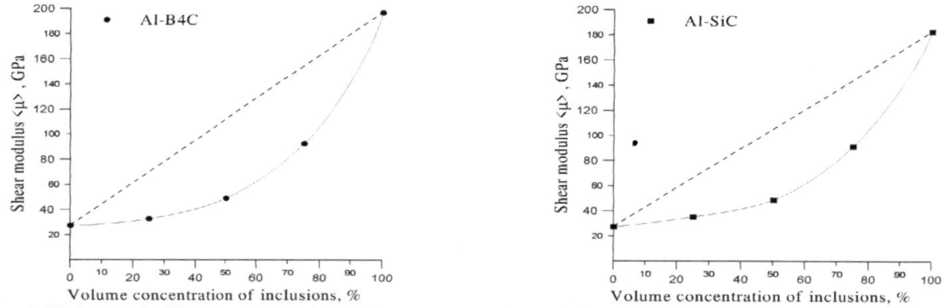

Figure 4. Values of effective shear modulus versus concentration of inclusions in composite materials. Solid lines exhibit the square approximation of the simulation results, dotted lines are linear approximation of component's properties of investigated composites

The influence of the concentration of ceramic inclusions on values of the effective Hugoniot elastic limit is also studied.

Values of the effective Hugoniot elastic limit in dependence on volume concentration of ceramic inclusions are presented in figure 5.

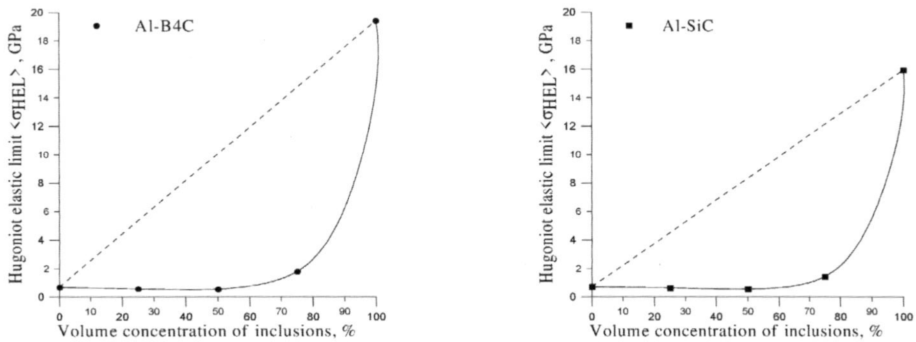

Figure 5. Calculated values of the effective Hugoniot elastic limit of composites as a function of volume concentration of ceramic inclusions. Solid lines exhibit the approximation of the simulation results, dotted lines exhibit linear approximation of component's properties of investigated composites

The effective Hugoniot elastic limit increases non-uniformly with growing concentration of inclusions, and its increment is maximal at high concentrations of inclusions.

4. Conclusion

The simulation results show that values of effective mechanical properties of the studied composites under shock-wave loading depend strongly on volume concentration of ceramic inclusions. The dependence of values of effective bulk and shear moduli of metal-ceramic composites on volume concentration of reinforcing ceramic inclusions is nonlinear and monotonically increasing.

The effective Hugoniot elastic limit demonstrates a weak dependence on the concentration of reinforcing inclusions up to 70%. This is explained by the fact that inelastic deformation develops at the shock front due to plastic flow of the aluminium matrix. At the inclusion concentration above 70% composites demonstrates a framework structure characterized by a negligible thickness of matrix layers or direct contact between particles. In this case, plastic deformation of the matrix is unable to ensure relaxation of increasing shear stresses. This causes the effective Hugoniot elastic limit to increase significantly, but not higher than the elastic limit in the corresponding polycrystalline ceramics.

References

[1] L A Gömze and L N Gömze 2010 *Építőanyag-JSBCM* **62** (4) 98 http://dx.doi.org/10.14382/epitoanyag-jsbcm.2010.18
[2] L A Gömze and L N Gömze 2013 *IOP Conf. Ser.: Mater. Sci. Eng.* **47** 012033 http://dx.doi.org/10.1088/1757-899X/47/1/012033
[3] Li Y, Ramesh K and Chin E 2000 *Int. J. Solids Struct.***37.51** 7547
[4] Zhou Z, Wu G, Jiang L, Li R and Xu Z 2014*Mater. Design* **63** 658
[5] Platova T M, Skripnyak V A and Karakulov V V 1995 *Vychislitel'niye Tekhnologii* **4**(1), 200
[6] Skripnyak V A and Karakulov V V 2004*Fiz. Mezomekh.*7(Spec.Iss., Part I) 329
[7] Skripnyak V A and Karakulov V V 2005*Vestnik Tomsk. Gos. Univer.* **50** 7
[8] Karakulov V V, Smolin I Yu and Skripnyak V A 2013*Vestnik Tomsk. Gos. Univer. Mat. Mekh* **4** 70
[9] Karakulov V V, Smolin I Yu and Skripnyak V A 2014 *AIP Conference Proceedings* vol 1623, ed V E Panin *et al* (Melville, NY: AIP Publishing) p 237

Thermodynamic parameters of mixtures with allowance for phase transition components under shock-wave loading

S A Kinelovskii and K K Maevskii

Lavrent'ev Institute of Hydrodynamics, Siberian Branch, Russian Academy of Sciences, 630090Novosibirsk, Russia

E-mail: skin@hydro.nsc.ru

Abstract. The shock-wave synthesis and compaction using powder mixtures are the one of perspective directions of new materials creation. The results of numerical experiments on modeling of shock wave loading of mixtures with allowance for phase transition components in their composition are presented. The significant change in volume in the region of phase transition components included in the mixtures allows us to expand the range of variation of thermodynamic parameters of the mixtures under shock wave loading. The calculation model is based on the assumption that all components of mixture under shock-wave loading are in thermodynamic equilibrium (model TEC). The model TEC allows us to describe the region of the polymorphic phase transition, considering the material in the region of phase transition as a mixture of low-pressure phase and high-pressure phase. The good agreement of these model calculations with the data of different authors defined on the basis of experiments is obtained. Thermodynamic parameters of the nitrides mixture, solid and porous mixtures with quartz as component were reliably described. This model is useful for determining the compositions and volume fractions of the components of the mixture to obtain the specified parameters of solid and porous materials under shock-wave loading.

1. Introduction

The possibility of changes in the structure and phase composition of mixtures under the action of shock waves is of interest for the creation of materials with the desired properties. Materials experiencing a phase transition under shock wave loading are of particular interest when solving this task. On the one hand, such materials, e.g., quartz, graphite, bismuth and others have a high compression ratio that allows you to expand the field of the feasible thermodynamic parameters [1, 2]. This gives the opportunity to purposefully achieve the conditions necessary for the production of ceramics with the given properties. On the other hand, many materials with unique properties are also experiencing a phase transition, which further stimulates interest in such research. There are different methods for modeling the thermodynamic parameters of shock-wave loading of mixtures [3], which, however, does not describe the data obtained on the basis of the experiment in the whole range of parameters, in contrast to the model TEC [4].

The interest in the research of compressibility of quartz mixtures is related to the properties of quartz and the possibility of creating materials with the required properties (heat-resistant and high-strength ceramics). The wide distribution of quartz in nature has led to much attention to it. The mass fraction of quartz is more than 60% taking into account mix and silicates in the earth's crust; the free quartz content is 12 %. As a consequence, there is the large base of experimental data on shock-wave loading on quartz and mixture with it that allows you to check the accuracy of the simulation.

Nitrides are often refractory and resistant to stress at high temperatures substances. Nitride coatings are used in energy, space technology, imparts hardness and corrosion resistance of products. The advantage of aluminum nitride AlN over other materials is the unique combination of its physical and electrical characteristics: such as the high thermal conductivity, good insulating properties, the moderate coefficient of thermal expansion at relatively low cost. Recent interest in silicon nitride Si_3N_4 is associated with the discovery of its high-pressure modification with a cubic structure (γ-phase) which has a considerable hardness comparable to that of stishovite—the high-pressure phase of SiO_2. Some researchers believe that stishovite takes the third place in this parameter, after diamond and cubic boron nitride. This feature of the γ-phase allows considering it as a promising technological material, though very high pressures more than 15 GPa and temperatures more than 1000 K required for the formation of γ-Si_3N_4 are a major obstacle not only to industrial use, but also to a detailed study of its properties. Therefore, the selection of optimal conditions under which it would be possible to synthesize relatively large amounts of this substance with a high output at acceptable values of external parameters (pressure and temperature) is a topical task. High temperature is important along with pressure in the formation of γ-Si_3N_4. The promising method of high-temperature shock compression is used to research phase transition in silicon nitride and achievements of γ phase [5, 6]. This method allows increasing the substance's temperature during compression, thereby promoting the transformation, and significantly reducing the residual temperature after the pressure drop, thereby ensuring quenching of the new phase. The effect is achieved by addition of components with high compressibility and specific heat, such as alkali halide salts, to the samples.

The model TEC that used in this work can reliably describe the thermodynamic parameters of the mixture under shock wave loading, including the region of phase transition of components. It should be noted, that the model TEC uses the parameters of the equation of state of the components and hence allows considering the interaction of mixture components, unlike the mixed method, which is mainly used for high-density mixtures [6].

2. Calculation model

The model TEC is used to describe the thermodynamic parameters of porous solids and mixtures of powders under shock wave loading [1, 4]. The equation of state of the Mie–Grüneisen type is applied when modeling the thermodynamic parameters of condensed phases. The initial internal energy and pressure of the substances are zero under normal conditions, taking into account the fact, that field of use of this model for pressures is over 5 GPa. Hence, the equation of state for condensed component has the form with the current and initial densities ρ and ρ_0, pressure P, specific heat c_v, and the current and initial temperatures T and T_0:

$$P(\rho,T) = A\left[(\rho/\rho_0)^k - 1\right] + \Gamma(T)c_v(T - T_0)\rho \tag{1}$$

The function $\Gamma(T)$ has the following form:

$$\Gamma(T) = \left\{[\Gamma(T_0) - \Gamma(T_\infty)]^{-1} + C(T - T_0)\right\}^{-1} + \Gamma(T_\infty) \tag{2}$$

$$C = \left\{[\Gamma(T_*) - \Gamma_\infty)]^{-1} - [\Gamma(T_0) - \Gamma_\infty]^{-1}\right\}(T_* - T_0)^{-1} \tag{3}$$

The parameters C, $\Gamma(T_\infty)$ are chosen from the condition of compliance of the calculated shock adiabats with known experimental results for every component, so that:

$$\Gamma(T) \rightarrow \Gamma(T_0) \text{ when } T \rightarrow T_0,$$
$$\Gamma(T) \rightarrow \Gamma(T_\infty) \text{ when } T \rightarrow T_\infty.$$

Here value of $\Gamma(T_0)$ is taken on the basis of known data at normal conditions. The coefficient C, which allows describing the experimental points upon moderate compression, is determined by the intermediate value $\Gamma(T_*)$ at temperature $T=T_*$. The asymptotic value $\Gamma(T_\infty)$ corresponds to the maximum temperatures. For the gas, we use the equation of state of ideal gas. We write the conditions of dynamic compatibility on the shock wave front, which are laws of conservation of the mass flux for each component of the mixture and laws of conservation of momentum and energy fluxes for the mixture as a whole [1, 4]. The resultant equations combined with the equation of state for each component are sufficient to find dependences of the type of $P(U)$ or $D(U)$, which can be treated as the shock adiabats of the multicomponent mixture. U and D are the mass and wave velocity, respectively. A, k, and Γ are the coefficients in the equations of state of the condensed component. For the mixture of n condensed components with the initial volume fraction μ_{n0} we can obtain the expression:

$$P=\left\{\sum_{i=1}^{n} A_i \frac{\mu_{i0}}{\sigma_i}\left[\left(h_i-\frac{k_i+1}{k_i-1}\right)\sigma_i^{n_i}+\frac{2k_i\sigma_i}{k_i-1}-h_i-1\right]\right\}\left[\sum_{i=1}^{n}\frac{\mu_{i0}}{\sigma_i}h_i+\left(\frac{h_g}{\sigma_g}\right)\left(1-\sum_{i=1}^{n}\mu_{i0}\right)-1\right]^{-1} \quad (4)$$

Here $h_i=2/\Gamma_i+1$, $i=1\ldots n$, $h_g=2/(\gamma-1)+1$. $\sigma_i=\rho_i/\rho_{i0}$, $\sigma_g=\rho_g/\rho_{g0}$ is the compression ratio of components, ρ_g, ρ_{g0} current and initial gas densities, $\gamma=1.41$ is the ratio of specific heat. In view of equal temperatures of all components, including gas, we finally have an $n+1$ equation for $n+2$ variables P, σ_i $(i=1\ldots n)$, σ_g, which allows us to construct a shock adiabat of mixture. A pure substance is considered as a mixture with one condensed component. The volume fraction of gas is assumed zero for calculation of parameters of solid material.

The model TEC may describe the behavior of the materials including the phase transition region. The material is considered as the mixture of low-pressure phase and high-pressure phase in this area. We write the conditions of dynamic compatibility on the shock wave front taking into account the phase transition. It is assumed that the volume fraction α of low-pressure phase passed into high-pressure phase in the region of polymorphic phase transition. The dependence of low-pressure phase, turning into the high-pressure phase, we can specify the following expression $\alpha=k\Delta E$ in the first approximation in the phase transition region. Here $k=(E_f-E_b)^{-1}$, E_b is internal energy at the beginning of the phase transition, E is current internal energy, $\Delta E=E_f-E_b$, E_f is internal energy at the finish of the phase transition, when there are full transition of low-pressure phase to high-pressure phase. The comparison of the calculations with the experimental data has shown that the value of k, defined for one of the porosity, allows one to describe the results for other values of the porosity. As the result, the three regions are set depending on the value of α that determines the volume of low-pressure phase and high-pressure phase of the shock adiabats for the materials with the phase transition [7]:

$$\begin{cases} E\le E_0 & \alpha=0 \\ E_0\le E\le E_k & \alpha=k\Delta E \\ E\ge E_k & \alpha=1 \end{cases}$$

3. Numerical results

The accuracy of the modeling materials used to create materials with desired properties, with the components experiencing a phase transition under shock wave loading, is confirmed, in particular, calculations of the shock adiabats for mixtures of quartz. Calculations of thermodynamic parameters of the mixtures of quartz with aluminum mass fractions wt. % Al (30) SiO_2 (70), Al (40) SiO_2 (60), Al (50) SiO_2 (50) are shown in figure 1. The calculation for pure quartz is given on the same figure for illustration. The possibility of accounting for the phase transition during shock-wave loading for the first time allows us to describe all the data obtained on the basis of experiments for all the considered mixtures.

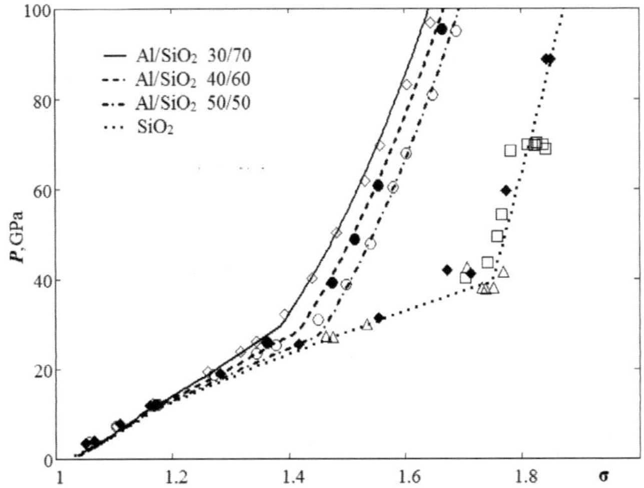

Figure 1. Shock adiabats in coordinates pressure – compression. The calculation and experimental results for the mixtures of quartz with aluminum of various compositions, and pure quartz. Experimental data from [8]

The model TEC allows describing the shock adiabats in the region of the phase transition of quartz mixtures with the significantly different components, in particular by density. Calculations of the mixtures of quartz with tungsten and quartz with teflon, as well as data obtained on the basis of the experiments [8] are shown in figure 2. The calculations were performed for the mixtures of following compositions: mixture of quartz with tungsten (density 10.19 g/cm^3), weight fraction wt.% W(88) SiO2 (12) and mixture of quartz with teflon wt.% SiO2 (55) Teflon (45), (density 2.38 g/cm^3). The calculations allow describing the data obtained on the basis of experiments with good accuracy.

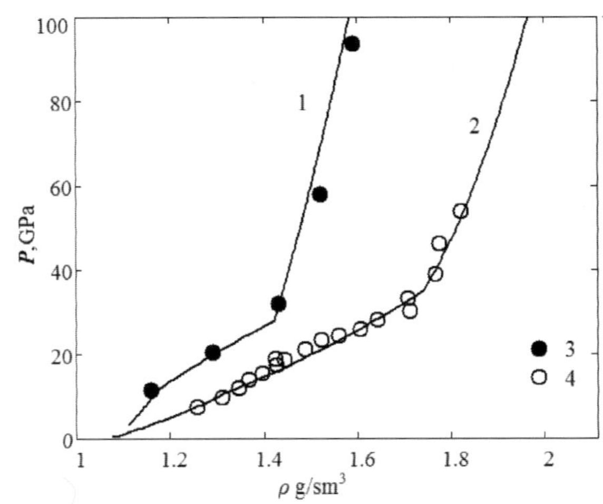

Figure 2. . Shock adiabats of mixture of quartz with tungsten (1), mixture of quartz with teflon (2)
Experimental data (3), (4) from [8]

The model parameters allowing reliable description of the thermodynamic parameters of pure nitrides under shock wave loading are used to simulate the parameters of the mixture of nitrides as components [7].

The experimental data [9] for aluminum nitride AlN (initial density 3.23 g/cm^3) and the data [10] for silicon nitride Si_3N_4 (initial density 3.44 g/cm^3) are used in the calculations of the shock wave loading of these materials with the phase transition. The shock adiabats are shown in figure 3 (a) with the shift for clarity. The initial pressures of the phase transition for AlN and for Si_3N_4 are prescribed at 20 and 30 GPa respectively. The transition of β-phase into highly-dense c-phase is reproduced in the calculations for Si_3N_4. The first three experimental points respective to the *a*-phase (corresponding to plastic deformation) did not take into account as it was done in [10]. The calculations show that the model TEC allows one to describe the unusual behavior of the shock adiabats [9] in the phase transition region (reduction of wave velocity with increasing of mass velocity).

The results for the mixture of AlN and Si_3N_4 with equal volume fractions and initial density 3.335 g/cm^3 are shown in figure 3 (b). Here the experimental data for pure AlN and Si3N4 [9, 10] are shown for the comparison. The beginning of the phase transition for AlN and Si_3N_4 in the mixture are determined at the same pressure as that for pure substances.

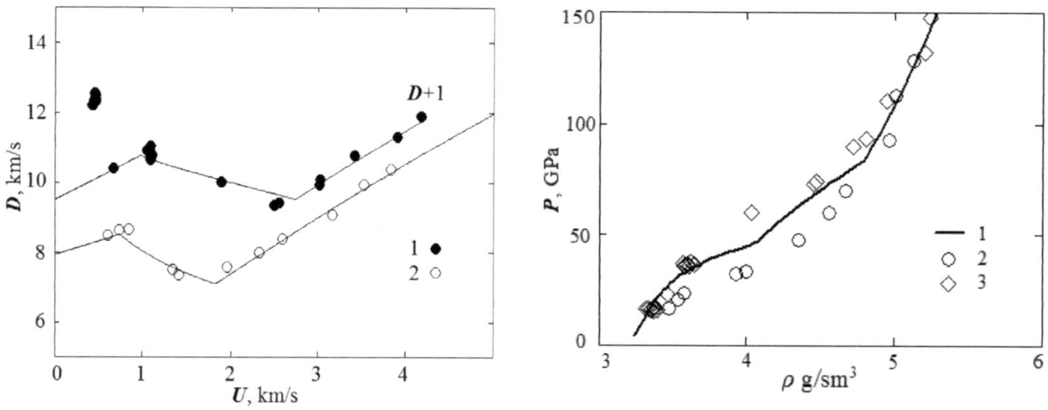

(a) (b)

Figure 3. Shock adiabats of silicon nitride (1) and nitride of aluminum (2) in the coordinates of wave and mass velocity (a). Shock adiabats of the mixture AlN and Si_3N_4 (1), experimental data of silicon nitride (2) [10] and aluminum nitride (3) [9] in the coordinates pressure and density (b)

The calculations for the mixtures of oxides are also performed on account of a large interest in silicon nitride and its compounds. Thermodynamic parameters are simulated for mixtures of Si_3N_4 with oxide, which were obtained on the basis of the experiments data [11]. The calculation for the mixture of Si_3N_4 and periclase MgO is shown in figure 4. MgO is used in industry for the production of refractories, including very fine abrasives for cleaning surfaces, in particular in electronics. The simulated mixture is characterized by the weight fraction wt % Si_3N_4 (95) MgO (5) and the density 3.164 g/cm^3

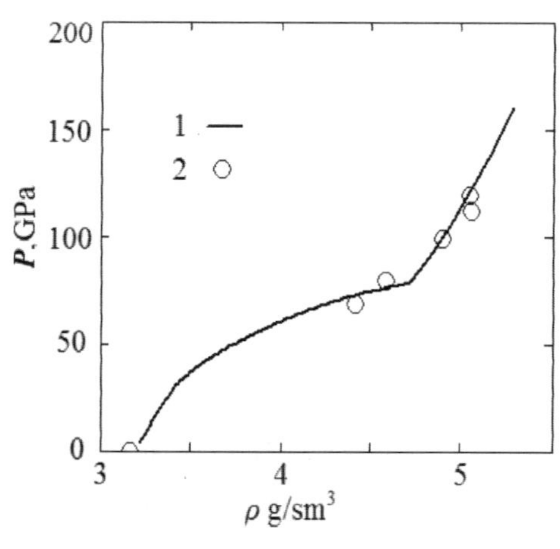

Figure 4. Calculation for the mixture of Si_3N_4 and MgO with the phase transition (1). Experimental data (2) [11]

The results show the accuracy of the description of thermodynamic parameters of the mixtures of this type under shock wave loading. The prospects of using materials with phase transition are shown taking into account their unique properties and possibility to change the values of pressure and temperature of mixtures in a wider range of values. This approach extends the range of achievable values of thermodynamic parameters.

4. Conclusion

The data of the model TEC calculations correspond well to the experimental data by various authors. The model parameters allow us to reliably describe the behavior of pure materials under shock wave loading were used to simulate thermodynamic parameters of mixtures with components experiencing a phase transition under shock wave loading. It was shown that the TEC model allows us to simulate thermodynamic parameters of mixtures under shock wave loading including mixtures that contain components experiencing phase transitions. The simulation results allow determining the ratio of mixture components with the aim of obtaining specified parameters of solid and porous materials after their exposure to shock waves for the synthesis of ceramics with desired properties.

References

[1] Kinelovskii S A and Maevskii K K 2016 *Technical Physics* **61**(8) 1244
[2] Kinelovskii S A and Maevskii K K 2016 *High Temperature* **54**(5) 675
[3] Oren E P and Francois X *J Shock Waves* 2010 20 73
[4] Kinelovskii S A and Maevskii K K *J. Appl. Mech. Tech. Phys.* 2013 **54**(4) 524
[5] L A Gömze and L N Gömze 2009 *Épitőanyag-JSBCM* **61** (2) 38 http://dx.doi.org/10.14382/epitoanyag-jsbcm.2009.7
[6] Yakushev V V, Zhukov A N, Utkin A V, Rogacheva A I and Kudakina V A *Combustion, Explosion and Shock Waves* 2015 **51**(5) 603
[7] Maevskii K K and Kinelovskii S A *AIP Conf. Proc.* 2014 **1623** 391
[8] Levashov PR, Khishchenko K V, Lomonosov I V and Fortov V E *AIP Conf. Proc.* 2004 **706** 87
[9] Bakanova A A, Bugaeva V A, Dudoladov I P and Trunin R F *Fiz. Zemli* 1995 **6**
[10] Hongliang He, T. Sekine, T. Kobayashi and H. Hirosaki *Phys. Rev. B* 2000 **62**(17) 11412
[11] S. P. Marsh (Ed.), LASL Shock Hugoniot Data, (Univ. California Press, Berkeley, 1980)

IOP Conf. Series: Materials Science and Engineering **175** (2017) 012031 doi:10.1088/1757-899X/175/1/012031

Effect of composition and calcination temperature of ceria-zirconia-alumina mixed oxides on catalytic performances of ethanol conversion

S G Chuklina[1], S A Maslenkova[1], A I Pylinina[1], L I Podzorova[2] and A A Ilyicheva[2]

[1]Department of Physical and Colloidal Chemistry, RUDN University 115419 Moscow, Russia
[2] Baikov Institute of Metallurgy and Material Science RAS, 119991 Moscow, Russia

E-mail: sofyaogan@gmail.com

Abstract. In the present study, we investigated the effect of preparation method, phase composition and calcination temperature of the (Ce-TZP) – Al_2O_3 mixed oxides on their structural features and catalytic performance in ethanol conversion. Ceria–zirconia–alumina mixed oxides with different (Ce+Zr)/Al atomic ratios were prepared via sol–gel method.Catalytic activity and selectivity were investigated for ethanol conversion to acetaldehyde, ethylene and diethyl ether.

1. Introduction

With increased availability and decreased cost, ethanol is potentially a promising molecule for the production of a variety of value-added chemicals [1]. Catalysts and catalytic processes for ethanol conversion to other chemicals such as hydrogen [2], ethyleneand small oxygenates like acetaldehyde is a subject of great interest [3], especially the research of suitable catalysts, supporting materials, operating conditions for high ethanol conversion, and hydrogen selectivity. In considering the reaction of dehydration: the climbing ethylene price and decreased bioethanol cost make bioethanol-to-ethylene conversion very attractive, especially from nonfood biomass feedstock [4].

Receiving ethylene and ether from alcohols on oxide supported catalysts is of particular interest because bio-ethanol, obtained by the fermentation of biomass, is more environmental friendly. This process is also open opportunities for utilization of renewable resources and by-products of chemical processes, such as glycerol. For increasing activity oxide catalysts for this reactions are usually promoted by active metals[5-7].The nature of the active site metal has a strong impact on the selectivity and activity of the ethanol conversion [8-11]

However, the expensive cost of noble metals still limited the widespread application of these catalysts. In this regard, mixed oxides are considered promising catalysts in dehydration and dehydrogenation of alcohols. Metal oxides and their combinations, such as ZrO_2, Al_2O_3 and CeO_2, were found to show high catalytic activity in acid-base and oxidation–reduction reactions, comparing to supported catalysts [12-13].For example supported and modified CeO_2–ZrO_2 and ZrO2-Al2O3 model catalysts have been extensively studied recently [14-16] as a tree way catalyst (TWC).

The aim of this work was the synthesis and study of the catalytic activity of(Ce -TZP)-Al_2O_3systemsin different composition and calcination temperature in the conversion of ethanol.

2. Materials and methods

Sample powders (Ce+Zr)/Al with 25, 50 and 75 wt.% atomic ratios (A10-CZ, A20-CZ, A50-CZ) were prepared by sol-gel method from the precursor solutions $ZrOCl_2$, $Al(NO_3)_3$, $Ce(NO_3)$. Precipitation was carried out with aqueous ammonia at 250°C for 120 minute. Oxides samples were obtained at 180°C and 500°C. The catalysts were characterized by XRD, ASAP and N2-adsorption. Microscopic study of the composites microstructure with/without modifying was carried out by atomic forcemicroscope "NEXT" (NTMDT) and scanning electron microscope «LEO1420" (CARL ZEISS).Ethanol conversion was carried out at 200°C-400°C with chromatography analysis of gas mixture (FID, helium as a carrier gas). Catalytic activity and selectivity were investigated for conversion of ethanol to acetaldehyde, ethen and diethyl ether (DEE).

3. Results and discussion

As it seen from the XRD pattern (Figure 1) on freshly prepared catalytic systems at 180°C crystalline phases are not identified. At T=180°C powder contains a substantially amorphous phase, with appearing of tetragonal ZrO_2. At 400°C begins the formation of the tetragonal pseudo-cubic phase. And at T= 500°C phase composition of the powder is represented by a crystalline phase of the solid solution based on t-ZrO_2 and amorphous Al_2O_3. This corresponds with the literature data [17-18].

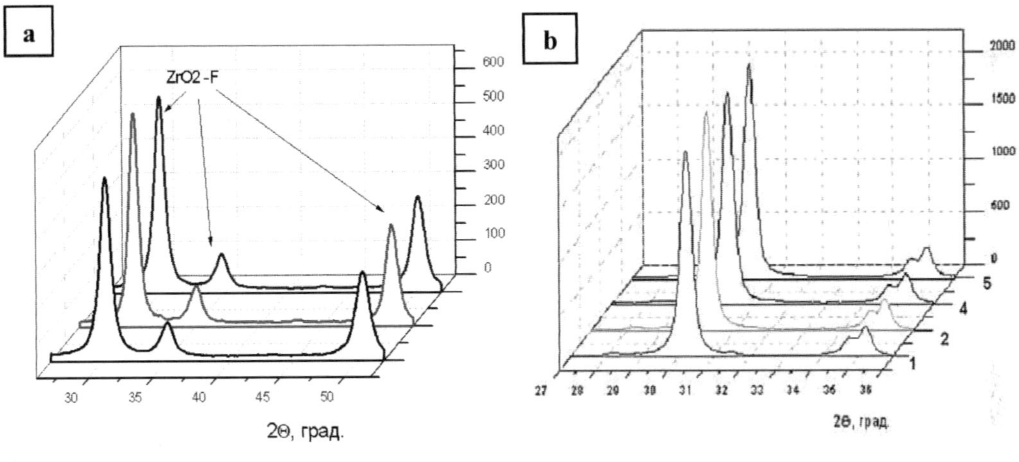

Figure.1 XRD patterns for oxide samples obtained at 180°C (a) and 500°C (b)

The observed differences in phase formation during heat treatment result in different grain size compositions in the microstructure of the compositesthat shown on figure 2(b) and [19-20].On the AFM images (Fig. 2) morphology of surface it could be identified 1-2 μm big grains of α - Al2O3 phase (corundum) with habit close to hexagonal drowned in solid solution of ZrO2 with smooth surface.

The main reaction was dehydration (as well as intramolecular dehydration) with selectivity for olefin and ether 60%. Alcohol dehydrogenation proceeds in parallel (less than 10% conversion). The selectivity is correlated with the catalyst composition and depends on calcination temperature.

Calcination temperature of the catalyst significantly affects the selectivity of the reaction of intermolecular dehydration. For example, for a mixed oxide system with composition A50-CZ selectivity after calcination increases 5 times, and for the sample with the smallest amount of aluminum A10-CZ the selectivity is 5 times less, as illustrated the diagram on Figure 3.

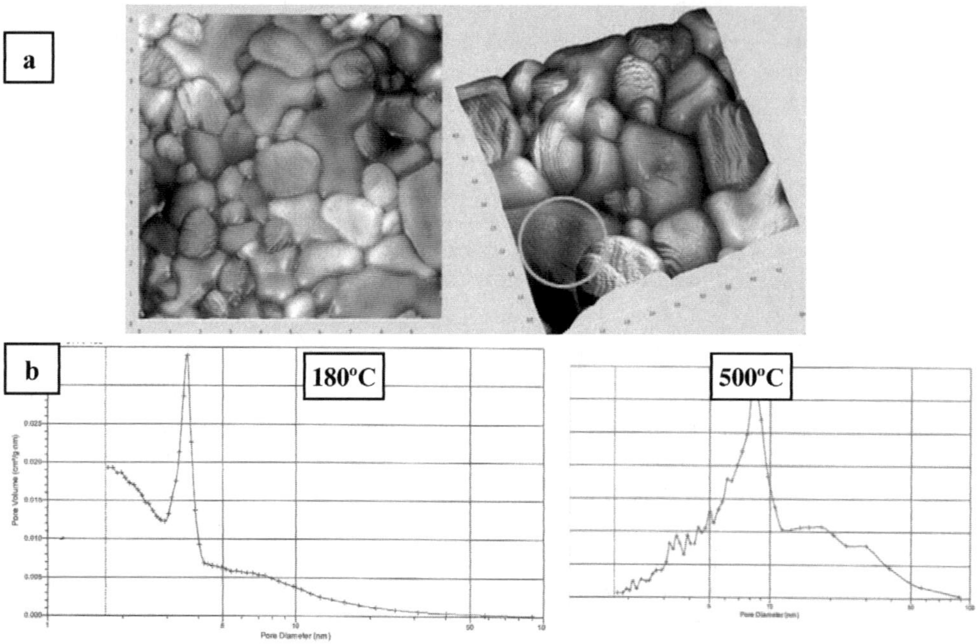

Figure. 2 a) AFM images of the composites (field of view - 10 μm) based on ACZ powders b) Pore volume and pore diameter distribution of amorphous (180°C) and crystalline (500°C) samples.

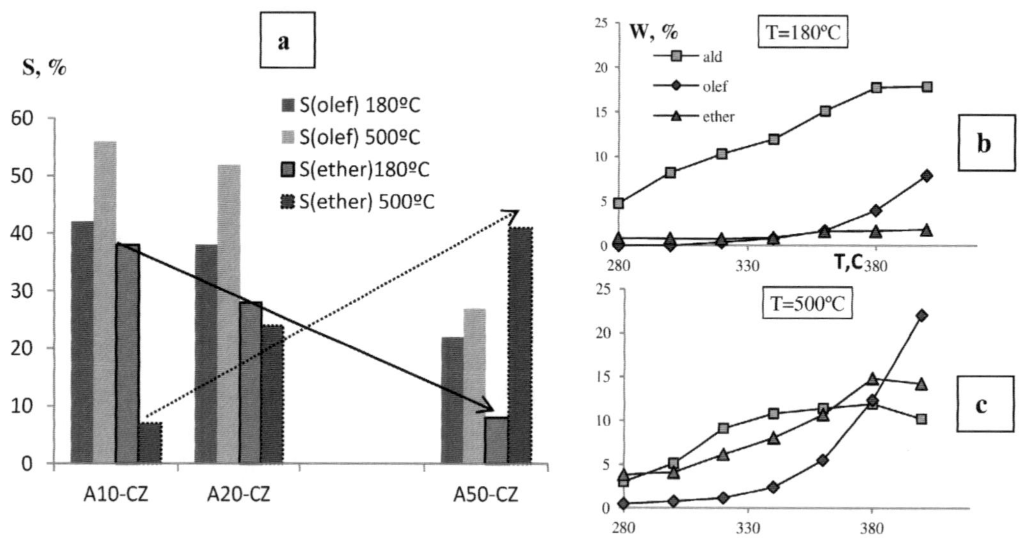

Figure. 3 a)% Selectivityof olefin and ether over samples with different compositions (% Al) obtained at 180°C and 500°C; b), c) Temperature dependences for sample A50-CZ obtained at 180°C and 500°C

4. Conclusions

Ethanol conversion reaction high selectivity on mixed oxides systems (Ce+Zr)/Al could be achieved by controlling not only the composition of nanopowders but also the calcined temperature.Hence, the preparation technique is of the highest importance for such catalytic systems design.

Acknowledgment
The authors are pleased to acknowledge the financial support of CRDF Global and Russian Academic Network Building, Education and Exposure Program (NBEE)

References

[1] J.Sun and Y. Wang 2014 *ACS Catal.,* **4(4)**1078–1090doi: 10.1021/cs4011343
[2] B.Cifuentes, M. Hernández, S.Monsalve2016 *Appl. Catal., A.*, **523**, 283-293 http://dx.doi.org/10.1016/j.apcata.2016.06.014
[3] K. Mudiyanselagea, I. Al-Shankitia, A. Foulisc2016 *Appl. Catal., B.*, **197**, 198–205 doi: 10.1016/j.apcatb.2016.03.065
[4] Hu Li, Zhen Fang, Jia Luo, Song Yang 2017 *Appl. Catal., B* **200** 182-191 http://dx.doi.org/10.1016/j.apcatb.2016.07.007
[5] T.A. Maia, J.M. Assaf, E.M. Assaf 2014*Fuel Process. Technol.***128** 134–145 http://dx.doi.org/10.1016/j.fuproc.2014.07.009
[6] Greluk, M., Rotko, M. & Machocki, A. Catal Lett (2016) 146: 163. doi:10.1007/s10562-015-1628-y
[7] Chuklina S.G., Pylinina A.I., 2016Russ. J. Phys. Chem. A **12 (90)**, 2370–76 doi:10.1134/S0036024416120074
[8] Elena I. Povarova, Anna I. Pylinina, Irina I. Mikhalenko 2014 *Physical Chemistry Communicatios*, **13** doi:10.1134/S2070205114030125
[9] L.V. Mattos, G. Jacobs, B.H. Davis 2012 *Chem. Rev.* **112 (7)**4094–4123 doi: 10.1021/cr2000114
[10] E.Varga, P. Pusztai, A. Oszko 2016 Langmuir**32 (11)** 2761–2770 doi: 10.1021/acs.langmuir.5b04482
[11] C.Wang, G.Garbarino 2016 ACS Catal., **6 (1)** 210–218 doi: 10.1021/acscatal.5b01593
[12] Ramasamy KK, MJ Gray, HM Job2016 *Catal. Today.***269**82-87 doi:10.1016/j.cattod.2015.11.045
[13] Ramasamy KK, MJ Gray, HM Job, 2016 Topics in Catalysis, **59(1)**46-54. doi:10.1007/s11244-015-0504-8
[14] Qiuyan Wanga, Zhenguo Lia 2011 J. Mol. Catal. A: Chem.**344(1–2)** 132–137 http://dx.doi.org/10.1016/j.molcata.2011.05.011
[15] I. Atribak, F. E. López-Suárez, A.Bueno-López 2011 *Catal. Today.***176**404-8 http://dx.doi.org/10.1016/j.cattod.2010.11.023
[16] LF Liotta, A Macaluso, 2003 Appl. Catal., A., **240 (1)** 295-307 http://dx.doi.org/10.1016/S0926-860X(02)00460-X
[17] Podzorova L.I., Iljicheva A.A., et al. 2002Neorg. Materials.**38 (12)**1455-61 doi:10.1023/A:1021319320332
[18] Podzorova L.I., Il'yicheva A.A., Pen'kova O.I., 2014 *Glass Phys. Chem***40 (5)**584-590 doi:10.1134/S1087659614050125
[19] Podzorova L.I., Ilicheva A.A., Shvorneva L.I. 2007*Inorg. Mater.* **43(9)** 972-975 doi:10.1134/S0020168507090117
[20] Zhukov I., Buyakova S.P., Kulkov S.N. 2016 *Épitőanyag-JSBCM* **68 (3)** 74-78 http://dx.doi.org/10.14382/epitoanyag-jsbcm.2016.13

IOP Conf. Series: Materials Science and Engineering **175** (2017) 012032 doi:10.1088/1757-899X/175/1/012032

High capacity MnO_x:ZrO_2 sorbent for elementary mercury capture: preparation, characterization and comparison to other sorbents

J Lakatos[1], C E Snape[2]

[1]Institute of Chemistry, Faculty of Material Science and Engineering, University of Miskolc, H-3515Miskolc, Hungary
[2]Department of Chemical and Environmental Engineering, Faculty of Engineering, University of Nottingham, NG7 2RD, UK

E-mail: mtasotak@uni-miskolc.hu

Abstract. Manganese oxide-zirconia type (MnO_x:ZrO_2) sorbents were prepared using the sol-gel technique by co precipitation $ZrO(NO_3)_2.xH_2O$ and $Mn(NO_3)_2$ xH_2O. The heat treatment below 500°C resulted a high surface area solid structure which consists of amorphous Mn_2O_3 (Bixbyite) and amorphous ZrO_2 phases. This material was found a high capacity oxidative sorbent for mercury removal from gas streams.

1. Introduction

Mercury (Hg) emissions have been of great environmental concern and are targeted under the Clean Air Act Amendments of 1990 [1]. Beside the elimination of mercury from the technologies which use it (thermometry, chlorine production etc.), great effort has taken to decrease the emission or removal mercury from the contamination sources [2]. Hg speciation is an important factor in the removal strategies since the elemental mercury removal require more complex process than the oxidised forms. Wide range of carbon base and non-carbon base sorbents were investigated. In order to improve efficiencies of active carbons for elemental Hg removal, investigation focused to surface modifications of carbons via sulphur and halogen (bromine and iodine) impregnation which resulted significant enhancement in the sorption capacity. Since the elemental Hg is sorbed in oxidised forms [4] the removal efficiency can be improved considerably if Hg is oxidised before/during sorption [5] Carbons itself can facilitate the Hg oxidation [6] as well as being mediated in flue gas by a combination of hydrogen chloride, SO_2 and fly ash/metal surfaces [6,7]. The oxidative capture of Hg is well documented in aqueous solution for analytical purposes, potassium permanganate in acidic solution is generally used for trapping Hg. Janssen et al.[8] used MnO_2 for collecting Hg from air. Granite et al. [9] compared a number of different inorganic substances, including MnO_2 supported on alumina, which, together with other metal oxides, e.g. Cr_2O_3, and metal sulphides, e.g. MoS_2, were found to exhibit moderate capacities for Hg capture among the candidates considered. The use of manganese oxides for Hg removal was the subject of a patent by Pahlman et al. [10], but the efficiency probable due to the low surface area of the sorbents (1-10 m^2 g^{-1}) restricted the widespread application. High oxidation power MnO_x phases can be built on properly selected supports. Manganese oxide-zirconia type (MnO_x:ZrO_2) sorbents were previously found to be efficient for NO_x capture[11]. Snape et. al.

[12] and Lakatos et al. [13-14] has demonstrated using the sol-gel technique a high capacity sorbent for oxidative capture for mercury can be produced. Recently Scala et al. [15,16] has prepared the MnOx active phase on high surface area γ alumina and produced it the honeycomb monolite form.

2. Materials and methods

The sorbent preparation used here was based on Eguchi and Hayashi's method [11]. $ZrO(NO_3)_2.xH_2O$ and $Mn(NO_3)_2$ xH_2O were dissolved separately. After mixing the solutions, the co-precipitated product was formed by adding ammonia. The solvent was then evaporated and the dried residue was calcined at 450°C using a tube furnace with a continuous flow of air. Finally, the sorbent was ground and the 75-250 μm fraction was used for the Hg sorption tests. The sorbent preparations were carried out at different Mn/Zr atomic ratio. The pore structure of the formed sorbents were characterised by gas adsorption measurement using N_2 sorption-desorption isotherms. The Mn-oxide, Zr-oxide and the MnOxZrO2 sorption isotherms are compared at fig. 1. In case of the cooprecipitated material Mn/Zr ratio close to the 1:1. The highest BET surface area and a well-developed mesopore structure was achieved, see Fig 1.

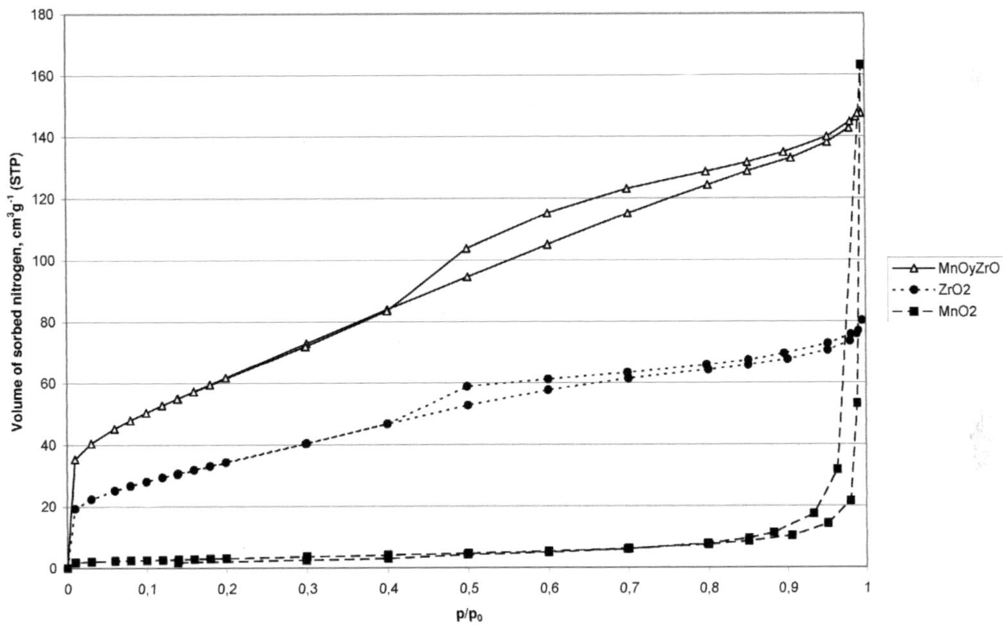

Figure 1. Nitrogen sorption isotherms (77 K) of different solids: Mn-oxide, Zr-oxide formed by precipitated alone and MnO_xZrO_2.formed by cooprecipitation. Each were prepared the same way

The sorbent was pelletized without binder and heat treated at 900 C for consolidate the pellet. The palletisation resulted only marginal effect on the sorbent surface area and pore structure (Fig.2).

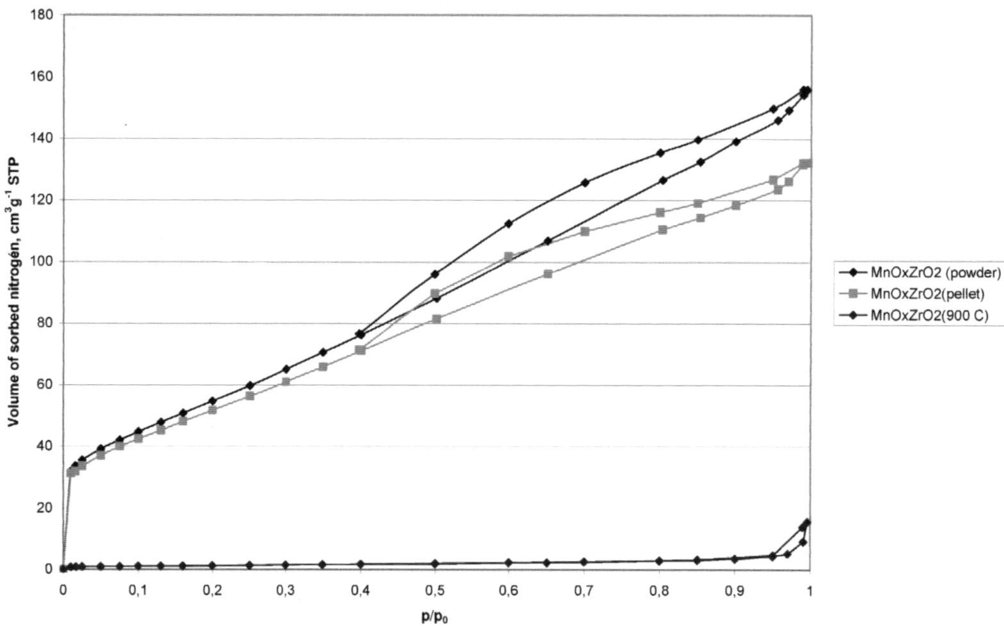

Figure 2. Nitrogen sorption isotherms (77 K) of powder form, pelletized, and palletised and 900C heat treated MnO_x ZrO_2 sorbent. The powder and pellet was activated by heat treatment at 450 C in oxidative atmosphere

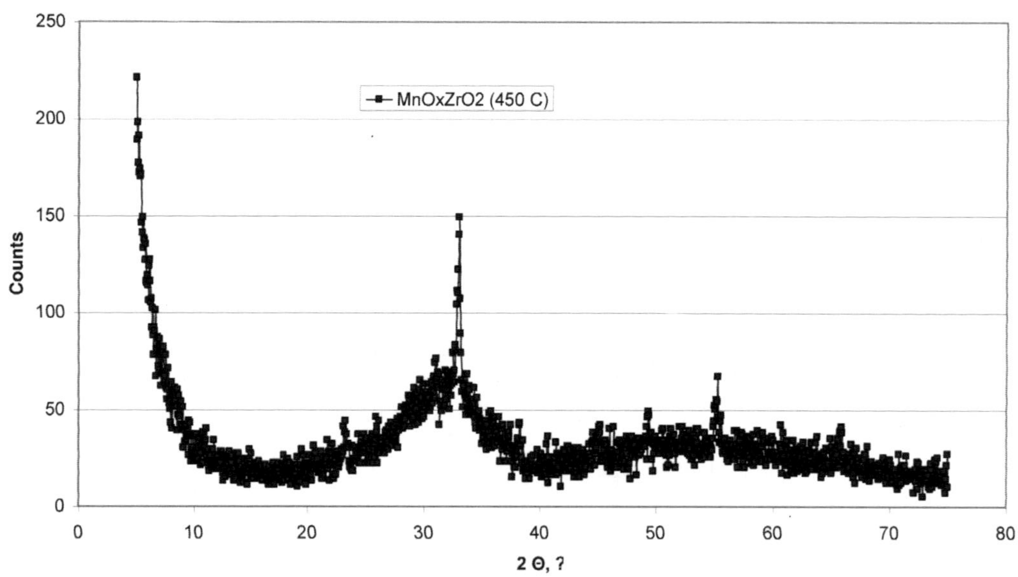

Figure 3. XRD pattern of the (MnO_x:ZrO_2) sorbents prepared by sol-gel technique and activated by heat treatment at 450 C in oxidative atmosphere

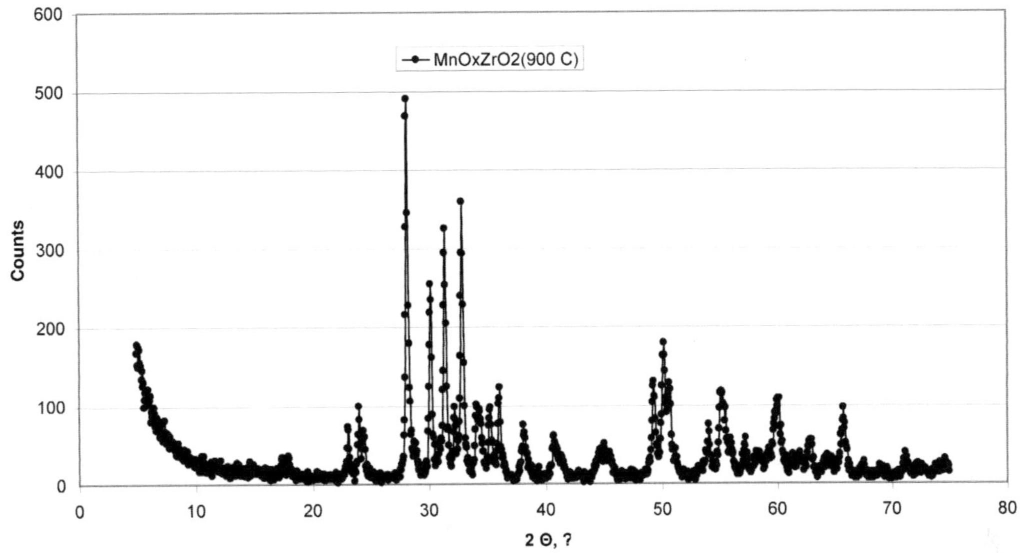

Figure 4. XRD pattern of the (MnO_x:ZrO_2) sorbents heat treated at 900 C

However the heat treatment at 900°C destroyed the amorphous structure Fig.3. since it has resulted nonporous crystalline phases Fig.4. The amorphous phase of MnO_x was identified by the database of XRD as Bixbyite (Mn_2O_3). During the heat treatment at 900 C this Bixbyite has transformed to Hausmannite (Mn_3O_4) while the amorphous zirconia has become Baddeleyite (ZrO_2).

3. Results and discussion

The Hg sorption tests were taken using the experimental set illustrated at the fig.5.[17]. The breakthrough curves for activated carbons an MnO_xZrO_2 sorbents are compared fig. 6. Due to the high capacities of the MnO_xZrO_2 sorbents, the Hg concentration in the test gas was kept approximately two orders of magnitude higher than it is typical in the flue gas (0.2 μg/m^3), and the sorbent has to be dilute with sand in the packed bed (1:1 - 1: 5 by volume) if the breakthrough like to be achieved in a reasonable time.

Table 1. Hg capture performance of MnO_xZrO_2 adsorbents and a comparison with commercial active carbon and manganese oxide

Mn/Zr atomic ratio	**BET surface area, m^2/g**	**Break-through time, g**	**Break-through capacity mg/g**
ZrO_2	127	< 0.2	< 0.02
0.1		0.55	0.06
0.5		98.6	24.9
0.86	228	>440	151
1		173	48.5
MnO_x	11	104	20.4
Mn O_2 ((Fischer Chemicals)		1:0	0.19
Activated carbon (Norit-Darco FGD(98% <45 μm)	600	1:1	0.06

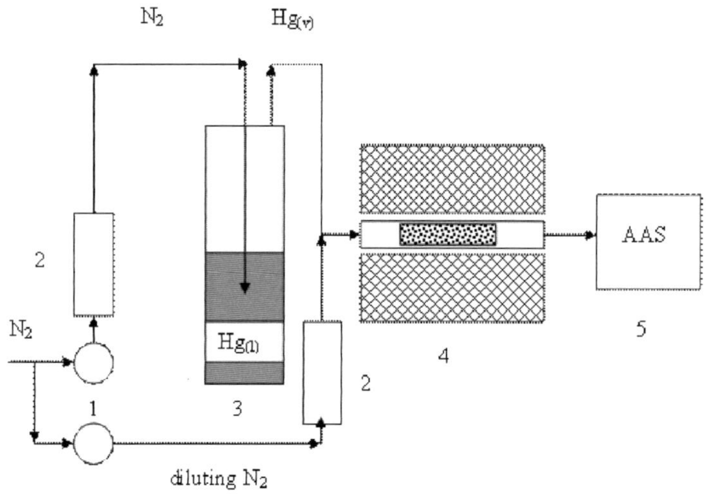

Figure 5. Experimental set up of test rig for Hg(0) sorption, (1) valves, (2) flow meter, (3) Hg vapour generator, (4) owen or thermostate with column filled with sorbent (preferred position is the vertical), (5)Hg detector(AAS (Atomic Absorption Spectrophotometer) or AFS (Atomic Fluorescence Spectrophotometer). Hg concentration calculated by the temperature and the saturation pressure data: CRC Handbook of Chemistry and Physics, 72nd Edn; CRC Press: Boca Raton, 1992; p 6-99.

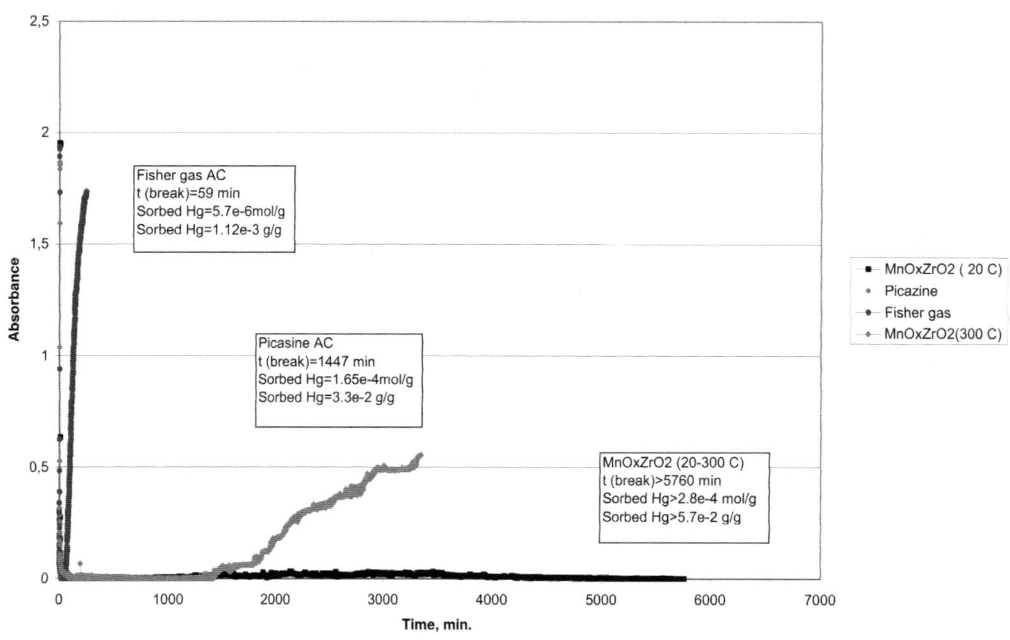

Figure 6. Break through curves of Hg(0) sorption on different sorbents. Column parameters: length: 50 mm, diameter: 5.5 mm, Flow rate: 0,2 ml/min. Mass of sorbents: m(Fisher gas):0,5995 g, m(Picasine):0,5074 g, m(MnO_xZrO_2): 1,1541g

The Hg(0) removal capacity by the MnO_xZrO_2 sorbents at different Mn/Zr atomic ratio and by the MnO_x or ZrO_2 formed the same process as the MnO_xZrO_2 sorbents were are given in Tab.1. The zirconia (support) displays negligible Hg uptake. The $MnO_x.ZrO_2$ sorbent prepared with 0.87 Mn:Zr atomic ratio achieved an exceedingly high breakthrough capacity of min. 15 m/m % Hg. The breakthrough capacity of 15 m/m % is over 200 times higher than that for activated Norit-Darco FGD carbon, which has been widely used for in direct injection trials for Hg capture from flue gas [5] The precipitated MnO_2 prepared in the same manner as the co-precipitated $MnO_x.ZrO_2$ sorbent was found have a much lower breakthrough capacity, as anticipated from the lower surface area. However, this is still considerably higher than for bulk manganese oxide represented in Tab.1 as Fischer chemicals. The preliminary investigation into the effect of temperature has proved that the sorption capacity does not fall until temperatures above 150°C. However, even at 300°C, the breakthrough capacity is still higher than 5% w/w Hg.

Conclusions

The higher than 15 m/m % Hg sorption capacity indicated that the MnO_x-based sorbents prepared by co-precipitation provided a considerable promise for Hg capture. These sorbents are much more effective than both bulk and the prepared manganese oxide alone and, also, the supported manganese oxides formed on a preliminary built support.

Acknowledgement

This research was partially carried out in the framework of the Center of Applied Materials Science and Nano-Technology at the University of Miskolc, Hungary.

References

[1] Keating, M. H.; Maxwell, M. H.; Driver, L.; Rodriguez, R. Mercury Study Report to Congress, USEPA Office of Air Quality Planning and Standards, EPA-452/R-97–010, December 1997.

[2] Krishnan, S. V.; Gullet B. K.; Jozewich, W. Mercury control in municipal waste combustors and coal-fired utilities. Environ. Progr. 1997, 16, 47-53.

[3] Lopez-Anton M. A.; Tascon, J. M. D.; Martinez-Tarazona M. R. Retention of mercury in activated carbons in coal combustion and gasification flue gas. *Fuel Proc. Technol.* 2002, *77-78, 353-358.*

[4] Huggins F. E.; Yap, N.; Huffman, G. P.; Senior, C. L. XAFS characterisation of mercury captured from combustion gases on sorbents at low temperatures. *Fuel Proc. Technol.* 2002, *8, 167-196*

[5] Pavlish J. H.; Sondreal E. A.; Mann, M. D.; Olson, E. S.; Galbreath, K. C.; Laudal, D. L.; Benson, S. A. Status review of mercury control options for coal-fired power plants. *Fuel Proc. Technol.* 2003, *82, 89-165.*

[6] Olson, E. S.; Miller, S. J.; Sharma, R. K.; Dunham, G. E.; Benson, S. A. Catalytic effect of carbon sorbents for mercury capture. *J. Hazardous Materials,* 2000, *74, 61-69*

[7] Dunham, G. E.; Dewall, R.; Senior, C. Fixed-bed studies of the interactions between mercury and coal combustion fly ash. *Fuel Process. Technol,* 2003, *82, 197*

[8] Janssen, J. H.; Van den Enk, J. E.; Bult, R.; De Grott, D. C. Determination of total mercury in air by atomic absorption spectrometry after collection on manganese dioxide. *Anal. Chim. Acta,* 1977, *92, 71-77.*

[9] Granite, E. J.; Pennline, H. W.; Hargis, R. A. Novel sorbents for mercury removal from flue gas. *Ind. Eng. Chem. Res.* 2000, *39, 1020-1029.*

[10] Pahlman, J. E.; Carlton, S. C.; Huff, R. V. et al. System and process for removal of pollutants from a gas stream. US Patent US2003157008, 2003.

[11] Eguchi, K.; Hayashi, T. Reversible sorption-desorption of NO_x by mixed oxides under various atmospheres. Catalyst Today 1998, 45, 109-115.

[12] C.E. Snape, C.Sun, J. Lakatos, R. Perry: Sorbent Composition WO patent: WO 2008/093137 A1.Publ. date: 7 Aug. 2008.

[13] J. Lakatos, Cheng-gong Sun, Ron Perry, Mark Kennedy and C. E Snape: Ultra high capacity co-precipitated manganese oxide sorbents for oxidative mercury capture. 237th ACS meeting Spring, Salt Lake City (2009)
[14] J.Lakatos, C. E Snape: Comparative study of elemental mercury removal from gas streams by carbon base and inorganic oxidative sorbents. Annual Word Conference of Carbon, Cracow, 2012. CD: ISBN 978-83-60958-99-5., Abst No. 732.
[15] F. Scala, C. Anacleria, S. Cimino: Characterisation of a regenerable sorbent for high temperature elemental mercury capture from flue gas. Fuel, 2013, 108, 13-18.
[16] F. Scala, S. Cimino: Elemental mercury capture and oxidation by regenerable manganese-based sorbent: The effect of gas composition Chem. Eng. Jurn. 2014, (article in press).
[17] J. Lakatos, C. E Snape: A simple testing rig and mercury vapour generator for investigation of mercury vapour removal characteristics of different sorbents. Material Sci. and Eng. Series II.2010, Vol 35. 95-107.

Influence of mechanical activation on the properties of natural zeolites from Tokaj Mountain

A Y Buzimov[1,2], S N Kulkov[1,2], E Kurovics[3], W Eckl[4] and S Pappert[4]
[1]Institute of Strength Physics and Materials Science SB RAS, 634055, Tomsk, Russia
[2]National research Tomsk State University, 634050, Tomsk, Russia
[3]Department of Ceramics and Silicate Engineering, University of Miskolc, H-3515, Hungary
[4]Fraunhofer Institute for Chemical Technology ICT, 76327 Pfinztal, Germany

E-mail: buzimov92@gmail.com

Abstract. This paper presents the results of a study of the influence of mechanical activation on morphology, specific surface area and phase composition of natural zeolite of Tokaj Mountain. During the mechanical activation of zeolites powders with specific surface area of 19-20 square meters per grams, significant changes in chemical and mineralogical compositions can be observed. The laboratory experiments had shown an intensive increase of specific surface area at the beginning of mechanical activation; a further relatively slow decrease and reduction of BET surfaces were observed. By increasing the mechanical activation time the amount of quartz, cristobalite-low, orthoclase mineral components were not stable, and their content have varied not so strongly as a decrease smectite 15 A, clinoptilolite, illite 2M1 or calcite. In addition, during the mechanical activation occurred amorphization, which was increased from 13% to 52%.

1. Introduction

The Tokaj Mountain region, located in the northeastern part of Hungary as part of the Carpathian Basin, is one of the most famous volcanic mountains of Hungary [1]. The interest of researchers towards natural zeolites is connected with their unique properties: an extremely high adsorption capacity, catalyzing action, thermal stability and resistance in different chemical environments [2].

The results of the study of structural, physicochemical properties of natural zeolites allow developing the theoretical bases for the directed change of useful properties of natural minerals that are required for sorption materials.

The specific surface area is a dominant parameter for zeolites [3]. To obtain the necessary magnitude of specific surfaces of materials depends not only on their chemical and mineralogical composition, but also on physical, mechanical and rheological parameters of the used materials [4, 5]. It is known that mechanical processing in ball mills can change the structure, morphology, specific surface area of zeolite-containing materials and occasionally improve their processing properties [6]. In spite of the numerous publications, the physicochemical properties of natural zeolites are still not sufficiently studied. The goal of this work was to study changes induced by mechanical activation on morphology, specific surface area, and phase composition of the natural zeolite Tokaj Mountain deposit.

2. Materials and Experimental Procedure

Natural zeolite from Tokaj Mountain was used as object of investigation. The mechanical activation of the powder was performed in a Retsch PM 400 planetary ball mill at 350 rpm for 600 minutes. The acceleration in planetary mill was 26.8 g. Ceramic grinding balls were used in planetary mill [7].
The crystal structure and phase composition of the zeolite before and after the mechanical activation were determined by X-ray diffraction analysis (XRD) using diffractometer Rigaku MiniFlex II with Cu-Kα (λ= 1.5418 Å) radiation in angular interval of 3–90° with steps of 0.04° [8]. The phases were determined by search method using powder diffraction data base file PDF2.

The morphology of the powder was analyzed using field emission scanning electron microscopy Zeiss Supra 55VP operated at an acceleration voltage of 3.5 kV. For the measurements, the samples were put on a carbon tape and covered with a gold layer.

The particle size was determined by using laser diffraction particle size analyzer Horiba LA-950 in the range 0.01 - 3000 µm.

Brunauer, Emmett, Teller (BET) surface area was determined using a micrometrics Tristar 3000 gas adsorption analyzer [9].

3. Results and Discussion

3.1. Initial of natural zeolite powder
On Fig. 1 it is shown the elemental analysis, SEM-pictures and X-ray patterns of natural zeolite powder.

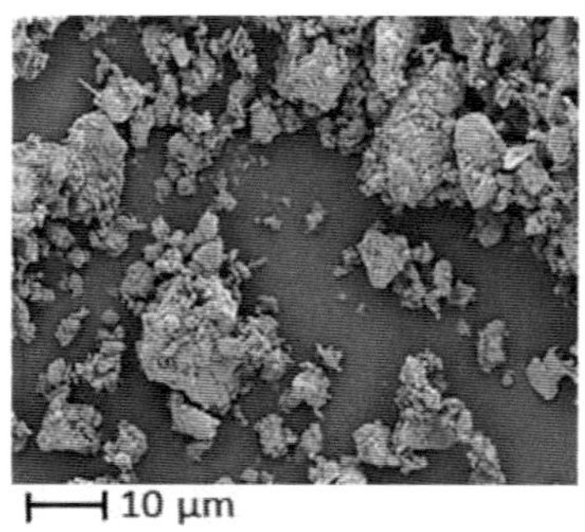

Element	Wt%	At%
OK	48.56	62.49
MgK	0.25	0.21
AlK	3.76	2.87
SiK	46.28	33.92
CaK	0.55	0.28
FeK	0.60	0.22

(a)

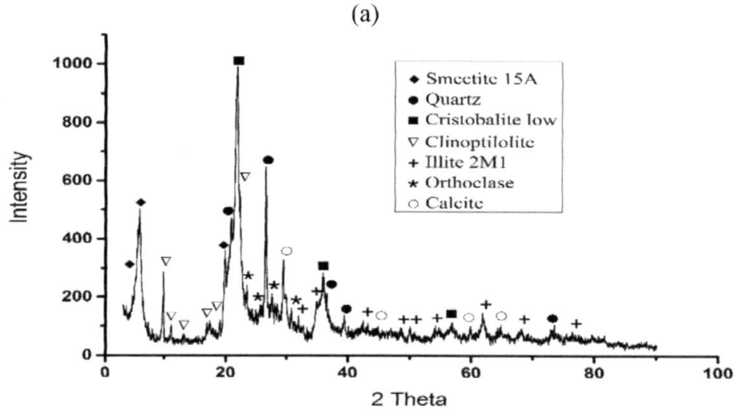

(b)

Figure 1. SEM-picture, elemental analysis(a), X-ray pattern(b) of natural zeolite powder

The results of elemental analysis showed that the zeolite powder in the initial state consisted of Al, O, Si, Ca, Mg, Fe. The SEM data showed that particles of zeolite powders has irregular shapes with an average particle size of 27 µm. Calculations of the X-ray pattern showed that the structure of zeolite consists of seven different phases: smectite 15A, quartz, cristobalite-low, clinoptilolite, illite 2M1, orthoclase, calcite with value of 20%, 8%, 15%, 14%, 15%, 9%, 6% respectively. XRD results showed that the presence of amorphous phase in initial zeolite was 13%.

3.2. Mechanical activation of natural zeolite powder

Initial zeolite powder has a specific surface area of 19 m^2/g. Experimental results regarding the change in BET specific surface area of zeolite samples with mechanical activation time in a planetary ball mill are shown in Fig. 2 (a).During the first 60 minutes, the specific surface area of zeolite powder increases, reaching 33 m^2/g. After 120 minutes, the specific surface slightly decreases and after 600 minutes of activation, there is a significant reduction to 20 m^2/g.

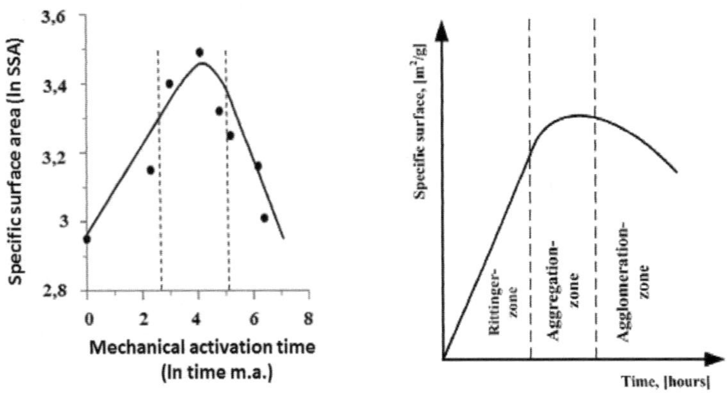

(a) (b)

Figure 2. The dependence of specific surface area values of zeolites from mechanical activations times on logarithmic scale (a), theoretical curve (b).

The changes of the specific surface area of the zeolites corresponding to the theoretical curve of the developed specific surface as function of the activation time is shown in figure 2 (b) – on the first stage one can observe rittinger-zone when the increase of the specific surface is proportional to the activation time; on the second stage there is an aggregation zone when the specific surface almost does not change, and the last stage - agglomeration-zone decreases the specific surface with the activation time [10-12].

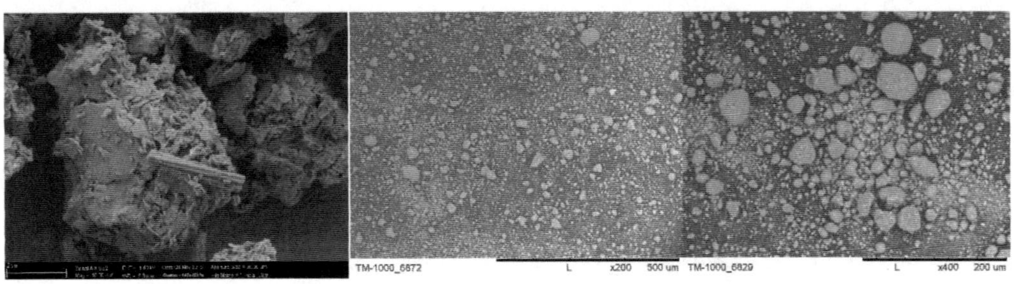

(a) (b) (c)

Figure 3. SEM-pictures, initial zeolite (a), after 20 minutes (b), after 600 minutes (c) of milling time.

With the increase in the mechanical activation time there is also a change in the particles size of zeolite powder, as shown in Fig. 3. The initial zeolite-average (value in microns) was 27 µm. After activation of 20 minutes, particle size was 5.5 µm and after 600 minutes of milling time, the particle size became 28 µm. Moreover, most particles have lost their initial shape and converted into a spherical shape during the mechanical activation.

The quantitative analysis of XRD patterns showed radical changes in mineralogical compositions, as shown in Fig. 4. The proportions of all mineralogical components of natural zeolite changed in dependence on the activation time. The most intensive changes can be observed in mineral components of smectite 15 A, clinoptilolite, illite 2M1 and calcite. After 60 minutes of activation time the content of smectite 15 A – decreased from 20% to 10%, clinoptilolite from 14% to 9%, meanwhile the amount of orthoclase – grew from 9% to 12% . In mineralogical composition, the considerable changes continued up to 600 minutes of mechanical activation. At 600 minutes of activation time the zeolite content of smectite 15 A, clinoptilolite, illite 2M1, and calcite reached its minimum value of 0.5%, 1%, 3%, 1% respectively, meanwhile the amount of quartz increased up to 17%. Under mechanical activation the amount of quartz, cristobalite-low, orthoclase mineral components were also not stable, but their content have varied not so strong. In addition, during the mechanical activation it occurred amorphisation, which increase from 13% to 52%.

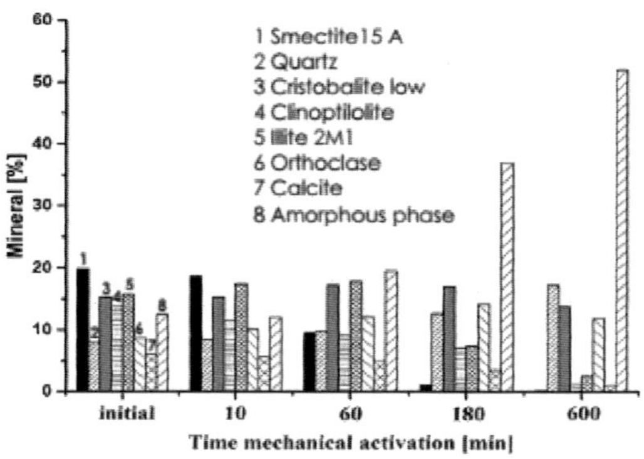

Figure 4. Mineralogical composition of natural zeolite during mechanical activation.

4. Conclusion

During the first 60 minutes, the specific surface area of zeolite powder increases, after 120 minutes, the specific surface slightly decreases and after 600 minutes of mechanical activation, there is a significant reduction of surface area. Most particles have converted into spherical shape during the mechanical activation.

During the mechanical activation of natural zeolite the most intensive changes observed in mineral components of smectite 15 A, clinoptilolite, illite 2M1 and calcite, and it was observed an increase of system amorphisation from 13% to 52%.

Acknowledgement

The research was conducted according to project RF Academy of Science № III.23.2.3.

5. References

[1] Viczián I 1997 *Clays Clay Miner.* **45** 114
[2] Akimkhan A M 2012 *Structural and Ion-Exchange Properties of Natural Zeolite* (Croatia:Intech)
[3] Olson E 2012 *J. of GxP Compliance* **16** 52
[4] Gömze L A, Gömze L N, Kulkov S N 2015 *Építőanyag–JSBCM* **67** (4) 143
http://dx.doi.org/10.14382/epitoanyag-jsbcm.2015.24

[5] Gömze L A 2003 *Épitőanyag–JSBCM* **55** (4) 133 http://dx.doi.org/10.14382/epitoanyag-jsbcm.2003.23
[6] Nikashina V A, Streletskii A N, Meshkova I N, Kolbanev I V, Grinev V G, Serova I B 2011 *Inorganic Materials* **47** 1341
[7] Geber R, Kocseha I, Gömze L A 2012 *Mater. Sci. Forum* **729** 344
[8] Song Z, Huang Y, Wang L, Li S, Yu M 2015 *Chem. Commun.* **51** 373
[9] Dedova E S, Shutilova E S, Geber R, Gömze L A, Kulkov S N 2016 *IOP Conf. Ser. Mater. Sci.Eng.* **140** 012007 http://dx.doi.org/10.1088/1757-899X/140/1/012007
[10] Gömze L A 2010 *Mater. Sci. Forum* **659** 19 *www.scientific.net/MSF.659.19*
[11] Juhász Z A, Opoczky L 2003 *Épitőanyag–JSBCM* **55** (3) 86 http://dx.doi.org/10.14382/epitoanyag-jsbcm.2003.16
[12] Gömze L A, 2016 APPLIED MATERIALS SCIENCE I., Compilation of Selected Scientific Papers, *Published by IGREX Ltd,* **ISBN 978-963-12-6600-9** pp 1-189

Synthesis and crystallization of titanium dioxide in supercritical carbon dioxide ($scCO_2$)

A Kaleva[1], S Heinonen, J-P Nikkanen, E Levänen

[1]Department of Materials Science, Tampere University of Technology, FIN-33101 Tampere, Finland

E-mail: aaretti.kaleva@tut.fi

Abstract. In this work, a simple and low-temperature method to synthesize titanium dioxide (TiO_2) particles with supercritical carbon dioxide is presented. The particles were synthesized by measuring 5 ml of tetra-n-butyl orthotitanate precursor to the supercritical chamber. The pressure was maintained at 15.0 MPafor all experiments. Reaction temperatures used were 50 °C or 70 °C. After reaching treatment parameters 10 ml of deionized water was introduced to the chamber with a co-solvent pump. A mixer was used inside the chamber to ensure proper mixing of water and precursor. Reaction times of 10, 60 and 300 min were used. Characterization of the particle crystal phase was determined by X-ray diffraction, differential scanning calorimetry and transmission electron microscopy. The specific surface areas were measured with nitrogen adsorption tests (BET). The results showed that the particles synthesized with reaction times of 10 and 60 min contained brookite as the crystalline phase. With longer reaction time of 300 min the phase shifted to anatase. In most experiments there was also significant amount of amorphous phase present. The specific surface areas varied between 274.3–566.6 m^2/g.

1. Introduction

Titanium dioxide is a material which exhibits high photocatalytic activity. It is used for example in antibacterial, self-cleaning, solar cell, water and air purification applications. The photocatalytic activity of titanium dioxide is affected by its particle properties such as crystalline phase, crystallinity, particle size and specific surface area. Titanium dioxide has three phases in nature: rutile, anatase and brookite. Among the three polymorphs of titanium dioxide anatase has proved to be more effective in photocatalytic applications than rutile or brookite. However, the photocatalytic activity of brookite varies between studies. The main reason for this is the challenges in the synthesis of brookite leading to fluctuation in its characteristic properties. Brookite is often synthesized in high pressures via hydrothermal route since it is reported to be a high pressure phase.[1, 2]The conventional synthesis methods for photocatalytic titanium dioxide require relatively high temperatures[3]. These methods also include the use of different solvents and catalysts which can be harmful for human and environment.

Supercritical carbon dioxide route is a processing method which is used in combination with sol–gel method to produce nanoparticles and also in other chemical processes [4]. When pure substances are pressurized and heated above their critical point they transform into supercritical fluids (SCF). SCFs have liquid-like solvation power and gas-like diffusion. Supercritical carbon dioxide is the most used supercritical fluid due to its low critical point of 31.2 °C and 7.38 MPa [5]. It is a green synthesis method because it is rather inert, non-toxic and a low-temperature process. Moreover, carbon dioxide can be recycled to prevent its release to the atmosphere. The reaction parameters, e.g., temperature and pressure, can be easily changed and accurately controlled to change the properties of supercritical fluids.[6]

In this work, titanium dioxide was synthesized by sol–gel methodat low temperature using supercritical carbon dioxide as a solvent. The effect of reaction temperature and reaction time to the properties of the synthesized particles was then studied.

2. Experimental

2.1. Synthesis procedure

All experiments were done with supercritical carbon dioxide apparatus (Thar RESS 250) which consists of high-pressure pump, co-solvent pump, high-pressure vessel, heating systems and a mixer. Synthesis precursors used were carbon dioxide (≥ 97%, AGA) tetra-n-butyl orthotitanate ($Ti(OCH_2CH_2CH_2CH_3)_4$, ≥ 97%,Sigma-Aldrich) and de-ionized water. The reaction pressure was 15.0 MPa in all experiments. Reaction temperatures were 50 °C or 70 °Cand the reaction times were 10, 60 or 300 min. The synthesis procedure was started measuring 5 ml of tetra-n-butyl orthotitanate to the reaction vessel and increasing the temperature and pressure to the preferred values. After reaching the reaction parameters 10 ml of water was pumped to the vessel by using the co-solvent pump with 3 ml/min speed. The precursors were mixed throughout the whole process using a stirrer inside the chamber. The reaction time was measured when water was first introduced to the chamber to the start of depressurization of the vessel since the depressurization time was between 2–5 min. The samples were then dried in 60°C for 1 h.

2.2 Characterization methods

Morphological characterization and particle size calculations were performed using emission scanning electron microscope (FE-SEM, Zeiss ULTRAplus. The BET specific surface areas were measured with nitrogen gas adsorption apparatus (Coulter Omnisorp 100CX). The crystal structures of titanium dioxide samples were determined with a Panalytical Empyrean Multipurpose Diffractometer. The crystallite sizes were estimated from XRD patterns using a Scherrer formula. Relative crystallinity between the samples were analysed from differential scanning calorimetry measurements (Netzsch Phoenix DSC 204 F1) by heating the samples to 500 °C with a 5 °C/min heating rate. Crystallite size was determined also from transmission electron microscope (JEM-2010 JEOL) pictures.

3. Results and discussion

3.1. Morphology, particle size and specific surface area

The SEM figures of the synthesized samples are presented in figure 1. The particles were agglomerated and the particle morphology was spherical. The spherical shape is expected to form when the pumped water enters the chamber and forms emulsion with supercritical carbon dioxide since they are only slightly soluble with each other. The dissolved tetra-n-butyl orthotitanate starts to react on the surface of the water droplets by hydrolysis and condensation reactions[4]. These reactions finally result in solid titanium dioxide particles. Some of the particles appeared to join together into irregular shapes which can be seen clearly from figure 1F resulting when the emulsion droplets fuse together while the precursors are reacting.

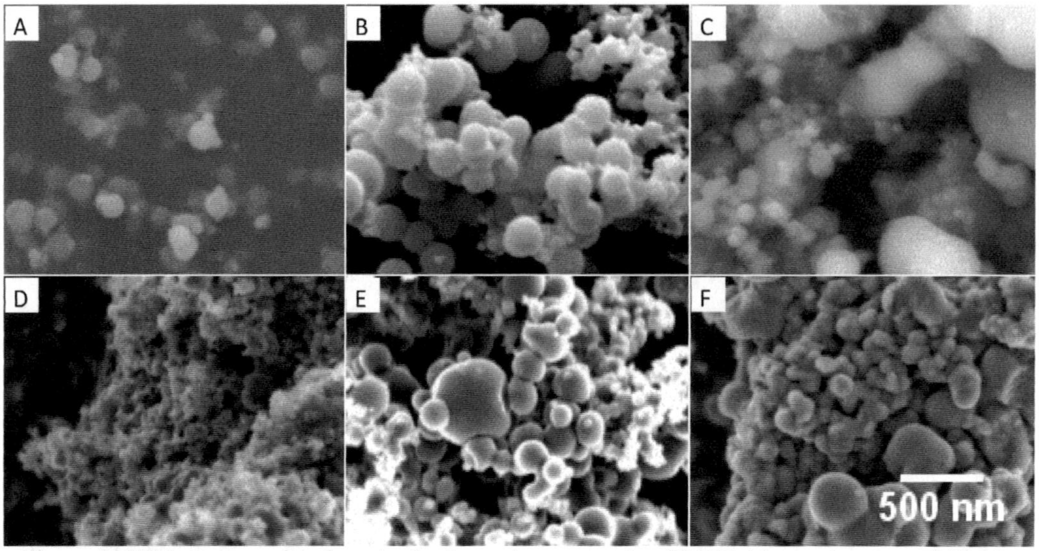

Figure 1. SEM micrographs of synthesized samples prepared at 50 °C with reaction time of 10 min (A), 60 min (B), 300 min (C)and at 70°C with reaction time of 10 min (D), 60 min (E), 300 min (F)

The particle sizes and specific surface areas of the particles are presented in Table 1. With the lowest reaction times the particle sizes are smallestcompared to longer reaction times from 0.148 to 0.303 µm with both reaction temperatures. This is due to the higher amount of amorphous titanium dioxide which is confirmed by the DSC results in the subsequent chapter. [7] Amorphous phase is present since the reaction time was too short to crystallize the particles. Particle size increases significantly when reaction time is increased to 60 min and it levelled out at 300 min reaction time. The particle size increases slightly also with the increase of reaction temperature. The particle size distribution is high as can be seen from the standard deviations of the particle size measurements. Specific surface areas for all treatments werevery high. The highest surface areas of over 500 m^2/g were obtained with the shortest reaction times due to the higher amount of amorphous phase. With 60 min and 300 min reaction times the surface areaswere between 274.3–319.8 m^2/g.With over 60 min reaction times the surface area decreased compared to the 10 min samples. Reaction temperature had a minor effect to the surface areas.

Table 1. Particle size averages deviations from all samples measured from SEM figures. BET specific surface areas from all samples

Treatment	Particle size average ±standard deviation (µm)	Specific surface-area (m^2/g)
150 bar 50 °C 10 min	0.303 ± 0.208	566.6
150 bar 50 °C 60 min	0.483 ± 0.437	319.8
150 bar 50 °C 300 min	0.488 ± 0.323	274.3
150 bar 70 °C 10 min	0.148 ± 0.076	527.2
150 bar 70 °C 60 min	0.629 ± 0.525	294.7
150 bar 70 °C 300 min	0.635 ± 0.404	305.9

3.2. Relative crystallinity

The relative crystallinities of the powders were analysed by differential scanning calorimetry. The results are presented in Figures 2 and 3 for samples treated at 50°C and 70°C, respectively. Three different peaks can be seen from the results. The endothermic peak at 100°C is due to the evaporation of water. One sample treated at 50°C with 10min reaction time has endothermic peak at 275°C which is due to the decomposition of organic residuals from precursor tetra-n-butyl orthotitanate. The only exothermic peak occurs at 390°C at which the crystallization of amorphous titania starts and forms anatase [7, 8]. The largest peaks indicate highest amount of amorphous phase while the smallest peaks higher crystallinity. The results show that the peak is biggest in 10 min samples in each treatment temperature which correlates well with particle size and specific surface area results. The samples treated at 50°C with 300 min treatment time and at 70°C with 60 and 300 min reaction times do not exhibit a peak at 390°C temperature which indicates high crystallinity. The results show that the increase in reaction temperature and reaction time increases the crystallinity as well.

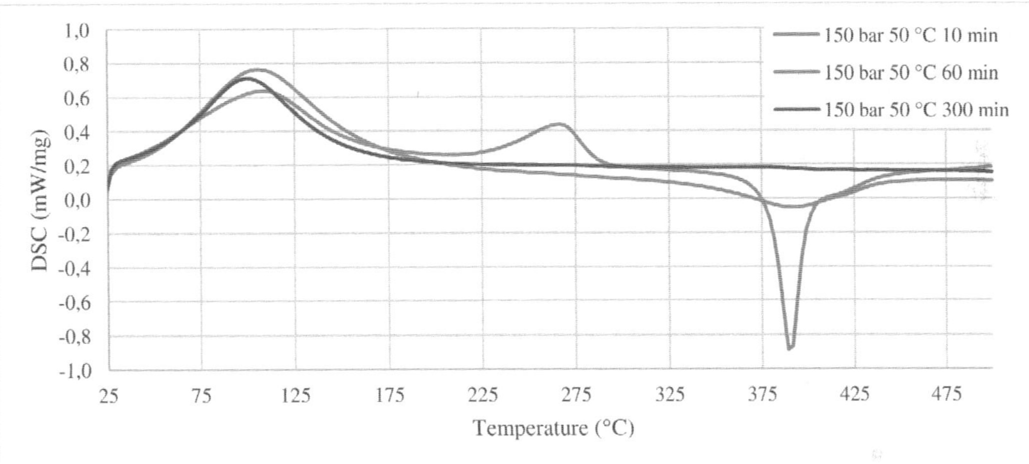

Figure 2. The DSC curves for samples treated in 50 °C with reaction times of 10, 60 and 300 min.

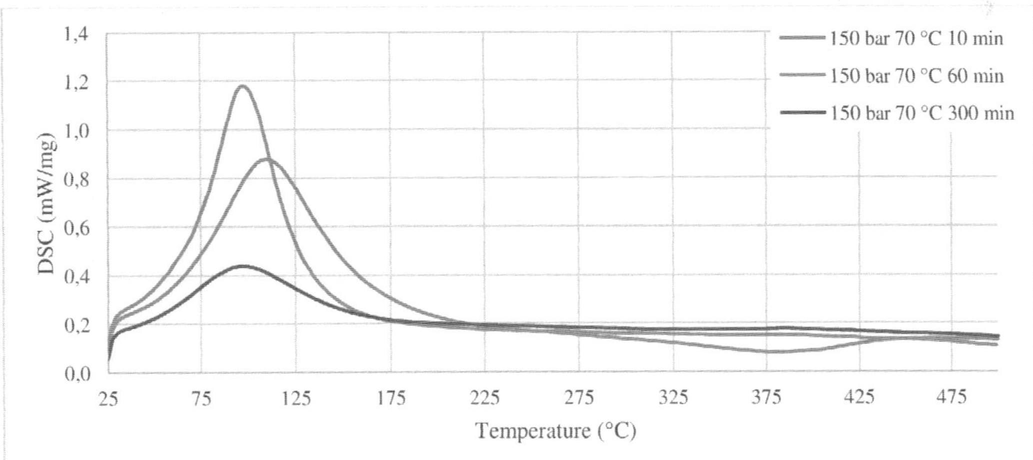

Figure 3. The DSC curves for samples treated in 70 °Cwith reaction times of 10, 60 and 300 min.

3.3. Characterization of crystalline phase and crystallite size

X-Ray diffraction was used to determine the crystalline phase of the particles and the results are shown in Figure 4. With 10 and 60 min samples in both temperatures the peaks were wide andslightly indistinct. There appeared to be a peak at 30°which would indicate the presence of brookite which has a characteristic peak at 30.8°[1].However, since the peaks are not so distinct and the amorphous phase is present in these sample, the dominate phase still remains uncertain. The formation of brookite could be due to the high pressure of the synthesis. With 300 min samples it was clear that the dominate crystallite phase was anatase in both temperatures. The crystallite size of anatase was determined with XRD by using the Scherrer equation to be approximately 3-4 nm with no clear change between different reaction parameters. In Figure 5, thereare TEM micrographs from two samples with 60 min reaction time in different temperatures. TEM micrographs were used as complementary data for the determination of the crystallite size of the particles. The measured crystallite size was 2.7–4.5 nm which supports the XRD data.

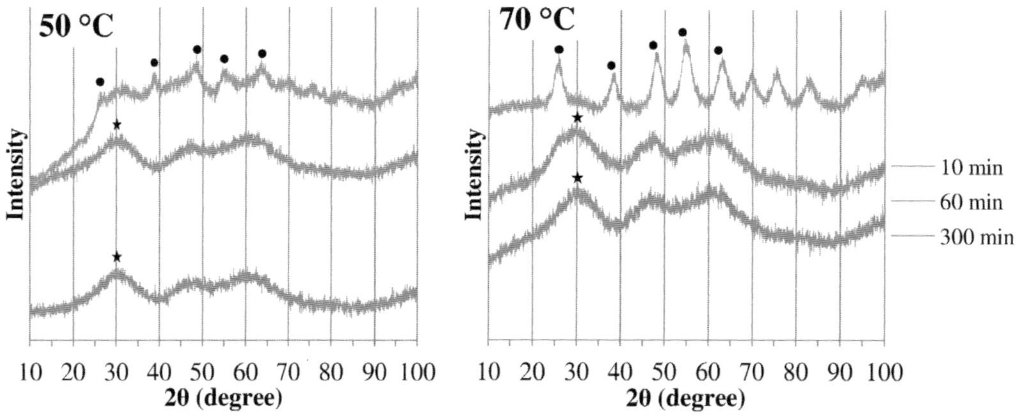

Figure 4. XRD patterns ofprepared samples. The characteristic peaks for anatase and brookite are marked as sphere and star, respectively

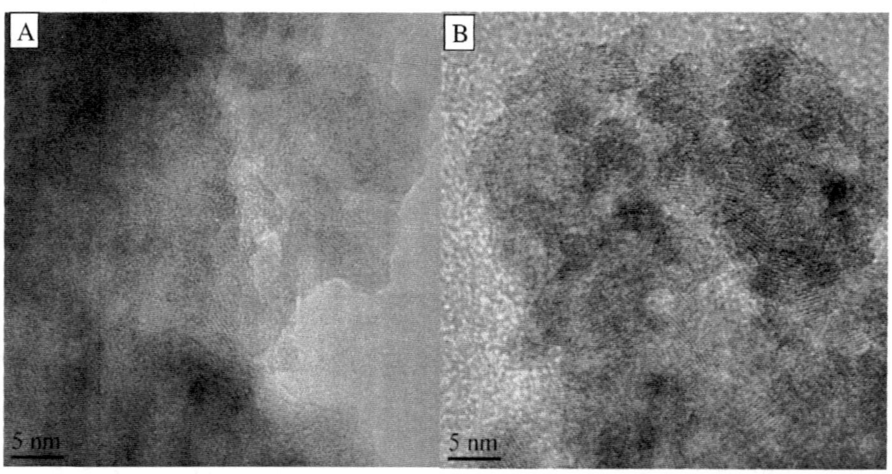

Figure 5. TEM micrographs of samples preparedin 15.0 MPa pressure for60minat50°C (A) and 70°C (B)

4. Conclusions

Titanium dioxide with high surface area and small crystallite size (approximately 4 nm)was successfully synthesized by using only supercritical carbon dioxide, water and tetra-n-butyl orthotitanate. The particle morphology was mostly spherical and particle sizes were between 0.15–0.63 μm. The smallest particle sizes formed with the shortest reaction times of 10 minand the particle size increased with increase in temperature and reaction time up to 60 min. The specific surface areas of all samples were very high ranging from 274–567 m^2/g.The highest surface areas formed with the shortest reaction times due to higher amounts of amorphous titanium dioxide. With the more crystallized samples the surface area was still high with no significant change by increasing temperature or reaction time above 60 min. The crystallinity of the particles increased by increasing temperature, reaction time or both. XRD results showed the formation of two titania phases: anatase and brookite. However, the amount of brookite phase in the synthesized powders remained uncertain and will need further studies.Moreover, the crystallization behaviour of titanium dioxide in supercritical carbon dioxide demands additional research, e.g., the effect of other reaction parameters like pressure to formation of titanium dioxide should be studied.

References

[1] Kandiel T, Robben L,Alkaim A, and Bahnemann D 2013 Brookite versus anatase tio2 photocatalysts: phase transformations and photocatalytic activities*Photochem. & Photobiol. Sci.***12**4 pp 602–609

[2] Nikkanen J-P, Heinonen S, Saarivirta E.H, Honkanen M and Levänen E 2013 Photocatalytically active titanium dioxide nanopowders: Synthesis, photoactivity and magnetic separation *IOP Conf. Ser. Mater. Sci. Eng.***47** 012066

[3] Fröschl T, Hörmann U,Kubiak P,Kučerová G, Pfanzelt M, Weiss C.K, Behm R.J, Hüsing N, Kaiser U, Landfester K and Wohlfahrt-Mehrens M 2012 High surface area crystalline titanium dioxide: potential and limits in electrochemical energy storage and catalysis*Chem. Soc. Rev.***41** p 5313

[4] Zhang X, Heinonen S and Levänen E 2014 Applications of supercritical carbon dioxide in materials processing and synthesis*RSC Adv.***4**105 pp 61137–52

[5] Sanli D, Bozbag S.E and Erkey C 2012 Synthesis of nanostructured materials using supercritical co2: part i. Physical transformations*J. Mater. Sci.***47** pp 2995-3025

[6] Beckman E.J 2004 Supercritical and near-critical co2 in green chemical synthesis and processing*J. Supercrit. Fluids***28** 2–3 pp 121–191

[7] XieH. Q, XiT. G, ZhangQ. H and WuQ. R2003 Study on the phase transformation behavior of nanosized amorphous TiO2*J. Mater. Sci. Technol.***19**5 pp 463–466

[8] Prefectural A and Nishishinwari G 2001 "Effects of heat treatment on photocatalytic property of sol–gel derived polycrystalline tio2," *J. Sol–Gel Sci. Technol.***22** pp 83–89

Influence of binding composition on the structure and properties of steel work-pieces obtained by injection moulding

E A Darenskaia[1], O Yu Vaulina[1], Yu V Myachin[1] and S N Kulkov[1,2,3]

[1]Tomsk Polytechnic University, 634050 Tomsk, Russia
[2]Tomsk State University, 634050 Tomsk, Russia
[3]InIstitute of Strength Physics and Materials Science of Siberian Branch Russian Academy of Sciences, 634055 Tomsk, Russia

E-mail: ivanovaea@tpu.ru

Abstract. The paper presents the results of the research analyzing the influence of binding composition on the structure and properties of the stainless steel samples obtained by injection moulding technique. It have been determined the tailored composition of binding, which provides sufficient feedstock viscosity, low porosity of work-pieces, etc. Three binding compositions polypropylene and paraffin wax have been studied: 1:6, 1:2, 2:1, respectively. Stearic acid has been used as a SAS (surface active substance).The results have shown that the binding compositions between 1:6 and 1:2 polypropylene and paraffin wax provides sufficient viscosity feedstock, dimensional stability during debinding and sintering, and the high density of the final product.

1. Introduction

Manufacturing of work-pieces using the method of powder injection moulding (PIM) is a prospective technology [1-4]. Raw material for this technology is powders (metal or ceramic) and a binding substance. The binder is one of the most important factors, which determines the success of the production of parts obtained by injection moulding [5-7].In most cases the binders are a mixture of organic components in which the main components are the natural waxesand synthetic polymers. Other substances may be added to modify the properties [8].In metal injection moulding (MIM) multi-component bindings are used, one of which is a low-molecular compound and another – a high-molecular one [9]. It provides sufficient mechanical strength to a work-piece after removal of a low-molecular component of a binding substance and good gas permeability for its subsequent removal. Practically, two-component bindings with additives of surface-active substance (SAS) have found wide application [10]. The ratio of binding components may be varied, and the content of a high-molecular component may stay in the range from 80 to 20% volume ratio [11].

The aim of this paper was toestablish an optimum ratio of components of binding for quality providing a feedstock and the green and sintered bodies.

2. Materials&Methods

Feedstock was produced from metal powder mixture (Table 1) and the binders with various compositions (Table 2) of high-molecular compound – polypropylene (PP) and low-molecular compound – paraffin wax.The binder № 1 contains 15% wt. PP, the binder № 2 – 35% wt. PPand the binder № 3 – 65% wt.

Contact angle of the binder was defined using formula$=\frac{\theta_{left}+\theta_{right}}{2}$, where $\theta_{left}$and θ_{right}– measured values of the superficial wetting angle on the left and right side.

Samples were produced in two stages: debinding– heating up to 550°C and holding at this temperature for two hours; sintering – heating to 1380°C and holding at this temperature for two hours.

Metallographic analysis of samples was conducted using LaboMet-I optical microscope. X-ray analysis was performed using a diffractometer with CuKa irradiation. Density of sintered samples was calculated by Archimedes method with weighing in ethyl alcohol.

Table 1. The chemical composition of the metal powder mixture

Chemical element	C	Cr	Fe
Concentration, % wt.	0.16 – 0.25	12 – 14	~84

Table 2. Three composition of binder

Number binder	**Content, % wt.**	
	Polypropylene	**Paraffin wax**
1	15	85
2	35	65
3	65	35

3. Results and Discussions

Figure 1 (a) shows that increasing of the amount of polypropylenein the binder led to a higher melting temperature, from 144 to 216°C. Binder № 3 has the highest meltingpoint, since it contains larger amount of polypropylene.

All considered binders for all selected temperatures have a contact angle less than 90 degrees (Figure 1 (b)) that means all of these binders have acceptable wetting ability.

The increase of the polypropylene content in the binder leads to growth of its melting temperature and decrease of its wetting ability due to an increasing of binder viscosity. Binders № 1 and № 2 had shown almost total wetting of steel plates at temperatures 1.2· melting point, but at such temperature values can begin burning of low-molecular compound (paraffin wax).

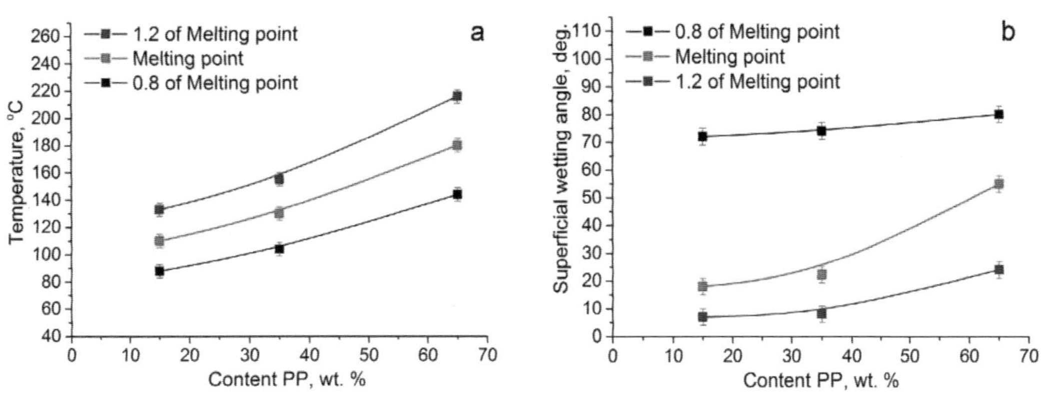

Figure 1. Effect of the polypropylene content in the binder on the melting point – *a* and superficial wetting angle –*b*

Samples produced from feedstock with the binder № 1 and 2 (15 and 35% PP) conserved the shape after debinding and sintering, whilethe sampleobtained using feedstockwith the binder № 3 (65% PP)

deformed after debinding and had porous.Obviously this is due to insufficient content of wax in the binder, which should generate channels for removing of polypropylene degradation products during at the initial time of debinding process [12]. Therefore we did not explore the sample sintered from feedstock with the binder № 3 (65% PP).

X-ray analysis showed that samples sintered with the binder № 1 and 2 (15 and 35% PP) has α-Fe structure (Figure 2). The lattice parameters α-Fe in the samples has a similar values: 0.2876 nm in the sample of the sintered feedstock with 15% polypropylene, and 0.2871 nm – 35% polypropylene.

The values obtained for the lattice parameters are compared with iron alloying with chromium: the value of the lattice parameter of pure ferrite 0.286 nm and ferrite Fe-Cr-alloy – 0.2876 nm.

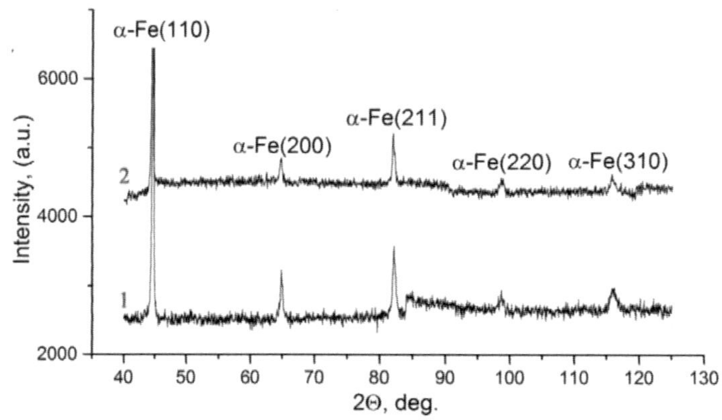

Figure 2. Diffractogram from the surface of the sample after sintering at 1380°C: *1* – binder № 1, *2* – binder № 2

The structure of the steel after sintering with the binder №1 (15% PP) has the form bainite (Figure 3(a)). The sample of the sintered feedstock with the binder №2 (35% PP) has a ferritic structure with grain sizes from 24 to 96 μm (Figure 3(b)). Data of the metallographic analysis confirm results of X-ray diffraction, that the structure is α-Fe.

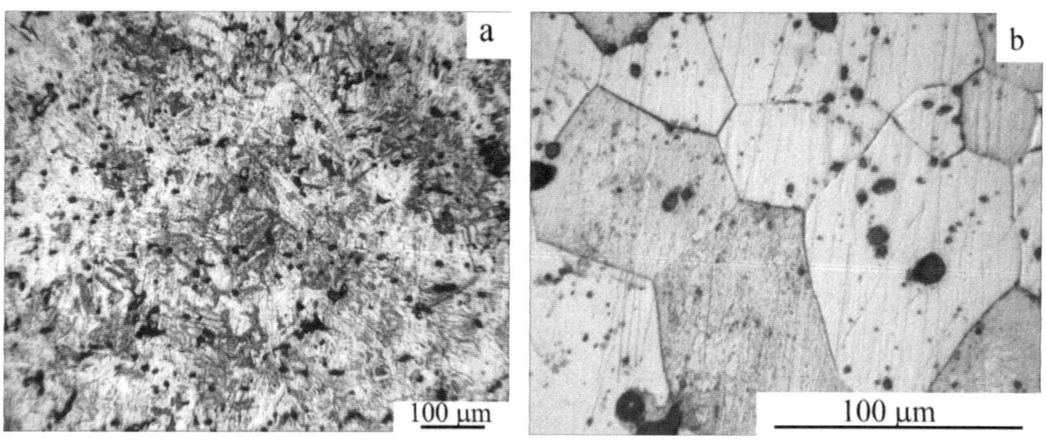

Figure 3. Microstructure of the steel after sintering at 1380°C: *a* – binder № 1, *b* – binder № 2.

Determination of density of sintered samples showed that samples, produced from the feedstock with the binder №1 (15% PP) and the binder №2 (35% PP) have densities 6.26 and 6.80 g/cm^3, respectively. This represents 81-88% of the density of cast steel with analogous chemical composition.

4. Conclusion

The wettingability binder depends on its composition - increasing of high-molecular compound in the binder leads to an increase in melting temperature and decreasing of wetting angle. The optimum ratio of the components of the binder is 15:85 and 35:65 polypropylene and paraffin wax, respectively. This composition of binder provides enough viscosity of feedstock for injection moulding, preserving the shape during debinding and sintering processes, and the satisfactory density of the final product.

Acknowledgements

The work was partially supported by Ministry of Education and Science of the Russian Federation, project № 14.578.21.0035-RFMEFI57814X0035.

References

[1] Rak Z S, 1999 *Powder Metallurgy and Metal Ceramics* **38** (3-4) 126
[2] Myachin Yu V, Vaulina O Yu, Darenskaia E A, Buyakova S P, Turuntaev I V and Kulkov S N, 2016 Structure and properties of powder steel obtained by metal injection molding *Perspective materials (Moscow: Limited Liability Company "Intercontact Science")* **7** 73 (in Russian)
[3] Isaev A, Grechishnikov V, Pivkin P, Ilyukhin Yu, Kozochkin M, Peretyagin P, 2016 *Építőanyag-JSBCM* **68** (2) 46 http://dx.doi.org/10.14382/epitoanyag-jsbcm.2016.8
[4] Egész Á, Gömze L A, 2014 *Építőanyag-JSBCM* **64** (4) 125 http://dx.doi.org/10.14382/epitoanyag-jsbcm.2014.23
[5] Bleyan D, Hausnerova B, Svoboda P, 2015 *Powder Technology* **286** 84
[6] Jin-Woong Lee, Suman Timilsina, Gi-Woo Kim, Ji Sik Kim 2016 *Powder Technology* **302** 187
[7] Ming-Shyan Huang, Hung-Chuan Hsu 2009 *Journal of Materials Processing Technology* **209** 5527
[8] Parhomenko A V, Samboruk A R, Ignatov S V, Kostin D V, Shul'timova A S, 2013 Development of binding substances in granulates for MIM-technologies *Vestnik of Samara State Technical University, Technical Sciences Series* **2** 91 (in Russian)
[9] Tirumani S Shivashankar, Ravi K Enneti, Seong-Jin Park, Randall M German, Sundar V Atre, 2013 *Powder Technology* **243** 79
[10] LI Yi-min, LIU Xiang-quan, LUO Feng-hua, WEJ Jan-lin 2007 *Trans. Nonferrous Met. Soc.* **17** 1
[11] German R M, Bose A, 1997 Injection molding of metals and ceramics *MPIF*
[12] Angermann H H, Yang F K, Van Der Blest O, 1992 *Journal of materials science* **27** 2534

IOP Conf. Series: Materials Science and Engineering **175** (2017) 012036 doi:10.1088/1757-899X/175/1/012036

The composite reinforcement production in digital manufacturing: experimental validation of the heat transfer and cure modeling results

I Kazakov[1], A Krasnovskii[1] and A Kutin[2]

[1]Department of Composite materials, Moscow State Technological University "STANKIN", 127055 Moscow, Russia
[2] Department of Mechanical Engineering, MSTU "STANKIN", 127055 Moscow, Russia

E-mail: aa.kutin@stankin.ru

Abstract. The experimental validation of the heat transfer and cure modeling results for 8-mm fiber-reinforced thermosetting composite reinforcement is reported in this article. The temperature and degree of cure of composite reinforcement are predicted using a two-dimensional heat transfer and curing model. The model uses the infrared radiant heating theory and takes into account the heat transfer between the composite rod and the surrounding air. The implicit finite difference method was used to solve the system of governing equations. The results obtained using mathematical model was compared to experimental data: the temperature field inside the composite reinforcement was measured by means of naked thermocouple; Differential Scanning Calorimetry (DSC) was used to measure the degree of cure of the final product. Calculated and measured temperature and degree of cure fields were in good agreement.

1. Introduction

The Fiber reinforced plastic (FRP) composite rods, presenting a high strength to weight ratio, resistant to repeated loading and not subjected to corrosion. Thanks to these features they are becoming more and more widespread in the construction industry as reinforcement for concrete structures [1], electrical insulators [2] and pump rods, pit-prop anchors in the mining industry, and area of their application is constantly expanding. The composite reinforcement is usually made of glass or basalt fibers and has special outer relief providing a good adhesion to concrete. The carbon fiber has higher characteristics but rarely used because of its high cost.

Among the several available techniques to produce composite reinforcement, nidltrusion process is becoming the most widespread and cost-effective continuous processing technique in Russia. One of the main attractions of this manufacturing method is the simplicity of tooling and low labor requirements. In this process, dry reinforcement fibers in the form of continuous strands (roving) are pulled through a resin bath for impregnation. After the wetting process, the reinforcement pack is collimated in the forming taper into a performed shape before entering the heating chamber where it cured. The forming taper is typically characterized by a conical convergent shape, in order to promote compaction of the material, the removal of the air and excess resin. The circle shape of the rod cross section is fixed by spiral protrusions formed by a helical winding device. The forming taper and helical winding devices are practically always combined in one. Due to the spiral protrusions the rod acquires some anisotropic characteristics [3], but their impact on the mechanical characteristics of the rod is usually neglected. The composite reinforcement is cured in the heating chamber without touching its walls in contrast to conventional pultrusion process. The heat provided by means of infrared (IR) heaters activates the exothermic cure reaction of the thermoset resin. Finally, outside the chamber, the composite reinforcement already polymerized and is pulled by a pulling system. Then a cut-off saw cuts the part into a desired length.

2. Theoretical modeling and numerical implementation

There have been several mathematical models [2, 4, 5] presented in the literature describing the cure process during pultrusion. However, these models are based on the assumption that the outer surface of the composite rod is in direct contact with the walls of a heated die. Therefore, in relation to the nidltrusion process such models are not appropriate.

Assuming the thermal conductivity along the radius is a constant, and is absent along the die due to the dimensions of composite rod in longitudinal direction is much greater than the dimensions in the transverse direction. Then, for the steady-state nidltrusion process heat transfer equation [4] can be rewritten as

$$\rho c_v U \frac{\partial T}{\partial x} = \rho_r \cdot (1 - v_f) \cdot H_{tot} \frac{d\alpha}{dt} + \frac{\lambda_1}{r} \frac{\partial}{\partial r}\left(r \frac{\partial T}{\partial r} \right), \tag{1}$$

where ρ - density of composite, c_v – specific heat, U is the pull speed, ρ_r - density of resin, H_{tot} - the total heat generated by the exothermic reaction of resin, λ_1 - heat-transfer coefficient, r - radial direction, v_f – fiber volume fraction.

The density of composite, ρ and resin density, ρ_r can be expressed as follows:

$$\rho = (1 - v_f) \cdot \rho_r + v_f \cdot \rho_f, \quad \rho_r = \alpha \cdot \rho_r^c + (1 - \alpha) \cdot \rho_r^u, \tag{2}$$

where ρ_f is the fiber density, ρ_r^u , ρ_r^c are uncured and cured resin densities correspondingly.

The specific heat capacities of composite material and resin are calculated in a similar manner:

$$c = (1 - v_f) \cdot c_r + v_f \cdot c_f, \quad c_r = \alpha \cdot c_r^c + (1 - \alpha) \cdot c_r^u, \tag{3}$$

where c_f is the specific heat of the fiber, c_r^u , c_r^c are uncured and cured resin specific heat correspondingly.

In the present investigation the well-established Kamal model has been adopted to describe the evolution of the cure reaction [6]:

$$\frac{d\alpha}{dt} = (k_1 + k_2 \cdot \alpha^m) \cdot (1 - \alpha)^n, \quad k_i = A_i \cdot \exp\left(-\frac{E_{ai}}{R \cdot T} \right), \quad (i = 1, 2), \tag{4}$$

where k_i is the reaction rate, A_i is the pre-exponential factor, E_{ai} is the activation energy, R is the gas constant, n,m are the equation superscripts.

Usually the heating chamber is equipped with IR heaters and the boundary conditions for the governing equation are:

$$T\big|_{x=0} = T_0, \quad -\lambda_1 \frac{\partial T}{\partial r}\bigg|_{r=R} = \gamma_1 + \gamma_2, \quad \frac{\partial T}{\partial r}\bigg|_{r=0} = 0, \tag{5}$$

where T_0 is initial composite temperature at the computational domain inlet.

The net heat transfer through radiation is calculated using the Stefan-Boltzmann law [7]:

$$\gamma_1 = \varepsilon_r \cdot C_0 \cdot \left(T^4\big|_{r=R} - T_{out}^4 \right), \quad \gamma_2 = \varepsilon_r \cdot C_0 \cdot \left(T^4\big|_{r=R} - T_H^4 \right), \tag{6}$$

where ε_r is the emissivity of the surface, C_0 is the Stefan-Boltzmann constant (5.67x10^{-8} W/(m^2·K^4)), T_{out} – the steady temperature of the air in the heating chamber sections, T_H – the IR heater temperature.

The implicit finite difference method was used to solve the system of governing equations (1-6). This method is based on discretizing the space and time domain through transformation into a finite difference form and solving the subsequent system of algebraic equations for the temperature and cure fields. For each time step, starting from the second (t), the unknown temperature is determined by using eq. (1, 4) taking into account the boundary conditions, temperature and the degree of cure at the previous time step (t-1). Then, the degree of cure is determined from the equation (4) for the current time step (t). Thus, the solution is determined in all nodes of finite-difference grid.

3. Experimental work and discussion of the results

This study is focused on the experimental validation of the heat transfer and cure modeling results for 8-mm fiber-reinforced thermosetting composite reinforcement.

The E-glass fibers were used as filler. The fiber volume fraction is v_f = 0.6.

An epoxy resin CYD-128 (bisphenol-A) was blended with Vestamin IPD (cycloaliphatic diamine) as hardener with the weight ratio 388:100. According with a mixture usually used for nidltrusion manufacturing process, several additives were added to reduce the product costs and improve the quality of the product. The DSC measurements were made in dynamic mode at METTLER TOLEDO using "STAR" software. The next cure kinetic parameters were obtained:

$$\frac{d\alpha}{dt} = \left(18\cdot\exp\left(-23820/R\cdot T\right) + 2850\cdot\exp\left(-38400/R\cdot T\right)\cdot\alpha^{0.85}\right)\cdot\left(1-\alpha\right)^{1.9},\quad H_{tot} = 456.44\ J/g.$$

The properties of components for nidltrusion simulation program are listed in Table 1.

Table 1. Values of physical quantities

Property	**Value**
Fiber density, g/cm^3	2.56
Uncured resin density, g/cm^3	1.06
Completely cured resin density, g/cm^3	1.27
Heat-transfer coefficient, λ_1, W/(m·K)	0.25
Specific heat capacity of the fibers, J/(g·K)	0.84
Specific heat capacity of uncured resin, J/(g·K)	1.69
Specific heat capacity of cured resin, J/(g·K)	1.36
The emissivity of the surface, ε_r	0.7
Inlet temperature, ^{0}C	24
The IR heater temperature, T_H, ^{0}C	300

The experiment to determine the temperature field of the composite reinforcement was performed using the heating chamber equipped with IR heaters (FSR 250 W, max operating temperature 400^{0}C). The heating chamber has 6 m length and consists of 8 sections. It was experimentally determined that the surface temperature of the heater is approximately 70^{0}C higher than the temperature setting for the corresponding section. Six IR heaters are usually placed in each section of the heating chamber and total 48 IR ceramic heaters are engaged in manufacturing process. During the experiment the naked thermocouple was placed inside the rod. All sections of the heating chamber were heated up to 230^{0}C (the temperature setting).

The simulated and experimental results are presented on Figure 1 (the points denote the experimental data). The simulation reproduce well the experiment up to 250^{0}C. Probably, at the temperature above 250^{0}C cured resin degradation occurs with the release of additional heat. Since the experiments to determine the exact value of the degradation temperature for the resin were not carried out, it was decided to accept this value as the temperature limit at which the developed mathematical model can be used with sufficient reliability.

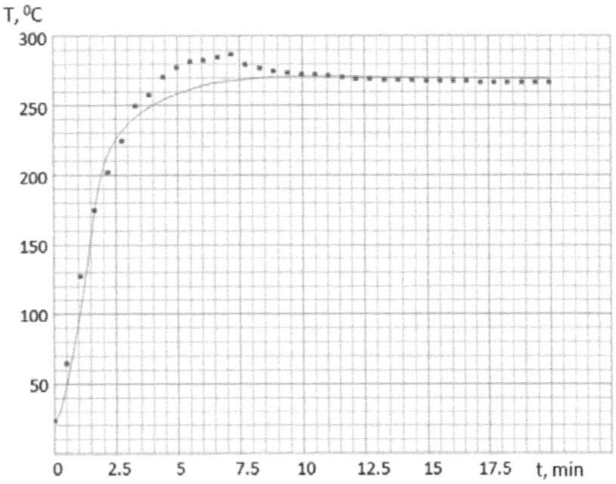

Figure 1. Temperature profiles within the rod along the heating chamber

The obtained degree-of-cure (DOC) curves are presented on Figure 2 and indicate the changes inside the composite reinforcement during the production. Estimated values of the DOC at the heating chamber output are 0.969 (for the nidltrusion speed of 2.5 m/min) and 0.942 (for the speed of 3 m/min).

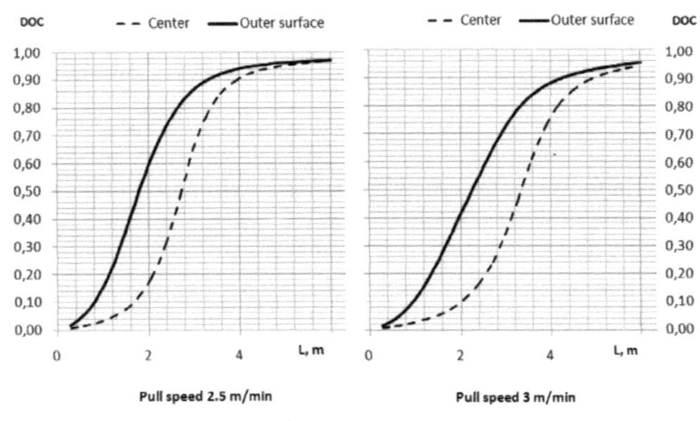

Figure 2. Degree-of-cure distributions within the rod

Several composite reinforcement rods were produced at the mentioned above temperature setting and pull speeds. After what, the samples 4x4 mm in size and weight of the composite material up to 15 mg were made by means of guillotine shears. The residual thermal effects of post-curing were determined using DSC to obtain the degree of cure of the composite reinforcement samples at the pulling speeds of 2.5 m/min and 3 m/min.

The DOC of the resin α was defined as the ratio of the difference between the thermal effects of curing the resin without filler (ΔHr) and after curing (ΔHc) of the sample to the original value of the thermal effect of the curing reaction of the resin ΔHr:

$$\alpha = \left(1 - \frac{\Delta H_c}{\Delta H_r \cdot C_c}\right), \qquad (7)$$

where C_c – the resin mass fraction in composite rod.

The table below shows the calculated and experimental values of the degree of cure.

Table 2. Calculated and experimental values of DOC

Pulling speed, m/min	ΔH_c, J/g	DOC (computed)	DOC (experimental)
2.5	2.715	0.969	0.973±0.03
3	6.25	0.942	0.939±0.03

4. Conclusion

The theoretical and practical work was presented regarding the mathematical modeling based on the mutual interactions between heat transfer, cure reaction and variation in the material properties for the nidltrusion process. Two-dimensional heat transfer and curing model is used to predict the temperature and degree of cure of composite reinforcement. Complex boundary conditions are considered, including prescribed temperature and heat radiation. The model uses the IR radiant heating theory and takes into account the heat transfer between the composite rod and the surrounding air.

The results obtained using mathematical model was compared to experimental data: the temperature field inside the composite reinforcement was measured by means of naked thermocouple; Differential Scanning Calorimetry (DSC) was used to measure the degree of cure of the final product. It was shown the calculated and measured temperature and degree of cure fields are in good agreement.

This thermal response model will be used in further studies of nidltrusion process. If the necessary material parameters are determined, this model can be further applied in different kinds of composite materials.

Acknowledgment

This research was made under financial support of Russian Ministry of education and science, project task N 9.2664.2014/K.

References

[1] Benmokrane B and Masmoudi R 1996 FRP C-Bar as Reinforcing Rod for Concrete Structures *The Canadian Society for Civil Engineering* **Advanced Composite Materials in Bridges and Structures 2nd International conference**

[2] Grigoriev S N and Kazakov I A and Krasnovskii A N 2016 The stress-strain state determination of a large diameter composite rod during pultrusion process *Proceedings of the International Conference on Civil, Architectural, Structural and Constructional Engineering, CRC Press* **66** Dong-A University, Busan, South Korea, 329–333 DOI: 10.1201/b19961-71

[3] Grigoriev S N and Krasnovskii A N and Khaziev A R 2012 *Mechanics of pultruded composite anisotropic solid rod. Plastics* **3** 18-25.

[4] Grigoriev S N and Kazakov I A and Krasnovskii A N 2015 Nidltrusion as the method for composite reinforcement production *Part I: Mathematical model of heat transfer and cure. Proceedings of the 2nd International Conference of Advanced Materials, Mechanical and Structural Engineering (AMMSE 2015) CRC Press* **42** Je-ju Island South Korea 207–210. DOI: 10.1201/b19934-43.

[5] Kutin A A and Krasnovskii A N and Kazakov I A 2014 The Fiber Orientation Angle Determination for a Composite Anisotropic Solid Rod in Pultrusion *Advanced Materials Research* **941(26)** 2273 - 2278 DOI: 10.4028/www.scientific.net/AMR.941-944.2273.

[6] Kamal M R 1974 Thermoset Characterization for Moldability Analysis *Polymer Engineering and Science* **14** 231-239.

[7] Bai Y et al. 2007 Modeling of thermal responses for FRP composites under elevated and high temperatures *Composites Science and Technology* **68** 47-56 DOI: 10.1016/j.compscitech.2007.05.039.

Research of products of high temperature synthesis flowing in the rotation conditions

G Ksandopulo[1], A Baideldinova[1], Y Riabikin[2], L Mukhina[1], E Ponomareva[1], N Vasilieva[3]

[1]Institute of Combustion problems, Almaty, Kazakhstan
[2]Institutr of Nuclear physics Almaty, Kazakhstan
[3] Chemical department of the Kazakh national university

E-mail: milabrega@yandex.ru

Abstract. The method of production of materials by out-furnace process of self-propagating high temperature synthesis (SHS), flowing in the conditions of action of centrifugal force, is developed presently. The primary purpose of working is achievement high level of generating of energy and use of it for forming of steady meta-stable crystalline phases with an uncommon set of physical and chemical properties.

1. Introduction

To study the Self-Propagating High-Temperature Synthesis (SHS) processes under the centrifugal forces effect there has been created a unique installation i.e. a high-temperature centrifuge [1]. Three steel reactors with internally installed quartz tubes stuffed with two-layer or multilayer powder reaction mixtures are rotated around a vertical axis. Ignition occurs from the centre after the engine is started and the installation is operated using routine mode. Further the combustion wave front spontaneously propagates along the reactor axis (Figure 1).

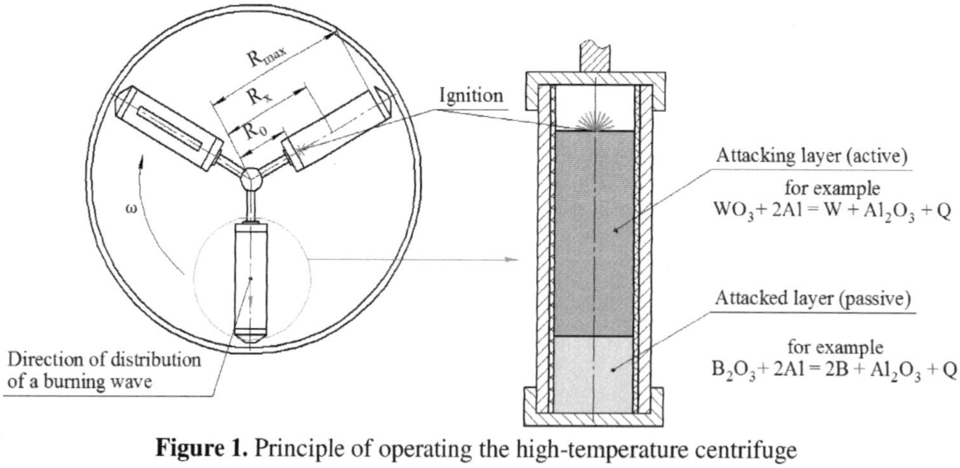

Figure 1. Principle of operating the high-temperature centrifuge

IOP Conf. Series: Materials Science and Engineering **175** (2017) 012037 doi:10.1088/1757-899X/175/1/012037

2 Experiments

The installation rotates with frequency ranging from 500 up to 3000 RPM corresponding to the centripetal acceleration up to 2000 g. According to (1), it is the angular velocity, quadratic dependence as a varying value, allows for modification of the experiment conditions in the best way. Therefore there is no point to change initial radius of the combustion wave front position. It is about four centimeters.

The total process time is 0.2-0.4 seconds, and temperature inside the reactor is 2500-3500 K.

$$F_C = m\omega^2 R \tag{1}$$

An attacking layer is ignited the first. There occur the processes of metal reduction from the oxide and formation of the adiabatic combustion wave [2]. This system can be considered adiabatic starting from the point when the heat release rate becomes considerably larger than the rate of heat loss to the environment. In the stationary conditions when the reaction mixture is affected only by the gravity force, molecules of the reduced metal oxide come together as particles and form an ingot outside the combustion wave front. Under the centrifugal forces effect the picture cardinally changes [3]. The metal particles come together to form larger clusters, they form an accelerated flow that overtakes the combustion front, and hits the nearest fresh mixture layer resulting in a heat-and-mass transfer to the pre-reaction zone (Figure 2). This effect is a key distinctive property of the described method. At the moment of collision kinetic energy of the clusters turns into the heat energy thus adding one more component to the heat balance of the system. Considering that the powder mixture is a porous medium and the metals clusters' temperature significantly exceeds melting temperature, non-viscous melted particles penetrate through the surface layer and go down the mixture to a certain depth. Once metal gets to the borders of the opposite charge powder grains it forms additional micro-centers of ignition. The combustion wave front is constantly extended and accelerated. All these factors lead to a continuous temperature increase before the combustion wave front, the rate of the latter's propagation along the reactor axis accelerates and the front is expanded. There takes place the so-called non-chain autoacceleration of the SHS wave. The figure below also illustrates velocity decrease at the border with the attacked layer.

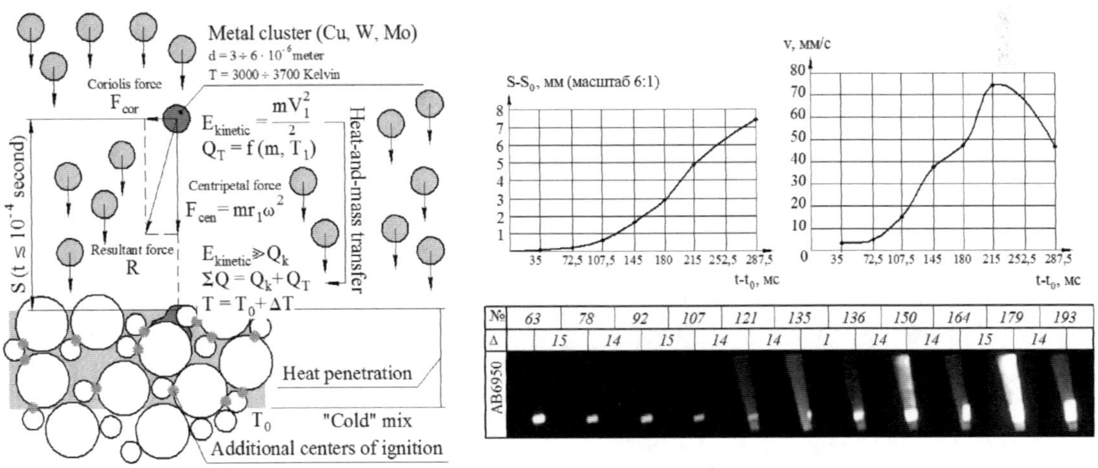

Figure 2. The SHS process mechanism within the centrifugal forces effect field
a – flow of accelerated centrifugal clusters of the reduced metal; b – expansion of the reaction wave front observed on the time-lapse snapshots, their velocity and relocation plots

3. Results and Discussions

When the active layer is fully burnt down the metal forms are ingot in the reactor bottom, while lighter slag (aluminum oxide) stays on the ingot surface and partly on the quartz tube walls. In the two-layer or multilayer system the total mass of the metal reduced in the attacking layer falls down onto the second - «cold» attacked layer. It is this thermo-mechanical effect that is a key purpose of the experiments. Attacked layers differ in their configuration. These can be low-calorie powder reaction mixtures that cannot be otherwise ignited, e.g. either mixtures based on the rare earth elements (REE), boron and barium oxides, or the oxides powders. One of the most attractive attacked layers is the carbon powder layer. The other versions of the attacked systems are various particle entrapping screens such as plexiglass, grain quartzite, and pressed graphite. If the reactor has a hole in its cover like a nozzle, the metal particles flow can be taken out to the external screen and then covered with any necessary substances. Figure 3 illustrates the results of the copper clusters flow effect on the installation parts. One can estimate its intensity by numerous burns and thermo-mechanical damage. In certain cases the metal is dispersed covering any available surfaces with a thinnest layer whilst in other cases it forms a film locally.

Striking and thermal strength of the attacking clusters flow is able to initiate highly-intensive high-temperature processes. The reaction duration at each cross-section does not exceed 10^{-4} seconds. Subsequently the adiabatic mode is interrupted, obtained products get cold rapidly, natural processes of the chemical reaction do not have any more time to be completed and there occurs instant "freezing" at the intermediate stages.

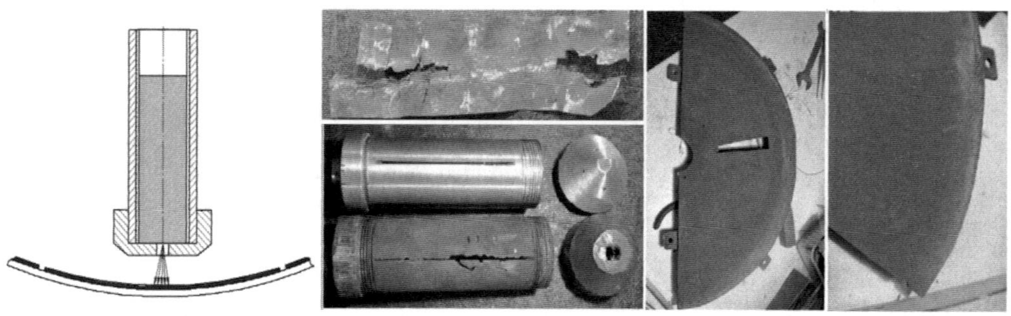

a b

Figure 3. External attacked layer
a – a plot of the combustion products outflow to the external screen; b – the result of the copper clusters flow effect outside the reactor.

This explains abundance of the free valency substances received during the experiments. Figure 4 illustrates a list of particularly interesting compounds that, according to the X-ray analysis, are contained in the food stuff.

There are found tungsten and molybdenum carbides, various silicides, borides, aluminates, and intermetallides; on the ERP most of these substances give peaks corresponding to the presence of the free valency, moreover most of them exist only in our laboratory. The most well-known ones are as follows: $CuBO_2$ – transparent conductive p-type oxide with high individual electric conductivity called delafossite, MgB_2 – magnesium diboride, a superconductor with high temperature transition (40 K), and $CuAlO_2$ – a semiconductor. Synthesis of these free valency substances and their extraction from the solid solutions gives a great variety of the possibilities. The most obvious possibility is to modify metals, as well as modify a crystal lattice during its solidification [4, 5]. Besides, the other possibilities are to provide additives for rocket fuel, high-temperature superconductors, superconductors [6], semi-conductors, paramagnetic materials, and affect the biological processes occurring with formation of the free valency.

IOP Conf. Series: Materials Science and Engineering **175** (2017) 012037 doi:10.1088/1757-899X/175/1/012037

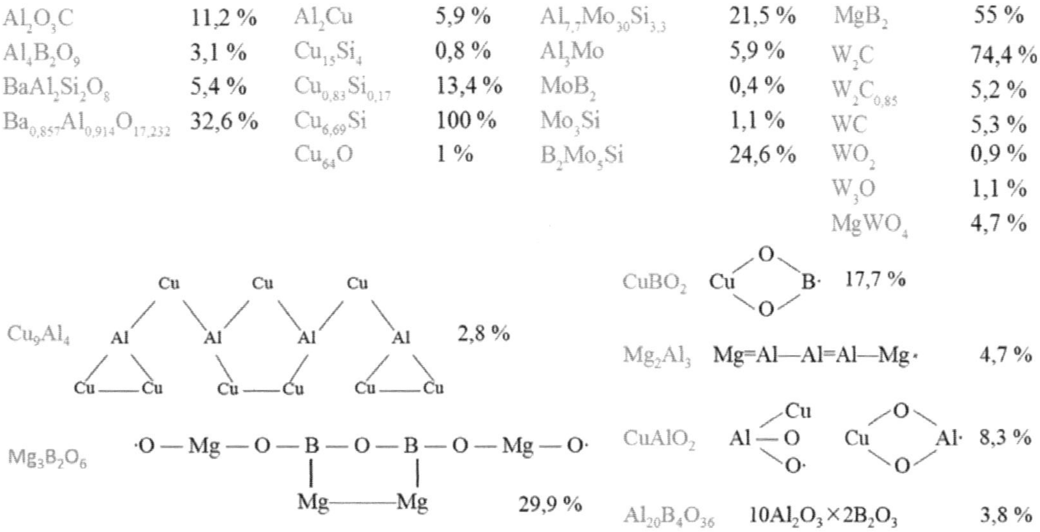

Figure 4. A list of nontrivial substances detected in the combustion products by the X-ray phase analysis

One more illustration of the possibilities given by this method is the result of attack of the common graphite powder by the copper clusters. Figure 5 illustrates a diagram of the reactor charging and a picture of the products. The picture shows that the product has a slag pellet in the bottom whilst on the top there is an ingot covered with a powder layer. According to the comparative analysis of the compensatory laser spectroscopy data, the copper ingot surface is covered with grapheme but not with graphite. Such experiments often reveal diamond carbon structures testifying to the presence of serious flow shock impacts, however researchers are much more interested in obtaining nanocarbon films.

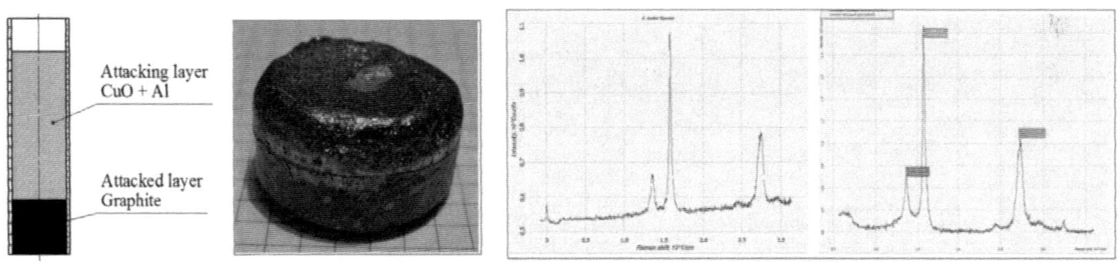

a b c d

Figure 5. Attack of a common graphite layer by the copper clusters
a – diagram of reactor charging; b – ingot picture; c – the result of the sample study from the ingot surface by the compensatory laser spectroscopy; d – the result of a similar study of the graphene sample

4. Conclusion

There is still a task to identify a bond between concentration of necessary substances in the combustion products and rotation frequency of the installation. Apart from that, it is important to find efficient methods to be used for extraction of these compounds in their pure form. To execute this project the chemical and analytical studies are being currently carried out and a higher intensity installation is being started-up.

References

[1] Ksandopulo G I, Shevchenko V N and Baideldinova A N 2010 High-Temperature Centrifuge / RoK Patent No 68317 as of 26.05.2010

[2] Ksandopulo G I, Baideldinova A N, Omarova K I, Ainabayev A M 2014 *Eurasian Chemical-Technological Journal* **16** (1) 49

[3] Baideldonova A, Ksandopulo G, Mukhina L 2016 *IOP Conf. Ser.: Mater. Sci. Eng.* **123** 012032 http://dx.doi.org/10.1088/1757-899X/123/1/012032

[4] Gömze L A, Gömze L N 2009 *Épitőanyag-JSBCM* **61** (2) 38 http://dx.doi.org/10.14382/epitoanyag-jsbcm.2009.7

[5] Gömze L A and Gömze L N 2011 *IOP Conf. Ser.: Mater. Sci. Eng.* **18** 082001 http://dx.doi.org/10.1088/1757-899X/18/8/082001

[6] Tolendiuly S, Fomenko S M, Abdulkarimova R G, Mansurovb Z A, Dannangoda G C, and Martirosyan K S 2016 *International Journal of Self-Propagating High-Temperature Synthesis*, **25** (2) 97

Influence of mechanical activation of steel powder on its properties

O Yu Vaulina[1], E A Darenskaia[1], Y V Myachin[1], I E Vasilyeva[1], S N Kulkov[1,2,3]

[1]Tomsk Polytechnic University, [2]Tomsk State University,
[3]Institute of Strength Physics and Materials Science of Siberian Branch Russian Academy of Sciences, Tomsk, Russia

E-mail: kolgay@tpu.ru

Abstract. It has been studied properties of stainless steel based powders after mechanical activation using planetary ball milling technique. It have been shown that after one minute mechanical activation porosity of sintered steel is less than 5%, which is less than the porosity of the sintered steel powder without mechanical activation. The sample without activation has austenite state, which changes after activation toaustenite and ferrite mixtures. X-ray analysis confirmed that the mechanical activation leads to a change in the phase state of the samples: the samples without activation of the FCC structure (γ-Fe), after activation - FCC (γ-Fe) and BCC (α-Fe). The hardness increases at increasing activation time from 800 MPa for the sample without mechanical activation to 1250 MPa for the sample with the activation time of 10 minutes.

1. Introduction

Mechanical activation is a way to accelerate the physical and chemical processes and is becoming more widely used due to action on materials a mechanical stresses **[1-4]**. Planetary milling is a high-energy activation process and may improve an initial state of powder mixtures before its using for preparing samples. Therefore one can needs to investigate the influences of such high-energy treatment on properties of powders and sintering body **[5-11]**.

The aim of this work is to investigate the influence of mechanical activation of stainless steel powder on the properties of sintered samples produced using injection moulding technology.

2. Materials and methods

Samples were prepared using powder metallurgy technology by mixing the initial powders with mechanical activation for 1, 5 and 10 minutes. The chemical compositions of samples are shown in Table 1.

Determination of samples porosity was carried out using "Analyzer of fragments microstructure for solid bodies SIAMS 700TM". The average size of pores was determined by randomly secants. Analysis of phase composition was carried out by X-ray with CoKα radiation.

Metallographic analysis was performed on"Labomet-I" metallographic microscope with system visualizationimage. Microhardness was measured with an indenter (diamond quadrangular pyramid) at a load of 50 g.

Table 1.-The chemical composition of the material

C, %	Cr, %	Ni, %	W, %	Fe, %
0,03	17	12	2	Base

3. Results and Discussion

Porosity

All samples have a residual porosity after sintering.Metallographic examination of the polished sample surface allowsevaluating the presence of pores, their number, shape, size and size distribution.On the polished surface one can see areas with a high concentration of pore and comparable size of the area where almost there are no pores.The pores are usually irregular shapes and different sizes. The results of theporosity size distributions are shownon the Figure 1.The distribution of pores in the sintered samples is uni-modal.

Size distributions are shown that the majority pore (up to 90%) for all samples have a size not more 10 micrometers. Samples with mechanical activation with 5 and 10 minutes have a larger pores size up to 70 micrometers.

It can be seen that the lower porosity (Figure 1) and a smaller average pore size (figure 1b) has sample after mechanical activationduring 1 min and the highest porosity values had samples with mechanical activation for 5 and 10 minutes.

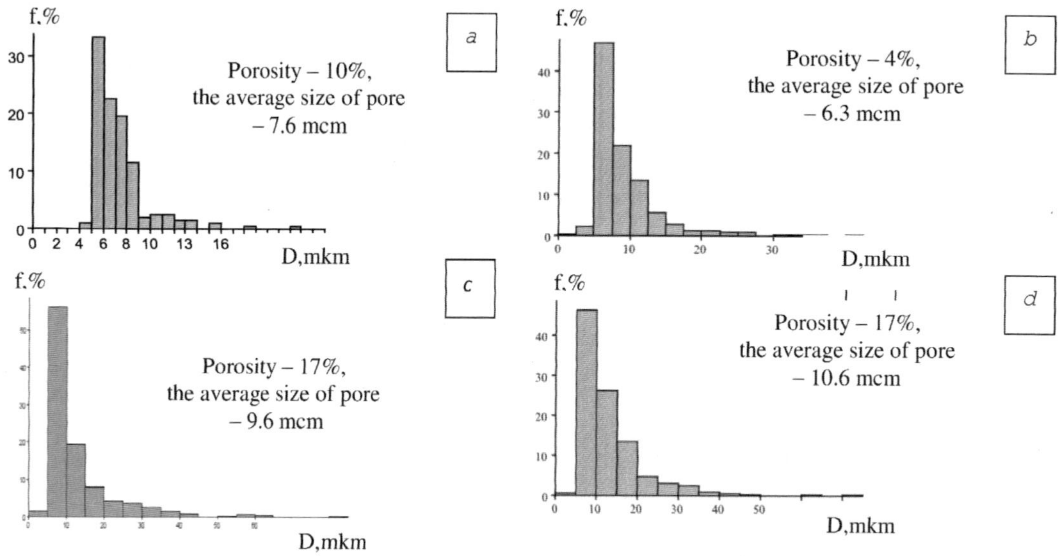

Figure 1 - Distribution of pore size for a sample a) in initial state without mechanical activation; after mechanical activation 1 (b), 5 (c) and 10 minutes (d)

Microstructure and Phase Composition

Metallographic studies after etching (Figure 2) are shown that sample without mechanical activation has an austenitic structure with a large number of twinsdue to adding of nickel to steel, which expands the range of γ-phase.

For samples after mechanical activation one can observe two areas first the ferrite – dark areas and second austenite - light areas. This also was confirmed by x-ray analysis.

X-ray analysis showed that the sample in initial state (Figure 3a) has a FCC structurewith lattice parameter 0.359561 nm. After mechanical activation (Figure 3b) were found two phases of the FCC structure(lattice parameters 0.36055, 0.3598, 0,35632nmfor 1, 5 and 10 min respectively) and BCC

IOP Conf. Series: Materials Science and Engineering **175** (2017) 012038 doi:10.1088/1757-899X/175/1/012038

lattice (0,28844, 0,28961, 0,28983 35632nmfor 1, 5 and 10 min respectively respectively). These lattice parameters slightly differ from pure γ-Fe (0,356 nm) and α-Fe (0,286 nm).

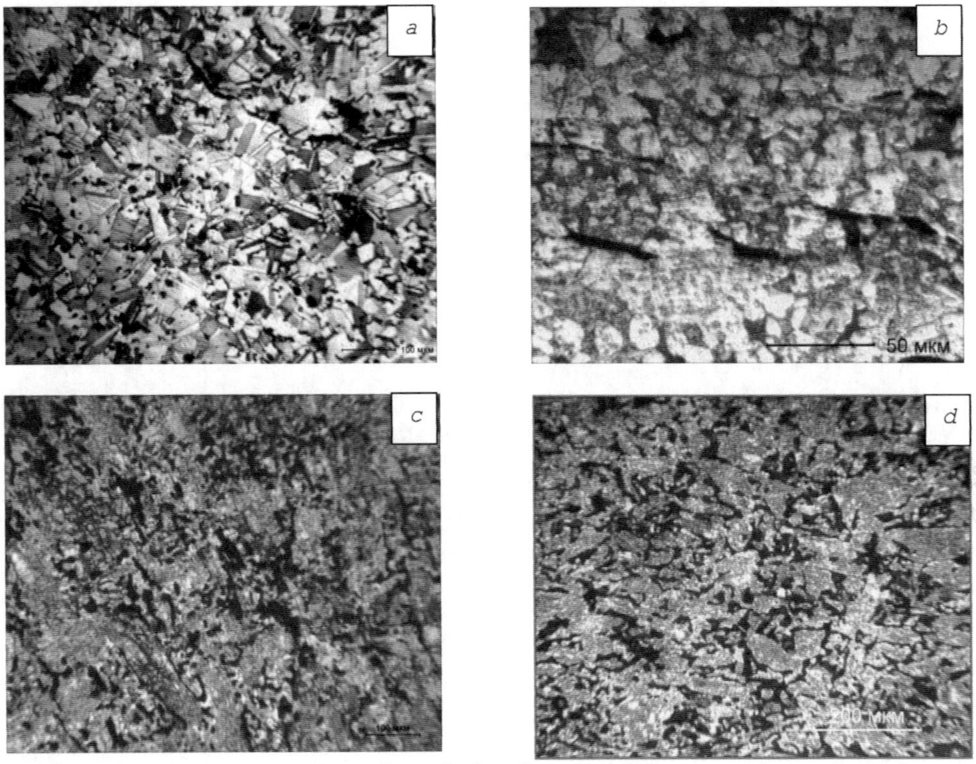

Figure 2 - The microstructure of samples etched surface: a) initial state and for mechanical activation 1 (b), 5 (c) and 10 minutes (d)

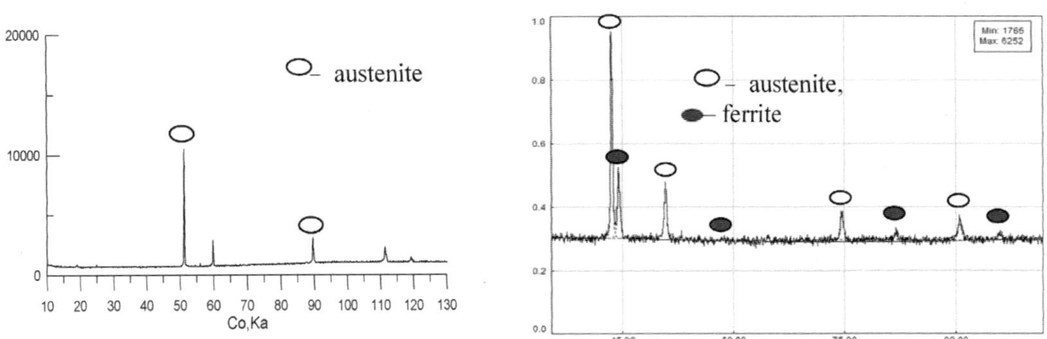

Figure 3 - Diffraction patterns on the surface of samples: a) without mechanical activation; b) a mechanical activation for 10 minutes

Microhardness

During an increasing the activation time hardness increases from 800 MPa(Table 2) in initial state and its increasing up to 1250 MPa for the sample with the activation of 10 minutes, which corresponds to the hardness of stainless steel. Despite the porosity, hardness corresponds to the hardness of the cast material.

Table 2. Microhardness value for samples with different activation time

Sample	**Microhardness**, MPa
without activation	815±11
1 min	794±50
5 min	1196±100
10 min	1265±120

4. Conclusion

After one minute mechanical activation porosity sintered steel is less than 5%, which is less than the porosity of the sintered steel powder without mechanical activation.

The sample without activation has austenite state, which changes after activation toaustenite and ferrite mixtures. X-ray analysis confirmed that the mechanical activation leads to a change in the phase state of the samples: the samples without activation of the FCC structure (γ-Fe), after activation - FCC (γ-Fe) and BCC (α-Fe).

The hardness increases at increasing activation time from 800 MPa for the sample without mechanical activation to 1250 MPa for the sample with the activation time of 10 minutes.

Acknowledgements

The work was partially supported by Ministry of Education and Science of the Russian Federation, project № 14.578.21.0035-RFMEFI57814X0035.

References

[1] Juhász Z A, Opoczky L 2003 *Épitőanyag–JSBCM* 55 (3) 86 http://dx.doi.org/10.14382/epitoanyag-jsbcm.2003.16
[2] Gömze L A 2010 *Mater. Sci. Forum* **659** 19 *www.scientific.net/MSF.659.19*
[3] Mohamad Nor N.H., Ismail M.H., Abu Kasim N.A., Muhamad N. and Taib M.A. 2014 *Applied Mechanics and Materials* **465-466**
[4] Mucsi G. 2016 *Épitőanyag-JSBCM* **68** (2) 56 http://dx.doi.org/10.14382/epitoanyag-jsbcm.2016.10
[5] Hueller M., Chernik G.G., Fokina E.L., Budim N. 2008 *Reviews on advanced materials science* **18** 366
[6] Bhimasena N.M., Berenika H., Tirumani S.S. 2015 *Powder Technology* **283** 24
[7] Gaffet E., Bernard F. 2002 *Mechanically activated powder processing* **27** 47
[8] Rogacheva A. S., Shkodicha N. F., Vadchenkoa S. G., Barasb F., Chassagnonb R., Sachkovaa N. V., and Boyarchenko O. D. 2013 *International Journal of Self Propagating High Temperature Synthesis* **22** (4) 210
[9] Myachin Y.V., Darenskaya E.A., Vaulina O.Yu., Buyakova S.P., Turuntaev I.V.,Kulkov S.N. 2016 *Advanced materials* **7** 73
[10] Pan Lei, Shparkovich A.A., Bolshunova A.B., Vaulina O.Yu., Turuntaev 2016 *Materials and technologies of new generations in modern materials science* 89-93
[11] Sosnovskaya A.A., Min Li, Darenskaya E.A. *Materials and technologies of new generations in modern materials science 97-99*

Transport Properties of Aliovalent Substitution Solid Solutions of the System (1-x)PbF_2-$x$$YF_3$-$SnF_2$

Y V Pogorenko[1], R M Pshenychnyi[1], V I Lutsyk[2,3] and A O Omel'chuk[1]

[1]Vernadskii Institute of General & Inorganic Chemistry of the Ukrainian NAS, 03680 Kyiv, Ukraine
[2]Institute of Physical Materials Science, SB RAS, 670047 Ulan-Ude, Russia
[3]Chemical Department, Buryat State University, 670000 Ulan-Ude, Russia

E-mail: vluts@ipms.bscnet.ru

Abstract. In polycrystalline solid solution of $Pb_{1-x}Y_xSnF_{4+x}$ ($0 < x \leq 0,17$) with a structure of β-$PbSnF_4$ in temperature 293 – 523 K the fluorine anions occupy three structurally nonequivalent positions which differ in the local environment and mobility. The concentration of mobility fluoride anions at 300 K is almost independent of the content of heterovalent substituent on YF_3 and at temperatures above 430 K increases with grow concentration of the YF_3. Fluoride ion exchange between nonequivalent subsystems increases with raising temperature, which causes an increase in conductivity. The electronic component of conductivity the synthesized samples by 2 orders of magnitude lower than the ion.

Keywords: solid fluoride-conducting electrolytes, heterovalent substitution solid solutions, lead, tin, and yttrium fluorides, conductivity, transport numbers

1. Introduction

The researchers are completed at recent time have been shown that a compound with a high unipolar conductivity that provided by the fluorine ions are interested for the development of new electrolyte and electrode materials of electrochemical devices for various purposes such as chemical power sources, sensors, analyzers, ion-selective electrodes, etc. [1, 2]. For many of them characterized by a high ionic conductivity (10^{-1} – 10^{-4} S cm^{-1}) and a small electronic component.

Among currently known solid electrolytes fluorine conductive particular interest are solid solutions based on PbF_2, SnF_2, LnF_3 (where Ln is rare-earth elements). In particular the attractive transport properties, in particular, has lead (II) tetrafluorinestannate (II), which has conductivity even at room temperature is approximately $3 \cdot 10^{-3}$ S cm^{-1}.

The replacement of Pb^{2+} or Sn^{2+} cations by heterovalent Zr^{4+}, In^{3+}, Al^{3+}, Ga^{3+}, and Na^{+} cations favored the formation of solid solutions, which generally have higher conductivity than β-$PbSnF_4$ [3]. As is known, some substituents (e.g., Y^{3+}) favor a decrease in conductivity, while others (Bi^{3+}) increase it [4, 5].

Despite the large amount of currently available information [4-7], the effect of heterovalent substituents on the character of the dependence of solid solutions based on lead(II) tetrafluorostannate(II) on the composition and temperature has not yet been determined unambiguously.

Therefore, it is of interest to reveal the relationship between the nature of the heterovalent substituent and the charge transfer in multicomponent fluoride systems from both scientific and practical viewpoints.

The present communication presents the results of our studies of the charge transfer in the polycrystalline samples of PbF_2–SnF_2–YF_3 solid solutions.

2. Experimental details

The heterovalent substitution solid solutions $Pb_{1-x}Y_xSnF_{4+x}$ were synthesized in two stages. At first, $Pb_{1-x}Y_xF_{2+x}$ solid solutions were obtained by solid-state synthesis [8]. PbF_2 and YF_3 were thoroughly ground, mixed in an appropriate ratio, and pelletized. The resulting pressed mixtures were heated to 973 K in quartz tubes under highly pure argon, kept at this temperature for 5 h, and cooled to room temperature in the switched-off furnace mode.

At the second stage, the resulting solid solutions were fused with an equimolar amount of tin difluoride in platinum crucibles under argon, kept at 773 K for 20 min, and cooled to room temperature in the switched-off furnace mode. This gave nonstoichiometric $Pb_{1-x}Y_xSnF_{4+x}$ phases.

The conductivity of the synthesized compounds was studied by impedance spectroscopy and bridge method using a P5083 DCAC bridge in the frequency range 10–70 kHz. The impedance spectra were recorded in a cell with platinum electrodes using an Autolab (Ekochemie) electrochemical module and a frequency response analyzer (FRA) in the frequency range 10^{-1}–10^6 Hz (at an output signal amplitude of 10 mV). Pelletized samples of the synthesized compounds with a diameter of 8 mm and a thickness of 2.0–3.0 mm were used. Platinum electrodes were used as current leads. The measurements were performed in an argon atmosphere at temperatures of 293–573 K in the cooling mode.

The ^{19}F NMR spectra have recorded on Bruker AVANCE 400 spectrometer at the temperature range 293 – 573 K. The chemical shift (δ in ppm) relative to a reference set C_6F_6 an accuracy of 1 ppm. The width (at half height ΔH) has measured in kHz with an error less than 1 %. Experimental data processed using computer programs MestReNova and MagicPlot.

3. Results and discussion

At the first stage of the synthesis in the concentration range 1.0– 29.0 mol % YF_3, $Pb_{1-x}Y_xF_{2+x}$ solid solutions of cubic crystal system formed, which had a fluorite structure as in the case of the starting β-PbF_2, but with smaller unit cell dimensions. For example, for the $Pb_{0.71}Y_{0.29}F_{2.29}$ solid solution, a = 5.886(1) Å, while for β-PbF_2 it is 5.940(1) Å.

At the second stage, $Pb_{1-x}Y_xSnF_{4+x}$ ($0 < x \leq 0.17$) solid solutions of the tetragonal crystal system isostructural with β-$PbSnF_4$ formed (Fig. 1). The replacement of some part of Pb^{2+} ions by Y^{3+} ions leads to an increase in the unit cell volume of the resulting compounds. The unit cell volume of the $Pb_{0.83}Y_{0.17}SnF_{4.17}$ sample was 103.301 Å^3, while for pure β-$PbSnF_4$ it is 101.51 Å^3. The dependences of the unit cell parameters of the synthesized compounds on the yttrium trifluoride content in them obey the Vegard and Rutgers rules [9].

Figure 1 presents the typical frequency and temperature dependences of complex conductivity of the synthesized polycrystalline samples of $Pb_{1-x}Y_xSnF_{4+x}$ solid solutions with different compositions in Nyquist coordinates.

At 293 K the impedance diagrams of all the samples contain only one deformed semicircle in the region of high frequencies, which transforms into a linear dependence in the region of low frequencies. This segment of the impedance diagram suggests polarization of the solid solution/blocking electrode interface due to the change in the ion balance in the near-electrode layer of the solid electrolyte under the action of the external electric field.

As the temperature and yttrium fluoride content in the solid solution increase, the radius of the deformed semicircles decreases, and they shift toward higher frequencies. The shape of the obtained impedance diagrams suggests that the total bulk conductivity of crystallites has no significant contribution of surface conductivity. To verify this assumption, the capacity of the complex conductivity of the samples at frequencies f_{max} corresponding to the maximum Z'' values of the deformed semicircles and the sites of hypothetical intersection with the abscissa axis was evaluated by the equation $2\pi f_{max} RC = 1$. The resulting values (1.0–4.0) $\times 10^{-11}$ F have a substantially lower order of magnitude than the capacity component that characterizes the surface conductivity of crystallites (10^{-9}–10^{-7}) F.

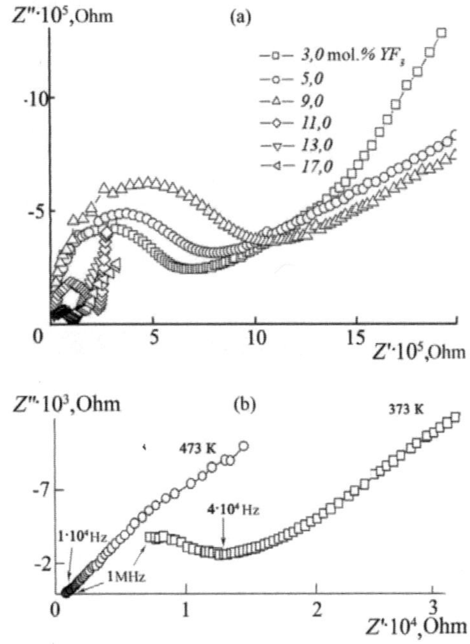

Figure 1. Impedance diagrams of the synthesized samples of (*a*) $Pb_{1-x}Y_xSnF_{4+x}$ solid solutions at 293 K and (*b*) of the $Pb_{0.87}Y_{0.13}SnF_{4.13}$ sample at 373 and 473 K

The temperature dependence of the conductivity of the synthesized compounds in the frequency-independent region of the impedance spectrum is satisfactorily approximated by the Arrhenius–Frenkel equation. For each sample including β-PbSnF4, the dependence has an inflection ("Faraday phase transition" [10]) in the range 435–475 K inherent in the majority of solid electrolytes with a fluorite and antifluorite structure (Fig. 2).

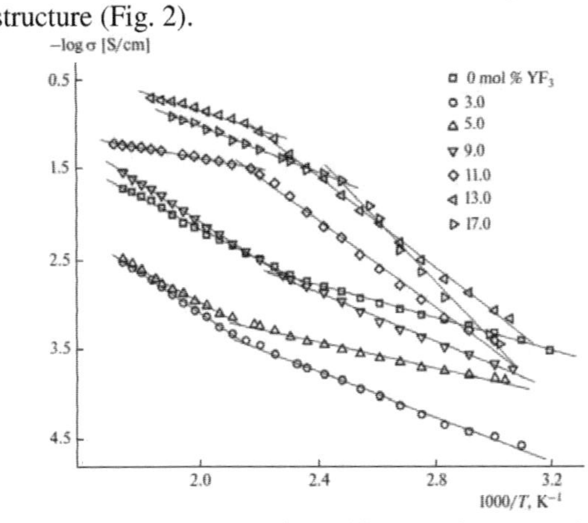

Figure 2. Temperature dependences of specific electric conductivity of the polycrystalline $Pb_{1-x}Y_xSnF_{4+x}$ solid solutions

The conductivity of the synthesized samples at low contents of yttrium ions ($x \leq 0.07$) is much lower than for the starting β-$PbSnF_4$. The $Pb_{0.91}Y_{0.09}SnF_{4.09}$ sample has almost the same conductivity as the stoichiometric β-$PbSnF_4$ phase in the high-temperature range (Fig. 2). In the samples containing more than 10.0 mol % YF_3, a considerable contribution to the total conductivity is from the structural changes in the solid electrolyte on heating. The activation energy in the high temperature range is higher than in the low-temperature one and increases with the concentration of the heterovalent substituent. The highest conductivity is inherent in the $Pb_{0.87}Y_{0.13}SnF_{4.13}$ sample. In the high temperature range, it is an order of magnitude higher than for β-$PbSnF_4$.

The conductivity parameters of the synthesized phases were estimated with the aid of the Arrhenius-Frenkel equation: $\sigma T = A\exp(-\Delta E_a / kT)$, where A is pre-exponential factor, ΔE_a is the activation energe of conductivity, k is Boltzmann constant (Table 1).

Table 1. Parameters of electric conductivity of $Pb_{1-x}Y_xSnF_{4+x}$ solid solutions

Sample	ΔE_a, eV	lg(A), S/(cm·K)	σ, S/cm	T, K
β-$PbSnF_4$	0,36 0,2	4,19 2,27	$9{,}02\cdot10^{-4}$ $1{,}88\cdot10^{-2}$	373 573
$Pb_{0.97}Y_{0.03}SnF_{4.03}$	0,48 0,25	4,49 1,92	$7{,}84\cdot10^{-5}$ $2{,}65\cdot10^{-3}$	373 573
$Pb_{0.95}Y_{0.05}SnF_{4.05}$	0,40 0,18	3,83 1,39	$2{,}47\cdot10^{-4}$ $3{,}34\cdot10^{-3}$	373 573
$Pb_{0.91}Y_{0.09}SnF_{4.09}$	0,47 0,30	5,42 3,32	$5{,}36\cdot10^{-4}$ $3{,}07\cdot10^{-2}$	373 573
$Pb_{0.89}Y_{0.11}SnF_{4.11}$	0,17 0,54	3,01 7,18	$1{,}67\cdot10^{-3}$ 0,0606	373 573
$Pb_{0.87}Y_{0.13}SnF_{4.13}$	0,24 0,53	4,28 7,42	$4{,}41\cdot10^{-3}$ 0,21	373 573
$Pb_{0.83}Y_{0.17}SnF_{4.17}$	0,30 0,72	4,7 9,99	$3{,}25\cdot10^{-2}$ 0,128	373 573

Below 300 K, the conductivity of the synthesized compounds is almost independent of the concentration of the heterovalent substituent (Fig. 3a). Above 400 K in the samples containing more than 7.0– 9.0 mol % yttrium trifluoride, the electric conductivity increases with the concentration of the hetetovalent substituent.

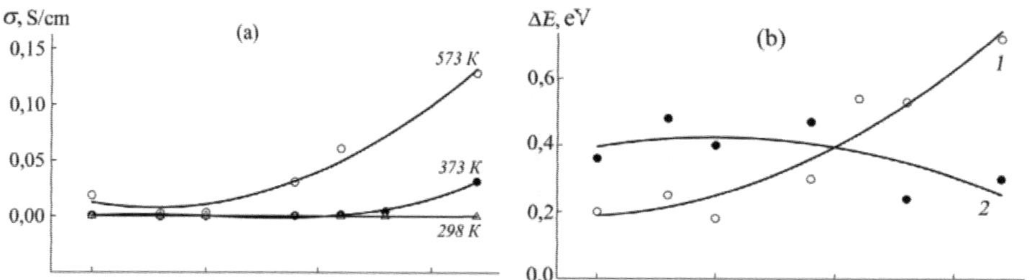

Figure 3. Conductivity isotherms (a) of the synthesized $Pb_{1-x}Y_xSnF_{4+x}$ samples and dependence of the activation energy of conductivity (b) of the samples of $Pb_{1-x}Y_xSnF_{4+x}$ solid solutions on the YF_3 content for the temperature range (1) before and (2) after the "Faraday phase transition" temperature

Figure 3*b* presents the dependence of the activation energy of conductivity of the synthesized samples before and after the “Faraday phase transition” on the concentration of the heterovalent substituent. Below 400 K at 9–10 mol % YF_3, the concentration is the dominant factor that determines the total electric conductivity. The activation energies of the synthesized samples are almost independent of the concentration of the heterovalent substituent and are close to the energy of the starting β-$PbSnF_4$ (table).

In the samples containing more than 10.0 mol % YF_3, a considerable contribution to the total conductivity is from the structural changes in the solid electrolyte on heating. The activation energy in the hightemperature range is higher than in the low-temperature one and increases with the concentration of the heterovalent substituent. The highest conductivity is inherent in the $Pb_{0.87}Y_{0.13}SnF_{4.13}$ sample. In the hightemperature range, it is an order of magnitude higher than for β-$PbSnF_4$.

The contribution of the electronic component of the total electric conductivity of the synthesized compounds was evaluated from the voltammetric characteristics of the(–)Ni/NiF_2+CaF_2|$Pb_{1-x}Y_xSnF_{4+x}$|Pt(+) electrochemical cell by the Hebb–Wagner method [11].The transport numbers of fluoride anions in the synthesized samples of solid solutions are almost independent of the concentration of the heterovalent substituent, and the electron component of conductivity exceeds the ion conductivity by more than two orders of magnitude.

Additional information about the nature of charge carriers in the system was obtained by high-temperature ^{19}F NMR spectroscopy. The results indicate that, like in the starting β-$PbSnF_4$, the fluoride anions occupy three structurally nonequivalent positions in the synthesized samples (F1, F2, and F3), which differ in the local environment and in the character of the Me–F bond, which agrees with the data of [12].

Unlike ^{19}F NMR spectra for tetragonal β-modification $PbSnF^4$ that at 300 K describes a narrow symmetric Lorentz function with a chemical shift of about 126 ppm and ΔH about 3 kHz the ^{19}F NMR spectra for solid solutions $Pb_{1-x}Y_xSnF_{4+x}$ are the set of broad lines (Fig. 4).

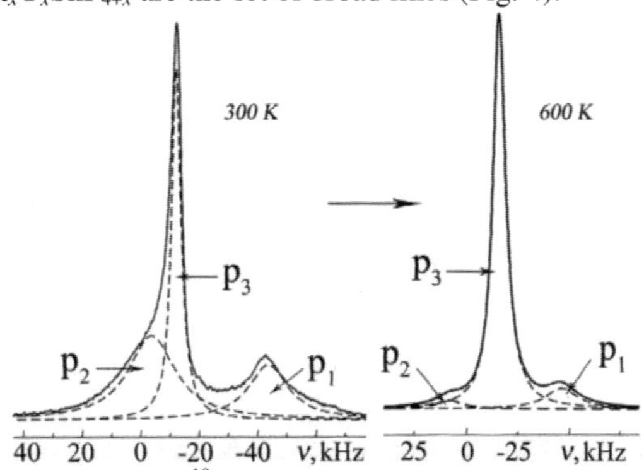

Figure 4. Evolution component of ^{19}F NMR spectra of solid solution $Pb_{0.89}Y_{0.11}SnF_{4.11}$ with increasing temperature

There is the redistribution of fluoride anions between places in the structure of synthesized samples with increasing temperature. Dominating the integrated intensity of superposition becomes narrow component (P_3) corresponding mobile ions.

The presence in the spectra of several components due to structural nonequivalence of fluoride ions, whose concentration can be measured by integrated intensities components. Contribution high mobility forms ions (component P_3) at 300 K greater than 32 % and increases with raise temperature, reaching values of 72–80 % at 523 K by attracting fluoride ions locally a mobile (P_2) and motionless (P_1) subsystems (Fig. 5). This largely reduces the area component P_2 and P_1 component in it remains unchanged until 400 – 450 K, and only further increase the temperature starts to drop, due to the sedentary growth rate of diffusion of fluoride ions.

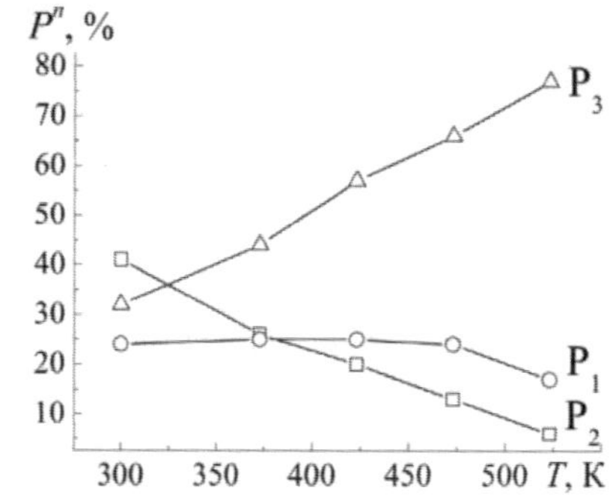

Figure 5. The distribution of fluoride ions localization in the crystal lattice $Pb_{0.89}Y_{0.11}SnF_{4.11}$ at different temperatures

The concentration of moving fluoride anions at 300 K does not depend on the concentration of heterovalent substituent YF_3, and at temperatures higher than 430 K increases with the concentration of the YF_3. Exchange of the fluoride ion between unequal subsystems are increases with increasing of the temperature and degree of substitution. This fact is reflected in the temperature dependences of electrical conductivity.

4. Conclusions

Heterovalent substitution solid solutions of the tetragonal crystal system isostructural with β-PbSnF4 are formed in the range of concentrations $0< x \leq 0.17$ in the $(1-x)PbF_2 - xYF_3 - SnF_2$ system. Like in β-$PbSnF_4$, the fluoride anions occupy three structurally nonequivalent positions. The mobile fluoride anions are mainly localized in the interstitial positions (F3). Their concentration exceeds 50% already at 300 K and increases with temperature. An inflection ("Faraday phase transition") was recorded in the temperature range 435–475 K for each composition of the synthesized samples, which is inherent in the majority of solid electrolytes with a fluorite and antifluorite structure. Below 300 K, the conductivity of the synthesized compounds is almost independent of the concentration of the heterovalent substituent. Above 400 K in the samples containing more than 7.0– 9.0 mol % yttrium trifluoride, the electric conductivity increases with the concentration of the hetetovalent substituent.

References

[1] Gurevich Y Y and Harkats Y I 1992 *Superionic conductors* (Moscow: Nauka)
[2] Sorokin N I and Fedorov P P 1997 *Inorganic Materials* **33** 1
[3] Donaldson J D and Senior B J 1967 *J. Chem. Sot. A.* 1821
[4] Ito Y, Koto K, Yoshikado S and Ohachi T 1986 *Solid State Ionics* **18–19** 1202
[5] Nikiforov A E, Zakharov A Yu, Chernyshev V A, Ugryumov M Yu and Kotomanov S V 2001 *Fiz. Tverd. Tela* **44** 1446
[6] Kanno R, Nakamura S and Kawamoto Y 1992 *Solid State Ionics* **51** 53
[7] Kavun V Ya, Ryabov A I, Telin I A, Podgorbunskii A B, Sinebryukhov S L, Gnedenkov S V and Goncharuk V K 2012 *Zh. Strukt. Khim* **53** 292
[8] Patwe S J, Balaya P, Goyal P S and Tyagi A K 2001 *Mater. Res. Bull.* **36** 1743
[9] Urusov V S 1987 *Theoretical Crystal Chemistry. Manual* (Moscow: Mosk. Gos. Univ.) (In Russian)
[10] Sorokin N I and Sobolev B P 2007 *Kristallografiya* **52** 870
[11] Wagner C 1956 *Z. Elektrochem* **60** 4
[12] Ito Y, Mukoyama T, Funatomi H, Yoshikado S and Tanaka T 1994 *Solid State Ionics* **67** 301-305

IOP Conf. Series: Materials Science and Engineering **175** (2017) 012040 doi:10.1088/1757-899X/175/1/012040

Micro-mechanical modelling of mechanical and electrical properties in homogeneous piezoelectric ceramic by using boundary integral formulations

M Biglar[1], F Stachowicz[1], T Trzepiecinski[1], M Gromada[2]

[1]Rzeszow University of Technology, Faculty of Mechanical Engineering and Aeronautics, 35-959 Rzeszów, Poland

[2]Institute of Power Engineering, Ceramic Department CEREL, Research Institute, 36-040 Boguchwała, Poland

E-mail: m_biglar@prz.edu.pl

Abstract. Recent experiments on polycrystalline materials show that microcrystalline materials have a strong dependency ona grain size. In this study, mechanical and electrical properties of polycrystalline materials in micro level were studied by using averaging theorems. To completely understand the size-dependency of polycrystalline materials, an integral non-local approach that can predict the stress-strain relations for these materials was presented. In microcrystalline materials, crystalline and grain-boundary were considered as two separate phases. Mechanical properties of the crystalline phase were modelled using crystalline brittle material and is composed of randomly distributed and orientated single crystal anisotropic elastic grains. For microcrystalline materials, the surface-to-volume ratio of the grain boundaries is low enough to ignore its contribution to the elastic deformation. Therefore, the grain boundary phase was not considered in microcrystalline materials and mechanical properties of the crystalline phase were modelled using appropriate integral non-local approach. Finally, the constitutive equations for polycrystalline materials were implemented into a boundary integral equation and the results and some examples are provided for piezoelectric ceramic.

1. Introduction

During recent decades, application of piezoelectric ceramics such as actuator, sensors and transducers have been increased due to their efficiency in converting mechanical loads to electrical ones and viceversa. However, the most important problem is that piezoelectric ceramics are too fragile due to their special microstructural structures. That is why, during recent years, lots of researches were done in order to investigate microstructural properties of piezoelectric ceramics [1, 2].

There are different methods to understand the properties and behavior of materials.One of the most favorite methods that is receiving attentions of researcher during last years is multi-scale modeling which is able to model the properties of materials simultaneously at different alternative scales. Modeling a specific material at different scales at the same time makes it possible to study the behavior of heterogynous materials actually [3-5].Due to the complexity of the material structure, there are no exact information about special characteristic of materials in a certain scale, therefore, one is able to determine these properties through a smaller scale. Thus, different classes of heterogeneous materials at different scales can be studied according to the same rules and the results can be compared in order to achieve exact and actual behavior of materials [6]. Another advantage of multi-scale that can be mentioned is that damage evaluation can be predicted via this method overall different scale [7-9].Damage models in multi-scale modeling are based on different theories. There are a lot of phenomenological and micromechanical models based on combination of quantities calculated at the macro-scale to evaluate damage.

2. Experimental

The macro-scale modeling is based on micro-continuum theories, however, paying attention to this point is necessary in order to save material integrity, no constitutive law or damage have to be considered. In the next step, by applying the boundary condition on a representative volume element (RVE) which was obtained from the calculation of macro scale, the cohesive laws can be brought up for modeling of crack initiation and propagation in the RVE. The RVE is a small volume of microstructure that has the general characteristics of the whole microstructure such as volume fraction, morphology and randomness of the phases and over which modeling of specific characteristics is carried out [10]. All information from the micro-scale can be sent to macro level to modify the results and model the next steps. In Fig. 1 it is shown that the calculation of micro-scale is able to provide boundary conditions for micro-scale, on the other hand, micro scale will provide some information by which the constitutive law can be modified and possible damage can be modeled for next steps as well.

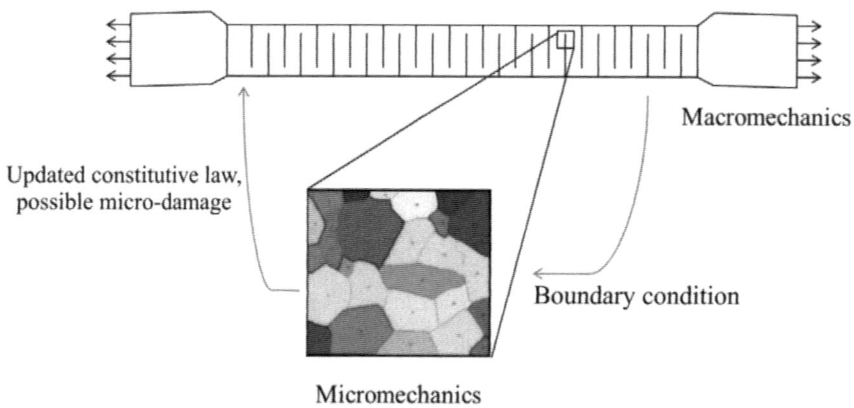

Figure 1. Schematic view of a multi-scale tension modeling

Via fundamental boundary element equation the displacement equation for each grain can be written as

$$C_{ij}^{H}(z_k')\tilde{u}_j^{H}(z_k') + \int_{S_{nc}^{H}} \tilde{T}_{ij}^{H}(z_k', z_k)\tilde{u}_j^{H}(z_k)dS_{nc}^{H} + \int_{S_c^{H}} \tilde{T}_{ij}^{H}(z_k', z_k)\tilde{u}_j^{H}(z_k)dS_c^{H}$$
$$= \int_{S_{nc}^{H}} \tilde{U}_{ij}^{H}(z_k', z_k)\tilde{t}_j^{H}(z_k)dS_{nc}^{H} + \int_{S_c^{H}} \tilde{U}_{ij}^{H}(z_k', z_k)\tilde{t}_j^{H}(z_k)dS_c^{H} \quad (1)$$

In equation (1), $\tilde{T}_{ij}^{H}(z_k', z_k)$ and $\tilde{U}_{ij}^{H}(z_k', z_k)$ are fundamental solution for traction and displacement which can be obtained via Stroh formulation for each grain. S_c^{H} and S_{nc}^{H} belong to boundary of internal grain and (contact boundary) and free boundary.

The overall volume average stress can be written according to averaging thorium as [11]:

IOP Conf. Series: Materials Science and Engineering **175** (2017) 012040 doi:10.1088/1757-899X/175/1/012040

$$\bar{\sigma}_{ij}^{m,t} = \frac{1}{V^m}\int_{S^m} x_i^m \dot{t}_i^m dS^m \tag{2}$$

Where x_i^m and $\dot{t}_i^m$ are the position vector of points which are located on RVE boundary and their traction respectively. And the similar way, the volume average strain can be written as [11]:

$$\bar{\dot{\varepsilon}}_{ij}^{m} = \frac{1}{2V^m}\int_{S^m} (u_{i,j}^m + u_{j,i}^m) dV^m \tag{3}$$

In order to avoid any localization of micro damage on macro scale results, a non-local approach has to be considered as below [11]:

$$\dot{\sigma}_{ij}^{D}(X') = D_{ij} C_{ijkl}^{M} \dot{\varepsilon}_{kl}^{M}(X') \tag{4}$$

where $\dot{\varepsilon}_{kl}^{M}$ is macro-strain, C_{ijkl}^{M} is fourth order elasticity stiffness tensor and D_{ij} can be obtained as follows:

$$D_{ij} = 1 - \hat{\bar{\dot{\sigma}}}_{ij}^{t}(X')[\hat{\dot{\sigma}}_{ij}^{e}(X')]^{-1} \tag{5}$$

$\hat{\bar{\dot{\sigma}}}_{ij}^{t}(X')$ is average stress which was already shown and $\hat{\dot{\sigma}}_{ij}^{e}(X')$ is elastic strain in the case of microdamage.

3. Results

In Fig. 2, the process of preparing RVE is shown. As it shows, the first step is clarifying the boundaries of grains via cad software, in the next step, image processing software was used in order to obtain the exact coordinates of grain automatically as well as separating grains. The benefit of this procedure is that the algorithm which was designed in this procedure is automatically able to propose exact information of every image which was abstained by SEM machines. So, boundary element method can be applied for huge ranges of SEM images without spending so much time.

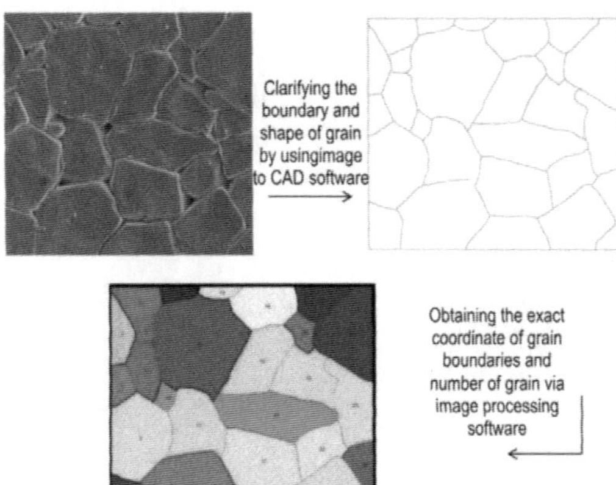

Figure 2. The workflow for generating the RVE of piezoelectric ceramics

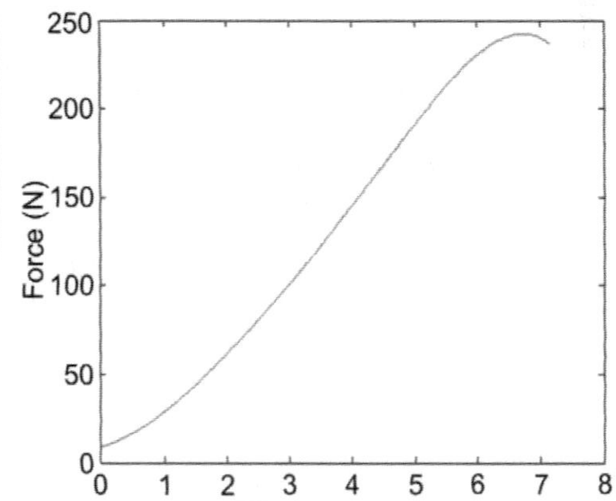

Figure 3. The reaction force versus displacement for the performed numerical simulations

Figs. 3 and 4 show the reaction forces and Potential jump for proposed grain. The reaction force can be obtained by multiplying average vertical stress on the length of that edge.

To study the influence of piezoelectric coupling on mechanical response, the calculations for two different cases with and without piezoelectric effect have been performed. The responses are represented in Figure. 5. From the responses it can be concluded that, according to the numerical procedures, the maximum stress can be hardly influenced by the coupling between mechanical and electrical properties.

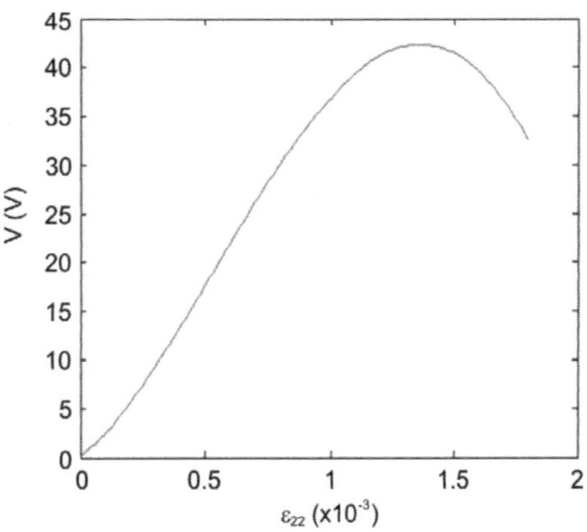

Figure 4. Potential jump V versus strainε_{22} for the performed numerical simulations

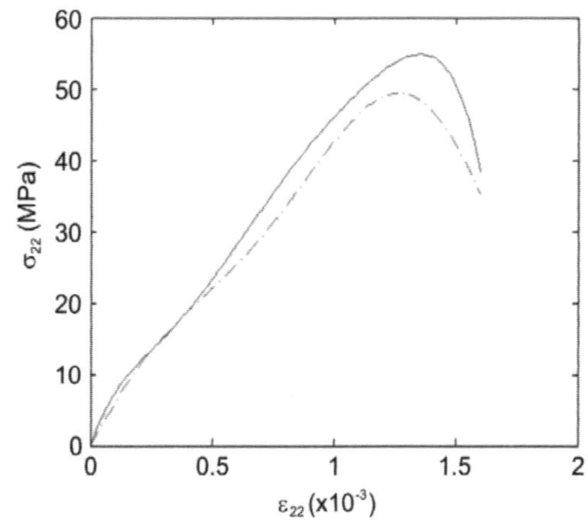

Figure 5. Stressσ_{22} versus strainε_{22} for the simulations of polycrystal with (solid) and without (dashed) piezoelectric effect

4. Conclusions

In this study, the mechanical and electrical response of piezoelectric materials were investigated through multi scale boundary element methods. The algorithm which was designed by using image processing software allows one to study a huge ranges of SEM photo automatically. By using one of the photos which was gotten in laboratory via SEM machine, the numerical calculationshave been done and the constitutive equations for polycrystalline materials were implemented into a boundary integral equation and the results and some examples were provided for piezoelectric ceramic.

Acknowledgments

The research leading to these results has received funding from the People Programme (Marie Curie Actions) of the European Union's Seventh Framework Programme FP7/2007-2013/ under REA grant agreement no PITN-GA-2013-606878.

References

[1] Tian W Y and Chen Y H 2000*Int. J. Solids Struct.***37** 7743
[2] Chen Y H andLu T J 2001 *Int. J. Solids Struct.***38**3233
[3] Khoei A R, Jahanbakhshi F and Aramoon A 2015 *Mech. Mater.***83** 40
[4] Sencu R M, Yang Z and Wang Y C 2016*Eng. Fract. Mech.***163** 499
[5] Kouznetsova V G, Geers M G D and Brekelmans W A M 2004*Comput. Meth. Appl. Mech.Eng.* **193**5525
[6] Silani M, Ziaei-Rad S, Talebi H and Rabczuk T 2014*Theor. Appl. Fract. Mech.***74** 30
[7] Espinosa H D and Zavattieri P D 2003 *Mech. Mater.***35** 333
[8] Espinosa H D and Zavattieri P D2003 *Mech.Mater.***35** 365
[9] Sfantos G K and Aliabadi M H 2007 *Int. J. Num. Meth. Eng.***69** 1590
[10] Amirmaleki M, Samei J, Green D E, van Riemsdijk I, and Stewart L 2016 *Mech. Mater.* **101** 27
[11] Sfantos G K and Aliabadi M H 2007 *Comput. Meth. Appl. Mech. Eng.***196**1310

Evolution of AC conductivity of wet illitic clay during drying

Š Csáki[1,2], I Štubňa[1], V Trnovcová[1], J Ondruška[1], L Vozár[1] and P Dobroň[2]

[1]Department of Physics, Constantine the Philosopher University, 949 74 Nitra, Slovakia

[2]Department of Physics of Materials, Charles University, 121 16 Prague 2, Czech Republic

E-mail: stefan.csaki@ukf.sk

Abstract. The evolution of the AC electrical conductivity during drying as well as the relationship between sample volume and moisture of green illite samples were investigated. The samples were prepared from illitic clay (80 mass % illite, 4 mass % montmorillonite, 12 mass % quartz and 4 mass % of orthoclase) and distilled water with initial moisture content 36 mass % and were freely dried in air. Conductivity was measured by the volt-ampere method with AC power supply of 5 V in the frequency range from 50 Hz to 10 kHz. The AC conductivity steeply increased with increasing moisture, up to 15 mass %. At higher values of the moisture, the AC conductivity was high and almost constant. The volume of samples increased with increasing moisture when the moisture was higher than 8 mass %. Below this value, the dimensions of samples do not significantly change. The dependence of the relative volume change on moisture is presented in a form of the Bigot´s curve.

Keywords: Illite, Electrical AC conductivity, Bigot´s curve, Drying.

1. Introduction

Illitic clay is very often used for manufacturing ceramic products by the "wet technology" in which ceramic body is shaped from a wet plastic mass. Therefore, the wet body consists of a porous solid skeleton and the water in pores and on crystallite surfaces. Water content can reach from 20 to 35 mass %. The illitic clay body expands due to adsorption of water molecules between individual clay crystals when the water content increases. The illite structure is built from an octahedral sheet located between two tetrahedral sheets forming a triple layer (figure 1). Ions K^+ are located between these triple layers, and in addition to them, water molecules can be also present here. To remove these water molecules, the sample must be heated up to 300 - 350 °C [1]. The surface of illite is formed from the tetrahedron layer with hexagonal patterns with hollows in their centers (figure 1) in which adsorbed polar water molecules are bound [2]. Large-radius K^+ ions are also present on the illite surface [3]. Water in clay body is commonly classified as 1) hygroscopic water, 2) capillary water and 3) free water [4].

The hygroscopic water is created from some layers: the first is extremely well- arranged monomolecular water layer directly and tightly bound to the electrically charged surface defects. The second layer (~1.3 nm in the case of illite [2]) is created from well- arranged water molecules. This layer gradually transforms to loosely bound adsorbed water; the thickness of this water layer is 2 – 6 nm [4]. The hygroscopic water is in equilibrium with the air humidity, i.e. the amount of the hygroscopic water in clay body is dependent on air temperature and pressure [4,5]. The hygroscopic water cannot be removed at room temperature [6].

The next progressively thickening water layer, so-called capillary water, covers the hygroscopic water. The capillary water is a part of the wet plastic mass that is prepared from illite powder and water, i.e. this water is introduced into the illite body during the technological process. The capillary water is held by surface forces in small pores, micropores, and capillaries. Molecules of capillary water are free and mobile. Due to this reason, it evaporates easily at room temperature. The capillary water separates the clay crystals, so its removal leads to a contraction of the clay body. Its amount is proportional to the porosity, i.e. the greater surface area the bigger amount of the hygroscopic water.

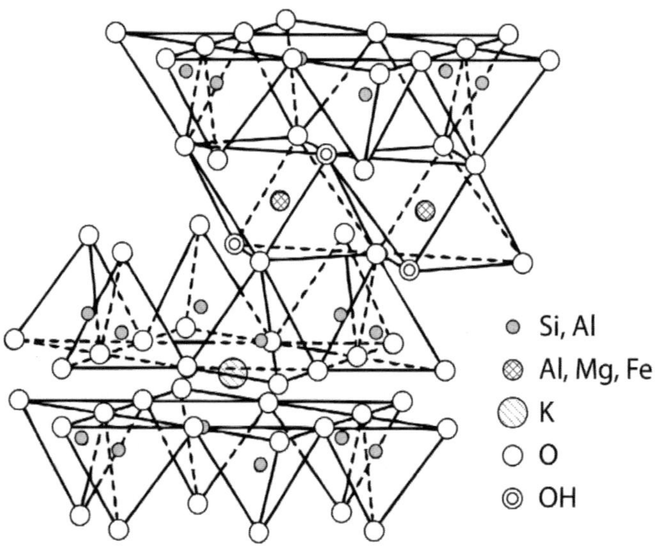

Figure 1. Structure of illite according to [7]

Electrical measurements are a useful, non-destructive tool for characterizing porous clays. The electrical conductivity of wet clay has two components - a volume conductivity and surface conductivity [8,9]. The volume conductivity of the wet clay results from conduction through the water occupying the pores. This conductivity is proportional to the amount of the hygroscopic and capillary water. The surface conduction occurs at the water-crystal interface. It represents the conduction in the hygroscopic water layer (which is the nearest to the crystal surface) and a conduction mechanism operating directly on the particle surface, probably associated with proton migration [10]. Mobile monovalent cations can easily move along surfaces of plate-like clay crystals and this movement results in electrical conduction under the applied electrical field [8,11].

The low-frequency AC electrical field may be used to study the conductivity of clays because polarization and relaxation processes typically affect responses at frequencies below 100 kHz [12]. Measurements at 1 kHz are reported in [9].

It should be pointed out that electrical properties of clays are still relatively unknown and published experimental data are limited. The articles published on the AC conductivity mostly investigate the effect of water content on this conductivity. Results for clays from three different sources show that the electrical resistivity of clay is low when the content of the water exceeds 15 – 20 mass %. As the clay becomes dryer, its resistivity increases considerably. Dependencies of the electrical conductivity on the moisture content can be met more often in works devoted to the study of soil [13,14]. In [15], the authors report on the relation between the electrical properties and the microstructure of clays.

The aim of this contribution is an experimental study of the AC electrical conductivity and shrinkage of the illitic clay during its drying. The AC conductivity was measured at different frequencies.

2. Experimental

Samples were made from illitic clay supplied by the mine in Füzérradvány, Hungary. The clay consists of 80 mass % of illite, 4 mass % of montmorillonite, 12 mass % of quartz and 4 mass % of orthoclase. The chemical composition of the clay is in Table 1.

Table 1. The chemical composition of illitic clay (mass %)

SiO_2	Al_2O_3	Fe_2O_3	TiO_2	CaO	MgO	K_2O	Na_2O	L.O.I.
58.0	24.0	0.6	0.05	0.38	1.70	7.85	0.10	7.3

Samples of dimensions 50×50×8 mm were made from the plastic mass obtained from 64 mass % of clay powder and 36 mass % of distilled water.

The relative amount of the water w in the sample was measured using the digital balance Kern EW with a sensitivity of 0.01 g. The relative moisture was calculated as

$$w = \frac{m(w) - m_d}{m_d}, \tag{1}$$

where $m(w)$ is the actual sample mass during drying and m_d is the mass of the dry sample after heating at 120 °C for 2 h.

The sample shrinkage was measured with the linear dial indicator with a resolution of 0.01 mm. The sample and linear indicator were placed on the balance (figure 2). The sample was put on supports to have a free access to the air from all sides of the sample.

The electrical resistance was measured with A-V method. The RC oscillator Hameg 8030-6 with AC voltmeter Hameg 8012 served as the voltage source and multimeter Fluke 289 served as AC ammeter. The AC conductivity was measured at frequencies 50, 200, 500, 1000, 5000 and 10000 Hz. We used two kanthal wires (∅ 0.3 mm) as electrodes. From the measured resistance R, thickness of sample b, radius of electrodes r and distance between the electrodes a, the clay conductivity can be calculated from the formula [16]

$$\sigma = \frac{1}{\pi Rb} \ln\left(\frac{a}{2r} + \sqrt{\frac{a^2}{4r^2} - 1}\right). \tag{2}$$

Equation (2) is valid for an infinite plane, but the area of the sample is 50×50 mm. As found in [16], if the side of the sample exceeds $6a$, the sample can be considered as infinite. Since a = 4 mm, this condition is fulfilled.

3. Results and discussion

The relationship between the relative moisture of the investigated clay sample and its contraction during drying, so called Bigot's curve, is depicted on figure 3. The initial mass fraction of the water in the sample is ~34 mass %. As the sample becomes dryer, its dimensions decrease down to the critical moisture. After reaching the critical moisture, which is ~13 mass %, the sample dimensions stay constant, in spite of the continuing drying. The linear contraction of the sample at the critical moisture is ~13 mass %.

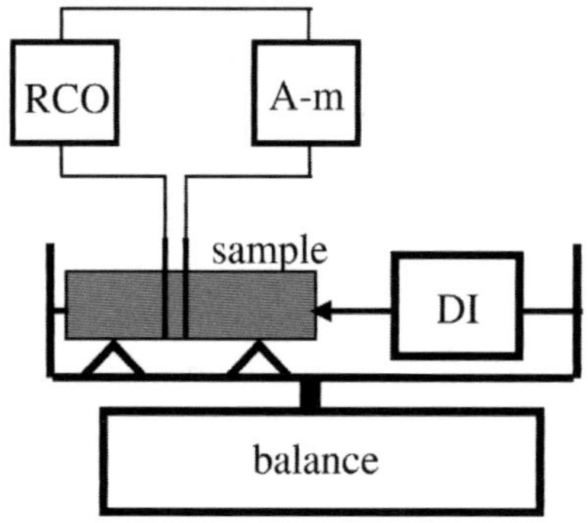

Figure 2. The scheme of the measurement unit. RCO – RC oscillator with V-meter, A-m – ammeter, DI – dial indicator

The initial volume of the sample is 2×10^4 mm^3 and its volume at the critical moisture is 1.32×10^4 mm^3, i.e. the relative volume reduction is 34 %.

The results of electrical measurements in which the relationships between the AC conductivity, measured at frequencies 50, 200, 500, 1000, 5000 and 10000 Hz, and the relative moisture calculated from equation (1) are shown in figure 4. As explained above, the water in the investigated sample occurs in two layers. The first layer, hygroscopic adsorption water, is bound to the crystal surface with electrostatic forces. The second layer is the capillary water introduced into the sample during technology.

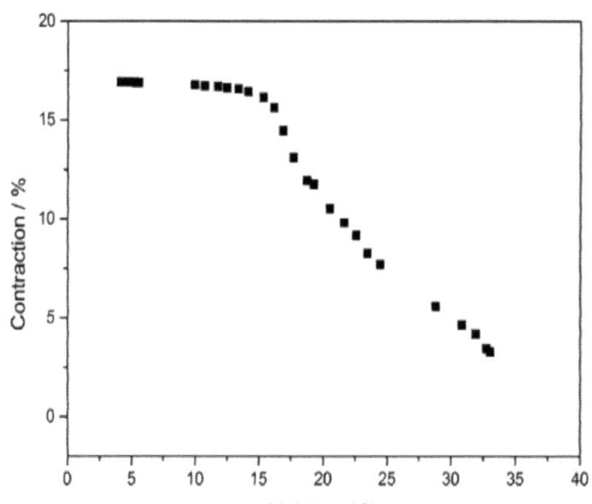

Figure 3. Relationship between the relative linear contraction and relative moisture

At the moisture w > 15 mass %, the capillary water is responsible for a good electrical conductivity. The conductivity very slightly decreases as the sample loses the capillary water from the initial moisture 34 mass % down to its critical value ~15 mass %. This behavior of the conductivity is a consequence of thinning the capillary water layers during drying. The only exception is the frequency 50 Hz at which a very flat maximum at $w \approx 25$ mass % is observed.

When the moisture content reaches the critical value, the AC conductivity is mainly determined by the capillary water, the content of which relatively rapidly decreases. Since a resistivity of the illite crystals is high, the current through illite crystals is negligible compared to the current through water layers. Since the thickness of the hygroscopic layer is very low and constant at room temperature, the conductivity of this layer can be considered as constant. As a consequence, the AC conductivity is mainly influenced by a rapid decrease of the capillary water.

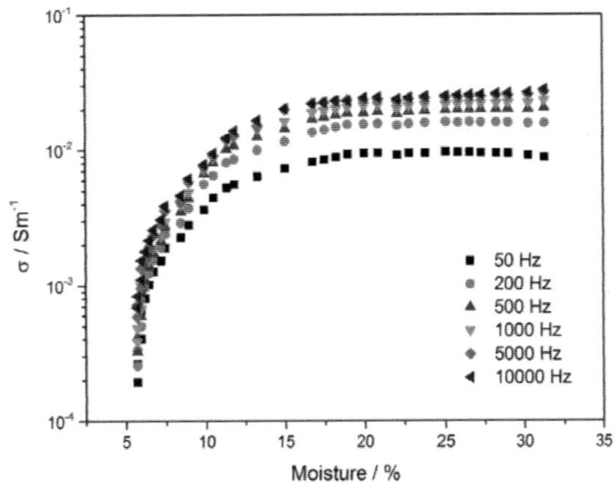

Figure 4. Relationships between AC conductivity and moisture measured at different frequencies

The frequency dependence of the AC conductivity for w = *const* can be obtained from figure 4. Such example is given for w = 20 mass % in figure 5. The relationship can be divided into two parts: 1) where the conductivity increases rapidly with the frequency (from 50 Hz to 1 kHz) and 2) where the conductivity changes slowly (from 1 kHz to 10 kHz). The moisture 20 mass % is relatively high and water layers continuously cover illite crystals; consequently, capillary water is the main environment for the AC current.

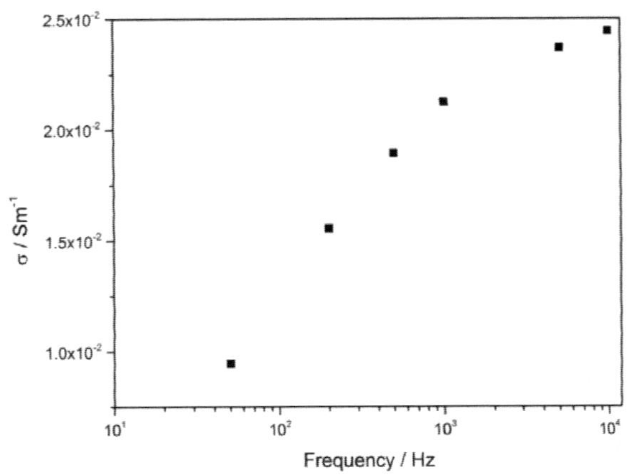

Figure 5. Frequency dependence of AC conductivity for w = 20 mass %

In water layers, the self-ionization of capillary water $H_2O + H_2O \leftrightarrow H_3O^+ + OH^-$ and the reaction $H_2O \rightarrow H^+ + OH^-$ occur. The created H^+ ions are very mobile charge carriers. The OH^- ions have dipole moments and can contribute not only to the electrical transport but also to the orientation relaxation. H^+ ions can also contribute to the relaxation by a localized hopping. This implies that complex picture of the ionic conductivity must be expected. A small contribution of K^+ ions (as they are components of illite) to the AC conductivity is also possible.

During drying, the majority of capillary water disappears from pores and crystal surfaces, the critical moisture is reached. The AC conductivity steeply decreases for $w < 13$ mass % for each frequency used in figure 4. Only hygroscopic water is present. Its molecules are relatively well oriented and the relative permittivity of this layer is about 80 [4].

4. Conclusions

The evolution of the AC conductivity during drying at different frequencies (between 50 Hz and 10 kHz) was monitored. To better understand the drying process, the contraction of the sample was also recorded. It was found, that:

- The contraction exhibits an approximately linear dependence on the moisture reduction until the critical moisture content is reached. Below the critical moisture content (15 mass %) the sample dimensions remain almost constant.
- The AC conductivity is relatively high above 15 mass % of the moisture content. Below this value, the conductivity steeply decreases as a result of the removal of the capillary water and decrease in the intercrystalline distances.
- At higher frequencies, the AC conductivity is higher and at the moisture content above 15 mass % it does not significantly depend on the moisture content.
- AC conductivity is built up of two mechanisms (according to frequency dependences) – bound water rotation combined with protonic jumps and grain conduction.
- The first mechanism predominates in the whole studied frequency range, at room temperature.

Acknowledgment: This work has been supported by the grant VEGA 1/0162/15 and UGA VII/1/2016 and SVV-2016-260322.

5. References

[1] Drits V A and McCarty D K 2007 The nature of structure-bonded H2O in illite and leucophyllite from dehydration and dehydroxylation experiments *Clays Clay Miner.* **55** 45–58

[2] Osipov V I 2012 Nanofilms of adsorbed water in clay: Mechanism of formation and properties *Water Resour.* **39** 709–21

[3] Siddiqui M K H 1976 The electric double-layer structure of illite and attapulgite clays *Clay Miner.* **11** 251–3

[4] Saarenketo T 1998 Electrical properties of water in clay and silty soils *J. Appl. Geophys.* **40** 73–88

[5] Martin R T 1960 Adsorbed Water on Clay: a Review *Clays Clay Miner.* **9** 28–70

[6] Levitt D G and Young M H 2007 Soils: Hygroscopic Water Content *Encyclopedia of Water Science, Second Edition* (CRC Press) pp 1136–9

[7] Grim R 1962 Applied Clay Mineralogy *Geol. föreningens i Stock. förhandlingar* **84** 533

[8] Tabbagh A and Cosenza P 2007 Effect of microstructure on the electrical conductivity of clay-rich systems *Phys. Chem. Earth, Parts A/B/C* **32** 154–60

[9] Wildenschild D 2000 Electrical properties of sand–clay mixtures: The effect of microstructure *Geophys. Res. Lett.* **27** 3085–8

[10] Revil A and Glover P W J 1998 Nature of surface electrical conductivity in natural sands, sandstones, and clays *Geophys. Res. Lett.* **25** 691–4

[11] Mojid M A and Cho H 2012 Effects of water content and temperature on the surface conductivity of bentonite clay *Soil Res.* **50** 44–9

[12] Kaya A 2001 Electrical Spectroscopy of Kaolin and Bentonite Slurries *Turk J Engin Env. Sci* **25** 345–54

[13] Auerswald K, Simon S and Stanjek H 2001 Influence of Soil Properties on Electrical Conductivity Under Humid Water Regimes *Soil Sci.* **166** 382–90

[14] Gong Y, Cao Q and Sun Z 2003 The effects of soil bulk density, clay content and temperature on soil water content measurement using time-domain reflectometry *Hydrol. Process.* **17** 3601–6314
[15] Kriaa a., Hajji M, Jamoussi F and Hamzaoui a. H 2014 Electrical conductivity of 1 : 1 and 2 : 1 clay minerals *Surf. Eng. Appl. Electrochem.* **50** 84–94
[16] Podoba R and Podolinčiaková J 2011 Needle electrodes for the measurement of DC and AC conductivity of ceramics *In: Didmattech XXIV 2011: Problems in teachers education* (Kraków: Instytut techniky TU) pp 31–4

IOP Conf. Series: Materials Science and Engineering **175** (2017) 012042 doi:10.1088/1757-899X/175/1/012042

Structured ZnO films: Effect of copper nitrate addition to precursor solution on topography, band gap energy and photocatalytic activity

S Heinonen, J-P Nikkanen, A Kaleva, L Hyvärinen and E Levänen

Department of Materials Science, Tampere University of Technology, FIN-33101 Tampere, Finland

E-mail: saara.heinonen@tut.fi

Abstract. ZnO is a widely studied semiconductor material with interesting properties such as photocatalytic activity leading to wide range of applications, for example in the field of opto-electronics and self-cleaning and antimicrobial applications. Doping of photocatalytic semiconductor materials has been shown to introduce variation in the band gap energy of the material.

In this work, ZnO rods were grown on a stainless steel substrates using hydrothermal method introducing copper nitrate into the precursor solution. Zinc nitrate and hexamethylenetetramine were used as precursor materials and the growth was conducted at 90 °C for 2 h in order to achieve a well-aligned evenly distributed rod structure. Copper was introduced as copper nitrate that was added in the precursor solution in the beginning of the growth. The as-prepared films were then heat-treated at 350 °C and band gap measurements were performed for prepared films. It was found that increase in the copper concentration in the precursor solution decreased the band gap of the ZnO film. Methylene blue discolouration tests were then performed in order to study the effect of the copper nitrate addition to precursor solution on photocatalytic activity of the structured ZnO films.

1. Introduction

Zinc oxide is an n-type semiconductor with band gap energy of 3.37 eV. [1]Due to its unique electronical and optical properties, ZnO has a wide range of applications for example in gas sensors and solar cells. [2]In addition to its other interesting properties, ZnO is believed to have higher efficiency in photocatalytic performance than TiO_2. [3] Among many methods to synthesize ZnO nanostructures, hydrothermal synthesis is a relatively cheap and simple method to produce structured high surface area powders and thin films [4].

Copper has many physical and chemical properties similar to Zn and optical and morphological properties of ZnO can be changed via incorporation of Cu into the lattice of ZnO nanostructures. [5]Many studies have demonstrated incorporation of copper leading to reduced band gap energy of ZnO. [6-9] The band gap narrowing can increase the visible light photocatalysis of the material. This has been explained with the Cu 3d levels forming a new energy level below the conduction band of ZnO. [9]Despite the enhanced visible-light absorption leading to better visible-light photocatalysis [10], these intra-band levels can act as recombination centres decreasing the photocatalytic activity of the material [3].

In this study, we have used hydrothermal method to produce structured ZnO films with high surface area synthesized in a solution containing copper nitrate to study the effect of copper nitrate addition to precursor solution on bandgap energy and photocatalytic activity of the films.

2. Experimental

2.1. Preparation and characterizationof zinc oxide films

Zinc oxide films were prepared by a hydrothermal method. Before the actual growth-step, a seed layer was formed on an AISI 304 stainless steel (SS) substrate (75mm x 25 mm) followed by the growth of ZnO nanorods. Before the process, substrates were cleaned by wiping with 2-propanol (C_3H_7OH, 99.5 %, VWR) and then rinsed with ethanol (C_2H_5OH, 99.5%, Altia Oyj). A solution for the seed layer was prepared mixing 2-methoxyethanol ($CH_3OCH_2CH_2OH$, ≥99,5%, Sigma-Aldrich) and ethanolamine ($NH_2CH_2CH_2OH$, ≥ 99.0%, Sigma-Aldrich) (96:4) and dissolving 0.23 M zinc acetate ($C_4H_6O_4Zn \cdot 2H_2O$, ≥ 98%, Sigma-Aldrich) into the mixture. Solution was then stirred at 60 °C for 2 h. After cooling down, the solution was spin coated (1500 rpm, 20 s) on the substrates and then the samples were heat treated at 350 °C for 20 min. The solution for growth of ZnO layer with tuned topography was prepared mixing hexamethylenetetramine ($C_6H_{12}N$, 99%, Sigma-Aldrich) and zinc nitrate ($Zn(NO_3)_2$, 98%, Sigma-Aldrich) volume ratio being 1:1. The concentrationsof both HMTA and zinc nitrate were 0.05 M. Copper nitrate tri hydrate ($Cu(NO_3)_2) \cdot 3H_2O$ was added to the solution. Two synthesis with varying copper concentration were conducted: in the first synthesis, the Cu/Zn molar ratio in the solution was 0.005 and in the second synthesis, the Cu/Zn molar ratio was 0.05. The synthesis temperature was90 °C. The plates were placed in the solution the seed layer side upside down and the growth was carried out for 2 hours. After removing the samples from the solution, they were rinsed with de-ionized water and dried in air. The samples were finally heat-treated at 300 °C for 30 min. The sample without Cu in the precursor solution is denoted as S1, the sample with lower Cu concentration in the precursor solution is denoted as $S1_{Cu0,5\%}$ and the sample with higher Cu concentration in the precursor solution is denoted as $S1_{Cu5\%}$.

The prepared zinc oxide films were characterized using field emission scanning electron microscopy (FESEM, Zeiss ULTRAplus) for surface morphology studies and X-Ray diffraction (XRD, Panalytical Empyrean Multipurpose Diffractometer) to define the crystal structure of the samples.The bad gap energies of the fabricated films were measured with UV-vis spectrophotometer (Shimazu UV-3600 UV/Vis/NIR spectrophotometer) in the wavelength range of 300–500 nm.

Photocatalytic measurements were performed in an aqueous solution of methylene blue (MB). The sample plates were placed in a decanter glass with 30 ml of 0.0085 mM MB solution. In order to reach adsorption equilibrium, the test solution was first kept in the dark for 60 min after which the ultra violet (UV)-lamp was turned on. Illumination was done from above with a 100 W/m^2 power using Ledia NIS330U-M UV-Gun (peak maximum at 365 ± 5 nm). The concentration of MB in the solution was followed by UV-VIS spectroscopy (Shimadzu UV-2501PC Spectrophotometer). At one-hourintervals, a 3 ml MB solution was taken for absorbance measurement. The absorbance was measured at the wavelength of 665 nm. A dark experiment without any irradiation but under otherwise identical conditions was also conducted for comparison. The discoloration of MB under UV irradiation but without sample plates was also tested to confirm that irradiation does not compose MB intrinsically.

3. Results and discussion

3.1. Morphology and crystal structure of ZnO films grown in different precursor concentration solutions and temperatures

The scanning electron microscopy images of the films prepared at different precursor solutions are shown in figure 1. The rod diameter was slightly increased with sample S1compared to the samples $S1_{Cu0,5\%}$ and $S1_{Cu5\%}$. Moreover, the rod diameter and density were equal for S1 $_{Cu0,5\%}$and $S1_{Cu5\%}$. The surface areas for the three samples were calculated from the SEM-images from the rod length, rod diameter and rod density for S1, $S1_{Cu0,5\%}$ and $S1_{Cu5\%}$. The received values were 109 cm^2, 134 cm^2 and 127 cm^2, respectively. The rod lengths, densities and surface areas are given in Table 1. It is notable that addition of copper into the precursor solution impactedclearly the rod alignment making the structure more aligned compared to the sample synthesized without copper nitrate.

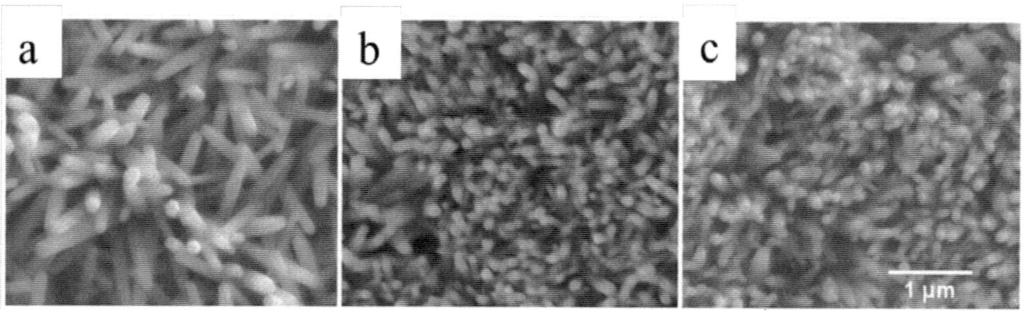

Figure 1 a–c. FESEM images of ZnO coatings prepared with a precursor solution a) without copper nitrate b) lower copper nitrate concentration and c) higher copper nitrate concentration

Table 1. Rod diameters, densities, lengths and surface areas of S1, S1Cu $_{0,5\%}$ and S1Cu $_{5\%}$

	S1	S1Cu0,5%	S1Cu5%
Rod diameter (nm)	141	101	104
Rod density (rods/μm^2)	10	25	24
Rod length (nm)	1270	891	840
Surface area (cm^2)	109	136	127

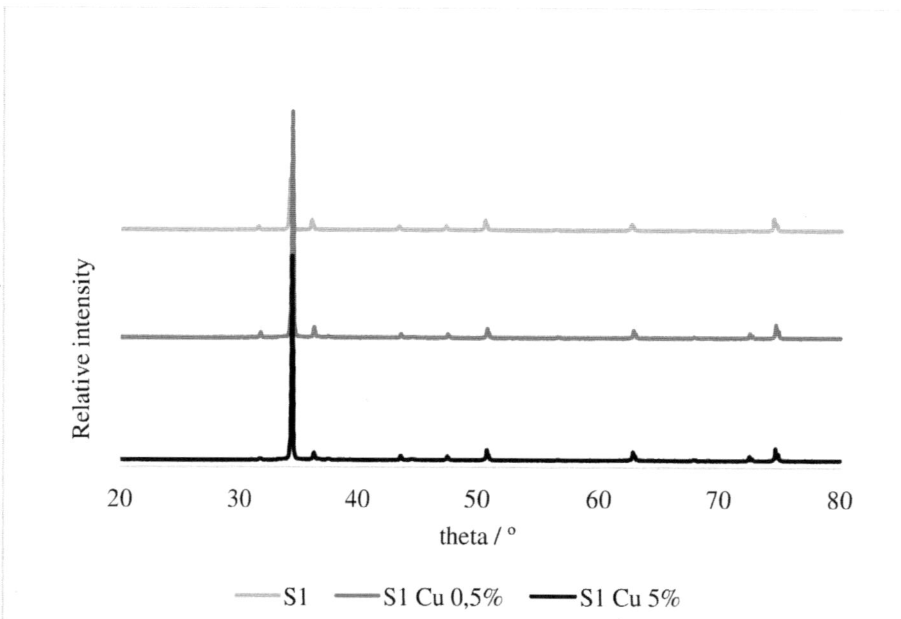

Figure 2. XRD analysis of the ZnO films with and without copper nitrate in the precursor solution

XRD analyses were conducted for all four coatings and all of them showed clear peaks of hexagonal wurtzite structure. Apart from peaks of wurtzite and iron from stainless steel substrate, no other peaks could be detected. The XRD patterns for each coating are shown in Fig. 2.

3.2. MB degradation results

The band gap energy values for samples synthesized without and with varying amount of copper nitrate in the precursor solution are listed in table 2. The increased amount of copper in the precursor solution was observed to decrease the band gap energy of the prepared films. The decrease of band gap energy could result from the increase of the lattice constant between Cu-O and Zn-O bonds. It can be assumed that copper forms a new energy level below the conduction band of ZnO which narrows the band gap.

Table 2. Band gap values of S1, $S1_{Cu0,5\%}$ and$S2_{Cu5\%}$

Sample	Band gap (eV)
S1	3.204
$S1_{Cu0,5\%}$	3.171
$S2_{Cu5\%}$	3.163

3.3. MB discolouration results

The dark tests were performed before the actual measurements for one hour during which the concentration of the solution did not change and thus the samples did not adsorb methylene blue. The MB solution was also irradiated without the samples and no change in the concentration of MB was observed, which means that irradiation did not intrinsically decompose MB. Based on MB degradation test results, the sample prepared without copper nitrate in the precursor solution (S1) had the highest photocatalytic activity although all the prepared coatings showed photocatalytic activity (Fig. 3).The sample $S1_{Cu0,5\%}$ with lower copper concentration in the precursor solution showed better photocatalytic performance compared to the sample $S1_{Cu5\%}$ that had higher copper concentration of the precursor solution. The lower photocatalytic activity of the samples prepared in the copper nitrate containing precursor solution could result from copper ions in the ZnO lattice acting as recombination centres. The reduced photocatalytic activity may also be due to the reduced band gap energies, which can be beneficial when only longer wavelengths of light are available. However, when artificial excitation is used the narrowing of the band gap may not be beneficial.

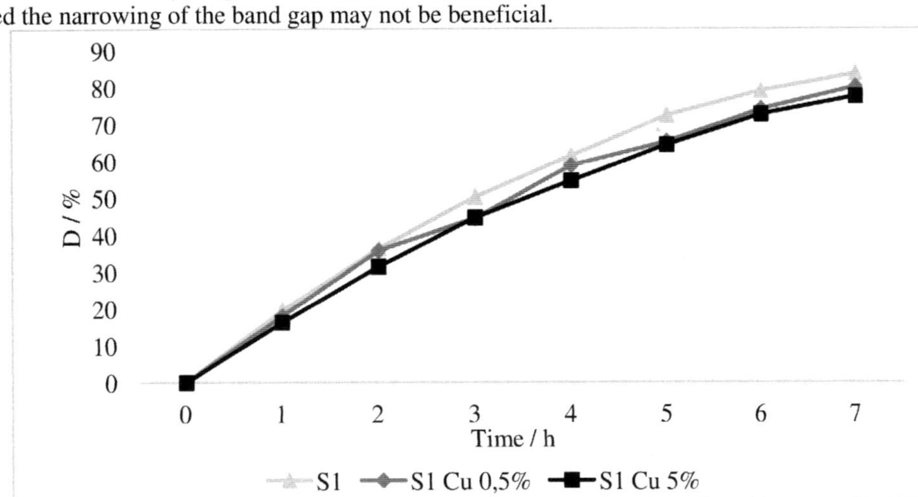

Figure 3. Percentage of degraded methylene blue as a function of time for samples S1, $S1_{Cu0,5\%}$ and $S1_{Cu5\%}$

4. Conclusions

In this paper, we prepared ZnO films with varying amount of copper nitrate in the precursor solution. It was shown that addition of copper nitrate resulted in well-aligned nanorod structure compared to the less aligned structure obtained via synthesis without copper nitrate. It was also demonstrated that addition of copper nitrate to the precursor solution of hydrothermally grown ZnO nanorods lead to decrease in the band gap energy of the coating. The photocatalytic activity of the coatings was also reduced when copper nitrate was used as a precursor despite the increased surface area that was obtained in the synthesis with copper nitrate. The reduced photocatalytic activity may have resulted from recombination centres formed by Cu but also from decreased band gap energy since artificial excitation was used in this study.

References

[1] Girija K G, Somasundaram K, Topkar A, Vatsa R K 2016 *J. Alloy. Compd.* **684** 15
[2] Joshi K, Rawat M, Gautam S K, Singh R G, Ramola R C, Singh F 2016 *J. Alloy. Compd* **680** 252
[3] Samadi M, Zirak M, Naseri A, Khorashadizade E, Moshfegh A Z 2016 *Thin Solid Films* **605** 2
[4] Saarenpää H, Sariola-Leikas E, Pyymäki Perros A, Kontio J M, Efimov A, Hayashi H, Lipsanen H, Imahori H, Lemmetyinen H and Tkachenko N V 2012 *J. Phys. Chem. C* **116** 2336
[5] Shabannia R 2016 *J. Mol. Struct.* **1118** 157
[6] Pung S Y, Ong C S, Mohd Isha K, Othman M H 2014 *Sains Malaysiana* **43** 273
[7] Zhao S, Bai Y, Chen J, Bai A, Gao W 2014 *Journal of magnetics* **19** 68
[8] Labhane P K, Huse V R, Patle L B, Chaudhari A L, Sonawane G H 2015 *J. Mater. Sci. Chem. Eng.* **3** 39
[9] Kanade K G, Kale B B, Baeg J-O, Lee S M, Lee C W, Moon S-J, Chang H2007 *Mater. Chem. Phys.* **102** 98
[10] Thennarasu G, Sivasamy A 2015 *Ecotox. Environ. Safe.* In press, DOI: 10.1016/j.ecoenv.2015.10.030

IOP Conf. Series: Materials Science and Engineering **175** (2017) 012043 doi:10.1088/1757-899X/175/1/012043

Preparation for Spherical particles of Praseodymium doped Perovskite Red Phosphor by Hydrothermal Reaction

T Kosaka[1], A Matsuda[1], M Mizunuma[1], and Y Tanaka[2]

[1]Department of Chemistry, Tokyo Gakugei University, Koganei, Tokyo 184-8501, JAPAN
[2]Department of Materials Science & Engineering, Tokyo Institute of Technology, Meguro, Tokyo 152-8550, JAPAN

E-mail: tkosaka@u-gakugei.ac.jp

Abstract. Spherical particles of 0.5mol%Pr^{3+}-doped $CaTiO_3$ were prepared by hydrothermal reaction on dissolved $CaCl_2$, $Pr(NO_3)_3$, and ammonium citratoperoxotitanate (IV) complex precursor solution with molar ratio of Ti/CA=1:2 and calcination in ambient atmosphere. The obtained particle exhibited red photoluminescence at 610nm. It is found that several particles have hollow structure. It is required that further investigation is needed to clarify the formation mechanism of these spherical hollow paricles.

1. Introduction

Pr^{3+}-doped perovskite-type titanates like $CaTiO_3$[1-3], $SrTiO_3$[4-6], and $BaTiO_3$[7,8] have been considered as attractive phosphors for field-emission display[4,9]since these materials exhibit closely ideal red emission.Previous studies with regard to Pr^{3+}-doped $CaTiO_3$ prepared by solid-state reaction or liquid phase reaction proposed that their red emission ascribe to the *f*-*f* transition of Pr^{3+}occurs via the band-gapand inter-valence charge transfer photo-excitation processes[10,11].

From the standpoint of powder preparation for the desired morphology and chemical uniformity, liquid phase process have many benefit. However, titanium tetrachloride or titanium isopropoxide, which are usually utilized as liquid titanium compounds, are susceptible to hydrolysis. Therefore, the preparation are required to conduct in a dry inert atmosphere. Recently, Professor Kakihana and his group developed water-soluble titanium complex (ammonium citratoperoxotitanate (IV)) and reported that the chemical formula is $(NH_4)_8\ [Ti_4(C_6H_4\ O_7)_4\ (O_2)_4]\cdot 8H_2O$ based on the results of single crystal x-ray diffraction [12].

Though it is known that the synthesis of uniform precipitate is relatively difficult by hydrothermal reaction due to the difference in hydrolysis rate of dissolved metal compound, citric acid as chelate ligands can act to yield a polyester network in which metal ion are dispersed uniformly [13]. Thus, hydrothermal reaction on precursor solution with dissolved citric acid as a chelating agent is a promising method for preparation of uniformprecipitates under specific conditions.

In this article, 0.5mol%Pr^{3+}-doped $CaTiO_3$red phosphor was prepared by hydrothermal synthesis on dissolved $CaCl_2$, $Pr(NO_3)_3$, and water-soluble titanium complex precursor solution and calcination in ambient atmosphere. We investigated the crystal structure, morphology, and photoluminescence of the powder obtained.

2. Experimental

Water-soluble Ti complex solution was prepared according to paper of Professor Kakihana's group [12]. 5×10^{-3} mol of Ti powder was dissolved into 20ml of 30% H_2O_2 and 5ml of 28% NH_3 solution to obtain a yellowish peroxy titanic acid solution with pH≈11. When the excess of H_2O_2 is decomposed, prescribed amount of citric acid (CA) was dissolved into this peroxy titanic acid solution with continuous stirring. After evaporative drying, yellow gel formed. This gel was dissolved into ion exchange water to make a water-soluble Ti complex solution with Ti concentration of 0.25mol/L.

Thereafter, $CaCl_2$ and $Pr(NO_3)_3$ solutions were added to the Ti complex solution and finally the precursor solution was prepared with molar ratio of Ti:Ca:Pr=1:1:0.005. The PTFE vessel introducing 20ml of transparent precursor solution was sealed in a stainless lined autoclave and hydrothermally heated at 423K for 24h. After the autoclave was cooled to room temperature, the precipitate was centrifugally separated and dried at 383K for 18h. Finally, the precipitate was calcined at 1023K for 1h in ambient atmosphere and then sample powder was obtained.

The X-ray diffraction (XRD) analysis of the obtained powder before and after calcination was carried out using CuKα radiation by Rigaku Ultima IV Multipurpose X-ray Diffraction system equipped with D/teXUltra-high-speed, position-sensitive detector. The morphology and chemical composition of the powder were investigated using Hitachi SU8020 field-emission scanning electron microscopy equipped with Horiba EMAX Evolution X-Max SDD detector. The photoluminescence measurement were performed at room temperature using Shimadzu RF-6000 spectrofluorophotometer with L42 long pass filter.

3. Results and discussion

Figure 1(a) shows XRD pattern of powder prepared from the precursor solution dissolving citric acid (CA) with Ti/CA=1:5 ratio and calcined at 1023K for 1h. XRD analysis revealed that perovskite ($CaTiO_3$) and anatase (TiO_2) phases dominate in the structure indicating calcium and titanium ions are inhomogeneousdistributed in the powder.

Figure 1(b) shows the SEM image of the as-prepared powder. The powder involved a spherical particle of 2-3μm in diameter. EDS analysis revealed that chemical compositionsof the spherical particle is 3.0at% Ca, 31.3at% Ti and 65.7 at% O. Considering Ca ion distributed other region in the powder, the excess amount of CA in the precursor solution lead to inhibit co-precipitation reaction with Ca ion stabilizingwater-soluble Ti complex. Therefore, it is indicated that the controlling for hydrolysis rate of Ti complex govern the chemical homogeneity of precipitate obtained by hydrothermal reaction.

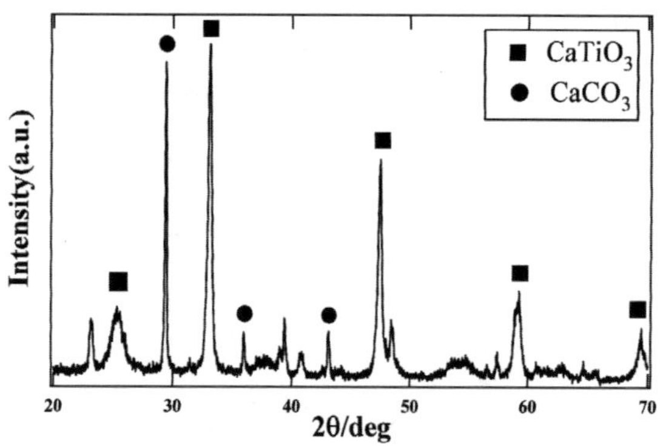

a.

b.

Figure 1.

Then, we conducted the preparation using the precursor solution with Ti/CA=1:2 ratio to stimulate the co-precipitation reaction with Ca^{2+} and Ti^{4+}. XRD revealed that the powder has amorphous-like structure after hydrothermal reaction at 423K for 24h and calcination at 1023K for 1h resulted in the formation of perovskite single phase as shown in Figure 2.

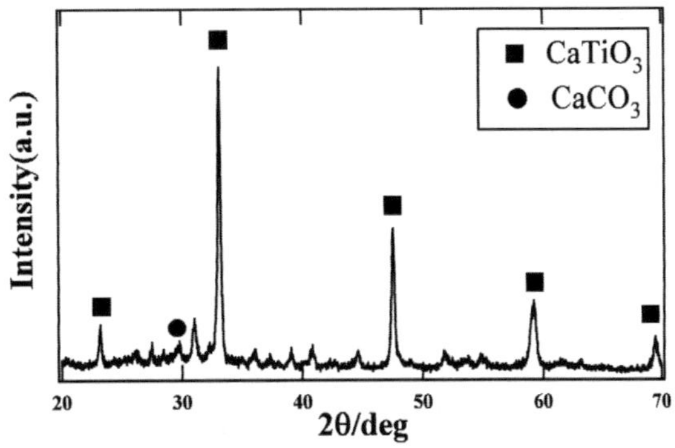

Figure 2.

SEM observation demonstrated that the powder contain considerable amounts of spherical particles with 2-3μm in diameter as shown in Figure 3 (a). EDS analysis confirmed that these spherical particles have chemical composition of 14.3at% Ca, 23.8at% Ti, and 61.9 at% O.These spherical particles exhibited red photoluminescence at 610nm, which is corresponded to the ones by previous works [14].

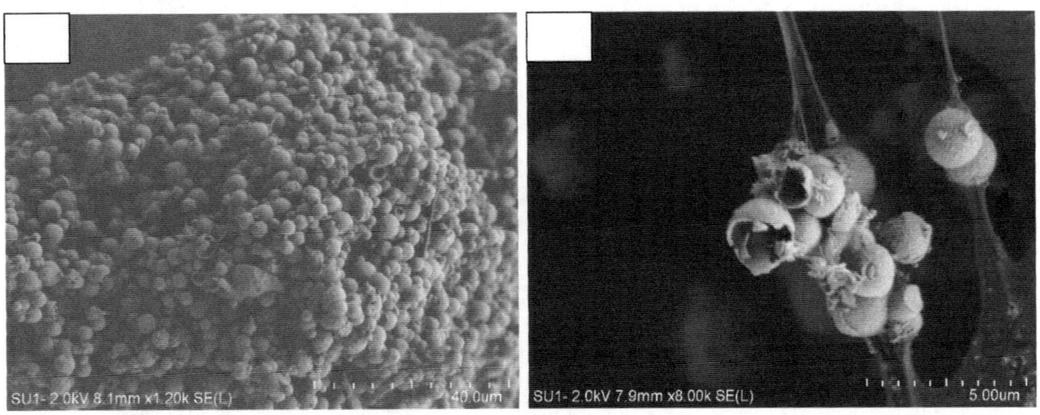

Figure 3. SEM images of (a) obtained spherical and (b) spherical hollow particles by hydrothermal reaction on the precursor solution with Ti/CA=1:2 at 423K for 24h and calcination at 1023K for 1h

We discovered spherical hollow particle in the powder during SEM observation as shown in Figure 3(b). Recently, one-pot synthesis for porous hollow $La_2O_2CO_3$microspheresusing hydrothermal treatmentof glucose and $La(NO_3)_3 \cdot 6H_2O$, and subsequent calcination was reported [15]. The authors proposed the formation process of the hollow sphere as follows. Firstly, carbonaceouspolysaccharide microspheres binding La^{3+} were formed via hydrothermal treatment of glucose/water/melt salt.Then, the sequential removal of carbon cores,densification, cross-linking, and phase transformation of La in thelayer via calcination result in the formation of the hollow spheres. Though polysaccharide like glucose isn't contained in our precursor solution, we speculated that formation process like mentioned above is proceededduring our hydrothermal reaction.

Figure 4 shows XRD patterns of obtained powders by calcination at 1023K for 1h after hydrothermal reaction at 423K for 6, 12, 18, and 24h. It is demonstrated that Ca rich phase dominatedin the powder in the early stage of hydrothermal reaction and $CaTiO_3$ ratio was increased as time course. These results indicated that hydrolysis reaction of metal ion isn't proceeded simultaneously even in present condition and chemically homogeneous particle are formed inducing resolve and precipitation reactions. However, we considered that further investigations are needed to clarify the formation mechanism of Pr^{3+} doped $CaTiO_3$ spherical hollow particles.

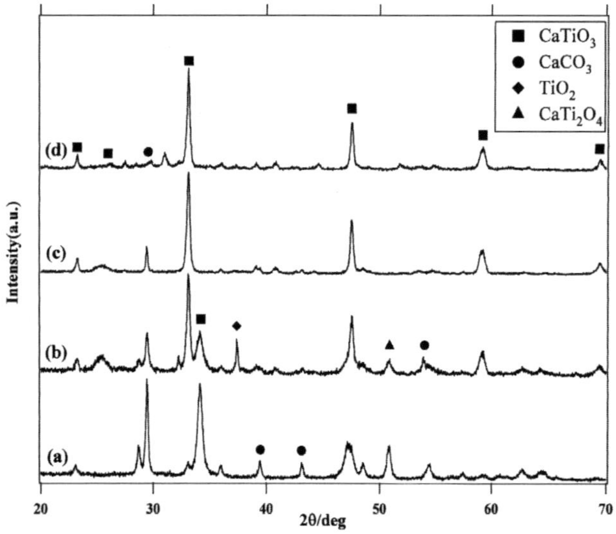

Figure 4. XRD patterns of obtained powders by hydrothermal reaction at 423K for (a)6h, (b)12h, (c)18h, and (d) 24h and calcination at 1023K for 1h.

4. Conclusion

The hydrothermal synthesis on dissolved $CaCl_2$, $Pr(NO_3)_3$, and ammonium citratoperoxotitanate complex precursor solution with Ti/CA=1:2 at 423K for 24hcaused to form precipitates which have amorphous-like structure and spherical morphology. The precipitates calcined at 1023K for 1h have perovskite single phase structure exhibiting red photoluminescence at 610nm. We proposed the formation mechanism of Pr^{3+} doped $CaTiO_3$ spherical hollow particles. However, further investigation is required.

References

[1] Diallo P T et al 2001 *J. Alloys Comd.***323-324** 218
[2] Okamoto S and Yamamoto H 2001 *Appl. Phys. Lett.***78** 655
[3] Boutinaud P, Pinel E, Dubois M, Vink A P, Mahiou R 2005 *J. Lumin.***111** 69
[4] Itoh S, Toki H, Tamura K, Kataoka F 1999 *Jpn. J. Appl. Phys.***38** 6387
[5] Yamamoto H, Okamoto S, Kobayashi H 2002*J. Lumin.***100** 325
[6] Inaguma Y, Nagasawa D, Katsumata T 2005 *Jpn. J. Appl. Phys.***44** 761
[7] Okamoto S and Yamamoto H 2002 *J. Appl. Phys.* **91** 5492
[8] Li J, Wu Y J, Kuwabara M 2005 *Jpn. J. Appl. Phys.***44** L708
[9] Cho S H, Yoo J S, Lee J D 1996 *J. Electrochem. Soc.***143** L231
[10] Inaguma Y, Tsuchiya T, Katsumata T 2007 *J. Solid State Chem.***180** 1678
[11] Boutinaud P et al 2010 *J. Lumin.***130** 1725
[12] Kakihana M et al 2001 *Inorg. Chem.* **40** 891
[13] Pan Y et al 2003 *J. Solid State Chem.***174** 69
[14] Diallo P T, Boutinaud P, Mahiou R, Cousseins J C 1997 *phys. stat. sol. (a)***160** 255
[15] Niu H et al 2011 *J. Alloys Comd.***509**744

IOP Conf. Series: Materials Science and Engineering **175** (2017) 012044 doi:10.1088/1757-899X/175/1/012044

Ceramic-like open-celled geopolymer foam as a porous substrate for water treatment catalyst

T Kovářík, T Křenek, M Pola, D Rieger, J Kadlec, P Franče

New Technologies - Research Centre, University of West Bohemia, 306 14 Plzeň, Czech Republic

E-mail: toko@ntc.zcu.cz

Abstract. This paper presents results from experimental study on microstructural and mechanical properties of geopolymer-based foam filters. The process for making porous ceramic-like geopolymer body was experimentally established, consists of (a) geopolymer paste synthesis, (b) ceramic filler incorporation, (c) coating of open-celled polyurethane foam with geopolymer mixture, (d) rapid setting procedure, (e) thermal treatment. Geopolymer paste was based on potassium silicate solution $n(SiO_2)/n(K_2O)=1.6$ and powder mixture of calcined kaolin and precipitated silica. Various types of ceramic granular filler (alumina, calcined schistous clay and cordierite) were tested in relation to aggregate gradation design and particle size distribution. The small amplitude oscillatory rheometry in strain controlled regime 0.01% with angular frequency 10 rad/s was applied for determination of rheology behavior of prepared mixtures. Thermal treatment conditions were applied in the temperature range 1100 – 1300 °C.

The developed porous ceramic-like foam effectively served as a substrate for highly active nanoparticles of selected Fe^{+2} spinels. Such new-type of nanocomposite was tested as a heterogeneous catalyst for technological process of advanced oxidative degradation of resistive antibiotics occurring in waste waters.

1. Introduction

Geopolymers are inorganicmaterials that form long-range, covalently bonded, predominantly amorphousnetworks [1].Geopolymers are a class of materials synthesized in alkaline medium with aluminosilicates sourced from calcined clays, natural minerals, industrial wastes etc. and/or from mixtures of these [2].The reactivity of raw aluminoslicates mainly depends on their chemical properties, surface functionalization, size/morphology of particles and structural composition [3]. In a strong alkaline medium the disaggregation process of Si–O–Si, Al–O–Al and Al–O–Si bonds on the surface of solid particles takes place. Disaggregation/dissolution of aluminates-silicates producesaluminate and silicate species and formation of hydroxylated complexes such as $Si(OH)_4^-$, $Al(OH)_4^-$ or $Al(OH)^{3-}{}_6$[4].Through the polycondensation reaction a free tetrahedral unites SiO_4 and AlO_4 are linked in an alternating manner to yield a three-dimensional amorphous matrix[5].

Geopolymers display excellent mechanical strength and resistance to attack by aggressive environment, their hardening kinetic is easily adjustable and ceramic-like structure and high melting point indicate structural stability at elevated temperatures [6].

Recently, study on geopolymers as a novel photocatalyst for degradation of dye from wastewater showed a synergetic excellent performance in cleaning process [7]. Therefore, porous geopolymer substrates having high specific area making them ideal candidates for catalyst supports.Many methods have been explored to synthesize reproducible porous substrates or foams with homogenous pore characteristics. Various processes were applied such as, gaseous method[8], suspension and solidification method through foaming agents[9] and freeze-castingtechnique [10]. Furthermore, little work has been devoted to the feasibility of making geopolymer-synthesized foam through polymeric sponge method. In this study, we investigated the optimal manufacturing process of the potassium based geopolymer open-cell porous substrate by the polyurethanesponge impregnation.

2. Experimental

The geopolymer paste was synthesized as a twocomponent system consists of alkaline potassium silicate solutionwith silica modulus 1.6and powder mixture of calcined kaolin clay and precipitated silica as a main source of solid silica and alumina as presented in Table 1 with particle size distributionas follows: d_{50} at 3.31μm, d_{90} at 6.15μm. The mixture was formulated by molar ratios: $SiO_2/Al_2O_3 = 5$, $SiO_2/K_2O = 3$, $Al_2O_3/K_2O = 0.6$, $H_2O/K_2O = 19.5$.The alumina, schistous clay and cordierite ceramic powders were used as granular fillerwith maximum grain size 100 μm and particle size distribution presented in Table 2.

Table 1. The chemical composition of solidcomponents (wt.%)

	SiO_2	Al_2O_3	CaO	MgO	TiO_2	K_2O	Fe_2O_3
Powder mixture	58.00	38.10	0.12	0.15	1.45	0.62	1.00

Table 2. Mean size and volume average particle distribution

Size distribution	**d_{50} [μm]**	**Dv10 [μm]**	**Dv90 [μm]**
Alumina	28.0	3.9	82.8
Schistous clay	39.9	7.3	82.1
Cordierite	53.0	13.8	99.4

The materials were proportioned by weight ratio and mixed in a laboratory automatic blender for 6 min (1000 rpm). After that, the portion of ceramic filler was added and mixtures were blended for an additional 5 min. The weight portion of different ceramic filler varied in relation to optimal workability of the mixtures. In virtue of this, dynamic rheological measurements were performed using a rotary rheometer (Ares G2, TA Instruments) operated in oscillation mode with small amplitude. Time dependent experiments were performed at an angular frequency 10 rad/sec (1.59 Hz) and strain rate 0.01% in plain-plain geometry. Consequently, polyurethane (PU) porous spongewas immersed into the geopolymer mixture and taken back several times. After that, it was heated in closed container at 65°C to accelerate rapid hardening mechanism. As a solid substrate the geopolymer/PUsponge was dried at room temperature for the next 24 hours and heat-treated in a kanthal furnace. The temperature was raised up to 600°C at 1°C/min in order to burn out the organic substrate entirely. Then the geopolymer substrate was sintered for 4 hours at various temperatures from 1100 to 1300°C.

3. Results and Discussion

3.1. Viscosity of the fresh mixtures

Our investigation of fresh mixtures used in this study indicated the interval of workability corresponding to a proportion of filler added and blended into the geopolymer paste. Based on trial test the maximum weight proportion of filler was defined for each mixture taking into account particle size distribution of ceramic powders. It was determined that geopolymer paste with alumina powder, labeled as a mixture A, consisted of 105 wt.% of ceramic filler. The weight of ceramic filler isexpressed as weight ratio to powder part, where 100 wt.% was powder part of geopolymer binder.

Based on these, geopolymer paste with schistous clay consisted of 210 wt.% (mixture B) and paste with cordierite powder consisted of 120 wt.% (mixture C).

Time dependent measurements confirmed the large interval of workability of potassium based geopolymer mixtures. In should be noted that, the time zero is taken as a + 10 min in virtue of preparation of the mixture for each measurement at 30 °C ± 0.1 °C. The pure paste in interval from 5 to 60 min presentedsteady state plateau in viscosity (η) between 18.9 Pa.s and 30.5 Pa.sas presented in Figure 1 (D curve).

The addition of ceramic filler increased viscosity rapidly in comparison with pure paste and emphasized the thixotropic properties of the system paste-filler, even before structural reorganization into three-dimensional aluminasilicate network.The mixtures A, B, C showed concurrent time dependent increase in viscosity after ~ 3 min. in measuring cell, which can be interpreted as a residual stresses relaxed prior to each rheological measurement.

The mixture B showed more increase in viscosity after 10 min of measurement with values over 33 kPa.s. The mixtures A and C revealed similar trend, after 10 min their viscosity ranging between 15 ~ 23 kPa.s and gradually raised up to 60 ~ 65 kPa.s after 60 min.

As showed, the proportion of ceramic filler markedly affected ultimate viscosity values including viscosity rate in timedependent experiment. Mixture B showed an intensive increase in viscosity with respect to highest proportion of granular filler (210 wt.%) in final mixture. The viscosity of mixture with schistous clay powder reached ~ 99 kPa.s after 60 min.

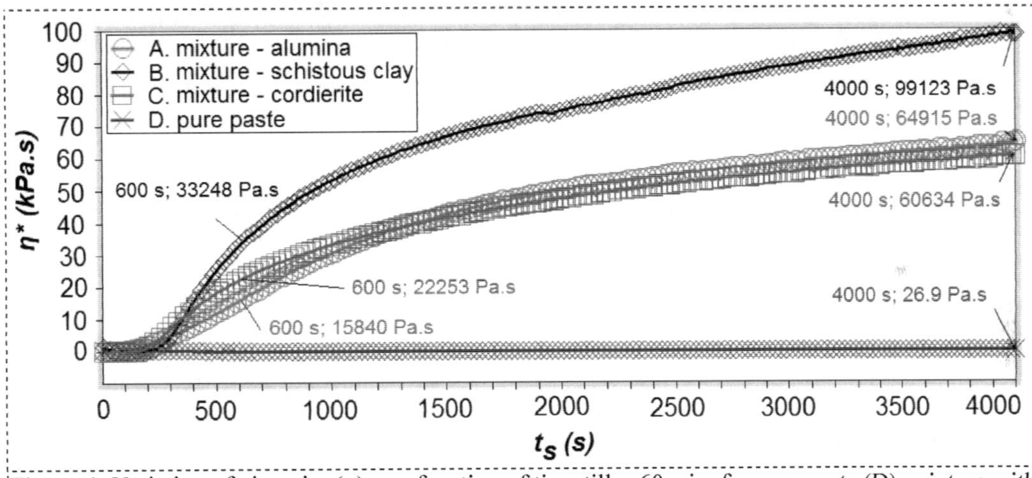

Figure 1. Variation of viscosity (η) as a function of time till ~ 60 min. for pure paste (D), mixture with alumina powder (A), mixture with schistous clay (B) and mixture with cordierite powder (C)

3.2. Microstructural and mechanical properties

It can be concluded that by comparing of three types of ceramic filler in geopolymer paste, after thermal treatment at 1300°C, microstructural properties vary in relation to the surface quality and structural integrity. Mixture A with very fine particles of alumina powder (Dv10 = 3.9μm) partially coagulated during blending process and even at lower viscosity (see Figure 1) non-homogenously covered PU sponge during impregnation. Geopolymer substrate after thermal treatment become fragile structure with partially connected struts in solid body of substrate. Compression strength ranging between

0.15 – 0.30 MPa in relation to thermal treatment at 1100 up to 1300°C. A similar results were obtained through mixture C where cordierite powder became fractional sediment after blending process and non-homogenously covered PU substrate due to sharp-edge geometry of the particles.

The homogenous-thick coating of the slurryon the surface of PU sponge revealed mixture B with 210 wt.% of schistous clay powder.Figure 2. shows heat-treated substrate at 1300°C for 4h, partially sintered grains and fully covered amorphous phase on the surface of sponge body with open pore size 0.5~2 mm (a) and hollow strut of geopolymer sponge in detail (b). Compression strength values, on experimental specimens 20x20x20 mm, reached 0.75 MPa after thermal treatment at 1300°C.

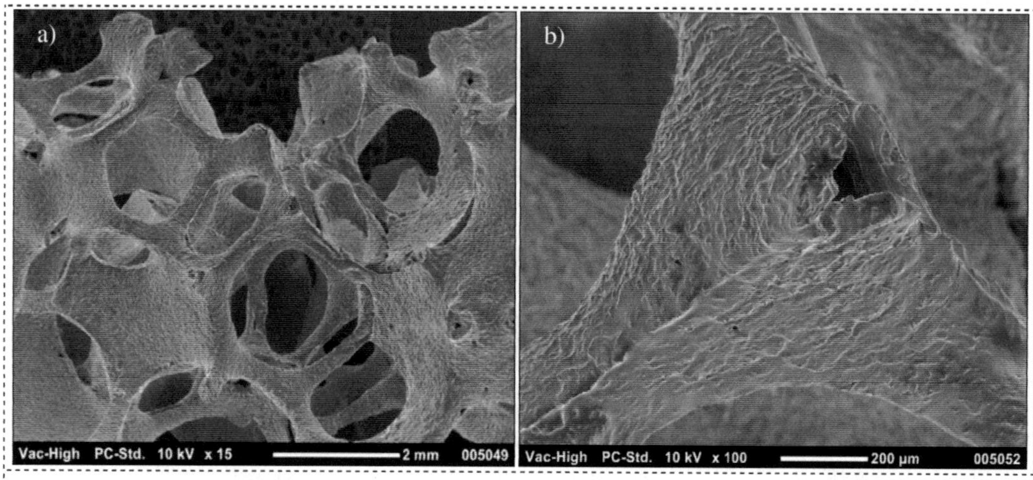

Figure 2.SEM photos of geopolymer sponge foam, based on mixture B, heat-treated at 1300°C.

4. Conclusion

In this investigation, the mixture B showed an intensive increase in viscosity with respect to highest proportion of granular filler in final mixture and revealed broad interval for workability of thegeopolymer slurry. The best conditions for homogenous-thick coating was mixture B with schistous clay filler resulted in compression strength 0.75 MPa after thermal treatment at 1300°C and uniform open pore size 0.5 ~ 2 mm.

Acknowledgements

The result was developed within the CENTEM project, reg. no. CZ.1.05/2.1.00/03.0088, cofounded by the ERDF as part of the Ministry of Education, Youth and Sports OP RDI programme and, in the follow-up sustainability stage, supported through CENTEM PLUS (LO1402) by financial means from the Ministry of Education, Youth and Sports under the National Sustainability Programme I.

References

[1] DavidovitsJ1989 *J. Thermal Anal.***35**429
[2] Srinivasula ReddyM,DinakarP,Hanumantha RaoB 2016*Microporous Mesoporous Mater.***234**12.
[3] KovářikT,Bělský, P, Novotný P, Říha, J, Savková J, MedlínR, RiegerD, Holba P 2015 *Constr. Build. Mater.***80**98
[4] Shi C,Feznández-JiménezA,Palomo A 2011 *Cem. Con. Res.***41** 750
[5] Yun-Ming Liew, Y-M,HeahCh-Y,Mustafa A B MKamarudin H 2016 *Prog. Mater. Sci.***83**595
[6] Duxson P, Lukey G C, Van Deventer J S J 2007 *J. Non-Crys. Sol.***353** 2186
[7] ZhangYJ,Liu L2013*Particuology***11**353
[8] Prud'homme E, Michaud P,Joussein E,Clacens J-M,Rossignol S 2011 *J. Non-Crys. Sol.***357** 1270
[9] Tang Q,Ge, Y-Y,Wang K-T,He Y, Cui X-M 2015 *Mater. Des.***88** 1244
[10] Papa E,Medri V,Benito P,Vaccari A,Bugani S,Jaroszewicz J,Swieszkowski W,Landi, E 2015 *Microporous and Mesoporous Mater.***215**206

Properties of oxide-hydroxide sintered ceramics

R V Levkov[1,2], **S N Kulkov**[1,2]

[1]Institute of High Technology Physics, Tomsk Polytechnic University, 634034Tomsk, Russia
[2]Laboratory of Physics of Nanostructured Functional Materials, Institute of Strength Physics and Materials Science SB RAS, 634055 Tomsk, Russia

E-mail: Levkov.r.v@mail.ru

Abstract. In this paper the study of porous ceramics obtained from aluminum hydroxide with gibbsite modification is presented. It was shown that aluminum hydroxide may be used for pore formation and pore volume in the sintered ceramics can be controlled by varying the aluminum hydroxide concentration and sintering temperature. It was shown that compressive strength of alumina ceramics increases by 40 times with decreasing the pore volume from 65 to 15%. Based on these results one can conclude that the obtained structure is very close to inorganic bone matrix and can be used as promising material for bone implants production.

1. Introduction

The development of material based on porous oxides with high strength is a fundamental challenge to modern materials [1-5]. Ceramics with a porous structure can be used to produce filters, catalyst supports, membranes etc. [6]. Porous ceramics is also interesting for medical applications due to its properties are close to bone tissue.

The traditional method of creating highly porous materials is based on “burn-pore-forming” additives [7]. On the one hand this method is effective in terms of management of the pore volume, but on the other hand the possible formation of carbon on the inner surfaces of pore is unacceptable for the manufacture of bio-implants and can lead to rejection of the implant [8]. To form a multi-level pore structure without addition of pore-former one can use the method of component decomposition, for example, the hydroxides to oxides, accompanied by the release of gas and, as a consequence, the formation of porosity [9]. This method based on the decomposition of aluminum hydroxide to the oxide has been known for a long time [10-12], but there is small amount data on the regulation of the pore space volume and the dependence of the pore space formation sintered at different temperatures, particle size of initial powders, as well as no data about dependences of the strength of the ceramic with changing of porosity [13].

The aim of this work is to study formation of porous structures in ceramic synthesized from aluminium hydroxide due to decomposition of hydroxide to the oxide.

2. Materials and experimental procedure

Aluminum hydroxide with gibbsite modification $Al_2O_3*4H_2O$ obtained by the decomposition of the aluminates solution was used as a starting component. The average particle size of the powder was 2.6 μm. Cylindrical form samples were pressed at 20 kN load, sintered in air at 1300, 1400 and 1500°C, with isothermal one hour holding at these temperatures. The structure, morphology, pore space, grain size and phase composition were investigated. The volume of the pore space was estimated by the formula $\theta = (\rho_{th} - \rho_e) / \rho_\kappa * 100$ %, where ρ_{th} the theoretical density material, ρe experimental density of the material. The pore structure of ceramics and morphology of the powder was investigated using scanning electron microscope (SEM) Philips-SEM 515 and optical microscope. The mean pore size of ceramics and the powder particles were calculated by the secant method [10].

X-ray diffraction studies were carried out on X-ray diffractometer with filtered CuKα radiation. Diffraction spectra were obtained in the range of angles $2\theta = 20 - 80$ degrees with a step $\Delta(2\theta)=0.05$, the exposure time was selected to achieve accuracy better when 0.5%. Compression testing of ceramics samples was carried out on universal test machine "Devotrans" with strain rate $2*10^{-4}$ s^{-1} [9].

3. Results and Discussion

On the XRD pattern of aluminum oxide samples peaks have a large width and only three peaks are clearly visible. Their angular positions correspond to the strongest reflections of the hydroxide of aluminum oxide of the $Al_2O_3*4H_2O$ composition [14-16]. X-ray peaks broadening could be caused by a small size of the coherent diffracting domain (CDD), or concentration in homogeneity by hydration degree of Al_2O_3 powder. These factors affect the specific surface - the specific surface of the powder is 173 m^2/g, the similar data were obtained in [17, 18].

According to the EDAX analysis the samples obtained contain 59,1 ± 2,3% (at.) of oxygen and 40,9 ± 1,3% (at.) aluminum, which corresponds to the chemical formula of aluminum oxide. No impurities in ceramics were detected.

The structure of sintered ceramics is presented in Fig. 1. As one can see grains of aluminum oxide have irregular shape and all samples are characterized by the presence of interparticle porosity regardless the sintering temperature. It was also observed that the increasing sintering temperature leads to a decrease in the interparticle pore size. The pore sizes for samples with increasing sintering temperature are decreasing (Figure 2(a,b,c)).

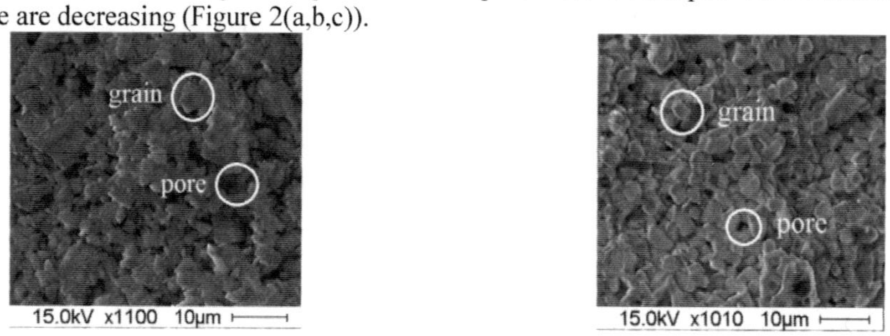

Figure 1. Ceramic structure after sintering at 1300°C and 1500°C

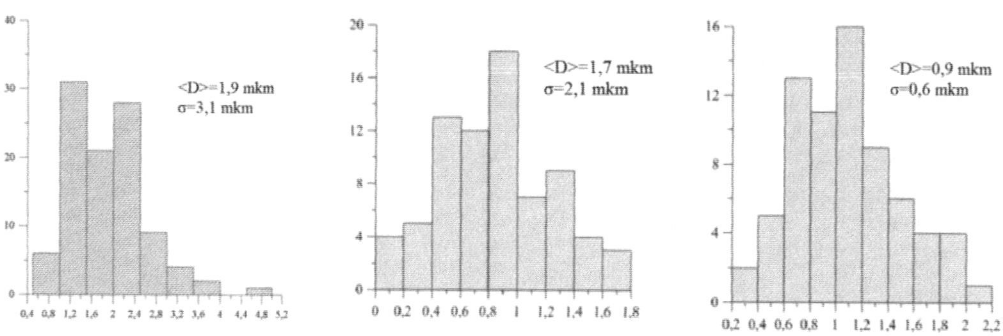

Figure 2. Pore size distribution, the sintered ceramics at the temperatures: (a) 1300°C, (b) 1400°C, (c) 1500°C

In Figure 3(a) the results of shrinkage and porosity of the samples are shown. Shrinkage has a linear dependence on sintering temperature. Ceramics sintered at 1300 °C demonstrates open porosity and at 1500 °C porosity is isolated.

The results of mechanical testing are shown in Figure 3(b). One can observe a significant increase in the compressive strength with increasing sintering temperature associated with the decrease in the pore space volume.

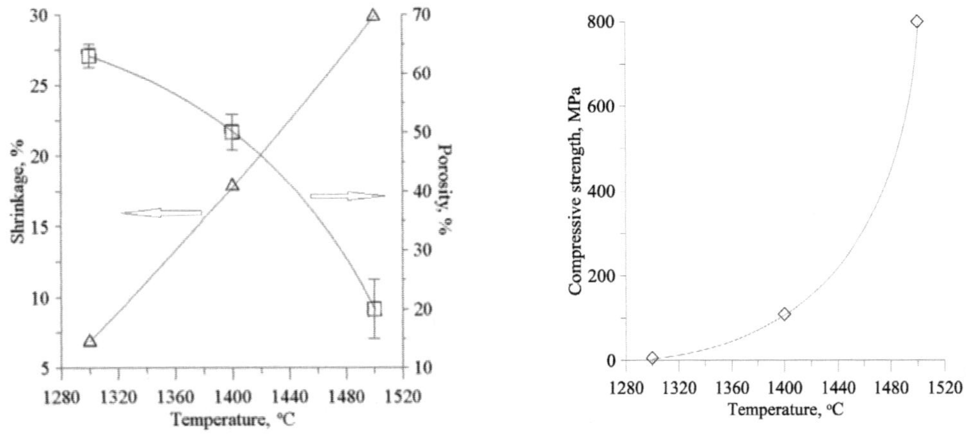

(a) (b)

Figure 3. (a) Dependency of shrinkage porosity on sintering temperature, (b) the dependence of the strength of the sintering temperature.

4. Conclusions

In the work presented it was shown that average pore size of aluminum oxide ceramics decreases from 1,9 to 0,9μm with increasing sintering temperature. Shrinkage decreases from 28 to 8 % with increasing sintering temperature. It was shown that the compressive strength increases with increasing sintering temperature, minimum compression strength is 20 MPa for sintering temperature 1300°C and the maximum 800 MPa for 1500°C.

5. Acknowledgments

This work was partially financially supported by Grant No 14.607.21.0069-RFMEFI60714X0069 of Ministry of Science and Education of RF.

References

[1] L A Gömze and L N Gömze 2009 *Építőanyag-JSBCM* **61** (2) 38 http://dx.doi.org/10.14382/epitoanyag-jsbcm.2009.7
[2] L A Gömze and L N Gömze 2010 *Építőanyag-JSBCM* **62** (4) 98 http://dx.doi.org/10.14382/epitoanyag-jsbcm.2010.18
[3] L A Gömze and L N Gömze 2011 *IOP Conf. Ser.: Mater. Sci. Eng.* **18** 082001 http://dx.doi.org/10.1088/1757-899X/18/8/082001
[4] L A Gömze and L N Gömze 2013 *IOP Conf. Ser.: Mater. Sci. Eng.* **47** 012033 http://dx.doi.org/10.1088/1757-899X/47/1/012033
[5] R. Ahmad, J. H. Ha, & I. H. Song, Synthesis of open-cell particle-stabilized Al2O3 foam using $Al(OH)_3$ particles. ScriptaMaterialia, 76, 85-88, (2014).
[6] Y. Heo, H. Im, J. Kim, &J. Kim. The influence of Al(OH)3-coated graphene oxide on improved thermal conductivity and maintained electrical resistivity of Al_2O_3/epoxy composites. Journal of Nanoparticle Research, 14(10), 1-10, (2014).

IOP Conf. Series: Materials Science and Engineering **175** (2017) 012045 doi:10.1088/1757-899X/175/1/012045

[7] A. D.Souza, C. C.Arruda, L.Fernandes, M. L.Antunes, P. K.Kiyohara &R.Salomão, Characterization of aluminum hydroxide (Al(OH)3) for use as a porogenic agent in castable ceramics. Journal of the European Ceramic Society, 35(2), 803-812. (2015).

[8] Farhadi, Fatollah, and Masoud Bahrami Babaheidary. "Mechanism and estimation of Al (OH) 3 crystal growth." Journal of Crystal Growth 234.4, 721-730, (2002).

[9] M. Goudarzi, D. Ghanbari, M. Salavati-Niasari,&A. Ahmadi, Synthesis and Characterization of Al(OH)3, Al2O3 Nanoparticles and Polymeric Nanocomposites. Journal of Cluster Science, 27(1), 25-38, (2016).

[10] N. L. Savchenko, I. N. Sevostyanova, T. U. Sablina, L.Molnar, R. Geber, L. A. Gömze, S. N.Kulkov, L. N. Gömze 2014 *Épitőanyag-JSBCM* **66** (2) 44 http://dx.doi.org/10.14382/epitoanyag-jsbcm.2014.9

[11] Deng, Zhen Yan, et al. "High Surface Area Alumina Ceramics Fabricated by the Decomposition of Al(OH)3." Journal of the American Ceramic Society 84.3, 485-491, (2001).

[12] Li, Shujing, and Nan Li. "Effects of composition and temperature on porosity and pore size distribution of porous ceramics prepared from $Al(OH)_3$ and kaolinite gangue." Ceramics international 33.4, 551-556, (2007).

[13] L A Gömze, L N Gömze, N. S. Kulkov, L. I. Shabalin, I. Gotman, F. Fedraza, G. L. Lecomte, T. Mayorova, E. Kurovics and A. Hamza 2015 *Épitőanyag-JSBCM* **67** (4) 143 http://dx.doi.org/10.14382/epitoanyag-jsbcm.2015.24

[14] V. V. Roerich, A. V. Krut'ko, A. D. Lastevsky, D. A. Kozlov, A. R. Avetisyan, A. M. Aronov N. A. Rychkova, B. B. The ventral interbody fusion in the lower cervical spine porous bioceramic implant, Surgery, 12(1), 63-68, (2015).

[15] Y Yang, S Shimai, Y Sun, M Dong, H Kamiya and S Wang 2013 *Journal of Materials Research* **28** 15

[16] Jing, L., Zuo, K., Fuqiang, Z., Chun, X., Yuanfei, F., Jiang, D., & Zeng, Y. P. The controllable microstructure of porous Al_2O_3 ceramics prepared via a novel freeze casting route. Ceramics International,36(8), 2499-2503, (2010).

[17] S P Buyakova, Han Vey, Lee Dong, Zhen Hayyun, T. Yu. Sablina, A. G. Melnikov, S. N. Kul'kov. The mechanical behavior of porous zirconia with active compression deformation, Writing in the Journal of Technical Physics. 44-48, (1999).

[18] S N Kulkov and S P Buyakova 2015 *Épitőanyag-JSBCM* **67** (4) 155 http://dx.doi.org/10.14382/epitoanyag-jsbcm.2015.26

Optimization of activator solution and heat treatment of ground lignite type fly ash geopolymers

Z Molnár[1], R Szabó[1], Á Rácz[1], J Lakatos[2], Á Debreczeni[3], G Mucsi[1]

[1]University of Miskolc, Institute of Raw Material Preparation and Environmental Processing, Hungary
[2]University of Miskolc, Institute of Chemistry, Hungary
[3]University of Miskolc,Institute of Mining and Geotechnics, Hungary

E-mail: ejtmolnar@uni-miskolc.hu

Abstract. Geopolymers are inorganic polymers which can be produced by the reaction between silico aluminate oxides and alkali silicates in alkaline medium. Materialscontaining silica and alumina compounds are suitable for geopolymer production. These can beprimary materials or industrial wastes, i. e. fly ash, metallurgical slag and red mud.

In this paper, the results of the systematic experimental series are presented which were carried out in order to optimize the geopolymer preparation process. Fly ash was ground for different residence time (0, 5, 10, 30, 60 min) in order to investigate the optimal specific surface area. NaOH activator solution concentration also varied (6, 8, 10, 12, 14 M). Furthermore, sodium silicate was added to NaOH as a network builder solution. In this last serie different heat curing temperatures (30, 60, 90°C) were also applied. After seven days of ageing the physical properties of the geopolymer(compressive strength and specimen density)were measured. Chemical leaching tests on the rawmaterial and the geopolymers were carried out to determine the elements which can be mobilized by different leaching solutions. It was found that the above mentioned parameters (fly ash fineness, molar concentration and composition of activator solution, heat curing) has great effect on the physical and chemical properties of geopolymer specimens. Optimal conditions were as follows: specific surface area of the fly ash above 2000 cm^2/g, 10 M NaOH, 30°C heat curing temperature which resulted in 21 MPa compressive strength geopolymer.

1. Introduction

Geopolymers can be produced by the reactions between alumina silicate oxides and alkali silicates, aluminates in high alkaline conditions. A possible way is to use fly ashmixed with NaOH solution. The reaction products might have different mechanical properties based on the type of fly ash (class C and F), the origin of the coal used and the applied combustion technologies, since they resulted difference in the raw material properties(composition, amount of glassy phases etc…) and the fly ash reactivity is also different. Activation state of fly ash can be enhanced by mechanical and/or chemical activation. The mechanical activation can be carried out by grinding in conventional ball mills or in high energy density mills (vibratory, stirred media mills).

Kumar et al.[1] activated the fly ash in attrition and vibratory mill. They stated that better mechanical properties of geopolymer reached by mechanical activation. Other authors investigated the effect of mechanical activation of fly ash on geopolymer properties in various mills [3-6]. Kumar et al.

[1] and Álvarez-Ayuso et al.[2] investigated the effect of NaOH molar concentration. They concluded that increasing of NaOH solution concentration better mechanical properties can be reached with different types of fly ash.

Álvárez-Ayuso et al.[7] use different heat curing temperature and curing time. The increasing curing time and curing temperature has positive effect on mechanical properties on the resulted products.

Leaching characteristics are very important from the point of view of practical applications of fly ash based geopolymers due to its environmental effect. According to Provis et al. [7] many elements from periodic table can bounded in geopolymer framework. The following elements bound with limited success or increase its availability by geopolymerisation: Mo, Ni, As, Se. Izquierdo et al. [8] investigated leaching properties of fly ash based geopolymers. They concluded that geopolymers suitable for immobilization of many hazardous elements from fly ash like Be, Bi, Cd, Co, Cr, Cu, Nb, Ni, Pb, Sn, Th, U, Y, Zr and rare earth elements, but more elements was mobilized like As, B, Se, V, W due to oxianionic mobilizing in high alkaline conditions. This confirmed the fact that an optimal dosage of the compounds, the synthesis and the curing conditions are crucial for the environmental properties of the final product.

2. Materials and methods

The fresh fly ash used as main component of geopolymers originated from Mátra power plant, Hungary has the following physical and chemical properties. Moisture content was 0.27 m/m%. The bulk density of the raw fly ash was 0.73 g/cm^3, particle density was 1.93 g/cm^3. Particle size distribution and specific surface area measured by Horiba LA950 V2 type laser particle size analyzer; the characteristic particle size values were x_{10}=10.77 μm; x_{50}=52.04 μm; x_{80}=119.32 μm, specific surface area: 1152.07 cm^2/g.Based on XRF analysis the chemical composition of the fly ash is given in Table 1. From the analysis the SiO_2/Al_2O_3 ratio was found to be 3.34 and the sum of SiO_2, Al_2O_3 and Fe_2O_3 content of the fly ash was 73.49 wt.%.

Table 1. Chemical composition of the lignite type fly ash (Visonta)

Component	**Visonta fly ash, Mass concentration, wt.%**
SiO_2	48.1
Al_2O_3	14.42
Fe_2O_3	10.97
Na_2O	0.37
K_2O	1.66
CaO	11.76
MgO	3.34
TiO2	0.492
P_2O_5	0.264
MnO	0.171

For geopolymer production NaOH solution in various concentrations was used as activator and Na-K-silicate (Betol SB) was used as network builder additive. Its chemical composition was as follows: K_2O=2.7%; Na_2O=13.7%; SiO_2=25.3%.

Fly ash was ground in tumbling mill for 5, 10, 30 and 60 minutes in order to mechanically activate the raw material (improving its reactivity), with e=80% of critical revolution number and φ=110 % mill filling ratio.

Geopolymer specimens made by mixing of raw and ball milled fly ash and alkaline activator in the ratio of 40 m/m % (L/S=0.67) and 45 m/m % (L/S=0.82) to fly ash. The higher sludge concentration was necessary in order to increase the workability of the geopolymer paste. Mixture was placed into pre-oiled moulds and compacted by vibration. The compacted mixture was kept in moulds for 24 hours insulated at ambient temperature. Then heat curedfor 6 hours. After heat curing the specimens were cooled down to ambient temperature. The compressive strength tests were carried out usinga Compression Testing Machine at the age of 7 days.The fly ash fineness, the NaOH concentration of activator solution (6, 8, 10, 12, 14 M), the amount of waterglass mass ratio in activator solution (25, 50, 75, 100 m/m%) and the heat curing temperature (30, 60, 90°C) were optimized.

In addition, FTIR analysisin transmission mode (KBr pellet) was performed in the case of various alkali concentration geopolymers to monitor the structural rearrangements during geopolymerisation reactions.

Chemical leaching tests have been done on both: the raw ground ash and the produced geopolymers. The mobilisation of the elements by distilled water and different strength acids (1 M acetic acid, 2 M nitric acid) were determined.

3. Results and discussion

In this part the research results are presented focusing on the optimization of the characteristic particles size and specific surface area of fly ash due to grinding, alkaline activator composition and curing temperature. The Figure 1. shows the effect of mechanical activation on fly ash fineness.

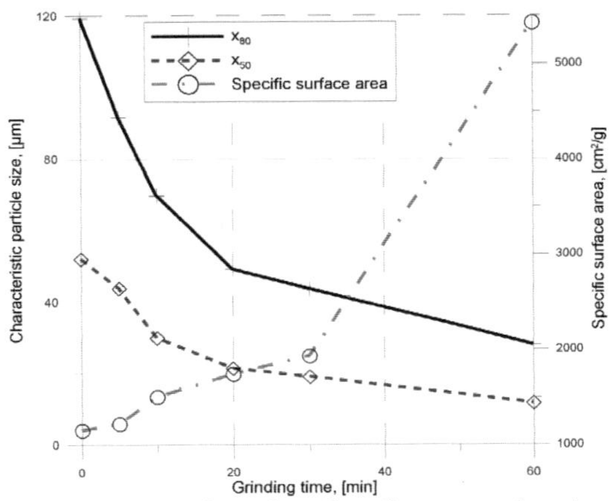

Figure 1. Changing of product fineness as function of grinding time

It was observed that the characteristic particle size became finer and specific surface area increased significantly by increasing grinding time even after 30 min residence time. The maximum fineness belongs to 60 minutes ground product (x_{50}=11.6 µm; x_{80}=27.9 µm; SSA=5425.55 cm^2/g).

The effect of grinding on the leachability of main structural elements: Si, Al, Fe is shown in Figure 2. An improving solubility can be reached for Si in alkaline solution as function of increasing fineness. These finding indicates that grinding breaking up the "spheres" of the fly ash resulting in the enhancement of soluble Si and Al which play important role in geopolymerisation reactions.

The effect of fly ash fineness on geopolymer mechanical properties (uniaxial compressive strength and specimen density) can be found in Figure 3.

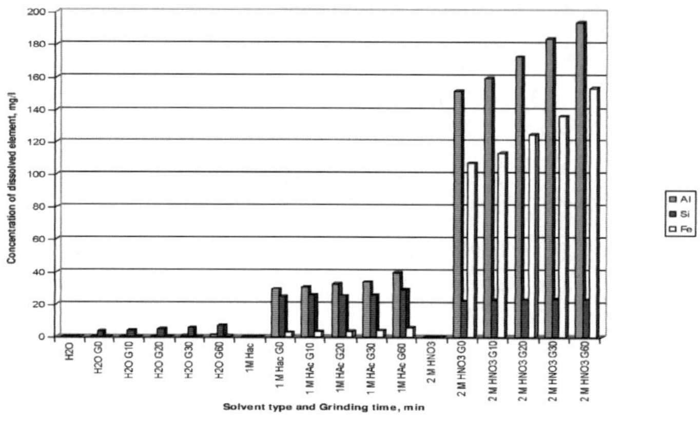

Figure 2. The effect of grinding on leachability of main structural elements of fly ash, Si, Al, Fe. Solid: Liquid ratio 1:50, contact time: 1 day

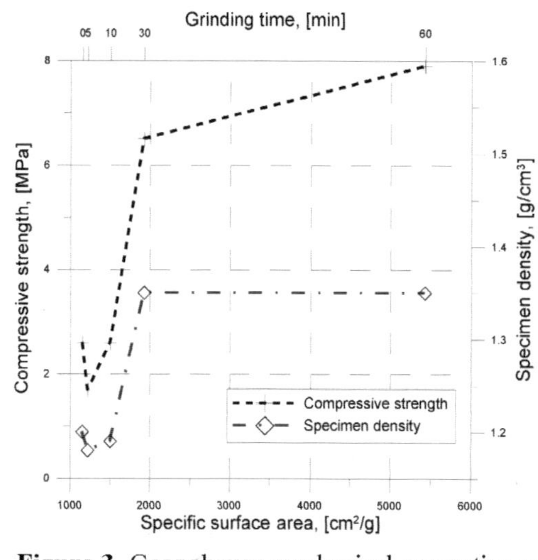

Figure 3. Geopolymer mechanical properties as function of fly ash fineness

The compressive strength improved by increasing specific surface area (or grinding time). The highest compressive strength (7.9 MPa) belongs to highest specific surface area (5425.55 cm^2/g). The tendency of specimen density change is similar to that of the compressive strength, good correlation achieved. The maximum density value was 1.35 g/cm^3.

Increasing the NaOH molarity of activator solution has a positive effect on geopolymer mechanical stability until 10 M (Figure 4.). After it the compressive strength decreased from the maximum 12.3 MPa (10 M NaOH) to 9 MPa (14 M NaOH), however the specimen density still increased (maximum value was 1.61 g/cm^3 belongs to geopolymer activated by 14 M NaOH solution). This observation is different than that of reported by other authors [3] therefore further structural investigation (FTIR) is necessary.

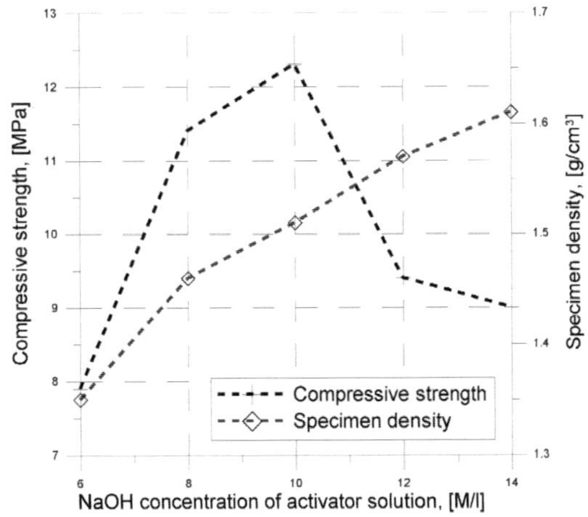

Figure 4. Effect of NaOH molar ratio on the geopolymer mechanical properties

Main peaks in FT-IR spectra (Figure 5) of geopolymer specimens corresponds to water molecules, O-C-O bonds and aluminosilicate bonds. In the case of geopolymer activated by 6 M NaOH solution the peak at 3353 cm^{-1} related to stretching vibrations of –OH, HOH bonds, while peak at 1646 cm^{-1} it is associated with HOH bending vibration. Peak at 1398 cm^{-1} related to O-C-O stretching vibration. Peak at 1012 cm^{-1} in case of fly ash shifted lower wavenumber (950 cm^{-1}; corresponds to asymmetric stretching vibration of Si-O-Si, Al-O-Si) which related to structural reorganization and formation of aluminosilicate gel phase associated with dissolution of fly ash amorphous phases in high alkaline condition. The peak at 950 cm^{-1} supplemented by a smaller peak at 905 cm^{-1} (in the case of geopolymers activated 10-14 M NaOH solution). Peaks at 797, 666, 615 and 561 cm^{-1} corresponds to Si-O-Si and Al-O-Si bonds symmetric stretching vibrations. It can be stated, that all of peak intensities became higher by increasing of NaOH concentration of activator solution.

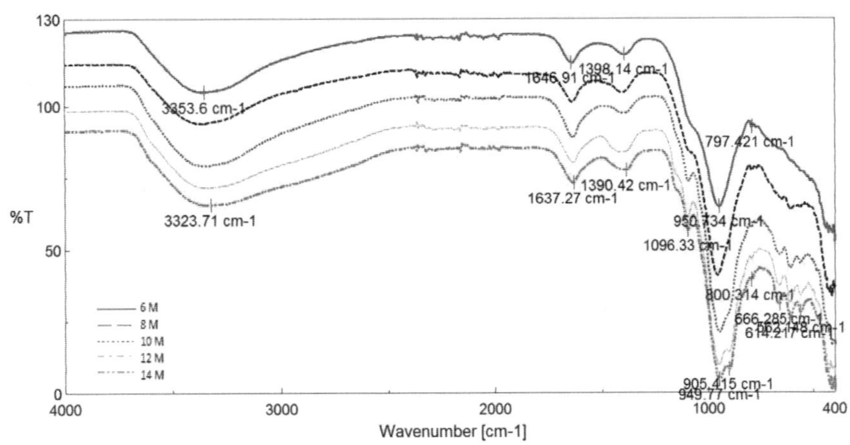

Figure 5. FT-IR spectra of geopolymers at various NaOH solution concentration

Based on FT-IR spectra it can be stated that, there are structural differences between geopolymer specimens, new peaks appeared in case of specimen spectra activated with 10-14 M NaOH solution in the range of 800-400 cm^{-1} and increased its intensity. In order to clarify the reasons of strength decrease, X-ray diffraction measurements are necessary in the future.

The effect of increasing amount of waterglass to NaOH solution on compressive strengthis shownin figure6. At 90°C the increasing waterglass ratio improved the compressive strength with a maximum value of 11.1 MPa belongs to 75 m/m% waterglass concentration. In case of lower temperature the highest compressive strength reached when activator solution was 100% Na-K silicate (13.1 MPa at 60°C, 21.3 MPa at 30°C). The lower temperature has positive effect on the compressive strength, which different from literature, where most of the results shows better mechanical properties in case of higher temperature [3]. The main cause presumably, that by increasing the amount of waterglass the molar ratios were changed and because the K based network formation became slower, therefore more timewould be needed to the diffusion of molecules and connect it. In case of higher temperature the reaction was so fast, that these diffusion processes couldnot be occurred and the formed structure will be disordered.

The effect of increasing ratio of waterglass to NaOH solution on specimen density is shown in Figure7. The results are correlated well with that of the compressive strength data.

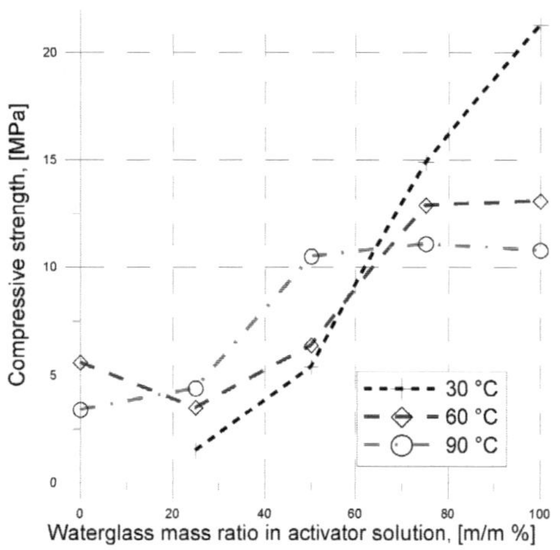

Figure 6. Waterglass addition to activator solution and its effect on compressive strength at various curing temperatures

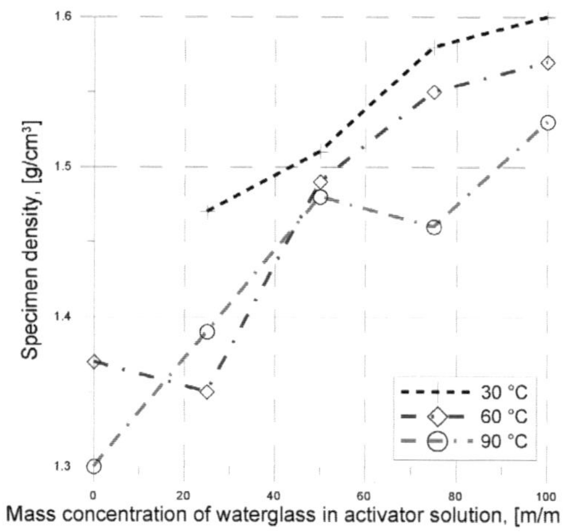

Figure 7. Waterglass addition to activator solution and its effect on specimen density at various curing temperatures

Leaching characteristics of geopolymers 1 M Acetic-acid for the main structuralelements are given in Figure 8. The leached Na concentration indicates the unconsumed NaOH. The other elements concentrations are also higher than it was in case of fly ash, which indicated that the new structure formed in the geopolymersis more sensitive than it was in the fly ash. However, it is important to note that other authors observed the leaching behaviour of various solvents resulted in more moderated leachability of structural as well as toxic components [9]. Moreover, it was found that toxic elements were mobilized in the geopolymer.

Based on the results presented in Figure 8 no significant effect of grinding fineness was found on the leachability of structural elements in strong acidic media.

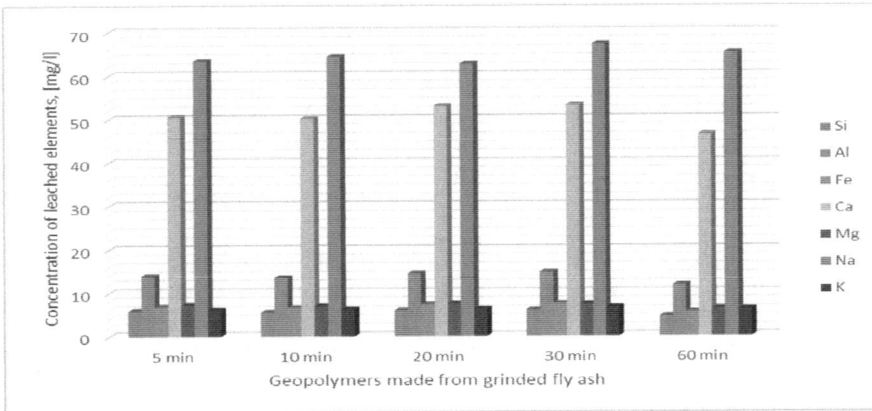

Figure 8. Dissolution of main elements from different geopolymers in 1 M Acetic-acid

4. Conclusions

Based on the experimental resultsthe following conclusions can be drawn:

- Increasing residence time in ball mill higher specific surface area, finer median particle size and x_{80}were reached.The finest product given by 60 minutes grinding time (SSA= 5425.55 cm^2/g, x_{50}=11.6 μm, x_{80}=27.9 μm).
- The fly ash fineness has positive effect on geopolymer physical properties. The finest fly ash (60 minutes ground) resulted the best compressive strength (7.9 MPa).
- Higher NaOH concentration increases the compressive strength to 12.3 MPa using 10 M NaOH solution. After it the compressive strength decreased.
- Na-K-silicate has positive effect on geopolymer compressive strength, 21.3 MPa reached when only Na-K-silicate solution was used.
- The compressive strength increased when lower temperature (30 °C) heat curing was used.
- FT-IR shows structural differences between geopolymers activated with different NaOH solutions.
- According to leaching experiments on ground fly ash it can be stated; grinding breaking up the "spheres" of the fly ash resulting in the enhancement of soluble Si and Al.
- No significant effect of grinding fineness was found on the leachability of structural elements in strong acidic media.

Acknowledgement

The described work was carried out in the frame of Center of Excellence of Sustainable Resource Management, University of Miskolc, Hungary. Hereby we would like to express our acknowledgement for the financial support of the project "Market oriented research and development of an innovative, eco-friendly thermal insulator material using polystyrene waste", under the identification number of PIAC_13-1-2013-0124 which is realized as part of the "Új Széchenyi Terv". Additionally, this work was supported by the János Bolyai Research Scholarship of the Hungarian Academy of Sciences (MTA).

References

[1] Kumar S, Kumar R, Alex T C, Bandopadhyay A, Mehrotra S P 2007 *Adv Appl Ceram***106**, No 3 pp. 120-127

[2] Álvarez-Ayuso É, Querol X, Plana F, Alastuey A, Moreno N, Izquierdo M, Font O, Moreno T, Diez S, V´ázquez É, Barra M 2008*J Hazard Mater***154** 175–183

[3] Mucsi G, Lakatos J, Molnár Z, Szabó R 2014*9th International Conference "Environmental Engineering"*, Vilnius, Lithuania, 22–23 May 2014

[4] Mucsi G, Molnár Z, Papp A2014*XXVIII. microCAD International Multidisciplinary Scientific Conference*, University of Miskolc, 10-11 April 2014

[5] Mucsi G, Csőke B, Mádai F, Debreczeni Á, Rácz Á, Molnár Z 2013*XV Balkan Mineral Processing Congress* Sozopol, Bulgária, 12.06.2013-16.06.2013. Sofia, pp. 864-868. (ISBN: 978-954-353-218-6)

[6] Mucsi G, Csőke B, Molnár Z 2010*Proceedings of the XIIth International Mineral Processing Symposium* 6-8. October 2010. Cappadocia Nevsehir, Turkey ISBN: 978-975-491-295-1, pp. 1235-1242

[7] Provis J L, van Deventer J S J 2009*Geopolymers, structure, processing, properties, and industrial applications*. (Woodhead publishing in materials, New York), ISBN: 978-1-84569-638-2 (e-book) p. 421-436

[8] Izquierdo M, Querol X, Davidovits J, Antenucci D, Nugteren H, Pereira C F 2009 *J Hazard Mater* **166** pp. 561-566

[9] Mucsi G, Szenczi Á, Molnár Z, Lakatos J 2016 *J Environ Eng Landsc* **24** pp. 48-59

IOP Conf. Series: Materials Science and Engineering **175** (2017) 012047 doi:10.1088/1757-899X/175/1/012047

Investigation of thermal properties of raw materials of asphalt mixtures

R Géber, A Simon, I Kocserha

University of Miskolc, Institute of Ceramics and Polymer Engineering, Hungary

E-mail: femgeber@uni-miskolc.hu

Abstract. Asphalt mixtures are composite materials, which are made of different grades of mineral aggregates and bitumen. During the mixing process mineral materials were blended with bitumen at relatively high temperature (~200 °C). As the binding process come off in these higher temperature range, thermal properties of asphaltic materials are important.

The aim of this project is to reveal the thermal properties of raw materials. During our research two types of mineral aggregates were tested (limestone and dolomite) by different methods. Differential thermal analysis, thermal expansion and thermal conductivity were investigated at technologically important temperatures.The results showed that the structure of mineral materials did not change at elevated temperatures, expansion of samples was neglible, while thermal conductivity changed by temperature.

1. Introduction

Asphalt mixtures are the most used pavement materials all over the world. The mixtures consist of mineral aggregates and bitumen as binding material [1-5]. One of the most important components of asphalt are fillers (*d<0.063mm*) which are the finest part in mineral aggregate. Mixed together with the binder a so-called asphalt mastic is formed.The contact between fillers and bitumen is not exactly known, until today.

During the lifetime of asphalt mixtures and the compacted pavements heat will play a significant role [6]. Raw mineral aggregates are stored in cold aggregate bins. At the beginning of the production aggregates are heated up in the drying drum in order to evaporate the adsorbed water from their surface. Furthermore, raw materials need to reach the mixing temperature. Then aggregates are transferredby the hot elevator to the hot aggregate bin, then to the mixer where hot bitumen is also added from the bitumen tank.After mixing, asphalt mixtures are loaded into heated trucks and transported to the field to be paved. Compacting process and traffic temperature both affect the properties of asphalt pavements.

The effect of temperature on the properties of mineral materials is not well known till today. Only a few research dealt with this area [7, 8]. Devecseri [8] investigated the changes of physical properties (*weight change, particle size change and colour change*) of different asphalt aggregates (*effusive and sedimentary rocks with the fraction of11/16 mm*) by elevated temperatures (*240°C and 480°C*). She found that high temperatures (480 °C) cause mineralogical, physical and colour changes in aggregates. She also stated that during heating aggregates behave contrary to each other, therefore there is no exact relation between their thermophysical and mechanical properties.

Considering the above mentioned, the aims of this research is to investigate the thermal properties of raw materials (*mineral fillers*) for a better understanding of the cohesion between fillers and bitumen.

2. Materials and methodology

2.1. Sample preparation

Limestone (*Mexikóvölgy*) and dolomite (*Pilisvörösvár*)fillers were tested during the research. Both minerals are widely used in Hungary as asphalt fillers.

Samples were prepared for further tests using standard sieves. Fractions of mineral materials that are smaller than a given particle size (*d<0.063 mm*) were created. After sieving,fillerswere dried to weight-constancy then they were hermetically sealed in containersafter cooling down. The reason of this was to avoid any moisture uptake.

2.2. Testing procedures

Determination of particle size distribution of fillers were doneby the use of Horiba LA-950V2 instrument.Wet method, sodium-pyrophosphate addition and ultrasonic treatment was used.

Morphological tests were carried out on Carl Zeiss EVO MA10 scanning electron microscope (SEM).Electron micrographs were taken in different magnifications. The aim of this test was to observe morphology and geometrical features of fillers.

Differential Thermal Analysis were done on MOM Derivatograph-C type instrument in temperature ranges of 50°C–1000°C. Heating rate was 5°C/ min. Weightchange of fillers and theirendothermic or exothermic reactions can be revealedwith this method.

With differential scanning calorimetry enthalpy changes and heat capacity of fillers were determined by Mettler Toledo DSC 823 E type instrument. Heating and cooling rates of 10°C/min were used in ranges of 30°C–200°C. Mineral fillers were placed in a 40 μl aluminium crucible.

In order to investigate the change in length of fillers a heating microscope (*manufactured by Camar Elettronica*) and Linseis L75 laser dilatometer were used. Heating microscope was used to measure optically the change of height of compacted specimens. Compacted samples (with a diameter of 2 mm and a height of 4 mm) were heated up to 200 °C with a heating rate of 5 °C/min. Both fillers could be compacted for this test.

Relative length change was determined during heating and cooling runs with the use of laser dilatometer. For this test a sample with a length of 20 mm and a diameter of 7 mm was needed. Only limestone filler was tested, because of compaction difficulties.

3. Results and discussion

3.1. Test results of mineral fillers

Table 1. contains all the test results performed on mineral fillers.

Table 1. Main properties of fillers

Material property	unit	Limestone *d<0.063 mm*	Dolomite *d<0.063 mm*
Mineral composition	wt%	100 % $CaCO_3$	100 % $MgCa(CO_3)_2$
Bulk density	$g \cdot cm^{-3}$	2.717	2.842
BET specific surface area	$m^2 \cdot g^{-1}$	1.02	0.38
Hydrophilic coefficient	-	0.76	0.89

IOP Conf. Series: Materials Science and Engineering **175** (2017) 012047 doi:10.1088/1757-899X/175/1/012047

Figure 1. shows the particles size distribution curves (*cumulative passing and hystogram*) of fillers. According to the results, it was found that these fillers can be characterized as polydisperse systems and they contain fine and coarse particles in different quantity by fraction. Mean particle diameter *(d_{50})* of limestone is 16.82 μm, and the mean particle size of dolomite is 39.84μm. 20 percent of limestone particles are under 10 microns, while in case of dolomite the amount of these fine particles are under 2 percent.

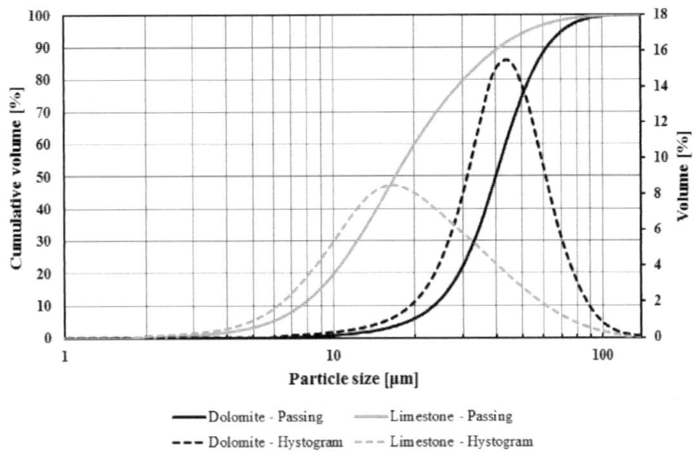

Figure1. Particle size distribution of fillers

Fig. 2. shows the microstructures of fillers (*magnification=200X and 1500X*).

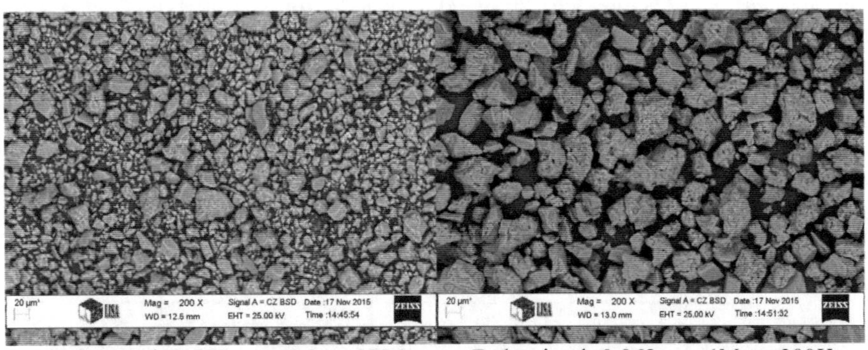

Limestone d<0.063 mm (*Mag=200x*) Dolomite d<0.063 mm (*Mag=200X*)

Limestone d<0.063 mm (*Mag=1500x*) Dolomite d<0.063 mm (*Mag=1500x*)

Figure 2. Scanning electron micrographs of fillers

IOP Conf. Series: Materials Science and Engineering **175** (2017) 012047 doi:10.1088/1757-899X/175/1/012047

The SEM micrographs confirm the results of particle size distribution tests. Comparing the micrographs, which were taken at a magnification of 200X, it can be observed that limestone filler is rich in fine particles while dolomite does not contain fines. This is the reason of the low BET specific surface of dolomite (*Table 1.*). Higher magnifications show well the morphology and the surfaces of particles. Limestone filler looks solid, on which submicronal particles are sticked. The shape of grains are various, mainly polygonal, because of the comminution processes. Dolomite particles are also polygonal and more porous than limestone. Fig. 3. shows results of differential thermal analysis.

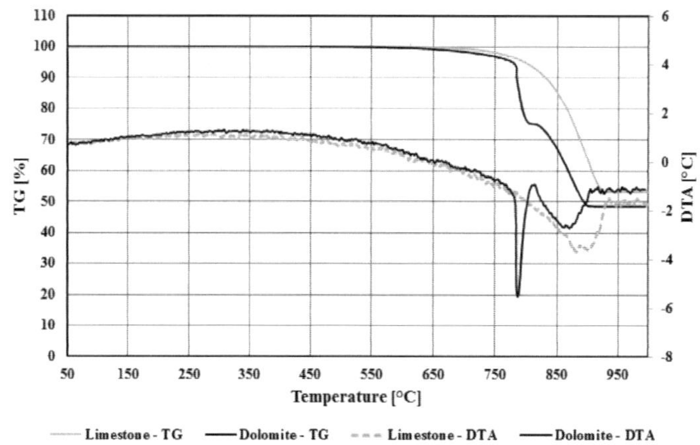

Figure 3. TG and DTA curves of fillers

TG (*thermogravimetric*) curves represent the weight change of fillers, while DTA (*differential thermal analytical*) curves show the temperature changes occurring within the materials.Limestone filler has a slow weight lossin 670–900 °C temperature range. This is due to the decarbonisation of $CaCO_3$. During this reaction CaO is formed while CO_2 is released. This phenomenon is an endothermic reaction, as the DTA curve shows a negative peak. TG curve of dolomite shows also a thermal decomposition, but in this case this process is a two-stage reaction. The first endothermic reaction belongs to the decomposition of dolomite, while the second endothermic reaction belongs to the decomposition of $CaCO_3$. Up to 200°C, fillers behave as inert materials (*in the viewpoint of asphalt technology*) because no reactions occur in this range. In Fig. 4. results of DSC analysis are summarized.

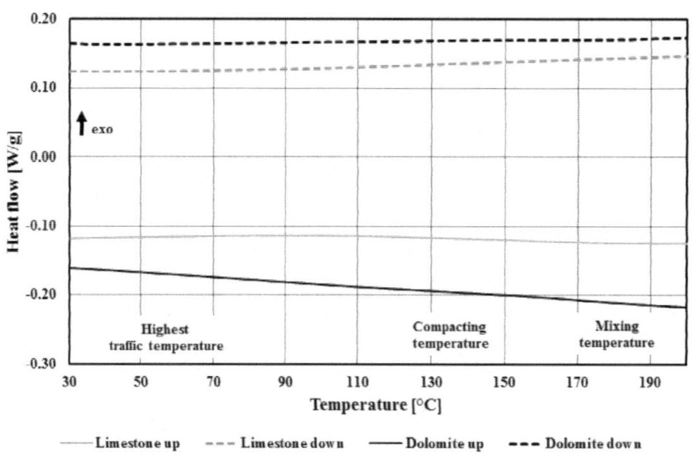

Figure 4. DSC curves of fillers

Heat flow curves recorded in the heating and cooling runs are shown. It can be concluded that there is no reaction, no significant enthalpy change occurs during the runs, no peaks are observed. As the diagram shows, no reaction occurred at the important temperatures.

Specific heat capacity can also be determined from DSC results, as Fig. 5. shows. As it can be seen dolomite has a higher specific heat capacity in the tested temperature ranges (*30°C–200°C*) than limestone. Specific heat capacity of fillers increased almost linearly in the tested ranges.

Heating microscope was used to measure the height change of compacted specimens during heating and it showed no change (*Fig. 6.*). The instrument is able to detect the silhouette of a lighted specimen by a digital camera. By image analysis the height change of the sample can be determined. In our case the height of the fillers was not changed during the test.

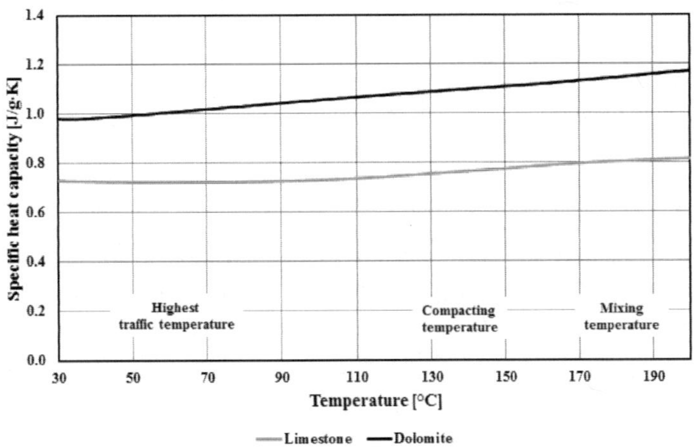

Figure 5. Specific heat capacity curves of fillers

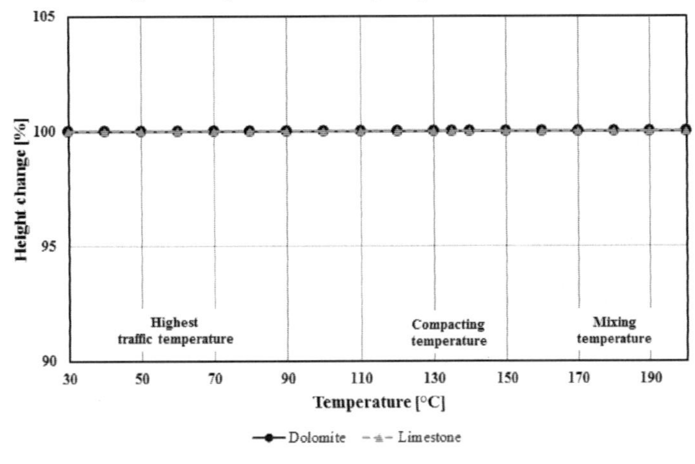

Figure 6. Height change of compacted fillers during heating

Dilatometric test by a precise laser dilatometer was also obtained. This method was used only on limestone filler, because dolomite was not able to compact. The reason of this is the particle size distribution of the mineral material (*Fig.1. and Fig. 2.*).

In Fig. 7. the relative length change curves (*heating and cooling runs*) are shown. At the maximum testing temperature (T_{max}*=200 °C*) the relative length change of the sample reached 0.15%. After cooling this change remained at 0.04 %, which is not a significant value. According to the dilatometric test and heating microscopic tests it can be concluded that neither filler change its dimensions during the heating process.

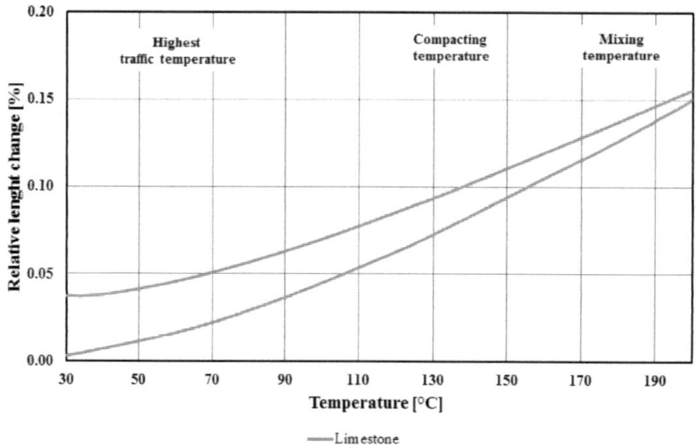

Figure 7. Relative length change oflimestone filler

Fig. 8. shows the results of thermal conductivity measurements. 4 testing temperatures were used (*20°C=room temperature, 60°C=highest traffic temperature, 135°C=compacting temperature and 180°C =asphalt mixing temperature*). Thermal conductivity values are increased by the increase of temperature. Dolomite has slightly higher conductivity values than limestone, but this difference is not significant (~0.03 W/m·K), except at room temperature. At this temperature the difference is almost two times higher.

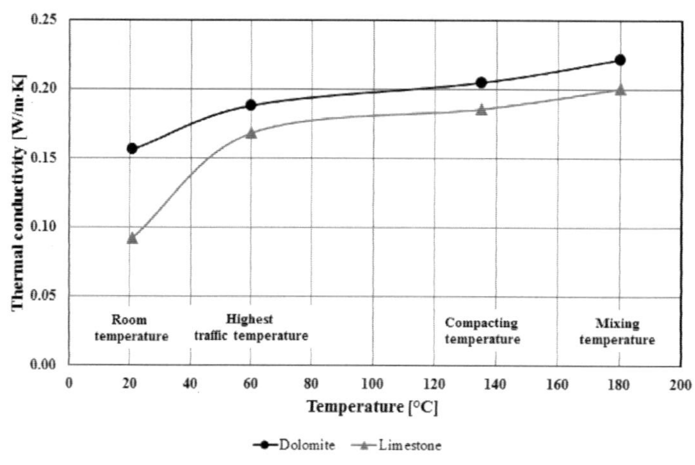

Figure 8. Thermal conductivity of fillers

4. Conclusions

During this research work two types of fine particle mineral materials were tested as asphalt fillers. Examinations of thermal properties of fillers were made by different techniques.

According to the particle size distribution, these mineral fillers can be characterized as polydisperse systems and they contain fine and coarse particles in different quantity by fraction.

IOP Conf. Series: Materials Science and Engineering **175** (2017) 012047 doi:10.1088/1757-899X/175/1/012047

The Authors have also observed the morphological features of the mineral materials. The morphology of fillers is various;dolomite has higher porosity in the particles, but does not contain much fine particles.

The samples were tested to reveal the thermal properties.According to the results, heat during the technological processes do not influence the properties of fillers (*no reactions, no change in length*). Difference was found only in specific heat capacity and in thermal conductivity. Dolomite has a higher specific heat capacity and thermal conductivity than limestone.

Acknowledgement
This work has been carried out as part of the TÁMOP-4.2.1.B-10/2/KONV-2010-0001 and TÁMOP-4.2.2/B-10/1-2010-0008 project within the framework of the New Hungarian Development Plan. The realization of this project is supported by the European Union, co-financed by the European Social Fund.

References
[1] J. Chen 1997 *Journal of Materials, Concrete Structures and Pavements* **36** 269 (in Chinese)
[2] A Kavussi, R G Hicks 1997 *Journal of Association of Asphalt Paving Technologists* **66** 153
[3] B M Harris, K D Stuart 1995 *Journal of Association of Asphalt Paving Technologists* **64** 54
[4] J Craus, I Ishai, A Sides 1978 *Journal of Association of Asphalt Paving Technologists* **47** 558
[5] L A Gömze and Á Kovács 2005 *Épitőanyag-JSBCM* **57** 34 http://dx.doi.org/10.14382/epitoanyag-jsbcm.2005.7
[6] T B Moghaddam, H Baaj 2016 *Construction and Building Materials* **114** 805
[7] U. Isacsson, H. Zeng 1997 *Construction and Building Materials* **11** 83
[8] G. Devecseri 2010 *Épitőanyag – Journal of Silicate Based and Composite Materials* **62 (1)** 23 http://dx.doi.org/10.14382/epitoanyag-jsbcm.2010.5

IOP Conf. Series: Materials Science and Engineering **175** (2017) 012048 doi:10.1088/1757-899X/175/1/012048

Plastic deformation effect of the corrosion resistance in case of austenitic stainless steel

F Haraszti and T Kovacs

Obuda University, Banki Donat Faculty of Mechanical and Safety Engineering – Material Science Department – 1081 Budapest – Hungary

E-mail: kovacs.tunde@bgk.uni-obuda.hu

Abstract. The corrosion forms are different in case of the austenitic steel than in case of carbon steels. Corrosion is very dangerous process, because that corrosion form is the intergranular corrosion. The austenitic stainless steel shows high corrosion resistance level. It knows that plastic deformation and the heat treating decrease it's resistance. The corrosion form in case of this steel is very special and the corrosion tests are difficult. We tested the selected steel about its corrosion behaviour after high rate deformation. We wanted to find a relationship between the corrosion resistance decreasing and the rate of the plastic deformation. We wanted to show this behaviour from mechanical and electrical changing.

1. Introduction

It knows that the most important corrosion forms in case of austenitic steels are the intergranular and the stress corrosion. The well-known susceptibility of austenitic stainless steels to intergranular corrosion after heat-treatment in the temperature range of 500°–800°C (sensitization) has long been attributed to depletion of Cr from regions of the alloy matrix adjacent to grain boundaries in which $Cr_{23}C_6$ had precipitated. Those regions of the steel in which the local Cr composition falls below about 12% have a diminished ability to form a passive film and hence corrode preferentially [1].

During the fabrication and welding technology on base of the microstructure changing become sensitised [2]. The precipitation showed in Fig. 1. The plastic deformation modifies the grain size and the welding technology causes heat input in practise.

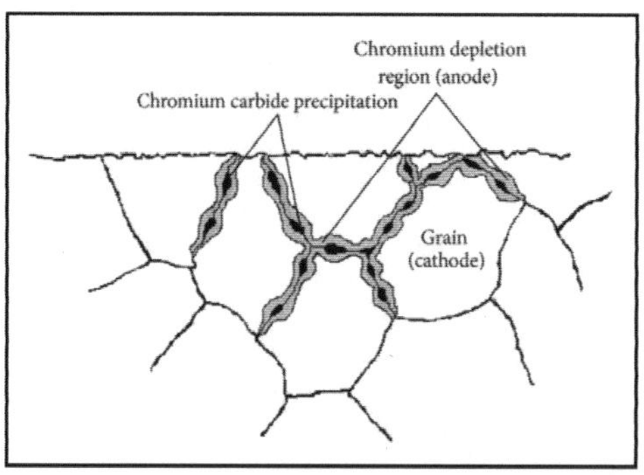

Figure 1. Schematic representation of the carbide precipitation at grain boundaries in austenitic stainless steel. *Source: http://www.hindawi.com/journals/ijelc/2013/970835/fig10/*

The measuring of the sensitization effect is very difficult experiment, usually use analytical or electron microscopy. In this way we need a test sample from the experimented part. We wanted to find a non-destructive experimental method to detect the sensitization. It known that the conduction of the metals depends on the microstructure and the quantity of precipitations of the material. Base of the experiments is that during the corrosion process the diameter of the sample and the conduction decrease. The relationship between the diameter and the conduction shows in (1) equation, where **R** is the conduction, **A** cross section area, **l** length and **ρ** resistivity [3-4].

$$R = \rho \frac{l}{A} [\Omega] \tag{1}$$

2. Materials and Experimental Procedures

2.1. Used materials
In case of our tests we used austenitic stainless steel (X15CrNiSi25-21, EN 1.4841, AISI 316) the chemical composition is given in Table 1. Number 3 and 4 sample were heat treated at 500°C duration range 1 hour in oven and after air cooling.

Table 1. Chemical composition of the used steel (at.%)

C%	Cr%	Ni%	Mn%	N%	Si
max. 0,20	24-26	19-22	max. 2,00	max. 0,11	1,50-2,50

2.2. Plastic deformation process
It is a cold metalworking process used to reduce the cross-section of a wire by pulling the wire through a dingle drawing die shows Fig.2. The cross-section reducing was 50%. By the way of this cold working process we wanted to increase the strength of the austenitic steel. The used austenitic steel recrystallization temperature is over than 550°C [5-6]. The cold working caused the increasing of the steel strength (R_m=1800 MPa) but on the used heat treating temperature the ultimate strength didn't decrease.

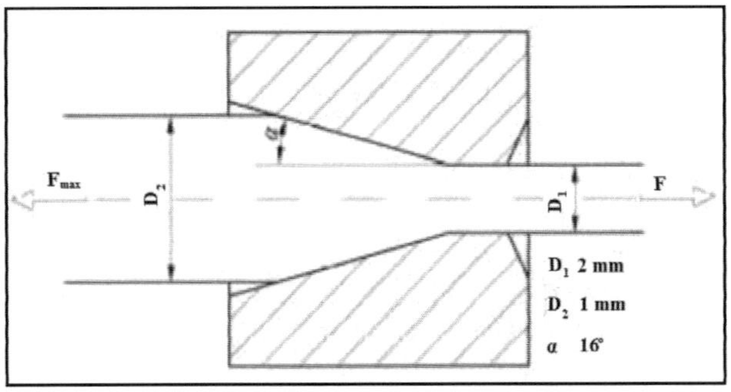

Figure 2. Metal working process

2.3. Corrosion tests
We used the Huey Test (ASTM A262) [7], used corrosive material was concentrated HNO_3 (Nitric acid) on 95°C, during 48 hours. By the way this test we can generate intergranular corrosion in the austenitic steel. The Fig.3. shows the sample number 3 surface SEM picture . On the surface of the sample we detected some cracks, assumable the cracks realized by the corrosion effect.

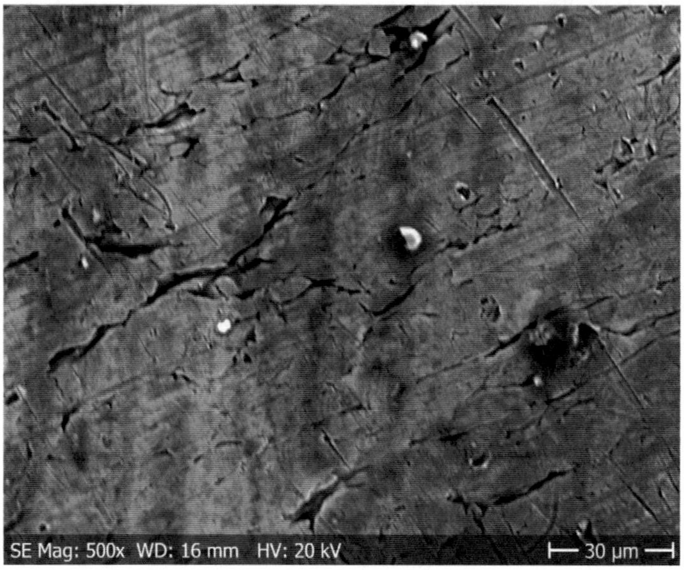

Figure 3. Surface of the number 3 sample (cold worked and heat treated)

3. Results and Discussion

Resistance testing

We tested the resistance range by the way of Wheatstone bridge.

Table 4. Resistance range in case of different treated samples

Number of sample	Samples	resistance (Ω) before corrosion test	resistance (Ω) after corrosion test
1	original sample without heat treating	0,1	0,3
2	Shaped (reducing 50%) sample without heat treating	0,3	0,6
3	Shaped (reducing 50%) sample heat treated in 500 °C	0,3	0,6
4	original sample heat treated in 500 °C	0,1	0,001

4. Conclusion

The experimental results give new information about the corrosion resistance range of the used austenitic steel.

I. After the Huey Test the resistance range changed in case of all samples.

II. In case of the number 1 sample the range of the resistance improved principally.

III. The cross section reduced samples (number 2 and 3) showed the same result about the measured resistance after the corrosion test. The resistance was the same, two times higher than before the corrosion test.
IV. In case of the sample number 4 the resistance decreased. This sample was heat treated in 500 °C without metal working.
V. We can conclude on base of our result that the used heat treating temperature (500 °C) in case of the AISI 316 austenitic steel samples doesn't caused relevant impact about the corrosion resistance when we accept the resistance range like the corrosion resistance representative.
VI. Our results are important about the corrosion resistance determination but we need more tests and result to understand this material process and base on the test parameters we can predict the corrosion behaviours. Also we need to validate our resistance tests results by the way of microstructure tests.
VII. The resistance test can be useful about the corrosion behaviour understanding because it is non-destructive test.

References

[1] C.S. Tedmon et all.: Intergranular corrosion of Austenitic stainless steel, Journal of the Electrochemical Society, 1971. Vol.118 Issue 2. 192-202
[2] M. Dománkova, E Kocsisová, I. Slatkovský, P. Pinke; The Microstructure Evolution and Its Effect on Corrosion Properties of 18Cr-12Ni-2,5Mo Steel Annealed at 500-900°C, Acta Polytechnica Hungarica, Vol.11, No.3, 2014. 125-137.
[3] F. Haraszti: The bases of corrosion's investigation EME kiadó, Kolozsvár, 2016, 185-188.
[4] F. Haraszti: Corrosion investigation of steel smaples EME kiadó, Kolozsvár, 2016, 189-192.
[5] T. Kovács-Coskun; B. Völgyi; I. Sikari-Nágl;Galvanic corrosion behavirour testing in case of aluminum and steel couple EME kiadó, Kolozsvár, 2013, 201-204.
[6] Reti T, Kovacs T: A phenomenological method for the prediction of damage accumulation processes under varying external conditions In: Materials Science Forum, Vol. 414-415: 2003, 317-322.
[7] ASTM Standard Practice in A 262 for Detecting Susceptibility to Intergranular Corrosion in Austenitic Stainless Steels

IOP Conf. Series: Materials Science and Engineering **175** (2017) 012049 doi:10.1088/1757-899X/175/1/012049

Researches of mechanical behaviour of the bone micro volumes and porous ceramics under uniaxial compression

T V Kolmakova, S P Buyakova and S N Kulkov

Institute of Strength Physics and Materials Science, Tomsk, Russia
Tomsk State University, Tomsk, Russia

E-mail: kolmakova@ftf.tsu.ru

Abstract. The research results of the mechanics are presented and the effective mechanical characteristics under uniaxial compression of the simulative micro volume of the compact bone are defined subject to the direction of the collagen-mineral fibers, porosity and mineral content. The experimental and computer studies of the mechanics are performed and the effective mechanical characteristics of the porous zirconium oxide ceramics are defined. The recommendations are developed on the selection of the ceramic samples designed to replace the fragment of the compact bone of a definite structure and mineral content.

1. Introduction

The long-term experience of the synthetic materials' use to repair the bone elements by means of endoprosthesis replacement points out that the biochemical compatibility of the implant with the body is an essential [1, 2], although insufficient, condition for functioning of the implant - bone system as a whole. Their biomechanical compatibility is an important condition. It is known that a substantial difference in the elasticity of the bone and implant can result in the loss of the latter due to the subsequent resorption of the bone tissue, which is in contact with the implant.

As regards the biochemical compatibility with the body, the preferred materials for the endoprosthesis replacement of the bone tissue are those referred to the ceramics class. Ceramics are identical to the inorganic bone matrix in the type of chemical bond.

The goal of research within the context of this paper is to develop the recommendations on the selection of the individual ceramic implants designed to replace the fragment of the compact bone of a definite structure and mineral content.

To achieve this goal in the course of the research the mechanical behavior is studied and the effective mechanical characteristics under uniaxial compression of the simulative micro volume of the compact bone and porous ceramics are defined; the experimental studies of the mechanical behavior of the produced porous zirconium oxide ceramics are performed and its effective mechanical characteristics are defined.

2. Materials and methods

The computer simulation of the mechanics of compact bone fragments and ceramic samples was performed in the software package ANSYS.

The geometrical model of the compact bone tissue micro volume (Figure 1) was based on a real image of the natural bone microstructure.

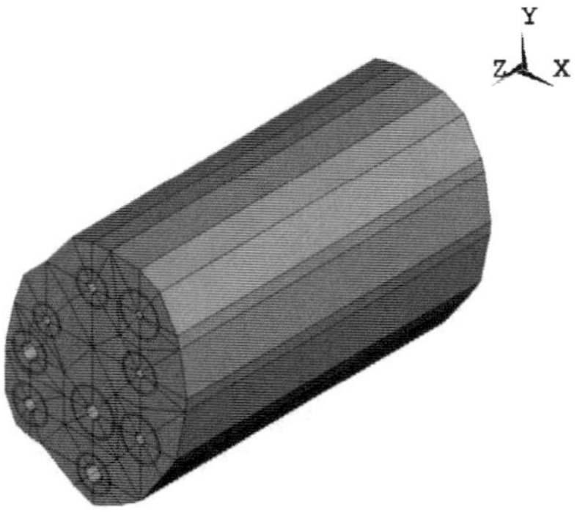

Figure 1. Geometrical model of compact bone tissue

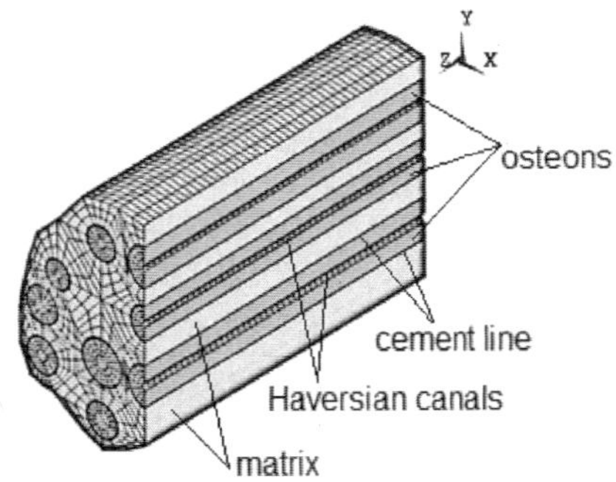

Figure 2. Finite element model of compact bone tissue micro volume

The micro volume was simulated as a reinforced composite material comprising the reinforcing structural elements, i.e. the osteons with the Haversian canals inside, the matrix and the cement streak separating the osteons and the matrix (Fig. 2). The total porosity of the simulative fragment of the compact bone P was derived by Haversian canals being explicitly simulated (Fig. 2), and Volkmann's canals being implicitly considered in the calculation of the effective mechanical characteristics of the structural elements of the simulative micro volume.

The material of the structural elements of the simulative micro volume of the compact bone was considered transversely isotropic. The effective mechanical properties of the matrix material and the reinforcing elements, osteons, were calculated subject to the volume ratio of the collagen and mineral components, the direction of the collagen- mineral fibers [3], the porosity due to the presence of Volkmann's canals.

The ceramics samples of porosity from 7 to 70 % were produced from the fine powder of zirconium oxide partially stabilized with yttrium oxide. The available pores in the ceramics structure on one side intend to ensure the ceramics integration with the bone tissue, and on the other side - to reduce the elasticity modules to the values comparable with elasticity modules of the bone. The adequate level of the ceramic samples porosity has been achieved both by using the pore-forming additives and without them. Figure 3 shows the histograms of pore size distribution of the obtained ceramic samples with porosity 7% and 60%. The uniaxial compression tests were performed on the universal testing machine "Instron-1185" at the constant loading rate of $3 \cdot 10^{-4}$ c^{-1}.

The geometric models of ceramic samples with porosity 7%, 40% and 60% were constructed based on the histograms of pore size distribution of ceramics obtained experimentally (Fig. 3). Figure 4 shows the finite element models of ceramic samples with different geometry of pore space.

The stress-strain state of the model samples of the compact bone tissue and ceramics was calculated using the method of the finite elements.The task was solved within the context of the linear elasticity theory.

3. Results and Discussion

Figure 5 presents the calculation results of the longitudinal elastic modules of the compact bone tissue samples with the different directions of the collagen-mineral fibers (I type - transverse fibers, I±45° type - fibers are arranged at an angle of ±45° in the neighboring osteon lamellas, II type – the fibers change their direction from perpendicular on parallel to a bone axis in the osteon lamellas, III type - longitudinal fibers) with different mass fraction of the minerals α and porosity P under axial compression as compared to the experimental results of P. Zioupos et al [4], X. N. Dong, X. E. Guo [5].

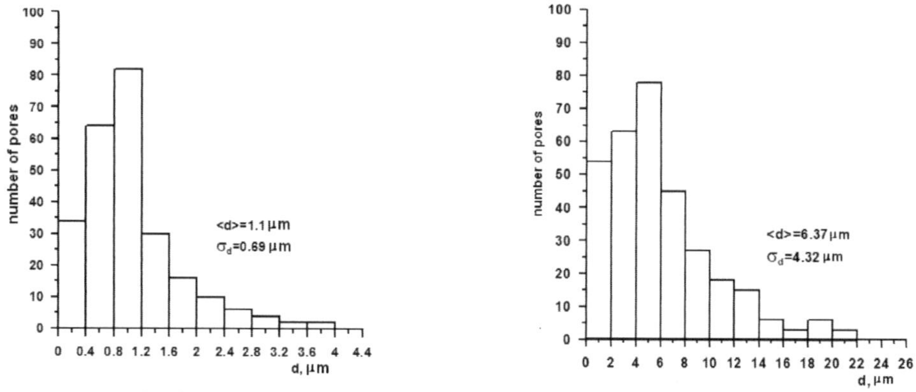

θ=7% (a) θ=60% (b)

Figure 3. The histograms of pore size distribution of the obtained ceramic samples

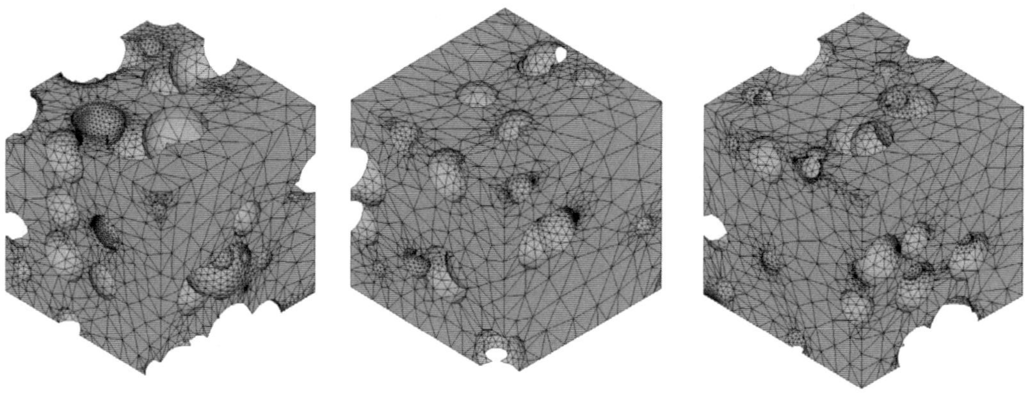

Figure 4. Finite element models of ceramic samples with different geometry of pore space (θ=7%)

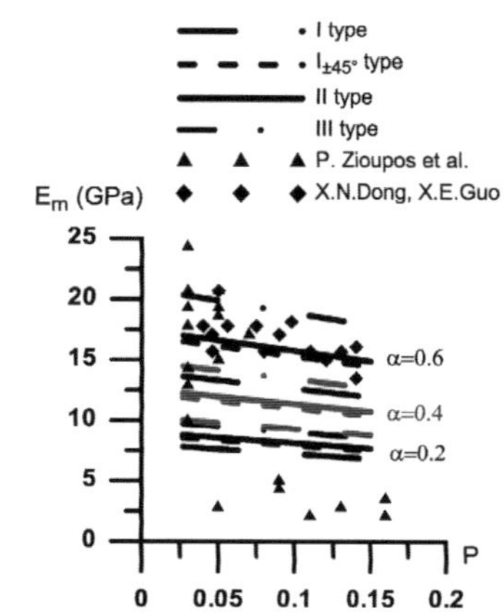

Figure 5. Dependence of longitudinal elastic modules of micro volumes of compact bone tissue with different mineral content (α) from porosity (P)

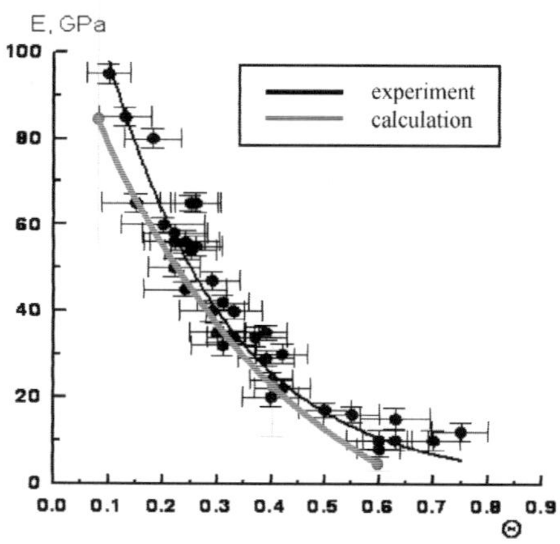

Figure 6. The dependence of the effective elastic modulus of the ceramics on the specific pore volume

The micro volume of the cortical bone of I ±45° type and the micro volume of type II, independent of their porosity, have the relatively equal longitudinal elastic modules (Fig. 5), though differ in the character of the stress and strain distribution, that is determined by the varied non-uniform deformation response of micro volume in the directions mutually transverse to the load application.

Calculations of elastic modules of model ceramic samples with different geometry of the pore space under uniaxial compression were conducted (Fig. 6). Results showed that the geometry of the pore space of the ceramic samples significantly influences the value of the modulus of elasticity.

As a result of the uniaxial compression tests of the zirconium ceramics samples $ZrO_2(Y_2O_3)$ there has been derived the dependence of the longitudinal elastic modulus (Fig. 6) and the ultimate compression strength of the ceramic samples on the specific volume of pores θ.

The comparison of the produced calculated elastic modules, the ultimate compression strength of the bone fragments, calculated using the expression of C. J. Hernandez [6], and experimentally produced mechanical properties of the ceramics (see the table below) demonstrated, that the ceramics with a specific pore volume of 50 to 60 % have the elasticity modules close to the compact bone with the mineral weight percent of 40 to 55 % and the porosity of 3 to 15 %.

Table 1. The calculated and experimental effective mechanical parameters of the compact bone tissue and ceramics

Ceramics $ZrO_2(Y_2O_3)$ (experiment)			**Compact bone (calculation)**				
θ	**E, GPa**	**σ_c, MPa**	**α**	**P**	**Type**	**Em, GPa**	**σ_c, MPa**
0.50	16.87	110.0	0.55	0.15	III	16.4	109.7
0.52	15.50	98.0	0.50	0.10	III	15.8	93.8
			0.45	0.027	III	15.5	81.2
0.54	14.17	87.3	0.45	0.15	III	13.7	62.7
			0.40	0.027	III	14.2	58.5
0.56	12.98	77.7	0.40	0.10	III	13.2	50.3
			0.40	0.15	III	12.6	45.1
0.58	11.89	69.2	0.45	0.15	II	11.6	62.7
			0.40	0.027	$I_{\pm45°}$	11.9	58.5
0.60	10.89	61.6	0.40	0.15	II	10.7	45.1
			0.40	0.15	$I_{\pm45°}$	10.5	45.1

4. Conclusion

The effective mechanical characteristics of the model micro volumes of compact bone and ceramic samples are defined. Results showed that the micro volume of the cortical bone of I ±45° type and the micro volume of type II, independent of their porosity, have the relatively equal longitudinal elastic modules, the geometry of the pore space of the ceramic samples significantly influences the value of the modulus of elasticity. The experimental studies of the mechanical behavior of the produced porous zirconium oxide ceramics were carried out. Its effective mechanical characteristics are defined. The recommendations were obtained on the selection of the ceramic samples designed to replace the fragment of the compact bone of a definite structure and mineral content. The ceramics with the porosity of 50-56 % may become a candidate to replace the compact bone fragment that has the collagen-mineral fibers arranged parallel to a bone axis. The ceramics with the porosity of 58 to 60 % may become a candidate to replace the compact bone fragment that has the collagen-mineral fibers, changing its direction from one to another lamellas or the one arranged at angles ±45°.

References

[1] L A Gömze 2016 APPLIED MATERIALS SCIENCE I. Compilation of Selected Scientific Papers, *Published by IGREX Ltd, Hungary* **ISBN 978-963-12-6600-9** pp 1-189

[2] S N Kulkov, S P Buyakova, M Chatzinikolaidou and I Kocserha 2015 *Építőanyag-JSBCM* **67** (4) 155 http://dx.doi.org/10.14382/epitoanyag-jsbcm.2015.26

IOP Conf. Series: Materials Science and Engineering **175** (2017) 012049 doi:10.1088/1757-899X/175/1/012049

[3] Bromage T *et al* 2003 *Anat. Rec.B: New Anat.* B **274** 157
[4] Zioupos P, Cook R B and Hutchinson J R 2008 *J. Biomech.* **41** 1961
[5] Dong X N and Guo X E 2004 *J. Biomech. Eng.* **126** 387
[6] Hernandez C J and Keaveny T M 2006 *Bone* **39** 1173

IOP Conf. Series: Materials Science and Engineering **175** (2017) 012050 doi:10.1088/1757-899X/175/1/012050

Alloying of steel and graphite by hydrogen in nuclear reactor

E Krasikov

Laboratory Head, National Research Centre «Kurchatov Institute», 123182 Moscow, Russia.

E-mail: ekrasikov@mail.ru

Abstract. In traditional power engineering hydrogen may be one of the first primary source of equipment damage. This problem has high actuality for both nuclear and thermonuclear power engineering. Study of radiation-hydrogen embrittlement of the steel raises the question concerning the unknown source of hydrogen in reactors. Later unexpectedly high hydrogen concentrations were detected in irradiated graphite.

It is necessary to look for this source of hydrogen especially because hydrogen flakes were detected in reactor vessels of Belgian NPPs.

As a possible initial hypothesis about the enigmatical source of hydrogen one can propose protons generation during beta-decay of free neutrons поскольку inasmuch as protons detected by researches at nuclear reactors as witness of beta-decay of free neutrons.

I. Introduction

It is known that in traditional power engineering hydrogen may be one of the first primary source of equipment damage [1]. This problem has high actuality for both nuclear and thermonuclear power engineering [2]. Particularly reactor pressure vessels (RPV) of the WWER-440/230 project were manufactured without stainless cladding that is were in contact with primary circuit water and accessible for hydrogen as a product of RPV wall corrosion. Analysis of the combined radiation-hydrogenation embrittlement of the 48TS type vessel steel was performed in [3] where at the mention of the American [4] and own data question concerning unknown source of hydrogen in metal that was irradiated in nuclear reactor in hermetic ampoules (was named as "irradiation-produced hydrogen" (IPH) was raised.

2. Materials and Methods

Table 1 lists chemical composition of the RPV steel used (48TS type). A-543 type US steel takes for comparison.

Table 1. Chemical composition of the RPV steels A-543 and 48TS (%%mass)

Type	C	Si	Mn	P	S	Cu	Cr	Mo	Ni
48TS	0,16	0,30	0,43	0,014	0,011	0,11	2,75	0,67	0,16
A543	0,14	0,18	0,20	0,011	0,015	0,07	1,60	0,50	3,01

4% solution of H_2SO_4 was used for additional electrolytic hydrogenation of the specimens (current density 0,1A/cm^2). Hydrogen concentration was determined by thermal degasation method at temperatures up to 1000°C with gas chromatograph (thermal conductivity detector) registration of gas released.

3. Experimental results and discussion

Determination of the hydrogen content in the irradiated steel fulfilled in the USA went to unexpected result: hydrogen content noticeably exceeded the quantity rated at (n,p) transmutation reaction: less than 0,1 ppm.

Results of the IPH concentration in steel analysis carried out in the USA are shown in Table 2 [4]. One can see that the greater the fast neutron fluence (FNF) the greater the hydrogen content.

Table 2. Dependence of the IPH concentration in steel versus FNF (E>1MeV)

FNF, $\times 10^{18}$cm^{-2}; $t_{irr.}$=225-300°C	0	7	200	400
IPH concentration, ppm; $t°_{degasation}$=1000°C	0,2	0,9	1,7	2,1

Ageing of the steel at 100-325°C during 48 hours revealed that IPH is not diffusible up to irradiation temperature that is IRH are in the irradiation produced traps. Inasmuch as IPH at temperatures of mechanical tests was immovable indicated values were subtracted from total quantity of hydrogen measured.

In I.V. Kurchatov Institute at several experiments was determined that steel specimens irradiated at relatively low (100-140°C) temperatures in sealed Ar contained ampoules hydrogen content was many times higher relatively initial content but was independent on FNF (Table 3) [3].

Table 3. Dependence of the IPH concentration in steel versus FNF (E>0,5MeV)

FNF, $\times 10^{18}$cm^{-2}; $t_{irr.}$=100-140°C.	0	100	170	190	270	450	500
IPH concentration, ppm; $t_{degasation}$=300°C.	0,2	2,9	4,3	13,3	24,9	9,8	3,1

Degasation kinetics are plotted in Figs 1, 2.

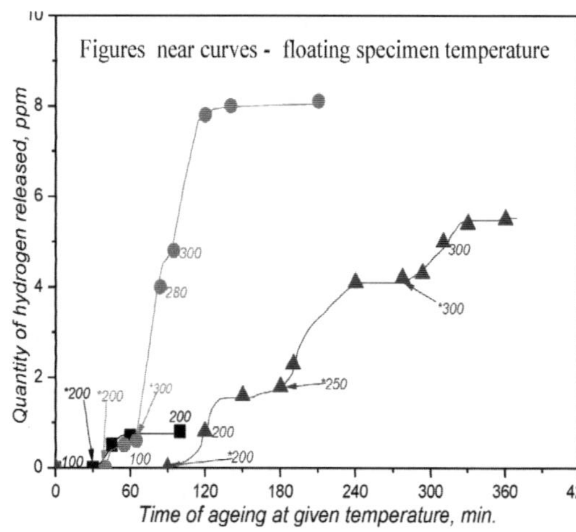

Figure 1. IPH degasation kinetics for irradiated steel ($4{,}5\times10^{20}$cm^{-2} at 140°C)

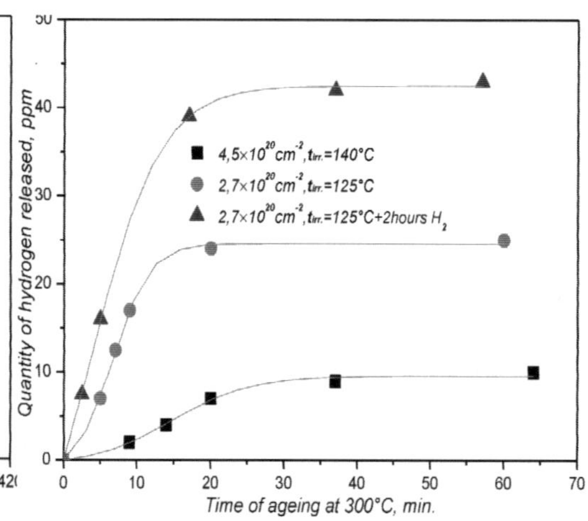

Figure 2. Hydrogen degasation kinetics for irradiated and irradiated+ hydrogenated steel

As one can see from Fig.1 that RIH discharge starts when heating temperature exceeds the irradiation temperature. It means that RIH is accumulated in radiation defects (traps).

IOP Conf. Series: Materials Science and Engineering **175** (2017) 012050 doi:10.1088/1757-899X/175/1/012050

Rather later data appear on unexpectedly high hydrogen concentrations in stainless steels irradiated in BWR type reactors (Fig.3 [5]) and high generations of hydrogen and helium in nickel [6].

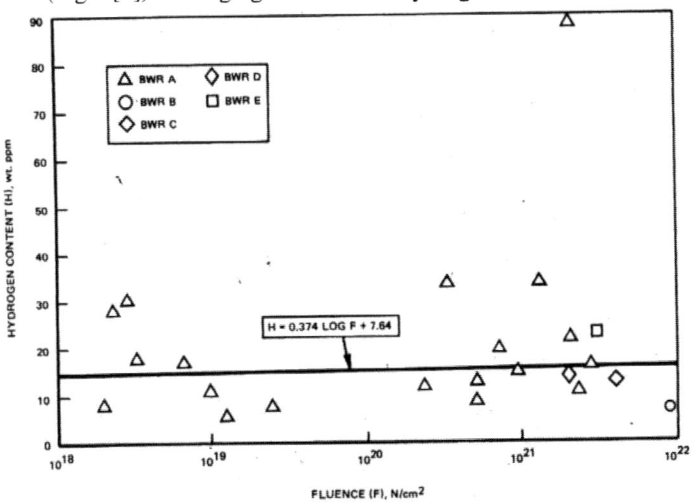

FIG. 2—*Hydrogen content of irradiated Type-304 SS as a function of fast* (E > *1 MeV) neutron fluence.*

Figure 3. IPH content in 304 SS versus FNF (E>1 MeV)

Surprisingly high hydrogen concentrations were revealed in irradiated graphite [7]. Table 4 presents the results of the thermal degasation of the GR-280 type graphite.

As it is seen hydrogen concentrations in irradiated specimens is one-two orders of magnitude higher than in unirradiated ones.

Table 4. Results of the thermal degasation of the GR-280 type graphite

FNF, cm^{-2} (E>0,5MeV)	Irradiation temperature, °C	Quantity of hydrogen released, ppm		
		Degasation temperature, °C		Total ppm
		1000	1800	
0		2,2	4,2	6,4
		1,5	2,9	4,4
1×10^{22}	500	20,8	-	20,8
		49,7	-	49,7
		41,2	1342	1383
		19,5	864	883
2×10^{21}	1100	-	369	369
		-	327	327

It is necessary to look for enigmatic source of hydrogen especially because in frame of inspections numerous flows were detected in the forged rings of the reactor pressure vessels in the Belgian nuclear power plants Doel 3 and Tihange 2 [8,9]. The owner Electrabel claimed that flaws were "most likely" hydrogen flakes.

One of the unobvious but probable initial hypothesis on enigmatic source of the hydrogen in operating nuclear reactor is generation of protons as a product of beta-decay of free neutrons (lifetime ~15 min.) [10].

References

[1] Vainman A 1990 Hydrogen embrittlement of the high pressure vessels. Kiev, Naukova dumka (in Russian)
[2] Alekseenko N 1997 Radiation Damage of Nuclear Power Plant Pressure Vessel Steels. ANS p 97
[3] Krasikov E 1974 Investigation of hydrogen embrittlement and hydrogen diffusion in irradiated steel, Ph.D Thesis, Moscow (in Russian).
[4] Brinkmann C 1970 Effects of Hydrogen on the Ductile Properties of Irradiated Pressure Vessel Steels. Report IN-1359, NRTS, Idaho Falls
[5] A.I. Jacobs 1987 Hydrogen buildup in Irradiated Type-304 Stainless Steel. *ASTM STP 956.* F.A. Garner, and N. Igata, Eds. ASTM, Philadelphia, pp 239-244
[6] Greenwood L, Garner F and Oliver D 2004 Surprisingly Large Generation and Retention of Helium and Hydrogen in Pure Nickel. *Journal of ASTM International,* April, vol.1, №4. Paper ID JAI11365, pp 529-539
[7] Biriukov A, Krasikov E 1998 Impact of neutron irradiation on graphite dehydrogenations. *VANT, ser. Thermonuclear fusion* №1-2 pp 3-8. NRC "Kurchatov Institute, Moscow (in Russian)
[8] Tweer I 2016 Flawed Reactor Pressure Vessels in the Belgian NPPS Doel 3 and Tihange 2. Comments on the FANC Final Evaluation Report 2015
[9] ORNL 2015 Evaluating of Electrabel Safety Cases for Doel 3/Tihange 2: Final Report (R1). ORNL/TM-2015/59349, Nov
[10] Mostoyoi Yu 1996 Neutron yesterday, today, tomorrow. *Successes of physics (UFN),* vol.166 №9 pp 987-1022 (in Russian)

Connection between diffusion coefficient and thermal conductivity of a metal matrix composite

M A Anisimova[1], A G Knyazeva[1,2] and I Sevostianov[3]

[1] Institute of High Technology Physics, National Research Tomsk Polytechnic University, 634050 Tomsk, Russia
[2] Institute of Strength Physics and Materials Science, 634055 Tomsk, Russia
[3] Department of Mechanical and Aerospace Engineering, New Mexico State University, Las Cruces, NM 88003, USA

E-mail: anisimova_mawa@mail.ru

Abstract. The paper discusses the calculation of the effective thermal and diffusion properties of metal matrix composites containing diamond particles. The effective properties are calculated using Maxwell homogenization scheme. We also establish cross-property connection between overall thermal conductivity and diffusion coefficient and illustrate it on example of Al\diamond composites.

1. Introduction

The paper focuses on the problem of the effective properties - thermal conductivity and diffusion coefficient - of metal matrix composites containing diamond particles. Composites of this kind are widely used for thermal management of electronic components and it is necessary to know their effective properties, such as thermal and electrical conductivity, thermal expansion coefficient, diffusivity etc. However, some effective properties, such as diffusion coefficient, are more difficult to measure experimentally than thermal or electrical properties. In this paper, we propose a method for evaluation of the effective diffusion coefficient through the thermal conductivity measurements.

2. Property contribution tensors

Property contribution tensors are used in the context of homogenization problems to describe contribution of a single inhomogeneity into the property of interest – it may be elastic compliance or stiffness, thermal or electrical conductivity, or diffusion coefficient [1].

2.1. Thermal conductivity problem

In the thermal conductivity problem, the key quantity is the conductivity contribution tensor that gives the extra heat flux produced by introduction of the inhomogeneity into a material subjected to otherwise uniform field of temperature gradient. We assume that the background material of volume V having the isotropic thermal conductivity k^0 contains an isolated inhomogeneity of volume V^1 of the isotropic thermal conductivity k^1. The limiting cases $k^1 = 0$ and $k^1 = \infty$ corresponds to an insulating and a superconducting inhomogeneities. Assuming linear relation between temperature gradient ∇T and the heat flux vector $\boldsymbol{q}$ per reference volume (Fourier law), for both constituents, the change in $\boldsymbol{q}$ in response to the presence of the inhomogeneity is given by

$$\Delta \boldsymbol{q} = \frac{V^1}{V} \boldsymbol{K} \cdot (\nabla T) \tag{1}$$

where the symmetric second-rank tensor $\boldsymbol{K}$ is the *conductivity contribution tensor* of the inhomogeneity.

This tensor for a spherical inhomogeneity is given by Sevostianov and Kachanov [2]

$$\boldsymbol{K} = 3\frac{k^0(k^1 - k^0)}{2k^0 + k^1} \boldsymbol{I} \ . \tag{2}$$

2.2. Diffusion problem

Diffusivity contribution tensor can be introduced by analogy with conductivity contribution tensors [3]. The homogeneous boundary conditions are assumed: the "remotely applied" concentration gradient, or flux of particles, would be uniform in absence of the inhomogeneity. Let, for example, the concentration gradient $\nabla c = \boldsymbol{G}^0$ be prescribed at the boundary of V. Then, the average over $V = V^0 \cup V^1$ flux and concentration gradient are

$$\langle \boldsymbol{J} \rangle_V = \frac{V^1}{V} \langle \boldsymbol{J} \rangle_{V^1} + \left(1 - \frac{V^1}{V}\right) \langle \boldsymbol{J} \rangle_{V^0}, \tag{3}$$

$$\langle \nabla c \rangle_V = \boldsymbol{G}^0 = \frac{V^1}{V} \langle \nabla c \rangle_{V^1} + \left(1 - \frac{V^1}{V}\right) \langle \nabla c \rangle_{V^0}. \tag{4}$$

Taking into account that $D^1 \langle \nabla c \rangle_{V^1} = \backslash \langle \boldsymbol{J} \rangle_{V^1}$ and $D^0 \langle \nabla c \rangle_{V^0} = \backslash \langle \boldsymbol{J} \rangle_{V^0}$, one can write

$$\langle \boldsymbol{J} \rangle_V = \frac{V^1}{V} D^1 \langle \nabla c \rangle_{V^1} + \left(1 - \frac{V^1}{V}\right) D^0 \langle \nabla c \rangle_{V^0} = \frac{1}{D^0} \boldsymbol{G}^0 + \frac{V^1}{V} (D^1 - D^0) \langle \nabla c \rangle_{V^1}. \tag{5}$$

Introducing tensor $\boldsymbol{A}$ that expresses $\langle \nabla c \rangle_{V^1}$ in terms of $\boldsymbol{G}^0$:

$$\langle \nabla c \rangle_{V^1} = \boldsymbol{A} \cdot \boldsymbol{G}^0 \tag{6}$$

the latter expression can be rewritten as

$$\langle \boldsymbol{J} \rangle_V = \left[D^0 \boldsymbol{I} + \frac{V^1}{V} (D^1 - D^0) \boldsymbol{A} \right] \cdot \boldsymbol{G}^0 . \equiv \left[D^0 \boldsymbol{I} + \frac{V^1}{V} \boldsymbol{H}^D \right] \cdot \boldsymbol{G}^0 . \tag{7}$$

The second term in the brackets represents the contribution of the inhomogeneity into overall diffusivity of the volume V and tensor $\boldsymbol{H}^D$ can be called the diffusivity contribution tensor.

For a spherical shape, tensor $\boldsymbol{A}$ can be written as

$$\langle \nabla c \rangle_{V^1} = \boldsymbol{A} \cdot \boldsymbol{G}^0 = \frac{3D^0}{D^1 + 2D^0} \boldsymbol{G}^0 \tag{8}$$

and the diffusivity contribution tensor for a spheroidal inhomogeneity has the following form:

$$\boldsymbol{H}^D = \frac{3D^0(D^1 - D^0)}{D^1 + 2D^0} \boldsymbol{I}. \tag{9}$$

3. Maxwell scheme

According to Maxwell's idea [4], we evaluate fields at far points in two different ways and equate the results. First, we evaluate this field as the one generated by a homogenized region Ω possessing the (yet unknown) effective properties. Secondly, we consider the sum of far fields generated by all the individual inhomogeneities within Ω (being considered as non-interacting ones). Equating the two quantities yields the desired effective property [5]. For the thermal conductivity problem, the result has the following form:

$$\boldsymbol{k}^{eff} = \boldsymbol{k}^0 + \left\{ \left[\frac{1}{V^\Omega} \sum_i V_i \boldsymbol{K}_i \right]^{-1} - \boldsymbol{P}^\Omega \right\}^{-1} . \tag{10}$$

where $\boldsymbol{P}^\Omega$ is the second-rank Hill's tensor for domain Ω [6]. For isotropic composite

$$P_{ij}^\Omega = \frac{1}{3k^0} \delta_{ij} , \quad \frac{1}{V^\Omega} \sum_i V_i \boldsymbol{K}_i = k^0 A \boldsymbol{I} . \tag{11}$$

where parameter A depends on the shape and properties of the individual inhomogeneities. Thus

$$k^{eff} = k^0 \left(\frac{1 + 2A\phi}{1 - A\phi} \right), \quad A = \frac{k^1 - k^0}{2k^0 + k^1} . \tag{12}$$

The effective diffusion coefficient is given by

$$\boldsymbol{D}^{eff} = \boldsymbol{D}^0 + \left\{ \left[\frac{1}{V^\Omega} \sum_i V_i \boldsymbol{H}_i^D \right]^{-1} - \boldsymbol{Q}^\Omega \right\}^{-1} . \tag{13}$$

where $\boldsymbol{Q}^\Omega$ is second-rank tensor that reflects the shape of domain Ω. For a spherical shape of Ω

$$\boldsymbol{Q}^\Omega = \frac{1}{3D^0} \boldsymbol{I} \tag{14}$$

$$D_{eff} = D^0 \left[\frac{1 + 2B\phi}{1 - B\phi} \right], \quad B = \frac{D^1 - D^0}{D^1 + 2D^0} \tag{15}$$

4. Cross-property connections

Comparison of equations (12) and (15) shows their complete identity. We now consider aluminum matrix composite with diamond particles in the context of hydrogen diffusion. For this system $k^0 = 236 W/mK$, $k^1 = 1500 W/mK$, $D^0 = 10^{-7} m^2/s$, and $D^1 = 0$. Solving (12) for the volume fraction of inhomogeneities and substituting it into (15), we can write the effective diffusion coefficient as a function of the thermal conductivity:

$$D_{eff} = D_0 + D_0 \left[\frac{3B(k_{eff} - k_0)}{A(2k_0 + k_{eff}) - B(k_{eff} - k_0)} \right] \tag{16}$$

Figure 1 shows dependence of $(D^{eff} - D^0)/D^0$ on $(k^{eff} - k^0)/k^0$.

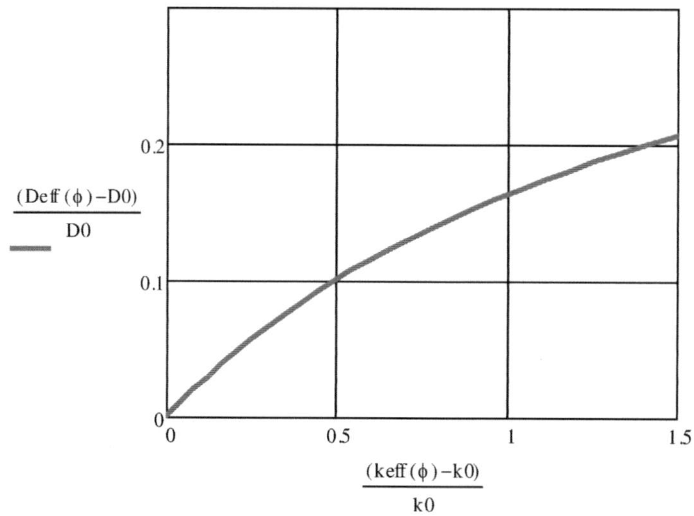

Figure 1. Cross-property connection between effective diffusion coefficient and effective thermal conductivity of Al/diamond composite

5. Concluding remarks

In this paper, we modelled effective thermal and diffusion properties of a particle reinforced composite. The effective properties are calculated using Maxwell homogenization scheme. We also established cross-property connection between effective thermal conductivity and diffusion coefficient and illustrate it on example of aluminum reinforced with diamond particles.

References

[1] Kachanov M, Sevostianov I 2005 On quantitative characterization of microstructures and effective properties *Int. J. Solids Struct.* **42** 309–36

[2] Sevostianov I, Kachanov M 2002 Explicit cross-property correlations for anisotropic two-phase composite materials *J. Mech. Phys. Solids* **50** 253-82.

[3] Knyazeva A G, Grabovetskaya G P, Mishin I P and Sevostianov I 2015 On the micromechanical modeling of the effective diffusion coefficient of a polycrystalline material *Philosophical Magazine* **95** (19) 2046-66

[4] Maxwell J C 1873 *A Treatise on Electricity and Magnetism* (Oxford: Clarendon Press)

[5] Sevostianov I, Giraud A 2013 Generalization of Maxwell homogenization scheme for elastic material containing inhomogeneities of diverse shape *International Journal of Engineering Science* **64** 23-36

[6] Markov K Z 2000 Elementary micromechanics of heterogeneous media *Heterogeneous Media: Micromechanics Modeling Methods and Simulations* ed Markov K Z, Preziozi L (Boston: Birkhauser) pp 1–162

IOP Conf. Series: Materials Science and Engineering **175** (2017) 012052 doi:10.1088/1757-899X/175/1/012052

The influence of metakaolin substitution by slag in alkali-activated inorganic binders for civil engineering

J Kadlec*, D Rieger, T Kovářík, P Novotný, P Franče, M Pola

New Technologies - Research Centre, University of West Bohemia, 306 14 Pilsen, Czech Republic

E-mail: jakadlec@ntc.zcu.cz

Abstract. In this study the effect of metakaolin replacement by milled blast furnace slag in alkali-activated geopolymeric binder was investigated in accordance to their rheological and mechanical properties. It was demonstrated that slag addition into the metakaolin binder can improve mechanical properties of final products. Our investigation was focused on broad interval of metakaolin substitution in the range from 100 to 40 volume per cents of metakaolin so that the volume content of solids in final binder was maintained constant. Prepared binders were activated by alkaline solution of potassium silicate with silicate module of 1.61. The particle size analyses were performed for determination of particle size distribution. The rheological properties were determined in accordance to flow properties by measurements on Ford viscosity cup and by oscillatory measurements of hardening process. For the investigation of hardening process, the strain controlled small amplitude oscillatory rheometry was used in plane-plate geometry. For determination of applied mechanical properties were binders filled by ceramic grog in the granularity range 0-1 mm. The filling was maintained constant at 275 volume per cents in accordance to ratio of solids in dry binder. The mechanical properties were investigated after 1, 7 and 28 days and microstructure was documented by scanning electron microscopy. The results indicate that slag addition have beneficial effect not only on mechanical properties of hardened binder but also on flow properties of fresh geopolymer paste and subsequent hardening kinetics of alkali-activated binders.

1 Introduction

Inorganic polymer cements can be synthesized by alkali-activation of a variety of materials including thermally activated clays, coal fly ash and blast furnace slag to produce a hardened material with mechanical and thermal properties potentially suitable for wide range of industrial applications [1, 2]. Alkali-activated pozzolan material was discovered by Purdon and Glukhovsky in the 1940s and systematically studied by them up to the 1960s [3]. Davidovits entitled this type of materials as "geopolymers" and introduced pioneering alkali-activated calcium-free systems based on calcined clays [4]. From a general point of view, geopolymers are classified as aluminosilicate binders activated by solutions of alkali metal hydroxides, carbonates and silicates. Sodium silicate solutions are widely used as geopolymer constituent controlling the dissolution processes of the solid phase via cleavage of the Si-O and Al-O bridging oxygen bonds. The dissolution rate depends on the pH, chemical composition, concentration, structural state, and distribution of silicate species in the system [5]. The following gel precipitation and solid phase formation is driven by polycondensation mechanism and this process is leading to the formation of amorphous aluminosilicate matrix [6]. The most used is kaolinite calcined to metakaolin which leads to increase of reactivity [7]. Not only calcined kaolinite, but many aluminosilicates, also containing alkaline earth oxides as CaO, can be hardened with this mechanism. Examples can be shales and other clay minerals, fly ashes, slags and theirs blends [8, 9].

Slag is used as an additive to geopolymers mixtures for improvement of mechanical strength and for decrease of setting time. It is expected that role of Ca^{2+} ions in geopolymerization process consists in incorporation into final structure as charge balancing ion and in precipitation of calcium silicate hydrates as nucleation sites which trigger rapid formation of geopolymer network respectively [10, 11].

This study describes the effect of metakaolin substitution by blast furnace slag on hardening rate and compressional and flexural strength of hardened material. For description of hardening rate the small amplitude oscillatory rheometry was used as promising method to detect the sol-gel transition of alkali activated geopolymer suspension to amorphous gel state [12,13].

2 Materials

In this study, a calcined shale (CS) as a metakaolin source and milled blast furnace slag (MS) were used as raw materials. For determination of mechanical properties calcined chamotte was used as filler. These materials were supplied by Ceske lupkove zavody a.s, Nove Straseci, Czech Republic and by Kotouc Stramberk a. s. in the case of milled blast furnace slag. The chemical compositions were determined by the XRF technique and are listed in Table 1.

Table 1. Chemical composition of raw materials [mass %].

Composition	SiO_2	Al_2O_3	CaO	MgO	TiO_2	Fe_2O_3	K_2O	LoI	Others
CS	53.02	39.9	0.11	0.12	2.33	1.22	0.58	2.05	0.67
MS	40.4	6.9	39.2	9.4	-	0.4	-	-	3.7
Activator	17.60	-	-	-	-	-	17.13	65.27	-

Granularity was determined by static light scattering technique and results are graphically presented in figure 1 b). The specific gravitiesof CS and MS were determined by Archimedes method to be 2.56 and 2.92 g/cm^3 respectively.This data was used for volume fraction substitutions in the range 0-60 wt.%. For alkaline activation an aqueous solution of potassium silicate with silicate module 1.61 and content of water soluble solids 34.73 % was used.

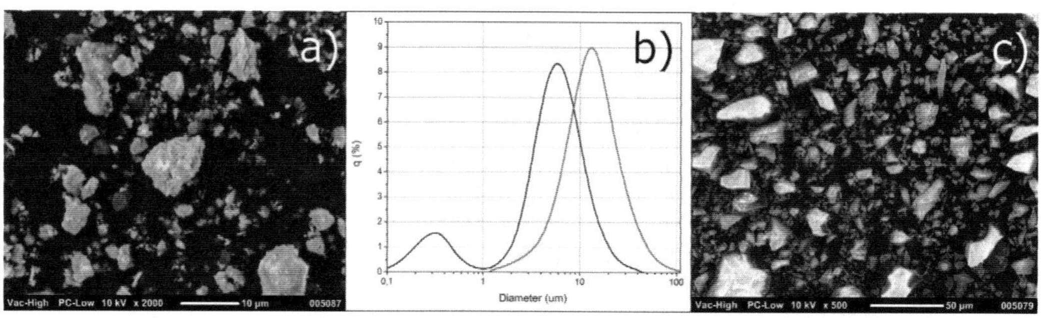

Figure 1. SEM images of calcined shale (a), particle size distribution (b) – black (CS) and red (MS) and SEM image of milled blast furnace slag (c)

3. Procedures and results

3.1Experimental concept

In this study, the experimental approach was as follows. The geopolymeric binder in the form of fresh paste was prepared by mixing of metakaolin with activator in weight proportion 1:1 for 5 minutes in laboratory vacuum mixer as a model binder and marked as 100-0. To determination of the substitution effect, metakaolin in this process was substituted by blast furnace slag in 10 vol% steps up to 60 vol% which was marked as binder 40-60. These binders were subsequently used for rheometrical measurements. These binders were filled by calcined chamotte with granularity range of 0-1 mm in accordance to Fuller-Thomson line of granularity for mechanical strength determination.

3.2 Rheological measurements

For description of flow properties and workability measurements on fresh pastes with Ford viscosity cups of diameters 4 and 6 mm were conducted. How it can be seen in figure 2 a) increasing metakaolin substitution leads to significant reduction of viscosity which is important to workability. This behavior can be attributed to increasing of overall particle sizes in individual pastes. The determination of setting time by small amplitude oscillatory rheometry was described earlier [13]. The measurements were performed in plain-plate geometry on an Ares G2 rheometer from TA Instruments with working gap of 0.5 mm and strain controlled to 0.01 %. After 15 minutes, isothermal heating at temperature of 30 °C was initiated for a period of time necessary for paste hardening. Subsequently, changes in position of maximum value on loss tangent curve were used for determination of hardening kinetics. The result for binder 70-30 is presented in figure 2 b) and overall results of loss tangent maxima clearly indicate that hardening of individual pastes is accelerated by increasing amount of slag. This effect is significant to the 40% substitution and it can be described as a fast dissolution of amorphous content of slag and subsequent construction of calcium silicate gel together with formation of geopolymer network which have effect on early age mechanical strength [11].

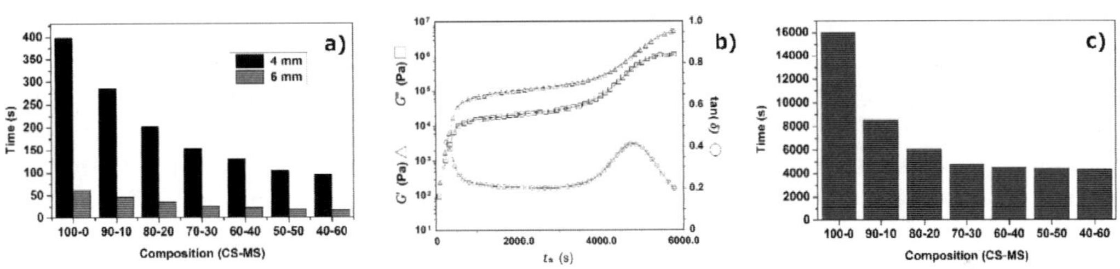

Figure 2. Results of flow measurements by Ford cups (a); Standard results of oscillatory measurement (binder70-30) where G', G" – storage and loss moduli in torsion, tan(δ) – loss tangent(b); results of individual oscillatory measurements(c)

3.3 Measurements of mechanical properties

For flexural and compression strength investigation, binder was prepared according to procedure described in section 3.1 in amount of 200 weight parts and mixed with 275 weight parts of filler. The mixing was performed in planetary mixer for a period of 10 minutes. Subsequently, the mixtures were cast in moulds with dimensions of 20×20×120 mm and 20×20×20 mm for flexural and compressional strength determinations. Before individual tests the samples were cured at ambient temperature in polypropylene bags to prevent humidity loss. Results for individual periods of time are presented in figure 3. The mechanical properties after 1st and 7thday clearly correspond with findings by rheometrical measurements up to the substitution of 30 %. Beyond this limit increasing the amount of slag is not further beneficial for mechanical properties. This is more apparent after 28 days where the best results are between 10 and 20% substitution. However, in all cases at the end of investigated time period the mechanical strength in compression and flexure exceeded 90 and 15 MPa respectively and thus all mixtures represent good binding properties.

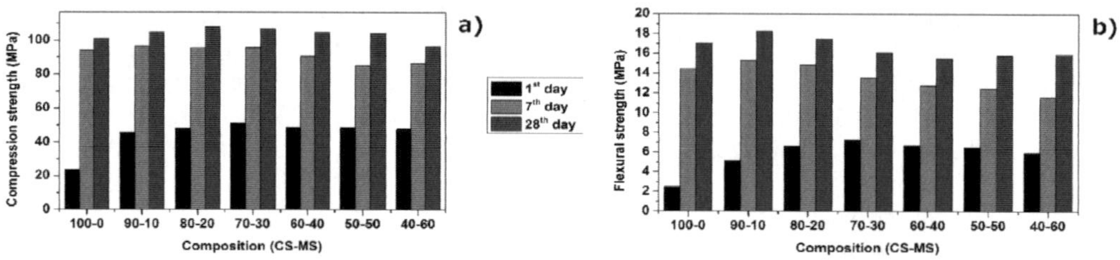

Figure 3. Results of strength measurements in compression (a) and flexure (b)

4 Conclusions

It can be summarized that the slag addition is beneficial mainly for fast hardening of geopolymeric binders where the 40% substitution of metakaolin by slag reduces setting time by 5 times. This behavior has beneficial effect also on early age mechanical strength up to the 30% substitution. On the contrary higher substitutions slightly decrease mechanical strength.Moreover,after 28 days the best results of mechanical strengths were achieved by 10-20% substitution of metakaolin by slag and it can be related to slower polycondensation of geopolymer network which results in less structural defects. However, in whole investigated range the mechanical properties are suitable for construction applications.Thus, it can be concluded that lower substitutions could be used for casting at construction sites and 30% and higher substitution can be advantage in manufacturing of prefabricated elements where short time of unmolding is significant parameter because of fast hardening and also where saving of natural resources is desirable.

Acknowledgments

The result was developed within the CENTEM project, reg. no. CZ.1.05/2.1.00/03.0088, cofounded by the ERDF as part of the Ministry of Education, Youth and Sports OP RDI programme and, in the follow-up sustainability stage, supported through CENTEM PLUS (LO1402) by financial means from the Ministry of Education, Youth and Sports under the National Sustainability Programme I.

References

[1] Pacheco-Torgal F, Castro-Gomes J and Jalali S 2008 *Const Build Mater***22** 1305-14
[2] Duxson P, Provis JL, Lukey GC and Van Deventer JSJ 2007 *Cem Concr Res***37** 1590-97
[3] Purdon AO The action of alkalis on blast-furnace slag 1940 *J Soc Chem Ind***59** 136-91
[4] Davidovits J 1991 *J Therm Anal***37** 1633–56
[5] Kouassi S S, Tognonvi M T, Soro J and Rossignol S 2011 *J Non-Cryst Solids***357** 3013-21
[6] Šesták J and Foller B 2012 *J Term Anal Calorim***109** 1-5
[7] Kovářík T, Bělský P, Novotný P, Říha J, Savková J, Medlín R, Rieger D and Holba P 2015 *Const Build Mater***80** 98-104
[8] Kim B-J, Yi C and Kang K-I 2014 *Const Build Mater***58** 206-13
[9] Rieger D, Čekalová M, Kullová L, Novotný P, Pola M 2015 *SGEM2015 Conference Proceedings Book4 (Albena)* vol 1
[10] Lee WKW and Van Deventer JSJ 2002 *Cem Concr Res***32**577-584
[11] Yip ChK, Lukey GC, Provis JL, Van Deventer JSJ 2008 *Cem Concr Res***38**554-564
[12] Steins P, Poulesquen A, Diat O and Frizon F 2012 *Langmuir***28** 8502-10
[13] Rieger D, Kovářík T, Říha J, Medlín R, Novotný P, Bělský P, Kadlec J and Holba P 2015 *Const Build Mater* **83** 26-33

IOP Conf. Series: Materials Science and Engineering **175** (2017) 012053 doi:10.1088/1757-899X/175/1/012053

The Influence of Surface Roughness on Biocompatibility and Fatigue Life of Titanium Based Alloys

S Major[1], P Cyrus[1] and M Hubálovská[1]

Department of Technical Education, Faculty of Education, University Hradec Králové, 500 03Hradec Králové, Czech Republic

E-mail: s.major@seznam.cz

Abstract. This article deals with the effect of treatment on the mechanical properties ofbiocompatible alloys. In the case of implants, it is desirable to ensure good biocompatibility. Generally, the environment in the body is very aggressive and implants can quickly degrade due the corrosion.The process of corrosion leads to the release of harmful particles into the body. Other reasons for rejection of the implants, is their coverage bacterial plaque. Another reason for the rejection of the implant may be a smooth surface. In some cases, the tissue does not adhere to the smooth surface of the implant, in this regionsoccurs an accumulation of body fluids.This problem can be solved with a rough surface. From the viewpoint of fatigue resistance, the rough surface containing grooves and holes has a negative influence on the fatigue resistance against mechanical loading. The rough surface can be produced by machining or asymmetric deposition of particles of oxides, nitrides or other particles on surface. In this work the formation and propagation of fatigue cracks in the material with granular surface is analysed. The formation and growth of fatigue crack originated from granular surface is simulated. Also, experimental studies were carried out.

1. Introduction

In the case of implants used in medicine, it is desirable to ensure good biocompatibility[1,2,3]. The biocompatibility is important property of materials used for construction of medical implants.Generally, the environment in the human body is very aggressive and implants can quickly degrade due the corrosion.The process of corrosion leads to the release of harmful particles into the body. For this reason, for example, titanium alloys with niobium content are preferred before alloys containing harmfulvanadium or nickel. The corrosion can be reduced through appropriate surface finishes.Another important problem isthe ability of the human body to accept implant. The body is trying to push out foreign objects. One of the reasons for rejection of the implants is their coverage bacterial plaque. Another reason for the rejection of the implant may be a smooth surface. In some rare cases (1% of cases in maximum), the tissue does not adhere to the smooth surface of the implant, in this regions occurs an accumulation of body fluids.In the area of fluid accumulation often occurs secondary inflammation.Experiments realized on animals shows that the body accepts better implants with a rough surface. This phenomenon can be explained by better adherence of tissue to the rough surface of implant. The rough surface can be produced by machining. From the viewpoint of fatigue resistance, grooves on the machined rough surface have a negative influence on the fatigue resistance against mechanical loading. The rough surface can be also produced by asymmetric deposition of oxide or nitride particles on implant surface.Currently we are studying the possibility of creating a rough surface by means of irregular and asymmetric deposition of particles on the smooth surface.

IOP Conf. Series: Materials Science and Engineering **175** (2017) 012053 doi:10.1088/1757-899X/175/1/012053

Particle distribution can be controlled for example by laser or electrical discharges. These particles then form granular surface that is well accepted by the body.Also in the case of surfacesprepared by this method may be reduced fatigue resistance of implant. Commonly realized surface finishes (for example machine parts with nitride smooth surface) are improving fatigue resistance[4].The improvement of fatigue resistance is caused by high compressive residual stresses introduced within the diffusion zone [4]. Since such a layer hinders the dislocation motion, the predominant failure mechanism in the high cycle fatigue is subsurface crack growth. In case the granular surface, granules are relatively large and they could act as crack initiation sites. The size, scatter and depth of granules on surface must be appropriately designed in order to prevent this.

2. Theoretical model of fracture process

For the strain-deformation analysis of the behaviour of the granules in the surface is used 3D model based on the finite element method FEM. This model is based on the assumption that the granules are spheres partially embedded in the surface disposed on the surface as shown in figure 1. Model parameters (such as granules diameter, see figure 1, the diameter of granule is from 0.5 to 1mm and its height is 0.4 mm at maximum) can be easily changed in the program. The FEM model itself uses symmetry, so that the model represents only one quarter of the granules and the surrounding material. The model assumes that the crack forms at the interface between the granules and surrounding materials. The emergence of the crack is more likely if there is a sharp difference between the mechanical properties (for example elastic modulus) of the granules and the surrounding material. The mechanical properties of material surrounding the granules are exponential functions of distance from surface. For further analysis, model modification of model proposed Navarro[5,6].was utilized. Main advantage of this model is that definition of boundary length of crack is not needed for distinction between the initiation and propagation phases of crack growth.

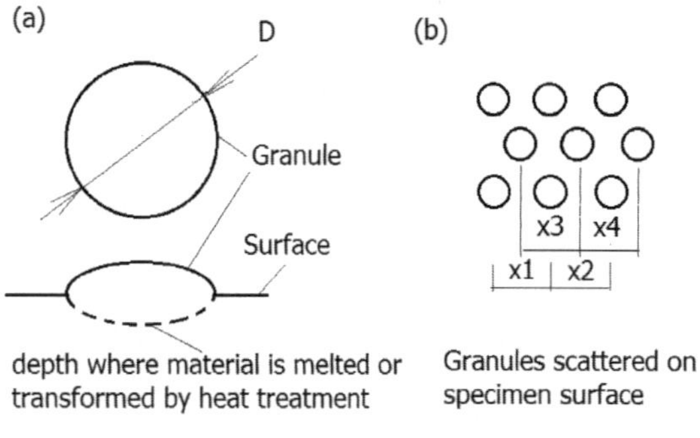

Figure 1. Granular surface:(a) Geometry of granules - the region of melted and transformed material is marked dashed line; (b) granules displacement on the surface

The fracture mechanics is used for calculation of the number of cycles needed to propagate a crack to length of final rupture a_{FR}. The following equation expresses the number of cycles:

$$N_i(\sigma_{eq},a)=N_{TR}(\sigma_{eq})-\int_a^{a_{FR}}\frac{da}{C\Delta K^m} \qquad (1)$$

where σ_{eq} is equivalent stress, $a$length of crack and N number of cycle. The constants C and m are known from Erdogan law. The subscript TR means the total number of cycles to rupture. Limits of integral are a – some crack length during fatigue process and a_{FR} – the length at the time of the final rupture.Initiation phase of the fatigue process can be described by the integration of equation:

$$\frac{da}{dN}=C\left(\Delta K^m-\left(\Delta K_{th,Long}\left(\frac{a^{P_{KT}}}{a^{P_{KT}}+a_0^{P_{KT}}-d_0^{P_{KT}}}\right)^{\frac{P_{KT}}{2}}\right)^m\right) \qquad (2)$$

Variables used in this equation are:$\Delta K_{th,Long}$- the growth threshold for long cracks;d_0 – the average distance to the first microstructural barrier; a_0- El Haddad parameter [7] . The P_{KT} factor is derived from the theoretical approximation of the Kitagawa–Takahashi diagram [8].

3. Specimens and the experimental procedure

3.1. Materials characterisation

Most of implants used in surgery are made from Ti-6Al-4V ELI alloy (Extra-Low Interstitials) currently. This alloy is very close to the alloy Ti-6Al-4V, which is intended for industrial use, but which may also arise in the construction of the implants. Ti-6Al-7Nb alloy or commercially pure titanium Grade 4 are being used as alternative materials. The surfaces studied in this work were made on specimens from these four titanium alloys.The mechanical properties of materials are shown inTable 1.

Table 1. Mechanical properties: Young modulus E, ultimate tensile strength σ_u, yield stress σ_y. The subscript *A* indicates the value after annealing of material.
Constants *C* and *m*are known from Erdogan law. These constants were obtained for annealed titanium, if these constants are used the crack growth da/dN is m/cycle and value of stress intensity factor is MPa.m$^{0.5}$.
Crack growth properties were measured according to ASTM E647.

	E[GPa]	σ_u [MPa]	σ_y[MPa]	$\sigma_{u,A}$ [MPa]	$\sigma_{y,A}$[MPa]	C	*m*
Ti-6Al 4V ELI	113.8	893	827	998	948	7.9 10^{-14}	4.9
Ti-6Al-4V	117	944	878	1027	978	8 10^{-14}	5
Ti-6Al-7Nb	105	920	842	1017	983	9.1 10^{-14}	4.8
PureTitanium	104.4	681	552	679	549	1.8 10^{-14}	5.13

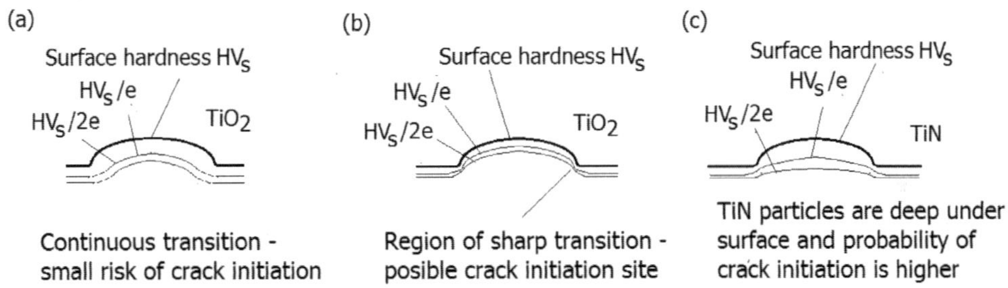

Figure 2. Profiles of surface layers on granules: (a) Plasma Immersion Ion Implantation; (b) Plasma spraying; (c) Nitriding.In figure is marked HV hardness and its decline to 1/e and 1/2e of its value on the surface

3.2. Surface treatment and surface layer

In the field of medical engineering, the most commonly surface treatment is deposition of TiO_2. While nitridation has positive effect on the fatigue life of steels [4], it is not generally case of titanium alloys. In case of granular surfaces, the spots on the sample surface are heated and locally melted by electric arc or laser.Nitriding takes place virtually simultaneously with the formation of the granules on the surface. Plasma nitriding was used making TiN surface. The TiO_2 surface was formedby two ways: Plasma Immersion Ion Implantation and Plasma spray.The TiO_2hasother good properties such as self-cleaning [9]. Profiles of surface layers on granules of different surface layers with regions of possible crack initiation are displayed in figure 2. In this figure is marked HV hardness and its decline. Hardness of surface layer corresponds to content of oxide or nitride particles in diffusion zone.

4. Results and discussion

In Figure 3 is shown theoretical curves for all four types of titanium and for three types of surfaces. From the figure it is clear that the best results giveTi-6Al-4V ELI withTiO_2 surface deployed by

Plasma Immersion Ion Implantation. This can be explained by the continuous distribution of particles on the granular surface. Worse outcomes for Plasma spraying method can be explained by presence of sharp transition, see figure 2. The worst results givenitridation of granuloid surfaces. The granules are throughout the volume nitride and between granules and smooth surface rise sharp transition stress.

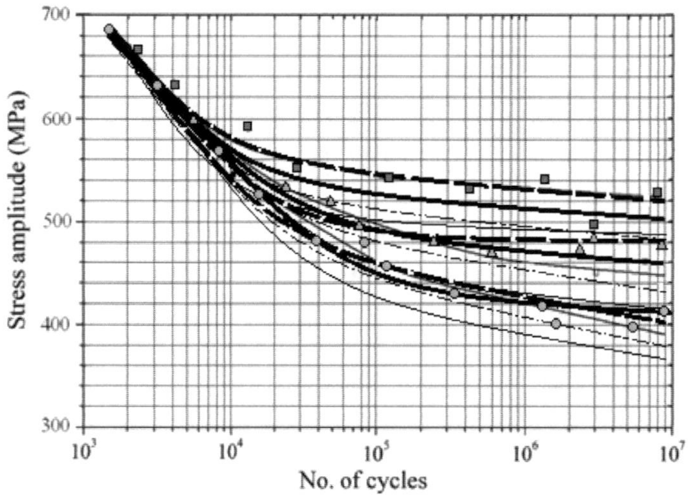

Figure 3. Experimental points: squares - Plasma Immersion Ion Implantation, triangles - Plasma spraying, circles-Nitriding; thick line - Ti-6Al-4V, dash thick line - Ti-6Al 4V ELI, thin line - Ti-6Al-7Nb, cash thin line - pure titanium

5. Conclusion

The best results giveTi-6Al 4V ELI with granuloid TiO_2 surface, this can be explained by continuous transition of stress between granules and smooth surface.

Acknowledgments

This research has been supported bySPR No. 2124 of University of Hradec Kralove.

References

[1] Major S, Kocour V, Cyrus P, 2016 *Frattura ed Integrita Strutturale*, 10 35 pp 379-388doi:http://dx.doi.org/10.3221/IGF-ESIS.35.43
[2] AyllónJ M, NavarroC, VázquezJ, Domínguez J, 2014 *Engineering Fracture Mechanics* 123 pp 34–43doi:10.1016/j.engfracmech.2014.03.011
[3] Griza S at al., 2012 *Engineering Failure Analysis* 25 pp 133–143doi:10.1016/j.engfailanal.2012.05.009
[4] Slámečka K at al., 2010 *International Journal of Fatigue* 32 pp 921-928
[5] Navarro C, Muńoz S, Domínguez J, 2008 *International Journal of Fatigue*pp 32–44doi:10.1016/j.ijfatigue.2007.02.018
[6] Navarro C, Vázquez J, Domínguez J, 2011 *Engineering Fracture Mechanics* pp 1590–601doi:10.1016/j.engfracmech.2011.01.011
[7] El Haddad M H at al., 1979 *Engineering Fracture Mechanics* 11 pp573–584doi:10.1016/0013-7944(79)90081-X
[8] Kitagawa H and Takahashi S, 1976 *Proceedings of the Second International Conference on Mechanical Behavior of Materials*. pp 627–31.
[9] Nakata K and Fujishima A, 2012 *J Photochem Photobiol* C 13. pp 169–189

IOP Conf. Series: Materials Science and Engineering **175** (2017) 012054 doi:10.1088/1757-899X/175/1/012054

Mechanical properties of non-woven glass fiber geopolymer composites

D Rieger*, J Kadlec, M Pola, T Kovářík, P Franče

New Technologies - Research Centre, University of West Bohemia, 306 14 Pilsen, Czech Republic

E-mail: DavidPJ@ntc.zcu.cz

Abstract. This experimental research focuses on mechanical properties of non-woven glass fabric composites bound by geopolymeric matrix. This study investigates the effect of different matrix composition and amount of granular filler on the mechanical properties of final composites. Matrix was selected as a metakaolin based geopolymer hardened by different amount of potassium silicate activator. The ceramic granular filler was added into the matrix for investigation of its impact on mechanical properties and workability. Prepared pastes were incorporated into the non-woven fabrics by hand roller and final composites were stacked layer by layer to final thickness. The early age hardening of prepared pastes were monitored by small amplitude dynamic rheology approach and after 28 days of hardening the mechanical properties were examined. The electron microscopy was used for detail description of microstructural properties. The imaging methods revealed good wettability of glass fibers by geopolymeric matrix and results of mechanical properties indicate usability of these materials for constructional applications.

1. Introduction

Alkali-activated aluminosilicate cements were discovered by Purdon and Glukhovsky already in 1940s [1] and are developed up to day. Davidovits entitled this type of materials as "geopolymers" because of polymer like chemical structure and introduced pioneering work on alkali-activated binders based on calcined clays [2]. Geopolymers are products of chemical reaction between aluminosilicate material and liquid alkaline environment where chemical cleavage of the Si-O and Al-O bonds in parent material leads to saturation of liquid solution and subsequent polycondensation of amorphous aluminosilicate matrix [3]. It is known that geopolymers can have high mechanical strength [4] and are inflammablebut by ceramic-like nature have brittle fracture character [5]. Using of fibers can change it into ductile one and prevent the crack propagation in material. Wide range of fibers is used for this purpose from organic natural and synthetic materials as PVC, PP, PVA and cellulose [5-7], inorganic carbon-based materials and mineral materials as basalt,glass, zirconia and others [8-10]. Fiber reinforcement is mainly used in form of short fibers for casting processes or as woven and non-woven composites prepared by impregnation.

This study focuses on mechanical properties of composites prepared by infiltration of geopolymeric binder into the non-woven glass fabric. The effect of different matrix composition and amount of granular filler on the mechanical properties of final composites are evaluated and documented by scanning electron microscopy. Furthermore, the hardening process of individual binder compositions was evaluated by the small amplitude oscillatory rheometry measurements [11, 12].

IOP Conf. Series: Materials Science and Engineering **175** (2017) 012054 doi:10.1088/1757-899X/175/1/012054

2. Materials

In this study, a calcined shale (CS) as a metakaolin source and calcined chamotte with granularity 0-0.1 mmas ceramic filler were used. These materials were supplied by Ceske lupkove zavody a.s, Nove Straseci, Czech Republic and by Kotouc Stramberk a. s. in the case of milled blast furnace slag. The chemical compositions were determined by the XRF technique and are listed in Table 1.

Table 1. Chemical composition of raw materials [mass %]

Composition	SiO_2	Al_2O_3	CaO	MgO	TiO_2	Fe_2O_3	K_2O	LoI	Others
CS	53.02	39.9	0.11	0.12	2.33	1.22	0.58	2.05	0.67
Activator	17.60	-	-	-	-	-	17.13	65.27	-

Granularityof CS and filler were determined by static light scattering technique and results are graphically presented in figure 1 b). For alkaline activation an aqueous solution of potassium silicate with silicate module 1.61 and content of water soluble solids 34.73 % was used.

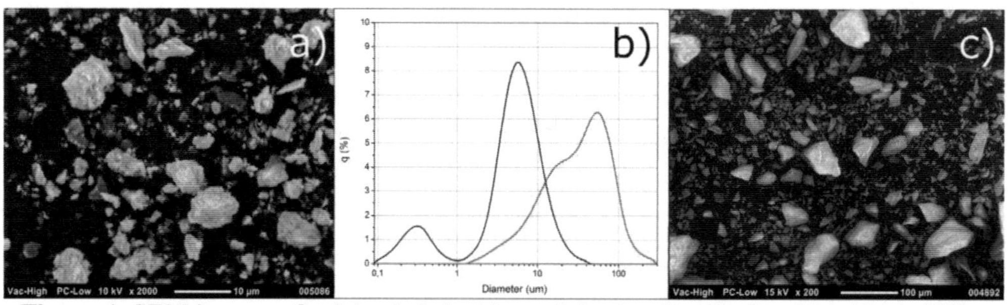

Figure 1. SEM images of calcined shale (a), particle size distribution (b) – black (CS) and red (filler) and SEM image of filler (c)

3. Procedures and results

3.1. Preparation of composites

In this study, the experimental approach was as follows. The non-woven glass fabric how it can be seen in figure 2 a) was incorporated by geopolymeric matrix with different composition in the range 100:70 – 100:100 wt.% of CS to activator solution respectively. Moreover, matrix 100:100 was filled by ceramic filler in range 20 – 100 wt.% in relation to CS for investigation of its impact on mechanical properties. Matrices were prepared by mixing of constituents in vacuum mixer for 5 minutes and incorporated into non-woven sheets with thickness of 4 mm by hand roller as can be seen in figure 2 a). The microstructure of non-woven fabrics is presented in figure 2 b). Five sheets were stacked layer by layer on each other for final specimen thickness and aged for 28 days in polypropylene bags.

3.2. Rheological measurements

Changes in complex viscosity were measured by small amplitude oscillatory rheometry. The measurements on unfilled pastes were performed in plain-plate geometry on an Ares G2 rheometer from TA Instruments with strain controlled to 0.01 %. Individual pastes were prepared by vacuum mixing and measured after 15 minutes at 30 °C isotherm.

Figure 2. Image of hand roller and non-woven fabric (a), SEM of non-woven glass fabric (b), tested specimens (c)

The changes of complex viscosity presented in figure 3 describe hardening rates of individual matrices and it can be deduced that increasing amount of activator leads to reduction of hardening rate by increasing liquid to solid ratio and delaying of saturationand polycondensation, but also to decreasing of matrix viscosity. The composition 100:100 was selected for filling experiment, based on obtained data where the initial viscosity reduction of two orders of magnitude was achievedin comparison with matrix composition 100:70.

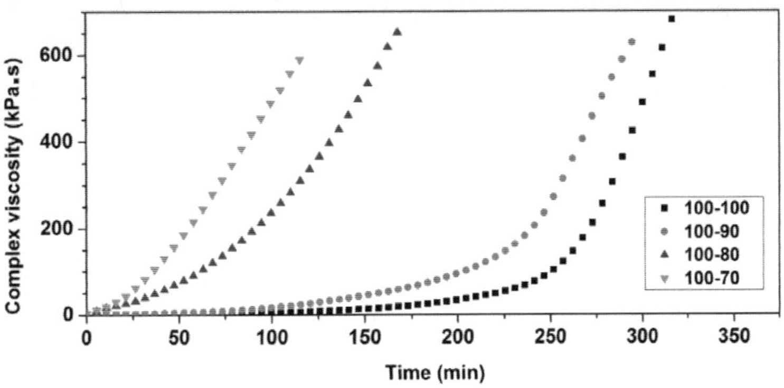

Figure 3. Complex viscosity of individual matrices

3.3. Measurements of mechanical properties

For flexural strength investigation, prepared non-woven sheets were divided by diamond saw to strips with dimensions of 100 x 10 mm which can be seen in figure 2 c) and were subsequently tested. Figure 4 summarizes flexural strengths of individual compositions measured after 28 days. How it can be seen the composition of pure matrix influences final strength and the best performance was achieved with composition 100:90:0. It can be attributed to sufficient hardening reaction of matrix and lower water content. In the case of matrices 100:100 increasing amount of filler improves the mechanical strength. However, the high viscosity of filled matrix at composition 100:100:100 prevents sufficient infiltration of matrix into the fabrics structure.

This effect is documented by scanning electron microscopy. In figure 5 a) the matrix 100:90 is fully infiltrated into the fabrics structure but in the case of matrices 100:100:80 and 100:100:100 the material stay also in between of individual sheets as can be seen in figure 5 b) and c) respectively. In last case this effect is highly significant and leads to decrease of overall flexural strength.

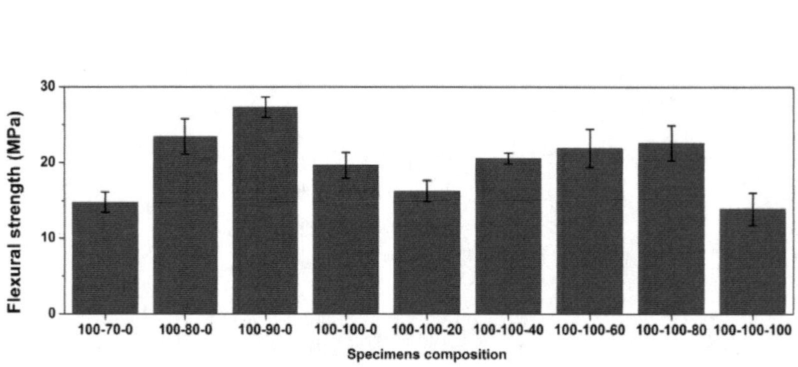

Figure 4. Results of flexural strength measurements

Figure 5. SEM micrographs of composites 100:90 (a), 100:100:80 (b) and 100:100:100 (c)

4. Conclusions

It can be concluded that in this study the best composition of matrix is 100:90 wt.% of calcined shale to activator solution. In the case of 100:100 matrixes the increasing amount of filler retains mechanical strength up to 80wt% of filler. Further increasing of filler amount leads to a considerable decrease in flexural strength. This effect is caused by remaining matrix in between of individual fabrics which is caused by increase of matrix viscosity. It can be described as main limitation of used impregnation process when high viscosity matrix cannot be completely expelled from stacked composite interspaces.

Acknowledgments

The result was developed within the CENTEM project, reg. no. CZ.1.05/2.1.00/03.0088, cofounded by the ERDF as part of the Ministry of Education, Youth and Sports OP RDI programme and, in the follow-up sustainability stage, supported through CENTEM PLUS (LO1402) by financial means from the Ministry of Education, Youth and Sports under the National Sustainability Programme I.

References

[1] Purdon AO The action of alkalis on blast-furnace slag 1940 *J Soc Chem Ind***59** 136-91
[2] Davidovits J 1991 *J Therm Anal***37** 1633–56
[3] Kouassi S S, Tognonvi M T, Soro J and Rossignol S 2011 *J Non-Cryst Solids***357** 3013-21
[4] Rieger D, Čekalová M, Kullová L, Novotný P, Pola M 2015 *SGEM2015 Conference Proceedings Book4 (Albena)* vol 1
[5] Toutanji H, Xu B, Gilbert J, Lavin T 2010 *Const Build Mater***24** 1-10
[6] Yan L, Kasal B, Huang L 2016 *Composites Part B***92** 94-132
[7] Alzeer M, MacKenzie K 2013 *Appl Clay Sci***75–76** 148-152
[8] Peigang He SY, Jia D, Yang Z, Duan X, Wang S, Zhou Y 2016 *Ceram Int***42** 7837-43
[9] Papakonstantinou CG, Balaguru P, Lyon RE 2001 *Composites Part B***32** 637-49

IOP Conf. Series: Materials Science and Engineering **175** (2017) 012054 doi:10.1088/1757-899X/175/1/012054

[10] Samal S, Thanh NP, Petríková I, Marvalová B, Vallons KAM, Lomov SV 2015 *Ceram Int***41** 12115-29
[11] Steins P, Poulesquen A, Diat O and Frizon F 2012 *Langmuir***28** 8502-10
[12] Rieger D, Kovářík T, Říha J, Medlín R, Novotný P, Bělský P, Kadlec J and Holba P 2015 *Const Build Mater* **83** 26-33

IOP Conf. Series: Materials Science and Engineering **175** (2017) 012055 doi:10.1088/1757-899X/175/1/012055

Influence of Sintering Time on the Structure Formation of Al-ZrW_2O_8 Pseudo Alloys

V S Shadrin[1,2], **S N Kulkov**[1,2]

[1]Strength and Design Department, Physical-Technical Faculty, Tomsk State University, 634050 Tomsk, Russia.
[2]Laboratory of Physics of Nanostructured Functional Materials, Institute of Strength Physics and Materials Science SB RAS, 634055, Tomsk, Russia

E-mail: kulkov@ms.tsc.ru

Abstract. Al – ZrW_2O_8 pseudo alloys were synthesized by free sintering of Al – ZrW_2O_8 powder mixture. Influence of sintering time on the structure formation of the pseudo alloys obtained was investigated. It has been shown that during sintering process zirconium tungstate decomposes into constituent oxides and re-synthesis of zirconium tungstate proceeds through intermediate stage – formation of WAl_{12} and $ZrAl_3$ intermetallic compounds.

1. Introduction

Zirconium tungstate ZrW_2O_8 with isotropic negative thermal expansion (NTE) behavior over a wide temperature range [1] is a perspective material for development of new class of metallic materials with controlled thermal expansion.

It is known that zirconium tungstate decomposes into its constituent oxides WO_3 and ZrO_2 upon heating and recent works [2] showed that decomposition of ZrW_2O_8 may begin already at 400 °C. Therefore the formation of alloycontainingZrW_2O_8particles will be strongly dependent on sintering time and one can expect some specific behavior of ZrW_2O_8 during interaction with metals. One of the easiest ways to observe these effects is to use aluminum as one of initial components due to low melting point of this metal. In addition aluminum based materials are lightweight, exhibit excellent mechanical properties, thermal and electrical conductivity and therefore are promising candidates for development of multifunctional Al – ZrW_2O_8 pseudo alloy with controlled thermal expansion.

The work presented is aimed at investigating the influence of sintering time on structure and phase composition of Al – ZrW_2O_8 pseudo alloys.

2. Experiment

Commercially pure aluminum powder ASD-6 and zirconium tungstate powder obtained by hydrothermal synthesis [3] were used as initial components. The amount of ZrW_2O_8 addition was 10 wt%. High energy ball milling for 30 seconds, 1 and 5 minutes using AGO-2 planetary ball mill with acceleration of 60g was performed to achieve homogeneous distribution of ZrW_2O_8 in aluminum powder and for mechanical activation of the components. Mechanically activated (MA) Al – ZrW_2O_8 powder mixtures were uniaxially cold pressed into ~ ⌀20mm×4mm cylindroids under 100 MPa and sintered at 600 °C in argon atmosphere for 1h and 5h. Surface of the specimens was etched in hydrofluoric acid. The microstructure of the specimens obtained was observed using TESCAN

Vega3 scanning electron microscope. Structure and phase composition were investigated using XRD analysis.

3. Results and Discussion

3.1. High energy mechanical activation of Al – ZrW_2O_8 powder mixtures

Aluminum powder is represented with spherical particles with average size of 3 μm. ZrW_2O_8 particles are rod-shaped with average length of 5 μm and are agglomerated, figure 1. Morphology and particle sizes of ZrW_2O_8 are in a good agreement with [3,4].

Figure 1. Morphology of ZrW_2O_8 powder

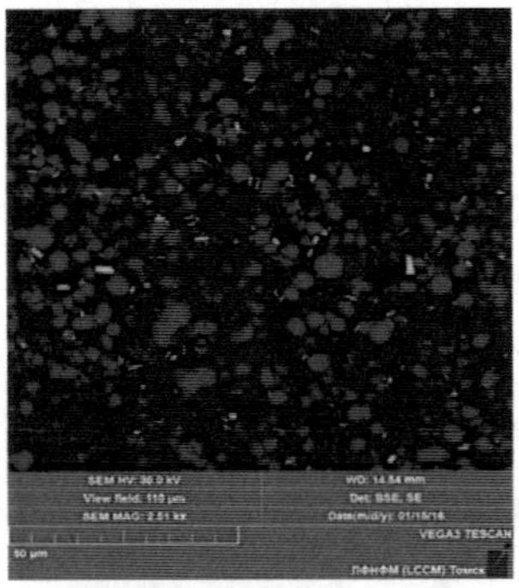

Figure 2. Morphology of Al - ZrW_2O_8 powder mixture after 1 min MA.

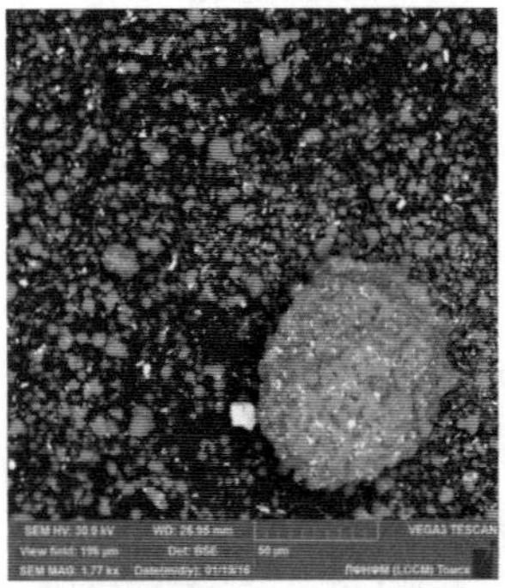

Figure 3. Morphology of Al - ZrW_2O_8 powder mixture after 5 min MA.

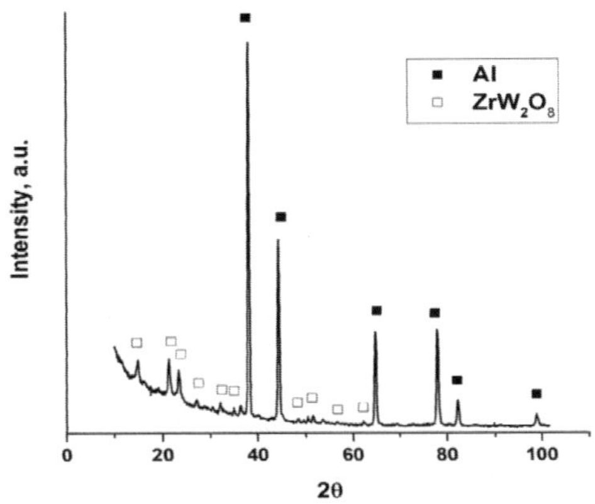

Figure 4. XRD pattern of Al – ZrW_2O_8 powder mixture after 1 minute of MA.

SEM analysis of the powder mixtures after high energy mechanical activation showed that 1 minute of MA time is enough to achieve homogeneous distribution of ZrW_2O_8 particles in aluminum powder, figure 2. Also during high energy mechanical activation there is a reduction in the length of rod-shaped particles to 1.5 μm, and their transverse size does not change. Increasing of the MA duration up to 5 minutes leads to formation of agglomerates, Figure. 3. However, even after aluminum powder agglomeration ZrW_2O_8 distribution remains homogeneous. XRD studies of the powder mixture after mechanical activation showed that the phase composition remains the same and there are no reactions between the mixture components during MA, Figure 4.

3.2. Sintered Al – ZrW_2O_8 samples

In figure 5 the structure of material after 1h of isothermal hold is presented. The structure is not homogeneous, one can observe elongated dense areas, porous matrix and white particles distributed both in the matrix and elongated areas. The content of white particles in matrix and in elongated areas is 3% and 20% correspondingly. The average size of white particles is 0.5 μm. obviously during the sintering process ZrW_2O_8 is redistributed in the matrix. Since after mechanical activation no irregularities in the initial powder mixture are found, one can assume that the formation of inhomogeneous structure takes place during the sintering process.

EDAX analysis showed that white particles consist of Al, W and Zr atoms. On the XRD pattern peaks corresponding to Al and $ZrAl_3$, WAl_{12}intermetallic compounds are presented, Figure 7. Calculated lattice parameters of intermetallics are in a good agreement with ASTM diffraction data cards: a=7. 5803 Å for WAl_{12} (8-331) and a=4.0130 Å for $ZrAl_3$ (2-1093). The lattice parameter of aluminum a= 4.0527 Å is increased in comparison with data card (4-787). The coherent diffraction domain (CDD) values (D) and lattice microdistortion (ε) for aluminum were calculated by Williamson-Hall analysis[5]: D=50 nm, $\varepsilon = 2.5 \cdot 10^{-3}$.

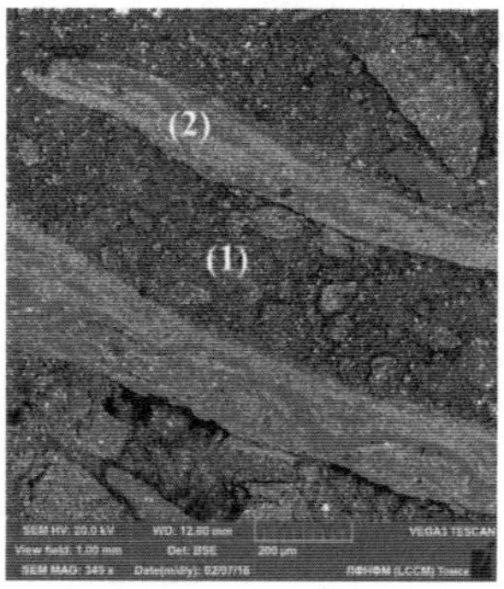

Figure 5. Etched surface of Al – ZrW_2O_8 sintered for 1h

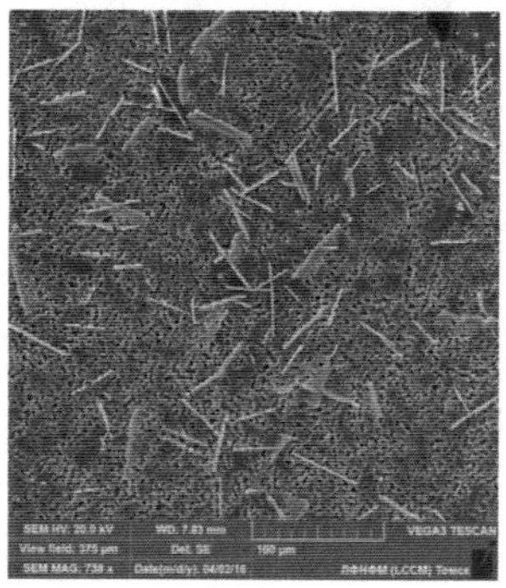

Figure 6. Etched surface of Al – ZrW_2O_8 sintered for 5h

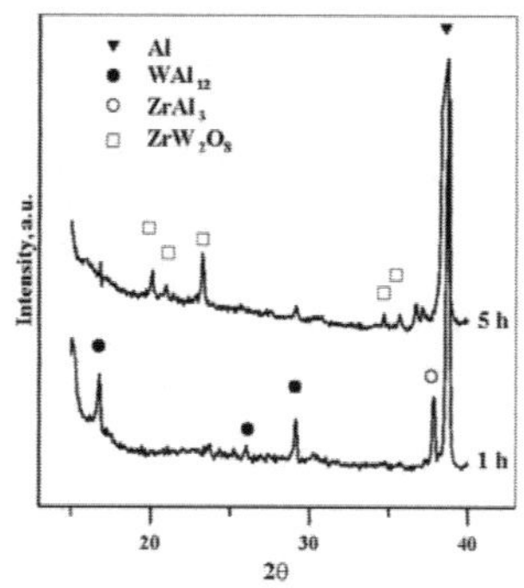

Figure 7. XRD patterns of Al – ZrW_2O_8 specimens

After 5h of isothermal hold the structure of the material is represented with microfibers with average length of 30 μmhomogeneously distributed in porous matrix, figure 6. On the XRD pattern, figure 6, one can observe peaks corresponding to Al and cubic ZrW_2O_8.Lattice parameter of aluminum increased in comparison with 1h isothermal hold, a= 4.0544 Å. ZrW_2O_8 lattice parameter a=9.2187 Å is increased in comparison with literature data [6]. CDD value and lattice microdistortions for Al are D=500 nm and $\varepsilon = 7 \cdot 10^{-3}$correspondingly.

The data obtained using scanning electron microscopy and XRD analysis allow us to suggest that formation of the material proceeds through decomposition of zirconium tungstate into constituent oxides WO_3 and ZrO_2 and synthesis of WAl_{12} and $ZrAl_3$ intermetallic compounds after 1 hour of isothermal hold. According to [7] WAl_{12} may be produced by exothermic*in situ* reaction between Al and WO_3. The generated heat may cause heating of aluminum higher than the melting point and subsequent cooling of the mixture leads to WAl_{12} crystallization. Formation of $ZrAl_3$ [8]goes on in a similar way.The further increase of isothermal holding time up to 5h leads to subsequent formation of ZrW_2O_8 microfibers. Lattice parameter and microdistortions of aluminum increase with increasing isothermal holding time, which may indicate the formation of internal stresses in the material.

4. Conclusions

It was shown that during sintering of Al - ZrW_2O_8powder mixture zirconium tungstate decomposes into its constituent oxides WO_3 and $ZrO_2$with subsequent formation of $WAl_{12}$and $ZrAl_3$intermetallic compounds. Increasing the isothermal holding time up to 5 hours leads to re-synthesis of the zirconium tungstate represented with elongated particles – microfibers. The formation of internal stresses in the material was shown using XRD and Williamson-Hall analyses.

5. Acknowledgments
This work was partially supported by agreement with the Ministry of Education 14.575.21.0040 (RFMEFI57514X0040)

References

[1] Evans J. S. O. Negative thermal expansion materials. The Royal Society of Chemistry, Dalton Trans., 1999, p. 3317 – 3326.
[2] Wu G., Zhou C., Zhang Q., Pei R. Decomposition of ZrW_2O_8 in Al matrix and the influence of heat treatment on ZrW_2O_8/Al–Si thermal expansion, Scripta Materialia, 2015, v 96, p. 29-32.
[3] Dedova E.S., Shadrin V.S., Gubanov A.I., Kulkov S.N. The Preparation and Structural Features of Zirconium Tungstate Possessing Abnormal Thermal Properties, Inorganic Materials: Applied Research, 2014, v. 5, No. 5, p. 471–475.
[4] Kulkov S.N., Dedova E.S., Pedraza F. Erdelyi J. The Influence of Temperature on the Properties of ZrW_2O_8, Epitoanyag - JSBCM, 2014, v.66, No. 2, p. 35-37.
[5] Williamson G.K., Hall W.H. X-ray line broadening from filed aluminium and wolfram, Acta Metallurgica, 1953, v. 1(1), p. 22-31.
[6] Mary T. A., Evans J. S. O., Vogt T., Sleight A.W. Negative thermal expansion from 0.3 to 1050 Kelvin in ZrW_2O_8. Science, 1996, v. 272, p. 90 – 92.
[7] Fan G.H., Wang Q.W., Geng L., Wang G.S., Feng Y.C. Preparation and characterization of aluminum matrix composites based on Al–WO3 system, Journal of Alloys and Compounds, 2012, v. 545, p. 130-134
[8] Laik A., Bhanumurthy K., Kale G.B. Intermetallics in the Al – Zr diffusion zone, Intermetallics, 2004, v. 12, p. 69-74.

Polycarbonate melt shear viscosity in wide range shear rate

G Toth[1], A Bata[1] and K Belina[1]

[1]Department of Materials Technology, GAMF Faculty of Mechanical Engineering and IT, Pallasz Athéné University, Kecskemét, Hungary

E-mail: togoagc@gmail.com

Abstract. In our work we present an investigation of polycarbonate melt shear viscosity in wide range shear rate. In the course of our research we aggregated more testing method results and mathematically determined viscosity equation of the tested material.

1. Introduction

The majority of the processing thermoplastic polymers is in melt condition. For example, extrusion, or injection moulding which are the two biggest plastic processing technology. Nowadays there is a need to simulate these processes. The simulations are help for the plastic product and mould designers. In these simulations one of the most important internal parameter the shear viscosity what depend on the temperature, pressure, molecular weight or structure.So easy to see how a top priority precise knowledge of the shear viscosity of plastic melts [1].

The shear viscosity basically is a rheological property. Some references are describing the rheology than sciencewhich describe the deformation of materials. For a more detailed descriptionthe rheology establishes the relationship between a body deformation and the tensionsand their changes over time. Also explore the dynamic (stress) and kinematics (deformation) variables linking the material characteristics of the status and structure of the material impact [2].In this paper we narrow our experimental to research rheology area.

The polymer materials science one of the youngest sciences, but it serves with more and more results for over 40 years thanks to a variety of polymers, material structure and chemical structure. Day by day there is continual development of new equipment, the researchers get more accurate, faster, more easily interpretable results. In parallel with the rheological studies show evidence for these processes are constantly adding new results.

In our work we tested polycarbonate (DOW Calibre 303 EP) melt shear viscosity in wide range of shear rate (0.05 s-1 – 240 000 s-1), with variable temperature steps. It is a thermoplastic water clear polycondensation type plastic.In the course of our research we summarized the results of several test methods are sections covering each other in good agreement. The procedures used were the rotational viscometry and measurement techniques based on the principle of capillary viscometry. After evaluating the results of those tests, information is givenabout the tested polycarbonate shear viscosity dependence on the shear rate and temperature.From the crude results we are defined the Cross viscosity equations (with three variables), on the temperature steps.The results are aggregated and summarized in a diagram. It showsviscosity curves in wide range shear rate around the processing temperature.

2. Method

Our results we got from three shear viscosity investigation methods, which covers a wide range of shear rate and temperature steps near the processing temperature. These were in the low shear rate range of the rotational viscometry (0,05 – 30 s^{-1}), in the middle-high range of the capillary viscometry (100 – 10.000 s^{-1}) and in the high range we used a special technique(10.000 – 240.000 s^{-1}) which was based on the capillary viscometry method.

2.1. Rotational viscometry

With the rotational viscometry we could measure torsion flow in polymer melt between parallel plates. The measuring technique conceptual layoutis shown in Figure 1. The CMT mean combined motor and transducer concept.

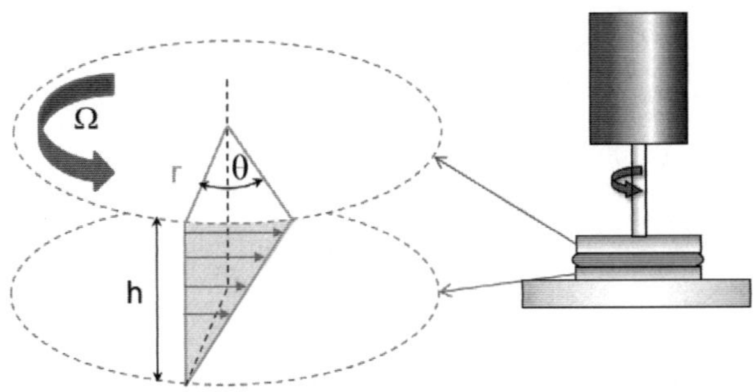

Figure 1. Rotation viscometry CMTconceptual layout and marking system [3]

From the marking system the stress, deformation and deformation rate components can be calculatingwith the equations described below (1, 2, 3).

$$\sigma = \frac{2}{\pi r^3} \times M \tag{1}$$

$$\gamma = \frac{r}{h} \times \theta \tag{2}$$

$$\dot{\gamma} = \frac{r}{h} \times \Omega \tag{3}$$

Where:

r	-	circle plate radius;
h	-	gap between the plates;
θ	-	driving engine rotation (rad);
Ω	-	driving engine angel speed (rad/s)
M	-	torque (μNm)

With the equations (1, 2, 3) the stress and deformation rate can be calculated form these considering the shear viscosity definable. It is a frequent measuring technique to determine shear viscosity for small molecular liquids, or oils and thermoplasts at low shear rate. With this method we can measure on higher temperature (to 600°C) and perception degradation process.

For our testing, we used SMT (Separate motor and transducer) conceptual TA Ares G2 type rotation viscometer with 25mm parallel circle plates and 0,4mm gap. In this type of rotation viscometer, the drive engine and the transducerare separated.

2.2. Capillary rheometer

The capillary rheometer is a measuring device in which we are able to measure thermoplastics shear viscosity.

50-10 000 s^{-1}shear rate ranges near to the processing temperature [4]. We used a Goettfert Rheograph 25 capillary rheometer. The measuring method is based oncontinuous melt flow in known geometry capillary(s) while the pressure drops measured between capillary inlet side and atmospheric pressure (Figure 2.).

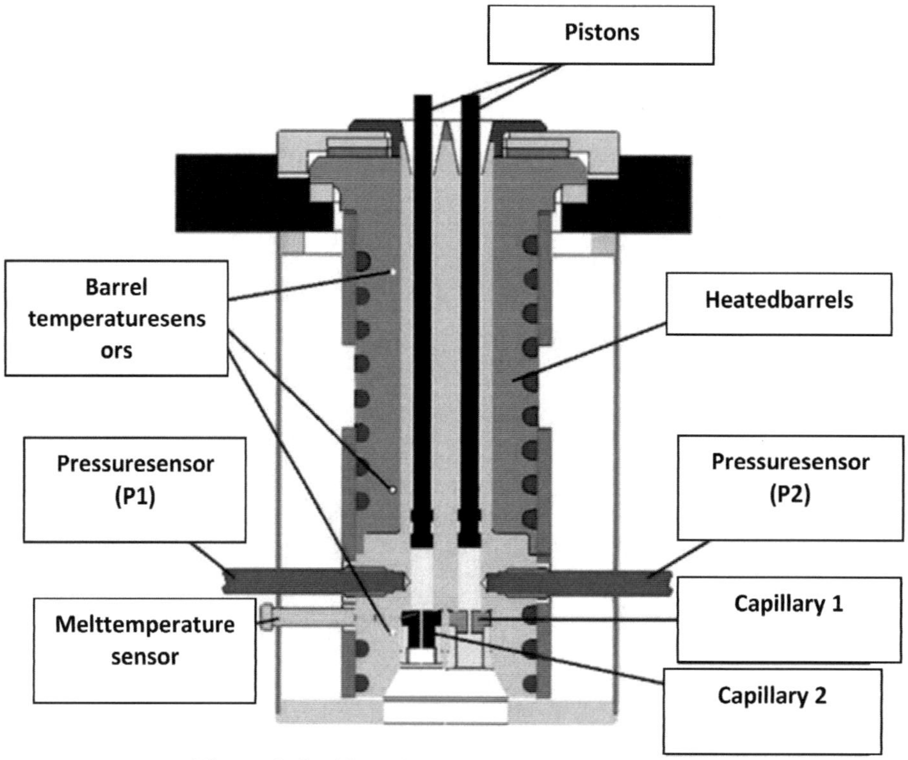

Figure 2. Capillary rheometerconceptual layout [4]

The variable parameters are the barrel temperature and piston speed (parallel with the deformation rate). In this case, the shear viscosity depend on temperature and deformation (shear) rate. In our tests the shear rate changed between 100-10.000 $s^{-1}$and the barrel temperatures changed in three steps around the material supplier recommended process temperature.The measuring method use two other geometry capillaries because of the entrance and exit effects (pressure drop). With this equipment in one step we can do the Bagley correction (correct the pressure drop) which gives the real shear viscosity.

2.3. Special capillary technique

This special capillary technique equations are the same as the capillary viscometry, but we used an injection machine's injection unit for the measurement, with changeable piston (screw) speed and barrel temperatures. An injection units piston can be faster than a measuring equipment and work on higher pressures. We used an ENGEL Victory 1050/300 injection-moulding machine and two special geometry capillary nozzles. In this case the injection machine was used as a capillary rheometer. With the injection-moulding machine repeatedly plasticating and injecting with varied injection settings through of the nozzles into the open air. Variables adjusted were for example the injection speed and the temperatures of the barrel zones. With this method the effect of temperature and injection speed on the viscosity can be shown.

The capillaries used were very important in terms of measurement because this gave us the possibility to calculate the corrected viscosity parameters. The main difference between the two capillaries were the length and diameter ratio. The diameter is unified (2mm) and the length was changed, one is very short just 2mm and another 20mm (Figure 3).

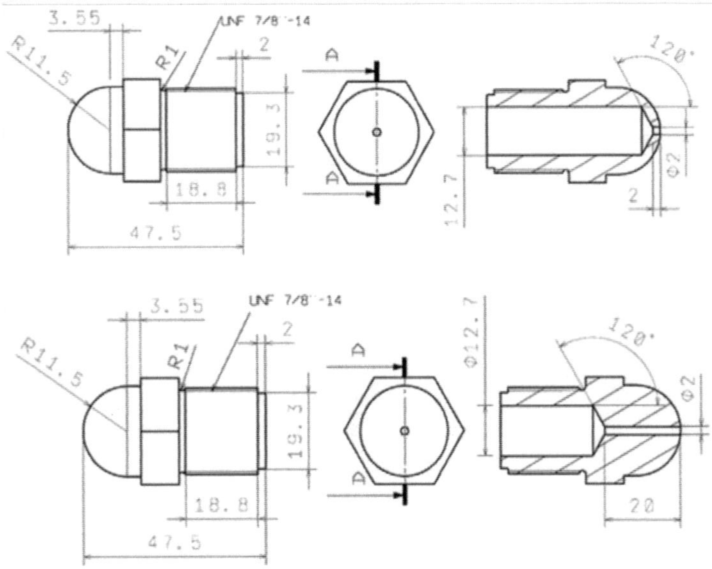

Figure 3. Geometry of the capillaries [5]

This difference is necessary for applying the following Bagley correction. During capillary rheometer tests, pressure is measured above the die inlet. The true pressure drop along the capillary is therefore 'hidden' by an additional pressure drop at the entrance of the die, where the flowing material goes from a wide reservoir (the main cylinder or barrel) to a narrow capillary, possibly also creating turbulences. Assuming that the same additional pressure drop takes place with different capillary lengths (but keeping constant the barrel and capillary diameters, and inlet shapes), it's possible to correct the pressure reading and estimate much more accurately the true pressure drop [7].

3. Results

Measurements were carried out at three different temperatures. Figure 4 shows the viscosity curves measured by different method. It can be clearly seen that the whole measured range covers almost eight scales.

The primer data were processed by the Cross equation [6]:

$$\eta = \frac{\eta_0}{1 + \left(\frac{\eta_0 \dot{\gamma}}{\tau^*} \right)^{[1-n]}} \tag{4}$$

where

η	-	viscosity (Pa*s);
$\dot{\gamma}$	-	deformation rate (s^{-1})
η_0	-	zero shear viscosity (Pa*s);

τ^* - stress limit between the Newtonian and non-Newtonian region (Pa)

$[1-n]$ - slope of the transient region (-)

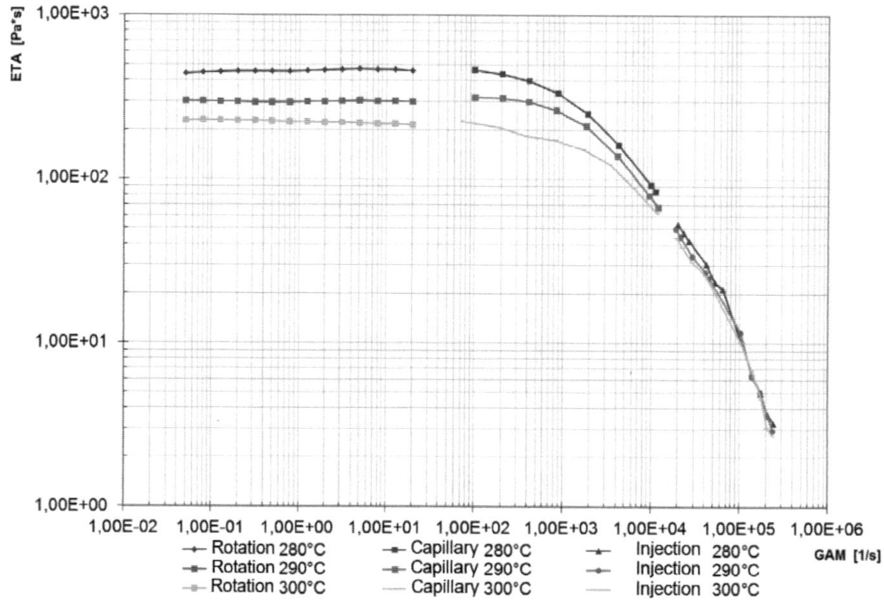

Figure 4. Shear viscosity curves separated by measuring methods

According to our results, the different methods can be merged, and the resulted viscosity curve describes the shear flow of a polymeric melt in very wide deformation rate range.

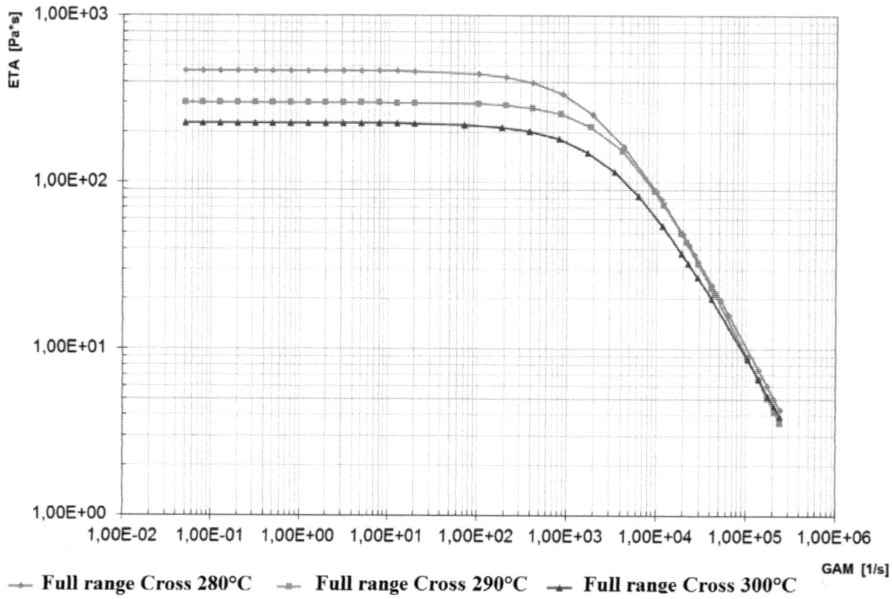

Figure 5. Shear viscosity curves calculated by Cross equation

The main advantage of this method is in the simulation. Using only one measuring method, the extrapolation gives around 15% deviation from the measured values. In our method the maximum deviation was only 5%.

4. Conclusion

Our measuring techniques can be used in the range 0.05 1/s – 240 000 1/s. Rotational viscosity covers the Newtonian region of flow [6-8], the capillary viscometry covers the transient region, the instrumented mould covers the non-Newtonian region. It was shown that the heat generated by the shear force has significant effect on the measuring temperature from around 50 000 1/s. Cross equation describes well the overall viscosity curve shows less than 5% deviation.

Acknowledgments

This project consumed huge amount of work, research and dedication. Still, implementation would not have been possible if we did not have a support of many individuals and organizations. Therefore, we would like to extend our sincere gratitude to all of them. We are also grateful to University of Miskolc, Department of Polymer Technology for provision of expertise, and technical support in the implementation. Without their superior knowledge and experience, the Project would lack in quality of outcomes, and thus their support has been essential. Nevertheless, we express our gratitude toward our families and colleagues for their kind co-operation and encouragement, which help us in completion of this project.

References

[1] T Czvikovszky, P Nagy and János Gaal 2004 A polimertechnika alapjai, *Műegyetemi Kiadó*, Budapest http://www.tankonyvtar.hu/hu/tartalom/tkt/polimertechnika-alapjai/adatok.html
[2] L Halasz, I Molnar and.I Mondva 1978 A polimerek feldolgozásának reológiai alapjai, *Műszaki Könyvkiadó*, Budapest (in Hungarian)
[3] TA instruments factory Ares G2 rotational viscometer equipment education material – Study of rheology theory and applications; http://www.tainstruments.com
[4] A Szűcs 2010 Műanyag ömledékek nagysebességű áramlásának tanulmányozása, PhD thesis, *University of Miskolc* (in Hungarian). http://www.kerpely.uni-miskolc.hu/downloads/docs/cikkek/szucsandras-ertekezes.pdf
[5] G Toth and T Szabó 2015 Polymer melt viscosity measuring by an injection machine *GARDUS*; **2** (2) ISSN: 2064-8014; 112 http://gradus.kefo.hu/index.php/gradus/article/download/203/218
[6] F.N.Cogswell 1981 Polymer Melt Rheology, London
[7] G E Baumann and S Steingiser 1963 *Journal of Plymer Sci. Part* A **1** 3395
[8] F Yang 1996 *Polymer Eng. and Sci.* **37** (1)

IOP Conf. Series: Materials Science and Engineering **175** (2017) 012057 doi:10.1088/1757-899X/175/1/012057

Micromagnetic investigation of low-symmetry 3D particles

T Blachowicz[1] and A Ehrmann[2]

[1]Institute of Physics – Center for Science and Education, Silesian University of Technology, 44-100 Gliwice, Poland
[2]Faculty of Engineering and Mathematics, Bielefeld University of Applied Sciences, 33619 Bielefeld, Germany

E-mail: tomasz.blachowicz@polsl.pl

Abstract. Investigating the anisotropies of magnetic nanoparticles is crucial for further development of magnetic data storage media, MRAM, magnetic logical circuits, or magnetic quantum cellular automata. Former theoretical and experimental examinations have revealed the possibility to gain highly symmetric nanoparticles with increased numbers of magnetic states per storage element. In a recent project, we have investigated low-symmetry T-shaped 2D and 3D particles from iron using the micromagnetic simulation software MAGPAR which is based on solving the Landau-Lifshitz-Gilbert (LLG) equation of motion for a mesh built from tetrahedral finite elements. To examine the influence of the reduced symmetry, simulations were performed on the 3D double-T particle with the field applied in different directions in the x-y base plane, ranging from 0 to 180° in 5° steps. Additionally, the external magnetic field was rotated laterally under different angles with respect to the x-y plane, i.e. 5°, 22.5°, and 45°. Similar simulations were executed for the 2D single-T particle. Our results show the strong impact of the shape anisotropy and the respective possibility to tailor magnetic anisotropies according to the desired behaviour by modifying the nanoparticles' form.

1. Introduction

Magnetic nanoparticles can be used for several applications, including magneto-electronic devices. In these low-dimensional structures, the shape anisotropy dominates over magneto-crystalline and magneto-elastic anisotropies, enabling tailoring the overall anisotropies by designing the geometry of the nanoparticle properly [1]. Magnetic nanostructures can also depict exotic magnetic states, such as flux-closed vortex states with significantly reduced stray fields in nano-rings [2,3]. Perpendicular rings can also exhibit stable onion states [4,5], while ellipses show onion, horseshoe, vortex, and domain wall states [6,7]. 3D-forms, such as half-spheres, tend to mixed in-plane and out-of-plane magnetization components [8-10].

Square magnetic nanoparticles show another special feature: Such systems can have stable intermediate states at remanence [11], as it is also known from exchange bias systems [12], allowing to use them as quaternary memory cells with 2 bits per storage position. This property depends on the geometrical dimensions of the nanoparticles, as shown in theory and experiment [13,14].

Since symmetric 2D nanoparticles with higher numbers of edges, such as sixfold or eightfold particles, only offer instable intermediate states which cannot be used for practical applications [15], it would be interesting to investigate asymmetric and three-dimensional magnetic nanoparticles. Our article thus aims at simulating two-dimensional T-shaped as well as three-dimensional double-T-shaped particles under different angles to get an insight into the impact of such symmetry breaks.

2. Methods

Magnetic nanoparticles were modelled using the micromagnetic simulation program Magpar, based on the finite element method [16]. Dynamical integration of the Landau-Lifshitz-Gilbert equation of motion is used in this program to simulate hysteresis loops.

For the simulation of the nanoparticles depicted here, mesh dimensions of maximum 2 nm were used. The dimensions of the 2D T-shaped nanoparticle are 250 nm x 250 nm x 10 nm; for the 3D particle, two of these particles were combined under an angle of 90°. Both systems are depicted in Fig. 1. As material, we chose iron (Fe) with an exchange constant of A_{Fe} = $2.0 \cdot 10^{-11}$ J/m, magnetic polarization at saturation J_{Fe} = 2.1 T and Gilbert damping constant α = 0.01. The y-axis is directed along the upper part of the T, the x-axis is directed along the base part of the T, while the second T in the 3D case is directed along negative z-direction. The out-of-plane angle θ is measured with respect to the surface normal (z-axis), while the in-plane angle ϕ is measured with respect to the x-axis.

The simulation was performed sweeping from zero external field up to +600 kA/m (with the external field in the sample plane or along an out-of-plane direction specified in the respective graph) and back up to -600 kA/m and to + 600 kA/m again, using a field sweeping speed of 10 kA/(m ns).

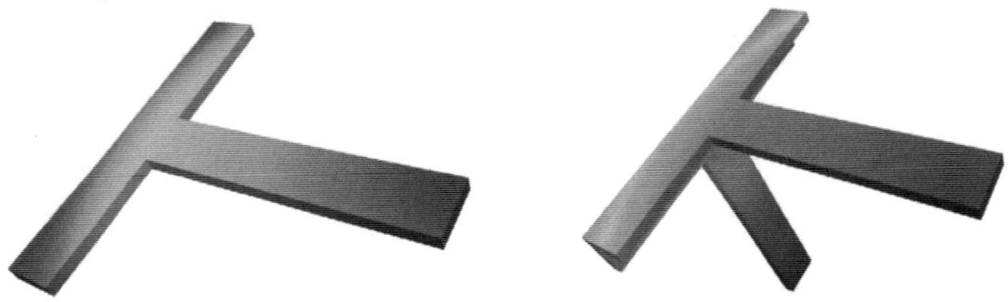

Figure 1. Geometries of simulated 2D T-shaped nanoparticle (left panel) and of 3D double-T-shaped nanoparticle (right panel).

3. Results

A necessary property of a hysteresis loop which shows four stable states at remanence is a step in the hysteresis loop, resulting from partly switched magnetization [11]. Fig. 2 depicts the results of in-plane measurements of the 2D nanoparticle for four different in-plane angles of the external magnetic field, measured with respect to the x-axis. Although a small step is visible in all cases at vanishing external magnetic field, there are only two stable states at remanence (those with smaller magnetization). Apparently in the 2D case the asymmetry of the T-shape does not support creation of additional stable states at vanishing external field.

In the 3D case, however, more degrees of freedom are available. Since optical as well as electrical detection methods allow measuring the magnetization in-plane or perpendicular to the plane with relatively simple means and intermediate angles necessitating more sophisticated methods, the most use-oriented measurement directions are out-of-plane (perpendicular to the sample plane) and in-plane along the magnetization direction, along the x- and the y-axis.

The results of these different evaluation methods is depicted in Fig. 3 for the double-T-shaped particle with the magnetization at an out-of-plane angle of 22.5° to the surface normal and an in-plane angle of 5°. Obviously, the highest magnetization is reached, approaching saturation, for the out-of-plane case, which fits to the external magnetic field being oriented strongly out-of-plane. However, in this hysteresis loop (blue line), no steps are visible which could indicate additional stable states at remanence. Such steps are only visible in the measurement along the y-direction. This shows that it may be helpful to examine different possible measurement directions to find special features.

For a "symmetric" out-of-plane angle of 45°, i.e. the magnetic field being oriented in the middle between both parts of the double-T, most in-plane angles do not reveal any indication for steps in the hysteresis loops. For an in-plane angle of 90° (Fig. 4), evaluations along the x- or y-direction as well as along the z-direction (not shown here) only show a small area where "ringing" occurs, a precession of the magnetization typical after switching processes. Evaluation along an intermediate angle (here $\phi = 45°$), however, shows clear steps stemming from partial magnetization reversal processes. Thus, the system could be used in this orientation to create quaternary memory cells.

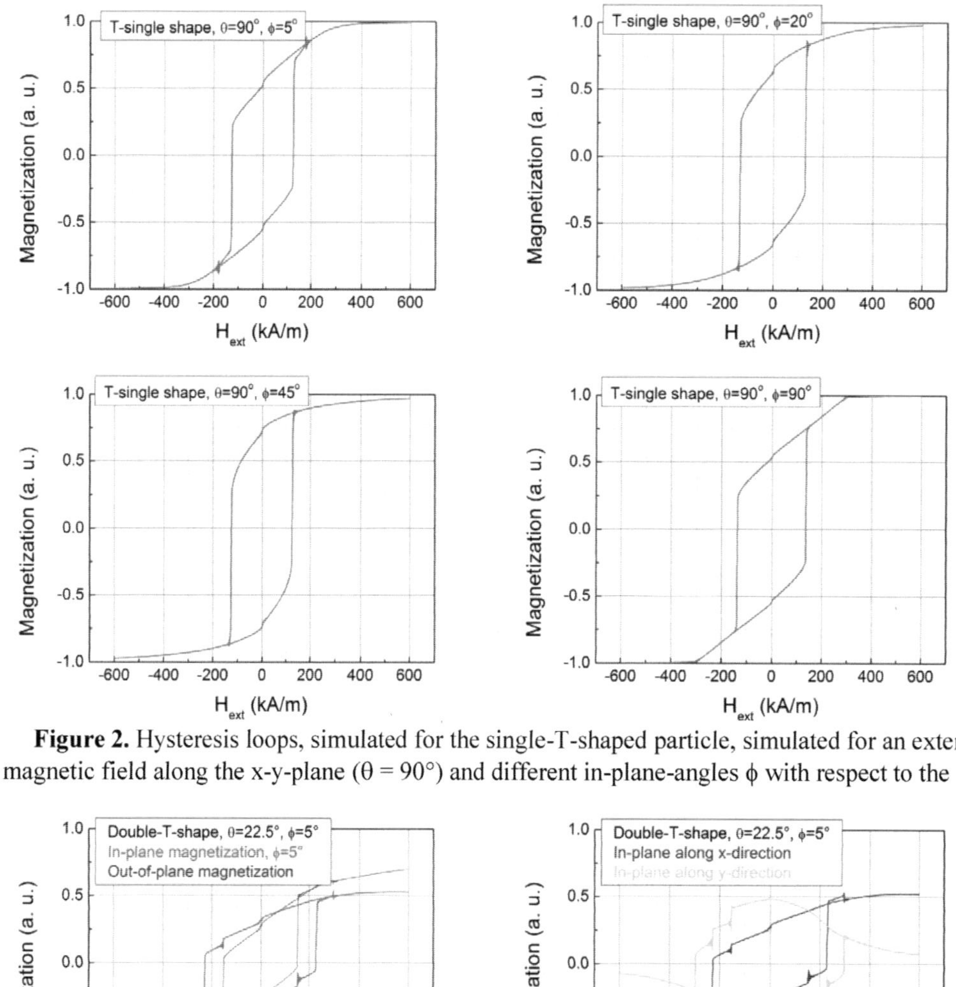

Figure 2. Hysteresis loops, simulated for the single-T-shaped particle, simulated for an external magnetic field along the x-y-plane ($\theta = 90°$) and different in-plane-angles ϕ with respect to the x-axis.

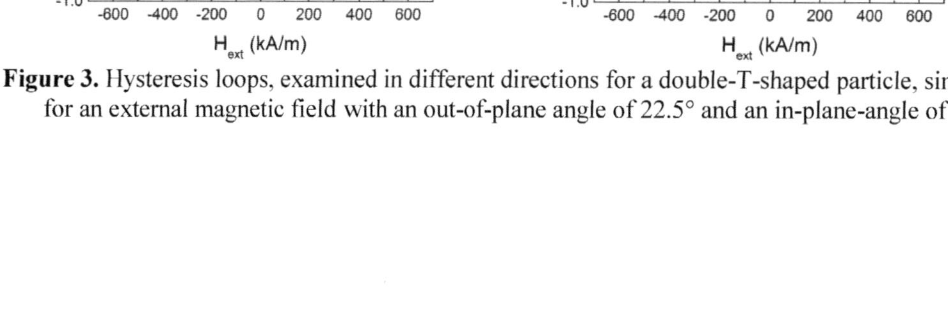

Figure 3. Hysteresis loops, examined in different directions for a double-T-shaped particle, simulated for an external magnetic field with an out-of-plane angle of 22.5° and an in-plane-angle of 5°.

While systems with one step can be created as 2D forms [11,14], 3D forms should offer a significant advantage to compensate the increased effort to create them. In Fig. 3, a system with two steps, i.e. 6 possible states at remanence, was depicted. All other systems, tested at out-of-plane angles of 5°, 22.5° and 45° with respect to the z-axis and in-plane angles of 0°, 5°, 20°, 45°, and 90° with respect to the x-axis, did not reveal signs of more than these two steps.

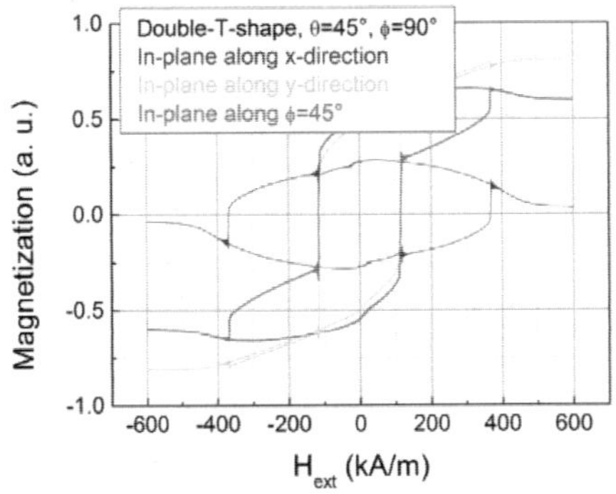

Figure 4. Hysteresis loops, simulated for the double-T-shaped particle, simulated for an external magnetic field with an out-of-plane angle of 22.5° and an in-plane-angle of 5°.

4. Conclusion

To conclude, we have shown that using asymmetric nanoparticles with 3D shapes can result in hysteresis loops with one or two steps on each side of the loop, corresponding with additional states at remanence. The stability of these additional states against small changes of the external field and erroneous shape modifications has to be tested carefully in future micromagnetic simulations.

Future simulations can investigate more 3D shapes to gain an insight into the mechanisms leading to additional stable states at remanence, finally resulting in a possibility to maximize the number of such stable states and thus the data storage capacity of a single nanoparticle.

5. Acknowledgements

This work was partially supported by the internal project BK-243/RIF/2016 of the Institute of Physics – CSE, Silesian University of Technology.

6. References

[1] Nogués J, Sort J, Langlais V, Skumryev V, Surinach S, Munoz JS, and Baro MD 2005 *Phys. Rep.* **422** 65
[2] Zhang W and Haas S 2010 *Phys. Rev. B* **81** 064433
[3] Zhu FQ, Fan DL, Zhu XC, Zhu JG, Cammarata RC, and Chien CL 2004 *Adv. Mater.* **16** 2155
[4] Subramani A, Geerpuram D, Domanowski A, Baskaran V, and Metlushko V 2004 *Physica C* **404** 241
[5] Wang J, Adeyeye AO, and Singh N 2005 *Appl. Phys. Lett.* **87** 262508
[6] Gao XS, Adeyeye AO, Goolaup S, Singh N, Jung W, Castaño FJ and Ross CA 2007 *J. Appl. Phys.* **101** 09F505
[7] Vavassori P, Grimsditch M, Novosad V, Metlushko V, and Ilic B 2003 *Phys. Rev. B* **67** 134429
[8] Soares MM, de Biasi E, Coelho LN, dos Santos MC, de Menezes FS, Knobel M, Sampaio LC, and Garcia F 2008 *Phys. Rev. B* **77** 224405
[9] Leong TG, Zarafshar AM, and Gracias DH 2010 *Small* **6** 792
[10] Amaladass E, Ludescher B, Schütz G, Tyliszczak T, Lee MS, and Eimüller T 2010 *J. Appl. Phys.* **107** 053911
[11] Blachowicz T and Ehrmann A 2011 *J. Appl. Phys.* **110** 073911
[12] Tillmanns A, Oertker S, Beschoten B, Güntherodt G, Leighton C, Schuller IK, and Nogués J 2006 *Appl. Phys. Lett.* **89** 202512
[13] Blachowicz T, Ehrmann A, Steblinski P, and Palka J 2013 *J. Appl. Phys.* **113** 013901

IOP Conf. Series: Materials Science and Engineering **175** (2017) 012057 doi:10.1088/1757-899X/175/1/012057

[14] Ehrmann A, Blachowicz T, Komraus S, Nees MK, Jakobs PJ, Leiste H, Mathes M, and Schaarschmidt M 2015 *J. Appl. Phys.* **117** 173903
[15] Blachowicz T, Ehrmann A 2013 *J. Magn. Magn. Mat.* **331** 21-23
[16] Scholz W, Fidler J, Schrefl T, Suess D, Dittrich R, Forster H and Tsiantos V 2003 *Comp. Mat. Sci.* **28** 366

Development and characterization of textile batteries

M Normann[1], T Grethe[1], A Schwarz-Pfeiffer[1], and A Ehrmann[2]

[1]Faculty of Textile and Clothing Technology, Niederrhein University of Applied Sciences, 41065 Mönchengladbach, Germany
[2] Faculty of Engineering and Mathematics, Bielefeld University of Applied Sciences, 33619 Bielefeld, Germany

E-mail: anne.schwarz-pfeiffer@hsnr.de

Abstract. During the past years, smart textiles have gained more and more attention. Products cover a broad range of possible applications, from fashion items such as LED garments to sensory shirts detecting vital signs to clothes with included electrical stimulation of muscles. For all electrical or electronic features included in garments, a power supply is needed – which is usually the bottleneck in the development of smart textiles, since common power supplies are not flexible and often not lightweight, prohibiting their unobtrusive integration in electronic textiles. In a recent project, textile-based batteries are developed. For this, metallized woven fabrics (e.g. copper, zinc, or silver) are used in combinations with carbon fabrics. The article gives an overview of our recent advances in optimizing power storage capacity and durability of the textile batteries by tailoring the gel-electrolyte. The gel-electrolyte is modified with respect to thickness and electrolyte concentration; additionally, the influence of additives on the long-time stability of the batteries is examined.

1. Introduction

Smart textiles have increasingly attracted interest in recent years by textile designers and technologists. Since they often contain electronic components, most of them need a power supply.

Electric energy can be stored in textiles using common rigid batteries or capacitors.However, due to the limitations imposed by integrated hard objects on textile fabrics in terms of haptics, drape or bendability, several approaches have been made to develop flexible textile-based energy sources. Besides textile capacitors and pseudo-capacitors, textile batteries are often investigated [1]. Capacitors and batteries can be created by layer systems coated on textile substrates [2] or in fiber-shape, the latter capable of being integrated into woven or knitted fabrics [3-5]. While early approaches concentrated, e.g., on silver electrodes combined with PEDOT as an electrolyte [6] to build rechargeable textile batteries, other researchers showed the advantage of carbon nanotubes as electrode material [7] or used similar materials as in common rigid batteries, such as lithium-iron-phosphate and lithium titanate [5].

Such textile batteries, however, usually lack the required energy densities and lifetimes. Additionally, as well-known from various research initiatives in the area of smart textiles, results can often not be reproduced with the necessary reliability. This is whythe recent project focuses on studying the influence of electrolyte thickness and concentration as well as the electrode materials as a base for further optimization of 2D textile batteries.

2. Experimental

A battery consists at least of two electrodes produced from different materials, separated by an electrolyte. The voltage between the electrodes depends on the difference between their redox potentials as given be the Nernst equation as well assecondary effects like over-potentials. This voltage is triggeredby a redox reaction in the electrolyte with different partial reactions near both electrodes. A current can flow when these areas are electrically connected.

Different metals can be combined as electrode pairs; however, the difference between their normal potentials should be high to enable high voltages between them. Some examples of normal potentials are $E_{0,Cu}$ = 0.34 V for copper, $E_{0,Ag}$= 0.79 V for silver, and $E_{0,Ni}$ = -0.23 V for nickel [8], showing that the combination of silver and nickel can result in higher voltage drops than copper-nickel cells Interestingly, carbon has also shown to be an interesting partner in combination with copper or nickel electrodes. As carbon will not take place in the redox reaction, it will only act as a pick-up electrode for a possible redox reaction in the electrolyte itself [9].In this research,copper coated polyester woven fabrics, marketed under the name "copper ripstop fabric" (Less EMF Inc.), silver coated polyamide fabrics, called "Shieldex® Zell" (Statex),and a carbon nonwoven fabric, type HP-U100/50C (HP-Textiles), are utilized.

As electrolyte, iodine-triiodide is used in the form of mixture of iodine and potassium iodide with a concentration of 5 % iodine in aqueous solution. Iodine-triiodide has proven to be more efficient than metal salts, such as copper sulphate or nickel sulphate [10].

This electrolyte can be used to form a gel-like electrolyte to decrease evaporation and drying which would result in a reduced ion conductivityof the electrolyte (Fig. 1). For the gel-electrodes, commercially available gelatine was mixed with the aqueous solution of iodine/potassium iodide at a temperature of 60°C using a stirrer. In the description of the single experiments, the iodine concentration in the aqueous solution is given, ranging from 0.25 % to 1.0 %. It was not possible to form uniform gelatine-gels with higher electrolyte concentration.

Electrolyte layers were produced using a squeegee or hand roller with 500 µm wet film thickness.The resulting dry film thicknesses are slightly increased.

Measurements are performed using adequate multimeters and a resistance decade to allow taking complete I-V-curves.

Figure 1. Textile battery consisting of carbon and copper electrodes, separated by an iodine-triiodide gel-electrode

3. Results

Fig. 2 depicts I-V-curves and P-V-curves of silver-carbon batteries with 0.75 % iodine in form of a gel. Different numbers of gel-electrolyte layers were introduced between both electrodes, resulting in different maximum voltages and currents. The maximum powers gained from these batteries increase with increasing gel-electrolyte thickness. This may be an indication for redox activity inside the electrolyte.

It should be mentioned, however, that increasing the electrolyte thickness is limited due to the increasing bending stiffness of the resulting textile battery. Thus, other parameters should also be taken into account to increase the battery power. Additionally, the increase of voltage and current with increasing electrolyte layer thickness is lower for higher gel-electrolyte thicknesses.

IOP Conf. Series: Materials Science and Engineering **175** (2017) 012058 doi:10.1088/1757-899X/175/1/012058

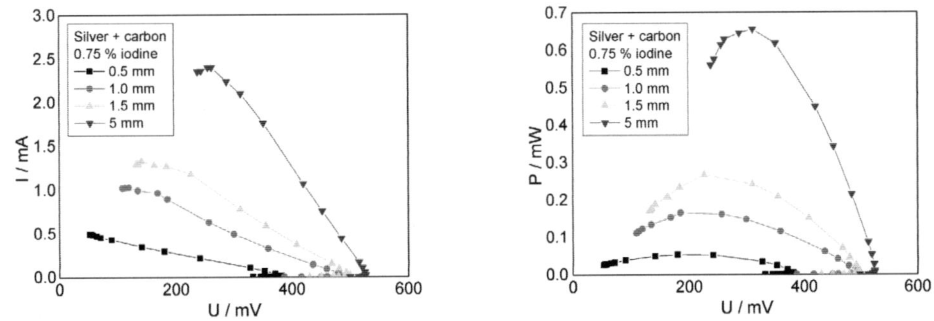

Figure 2. I-V-curves (left panel) and P-V-curves (right panel) of silver-carbon batteries with different gel-electrolyte layer thicknesses. All values are averaged over 5 batteries

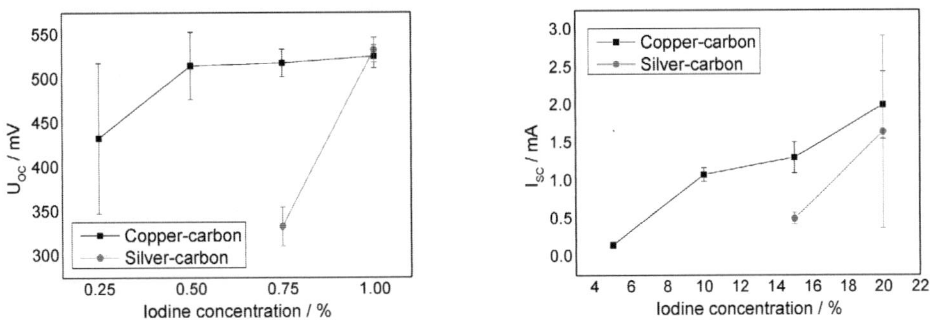

Figure 3. Open-circuit voltages and short-circuit currents, measured for different concentrations of iodine/triiodide in the gel of thickness 0.5 mm for copper-carbon and silver-carbon batteries. All values are averaged over 5 batteries

In Fig. 3, the open-circuit voltages and short-circuit currents for battery series with different parameters are depicted. The open-circuit voltages measured on copper-carbon batteries saturate around 0.5 % iodine in the aqueous solution), while for silver-carbon batteries, a higher concentration of 1.0 % iodineis necessary to reach higher voltages. The short-circuit current is increased for higher iodine-triiodide concentrations for both electrode combinations.Nevertheless, it must be mentioned that the partly large error bars indicate instabilities in the battery performance. While measured current values stabilize with increasing iodine-triiodide concentrations, the opposite effect is visible for the short-circuit currents. Apparently, a compromise must be found to create textile batteries with reliable performance.

Furthermore, increasing the amount of iodine-triiodide is limited to the values used in this research. Higher concentrationsof the electrolyte in the gel resulted in severe problems during gelling, making the layer mechanically instable.

The electrode potential of the cells is mainly influenced by the iodide concentration, since it is linked to the solubility product of AgI:

$$E = E_0 + \frac{RT}{F}\ln[Ag^+]$$

$$K_l = [Ag^+]\cdot[I^-]$$

$$E = E_0 + \frac{RT}{F}\ln\frac{[K_l]}{[I^-]}$$

with K_l: solubility Product, $[A]$: activity of species A, E_0: standard potential, R: gas constant, T: absolute temperature, and F: Faraday constant.

However, as depicted in Fig.2, the available amount of iodine has significant influence on the electric charge, delivered by the system. This gives rise to the idea that iodine also contributes to the electrochemical reaction. Therefore an oxidation of the metal electrodes by iodine can be proposed, but the system is limited by the solubility of the metal iodides.

4. Conclusion

To conclude, we have shown that textile batteries can be created using metalized or metallic electrodes, combined with a carbon electrode. As the electrolyte between both electrodes, iodine-triiodide was used which is quite common in flexible and textile solar cells, as well as batteries.

Gelling this electrolyte results in flexible, bendable textile batteries which can even be used without further treatment for short times, while long-term stability necessitates either encapsulation or addition of stabilizing agents

Future research will focus on further electrolyte materials and different possibilities to solidify them as well as other electrode materials forming iodides of higher solubility.

5. References

[1] Jost K, Dion K and Gogotsi Y 2014 Textile energy storage in perspective, *Journal of Materials Chemistry A***2** 10776-10787

[2] Jost K, Perez C R, McDonough J K, Presser V, Heon M, Dion G and Gogotsi Y 2011 Carbon coated textiles for flexible energy storage, *Energy Environ. Sci.* **4** 5060

[3] Gorgutsa S, Gu J F and Skorobogatiy M 2012 A woven 2D touchpad sensor and a 1Dslide sensor using soft capacitor fibers, *Smart Mater. Struct.***21** 015010

[4] Jost K, Stenger D, Perez C R, McDonough J K, Lian K, Gogotsi Y and Dion G 2013 Knitted and screen printed carbon fiber textile-supercapacitors for applications in wearable electronics, *Energy Environ. Sci.* **6** 2698-2705

[5] Liu Y, Gorgutsa S, Santato C and Skorobogatiy M 2012 Flexible, Solid Electrolyte-Based Lithium Battery Composedof $LiFePO_4$Cathode and $Li_4Ti_5O_{12}$Anode for Applicationsin Smart Textiles, *J. Electrochem. Soc.***159** A349-A356

[6] Bhattacharya R, de Kok M, Zhou J2009 Recharchable electronic textile battery, *Appl. Phys. Lett.***95** 223305

[7] Hu L2010 Stretchable, Porous, and Conductive Energy Textiles,*NANO Lett.***10**708-714

[8] Schwarz-Pfeiffer A, Normann M, Happe B, Ehrmann A 2015 Development of fibre-based energy sources, *Proc. of Aachen-Dresden International Textile Conference*

[9] Normann M, Herrmann A, Grethe T, Fiedler J, Schwarz-Pfeiffer A, Ehrmann A 2016 Energiegewinnung und -speicherung mittels Textilien, *15th Chemnitz Textile Technology Conference*

[10] Normann M, Kyosev Y, Ehrmann A, Schwarz-Pfeiffer A 2016 Multilayer Textile-Based Woven Batteries, in: Y. Kyosev (Ed.): *Recent Developments in Braiding and Narrow Weaving*, Springer International Publishing, 129-136

IOP Conf. Series: Materials Science and Engineering **175** (2017) 012059 doi:10.1088/1757-899X/175/1/012059

Metamaterial Absorber Based Multifunctional Sensor Application

Z Ozer[1,2], A M Mamedov[3,4*], E Ozbay[3]

[1]MTMs and Photonics Research Group, Mustafa Kemal University, Hatay 31200, TR
[2]Vocational School of Mersin, Mersin University, Mezitli, Mersin 33335, TR
[3] Nanotechnology Research Center (NANOTAM), Bilkent University, Ankara, TR
[4]International Scientific Center, Baku State University, Baku

E-mail: mamedov@bilkent.edu.tr

Abstract. In this study metamaterial based (MA) absorber sensor, integrated with an X-band waveguide, is numerically and experimentally suggested for important application including pressure, density sensing and marble type detecting applications based on rectangular split ring resonator, sensor layer and absorber layer that measures of changing in the dielectric constant and/or the thickness of a sensor layer. Changing of physical, chemical or biological parameters in the sensor layer can be detected by measuring the resonant frequency shifting of metamaterial absorber based sensor. Suggested MA based absorber sensor can be used for medical, biological, agricultural and chemical detecting applications in microwave frequency band.We compare the simulation and experimentally obtained results from the fabricated sample which are good agreement. Simulation results show that the proposed structure can detect the changing of the refractive indexes of different materials via special resonance frequencies, thus it could be said that the MA-based sensors have high sensitivity. Additionally due to the simple and tiny structures it could be adapted to other electronic devices in different sizes.

Keywords: Metamaterial, Absorber sensor, X-band, Finite integration technique.

1. Introduction

Metamaterial absorbers are more susceptible and easily obtainable materials for applications [1-2]. These artificially structures capable of manipulating electromagnetic waves in a desired frequencies represent a promising approach to meet this need in wide range of application areas [3-7].

In this paper, we proposed to evaluate numerically and experimentally density, pressure and marble type sensing capable of the metamaterial absorber (MA). The most important feature of designed MA structure is the resonant frequency that changes due to environmental influences. The resonant frequency shift happens depending on the some physical parameters such as density, pressure and dielectric constant. This resonant frequency shift allowing us to precise detection of parameter change. We have designed the proposed structure and chosen the substrates as well as the working frequency range. Other expectation of the designed sensor can be listed as low engineering cost, low loss factor, resolution, sensitivity and durability that we have taken into account for the proposed design.

2. Numerical Design &Fabrication Techniques

Firstly, as seen in fig. 1a we designed MA-based absorber sensor by FIT based 3D EM solver and simulated the MA-based absorber sensor. We used high frequency RT/duroid® 5870 laminates as a substrate which has 0.76 mm thickness and dielectric constant of 2.33. The MA-based sensor design consist of three layers which has copper plated split ring resonator (SRR) with RT5870 substrate, sensor layer and copper plated RT5870 substrate respectively as seen in Fig. 1. In numerical analysis, open boundary conditions applied in z direction and PEC boundary conditions applied in x-y plane for using this design to different sensing applications. The proposed design can be used in different applications at millimeter wave region in technology such as density, humidity, temperature and pressure. In this study we conducted and discussed density and pressure sensor applications as numerically and experimentally.

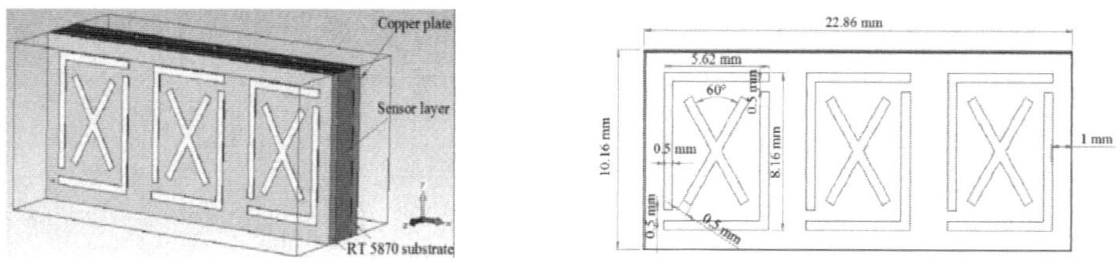

(a) **(b)**

Figure 1. (a) The proposed MA based sensor design (b) sizes of the proposed MA based sensor design

Fig. 1 b shows the dimensions of suggested MA based sensor design. Resonator layer consist of three similar SRRs has cross patch inside. Dimensions of the unit cell are suitable for X-band waveguide; resonators have 1 mm space with the side wall and 0.5 mm patch width.

3. Metamaterial Absorber Based Sensor Applications

This section weprovide details of density and pressure sensor application.In addition this design can be used in other sensor applications by the sensor layer is filled with different materials. According to sensing parameters, for example humidity or temperature, suitable material can be chosen when the environmental conditions affect the dielectric constant of the related material, this is going to change overall reflection constant which is required sensor applications.

3.1. MA Based Density Sensor

One of the distinguishing features of the materials which changes due to the used conditions are density and density sensing is one of the important applications in this study. For density sending application, Arlon 300, Arlon 350 and Arlon 450 materials which have densities 2.07 g/cm^3, 2.40 g/cm^3 and 2.45 g/cm^3 respectively are chosenin this part.Overall reflection coefficient of the unit cell changes by the inserted materials properties in the sensor layer.

Fig.2 shows numerically obtained absorptions values of the different materials calculated via $A(w) = 1 - S_{11}^2$. It's clear that resonance frequencies are linearly changed by inserted materials in sensor layer as 9.5, 9.42 and 9.3 GHz respectively. Fig.2. shows absorptions value is decreasing by the material density while there was a linear relation between the material density and resonance frequencies. The results show that the simulated values of the samples are good agreement of experimental obtained ones.

3.2. MA Based Pressure Sensor

Another sensor application is pressure sensor that we investigate the MA based sensor parameters for pressure sensor application. We assumed that the sensor layer filled with air and thickness of the sensor layer varies with the different pressure values. It's clear from the Fig. 3. the sensors resonant frequency shifts when the different pressure values acting on the sensor by changing layer thickness where changing the capacitance of structure.

Furthermore, marble type sensor application is realized and discussed. This kind of application could be popular in future especially in agricultural product sensor area. This study can be the basis for the development of multi-functional devices by adding extra feature. This structure can be developed for different potential applications such as portable detector devices.

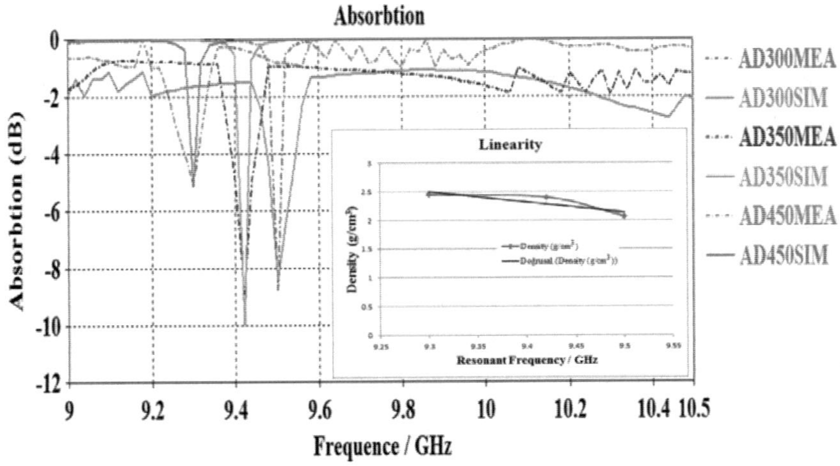

Figure 2. Numerical and experimental absorption values for density sensor application based on MA design

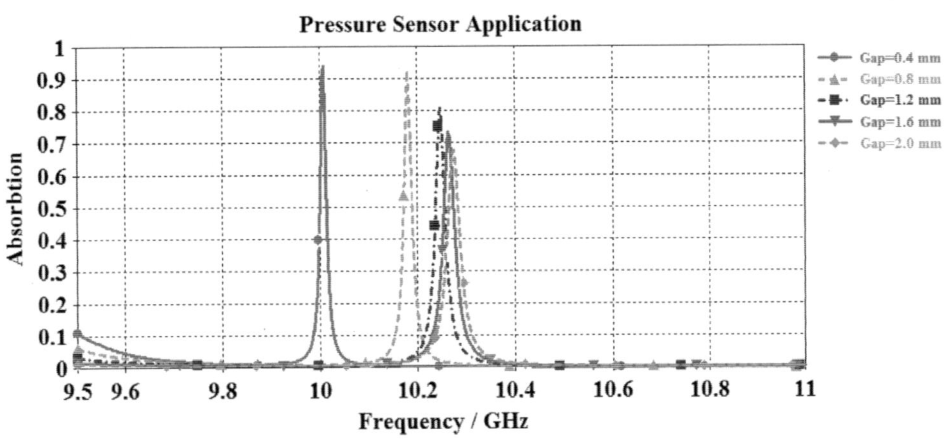

Figure 3. Pressure sensor application of MA based density sensor

4. Conclusion

In this study we investigate the different MA based sensor applications asdensity, pressure and marble type detection applications conducted for analyzing sensing capabilities of the suggested design. In order to show the sensing capabilities of the proposed design, density sensor application has been tested experimentally. It can be seen that experimental results are well consistent with the numerical results. The linearly shifting of the resonant frequency depends on the absorption level. . Simulation results show that proposed sensor can be used for different sensing applications when different kind of materials placed in the sensor layer. These materials that can be selected from linearly changes the results of a sensor design depending on the detection parameters such as humidity, temperature, etc. As an application addition to pressure and density sensor applications, marble type sensor application is performed and analyzed in detail. The result showed that MA based sensor can be used as marble type sensor.

References

[1] Yang A, Yan C C, Tian J B, Wang C, Han Y, Zhang D H, Li D D, Xu Z J 2014 *Appl. Phys. A*, **117** 537
[2] Dincer F, Karaaslan M, Unal E, Akgol O and Sabah C 2014 *Journal Of Electronic Materials*. **43** 11
[3] Sabah C, Dincer F, Karaaslan M, Unal E, Akgol O and Demirel E 2014 *Optics Communications*. **322** 137
[4] Kong H, Li G, Jin Z, Ma G, Zhang Z and Zhang C 2012 *J Infrared Milli Terahz Wave*. **33** 649
[5] Ozer Z, Dincer F, Karaaslan M and Akgol O 2014 *Optical Engineering*. **53(7)** 0751090
[6] Akturk C, Karaaslan M, Ozdemir E, Ozkaner V, Dincer F, Bakir M and Ozer Z 2015 *Optical Engineering*. **54(3)** 035106
[7] Grant J, Carranza I E, Li C, McCrindle I J H, Gough J and Cumming D R S 2013 *Laser Photonics Rev*. 7 1043

IOP Conf. Series: Materials Science and Engineering **175** (2017) 012060 doi:10.1088/1757-899X/175/1/012060

Evaluation of the shrinkage and creep of medium strength self compacting concrete

C J De La Cruz[1], G Ramos[2] and W A Hurtado[1]

[1]Universidad Nacional de Colombia, Facultad de Minas - Medellín, Colombia
[2]Universitat Politécnica de Catalunya, Barcelona, España

E-mail: cjcruz@unal.edu.co, gonzalo.ramos@upc.es and wahurtad@unal.edu.co

Abstract. The difference between self compacting concrete (SCC) and conventional concrete (CC) is in fresh state, is the high fluidity at first and the need for vibration at second, but in hardened state, both concretes must comply with the resistance specified, in addition to securing the safety and functionality for which it was designed. This article describes the tests and results for shrinkage and creep at some medium strength Self Compacting Concrete with added sand (SCC-MSs) and two types of cement. The research was conducted at the Laboratorio de Tecnología de Estructuras (LTE) of the Universitat Politécnica de Catalunya (UPC), in dosages of 200 liters; with the idea of evaluating the effectiveness of implementation of these new concretes at elements designed with conventional concrete (CCs).

1. Introduction

Self Compacting Concrete (SCC) is characterized by its high workability and resistance to segregation. In its basic composition it has the same materials as conventional concrete (CC), but their characteristics are different fresh. The SCC containing a larger amount of fine and maximum aggregate size is less than a CC; also for mixing water-reducing high range they are used.

It is important to note that concrete with high volumes of pasta and low volumes aggregate exhibit higher shrinkage strains (Neville, 1996), (R. Loser et al, 2009). Under the same environmental conditions, the retraction curve - Time of a SCC is very similar to that of a CC (Xiao-jie Liu et al, 2008), (De La Cruz, 2006); when resistance also remains constant, the properties (when hardened) of SCC are similar to a CC (De La Cruz, 2006), (Persson, 2001), (Bravo, 2004).

For the performance of medium strength self-compacting concrete (SCC-MS) in the cured state, it was initially decided to make the shrinkage and creep tests up to 180 days (DEMEC and gauges embedded points) (De La Cruz, 2006). Finally results were obtained up to 629 days to start the test; with the idea of meeting the deferred behavior for this type of concrete. With the completion of these trials (shrinkage and creep), they wanted to know the degree of dimensional stability over time, two (2) types of SCC-MS specifically SCC-MS with cements CEM II 32.5 R and CEM I 42.5R; for water / cement (W / C) 0.45 and 0.42; since both SCC-MS had the best features of self-compactability and compression resistance than expected, at lower cost, as De La Cruz (2006). All tests were performed at the *Laboratorio de Tecnología de Estructuras* (LTE) of the *Universitat Politécnica de Catalunya* (UPC).

2. Experimental details

Two (2) batches were done for each type of SCC-MS (with CEM II 32.5 R CEM I 42.5 R and 120 liters each), in the forced action mixer LTE capacity of 250 liters.

Each SCC-MS were performed two (2) of characterization tests in fresh, as are trials Slump and Extension Bars with ring (De La Cruz, 2006), (Bravo, 2004), to verify self- compactability of concrete; because they are the two (2) that define processes quickly and reliably the behavior in fresh SCC. Three (3) specimens of each type of control SCC-MS were also developed for evaluating the compressive strength after curing in a humid chamber to specification UNE 83-301 (Gettu et al, 2000), at a temperature of 23 ° C and humidity on 90%, tested at 28 days according to the UNE 83-304 (Gettu et al, 2000). Dosages and results of characterization tests are shown in Table 1. In the same batches 14 cylindrical samples, 150 x 300 mm were prepared; corresponding three (3) of each type of concrete for testing shrinkage and four (4) of each type of concrete for fluence.

Table 1. Dosages (Kg/m^3) and fresh characterization and compression resistance in specimens SCC-MS [MPa]

Concrete		SCC-MS	SCC-MS
Type of Cement		II 32.5 R	I 42.5 R
W/C		0,45	0,42
Cement		392	340
Water		196	163
Additives		4,11	8,84
Sand 0-2		651	680
Sand 0-5		434	453
Gravel 5-12		723	756
	Characterization Tests		
Slump	T50 [s] D_f	2	2
	[mm]	680	690
J. Ring	D_f	650	640
	C.B	0.8	0.8
Real U.W	[Kg/m^3]	2420	2420
Strength and C.V	[MPa] y [%]	30 2.9	42 2.2

These concretes were compared with a conventional concrete (CC) of the same strength. To evaluate deformations, the molds were provided with strain gauges (120 Q) of length 10 cm, arranged in the center of the vertical axis as shown in Figure 1. This is to capture the axial strain in the central third of the specimen.

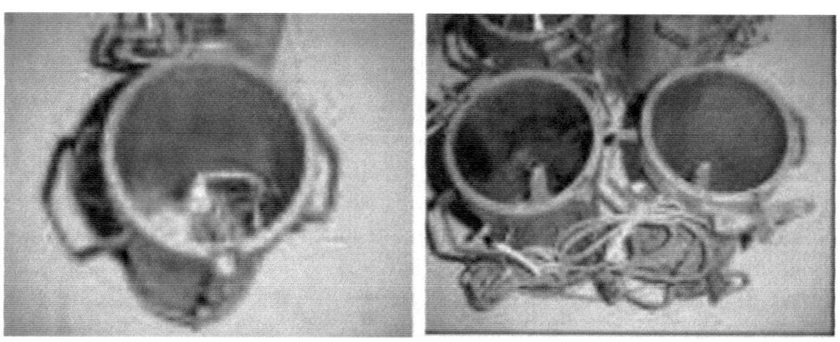

Figure 1. Molds for specimens duly provided with strain gauges (De La Cruz, 2006)

Processed specimens for the tests were demolded at 24 hours of being emptied, they are polished and immersed in water in a tank located in the climate chamber with relative humidity (RH) of 50% at a temperature of 19 ° C, for 28 days.

3. Shrinkage tests

Drying shrinkage is determined for three (3) specimens of each type of SCC-MS. Once the samples were immersed in water they were connected to data acquisition equipment. This, with the idea of tracking the processes of deformation of the SCC-MS during curing. In Figure 2, the process sequence once the specimens manufactured for trials presented. After 28 days of curing in water, each specimen was adhered external measurement points (DEMEC) located each on its side surface 120° as shown in Figure 3.

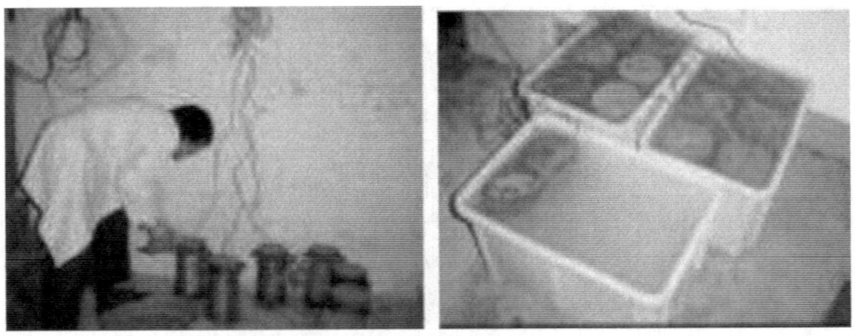

Figure 2. Procedure shrinkage and creep test on specimens of SCC-MS (De La Cruz, 2006)

Figure 3. Test tubes with strain measurement points on the surface and attached comparator DEMEC N°8865 lengths. (De La Cruz, 2006)

Other readings were completed 24 hours and continued to cure at 48 and 72 hours to 42 days. To have a better understanding and control of volumetric changes that could have the SCC-MS; in Figures 4 and 5, the behavior of specimens (e values) occurs from the time the curing has ended and be kept in a climatic chamber for 42 days (RH 50% and temperature 19°C). Deformation monitoring with electronic readings during the curing time, for both concrete showed an expansion (deformation negative values), which was expected given the curing conditions to which the specimens were undergoing.

So, given the performance of the test conditions (curing in water), endogenous contribution is negligible and, therefore, the measured strains are mainly to drying shrinkage. As shown in the graphs below, the securities electronic gauges in the early hours of drying, have a rapid growth of the strain, which tends to increase more gradually to 24 hours after the two SCC-MSs. Hence, the behavior is very similar during retraction measured for both SCCs-MS.

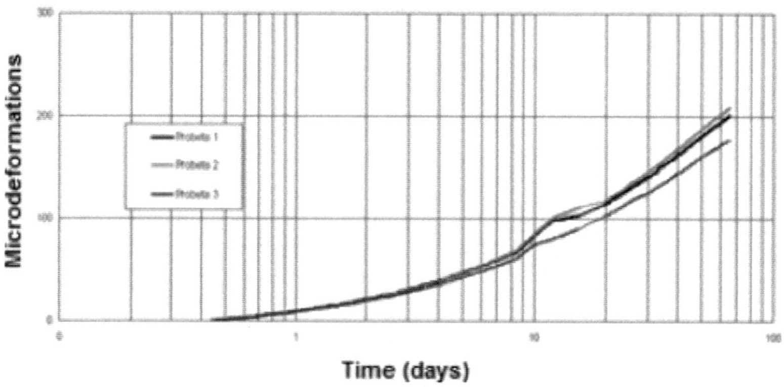

Figure 4. Evolution of the retraction SCC-MS with CEM II 32.5 R during drying (De La Cruz, 2006)

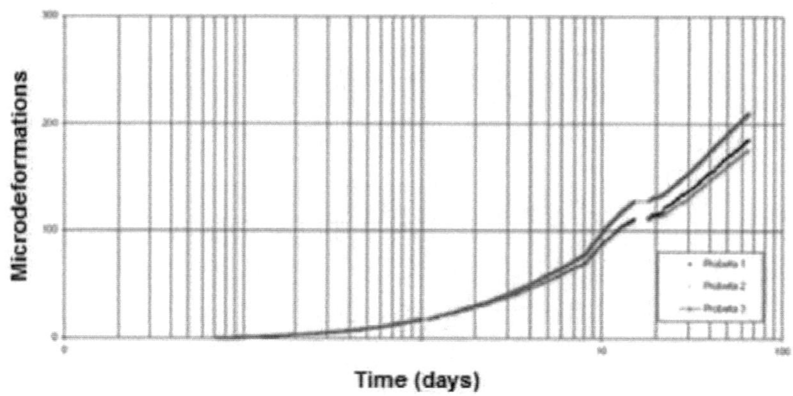

Figure 5. Evolution of the retraction SCC-MS with CEM I 42.5 R during drying (De La Cruz, 2006)

In Table 2, the results of deformation are summarized shrinkage both methods (points DEMEC and average of electronic values embedded gauge), which shows that the specific values (old) are approximately equal from 19 days.

The trend and behavior of each individual specimen and their average age (three (3) specimens for concrete) compared to the average electronic readings from the two (2) days of curing be removed. The differences in average readings deformation of the two (2) methods. For each type of SCC-MS at 28 and 42 days it is approximately 18% with CEM II 32.5 R and 25% with CEM I 42.5 R (De La Cruz, 2006).

It seems that at the age of 19 days of trial, the concrete specimens manage both internal and external balance. So the data obtained with both methods are becoming very similar, if not the same after this age. The differences are not comparatively high if one considers that the strain readings with DEMEC points, were performed manually with higher possibility of error (attached implement points- and strain measurements- sample), ànd electronic values are taken directly from the data acquisition equipment (embedded gauges). So, given the performance of the test conditions (curing in water), endogenous contribution is negligible and, therefore, the measured strains are mainly to drying shrinkage.

In Table 3, the shrinkage of SCC-MSs at ages 28 and 42 days and the results are presented as further research to conventional concrete (CCs), as is the Gettu et al. (2000) and Structural Concrete, EHE (2003).

IOP Conf. Series: Materials Science and Engineering **175** (2017) 012060 doi:10.1088/1757-899X/175/1/012060

Table 2. Values of drying shrinkage measured in single specimen. (De La Cruz, 2006)

Drying time (days)	SCC-MS CEM II 32.5 R (microdeformations)						SCC-RM CEM I 42.5 R (microdeformations)					
	DEMEC					GAUGES	DEMEC					GAUGES
	1	2	3	Average	C.V [%]	Average	1	2	3	Average	C.V [%]	Average
2	14	10	10	12	17	21	10	21	25	19	39	23
3	21	18	26	22	20	31	52	35	49	45	20	34
4	49	42	42	44	9	40	66	53	88	69	26	44
7	59	49	53	54	10	59	77	67	84	76	12	65
8	73	77	89	80	10	65	94	91	91	92	2	71
	10	10	10				12	12	14			
14	5	1	0	102	3	100	2	3	0	128	8	110
	13	12	12				14	13	15			
19	3	6	1	127	5	112	7	7	4	146	6	115
	14	15	13				15	15	16			
21	0	4	1	142	8	117	7	1	5	158	4	117
	16	18	15				17	17	17			
28	5	9	2	169	11	135	5	2	2	173	1	130
	19	21	17				20	20	20			
42	9	0	9	196	8	160	3	3	3	203	0	153

Table 3. Results of retraction points DEMEC and gauges in SCC-MS specimens (microdeformations) (Davies, 2006) (De La Cruz, 2006)

METHODOLOGY USED								
Type of concrete	**DEMEC**		**Gauges**		**By al. (CC)**	**Gettu et (2000)**	**By article (CC)**	**EHE 39º**
Days	28	42	28	42	28	42	28	42
SCC-MS CEM I 42.5 R	173	203	130	153				
SCC-MS CEM II 32.5 R	169	196	135	160	160	180	99	112

If the results established by shrinkage compared Gettu et al. (2000) (EHE), the EHE (Article 39, Section 39.7) for a CC at 42 days, it is observed that for SCC-MS with CEM I 42.5 R is higher by 27%, and for the SCC-MS with CEM II 32.5 R is increased by 30%. These percentage differences are relative, as in studies for CC by Gettu et al. (2000) (EHE), increased shrinkage is obtained at 15% that of the SCC-MSs and 38% higher than that proposed by the Article 39 of the EHE (Gettu et al, 2000). As described, the values obtained with DEMEC points and gauges, are intermediates to research and the article being compared, on the premise that these proposals are given for CC.

Besides measurements taken manually, can have more precision errors (sticking points and readings) with the embedded gauges, the results have not large percentage differences, suggesting verify the effectiveness of the measurements with both methods, being either I them suitable for determining shrinkage SCC-MSs.

One of the conclusions of De La Cruz (2006), regarding the shrinkage behavior of the SCC-MSs with different types of cement; for the same strength, they are similar. And also it notes that trends in the evolution of the shrinkage after 28 days of curing (considering zero (0) to read just after curing) for the two SCC-MSs are very similar during the 42 days of drying. In Figure 6, the evolution of the shrinkage occurring up to 530 days, the specimens SCC-MS with CEM II 32.5 R Reaching a maximum strain of 331 microstrain.

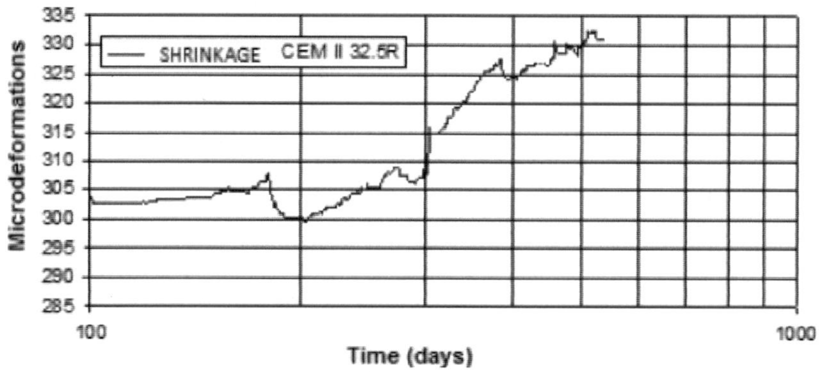

Figure 6. Evolution of drying shrinkage of SCC-MS with CEM II 32.5 R

Is fundamentally important to note that mentioned comparisons between concrete is based on mechanical and environmental properties are similar, regardless of the component materials.

4. Creep test

For the creep tests (ASTM C 512-87), they were prepared four (4) samples of each type of SCC-MS. The specimens were embedded strain gauges as shown in Figure 7, once demolded spot faced and were immersed in water for 28 days. As for the shrinkage test, once the SCC-MS (120 liters for each SCC-MS) obtained the procedure shown in Figures 1 and 2. After the curing is followed, a voltage was applied uniaxial compression equivalent to 40% of its compressive strength at the age of 28 days; that is for SCC-MS with CEM I 42.5 R of 17 MPa and with the SCC-MS CEM II 32.5 R of 12 MPa; oleopneumatic using racks of charge as shown in Figure 7.

In Table 4, results are summarized creep test deformation with both methods (DEMEC points and average values of the embedded electronic gauges).

Figure 7. Creep test in the LTE (De La Cruz, 2006)

Table 4. Values of creep measured in single specimen. (De La Cruz, 2006)

DAYS	SCC-MS CEM II 32.5 R (microdeformations)							SCC-RM CEM I 42.5 R (microdeformations)						
	1	2	3	4	AV.	C.V [%]	GAUGES	1	2	3	4	AV.	C.V [%]	GAUGES
2	214	175	186	196	190	2	530	305	25	228	186	186	17	650
3	249	227	301	284	260	7	590	343	409	340	238	330	9	700
4	420	325	322	399	370	6	620	466	480	357	322	400	9	730
7	406	392	403	427	410	3	690	504	550	413	381	460	9	800
8	476	462	525	539	500	5	670	595	578	574	448	550	7	830
14	637	633	642	630	640	1	830	742	648	644	606	660	6	940
19	700	805	766	756	760	5	840	886	731	749	690	760	8	950
21	721	822	781	798	780	5	840	900	742	819	728	790	7	960
28	819	910	942	896	890	5	890	1001	928	872	826	910	6	1020
42	956	1001	1117	1005	1020	6	1010	1120	1047	1001	917	1040	5	1080

If the deformations are compared with two (2) methods for each type of concrete, you can see that they are showing the same values at 28 days with cEm II 32.5 R and with a difference of 11 % with CEM I 42.5 R. A 42 days differences in the value of creep deformation between 4% and 6%. Electronic Observing the readings 28 and 42 days, the percentage difference between the two (2) SCC-MSs is 12%. It is older also the CEM I 42.5 R. That is consistent according to Mari's comments and Cladera (2003).

In this paper partial data and the evolution of creep deformation (under constant pressure) of the SCC-MSs occur during the 629 days of being subjected to load in the racks (I report given by the LTE (records sent to Colombia by the LTE (UPC) / 2008)). This evolution occurs both with electronic readings of gauges, as with DEMEC points.

To determine the creep deformation, it must be subtracted from the total deformation (£$_{total}$) (Eq. (1).) the initial deformation and shrinkage strain. The voltage- dependent deformation, at time t, for a constant voltage applied at t0, can be estimated according to the Article 39 of the EHE (2003) (Gettu et al, 2000), the following criteria:

$$\varepsilon_{total} = \sigma(t_0) \cdot \frac{\frac{1}{E_{0,t0}} + \varphi(0-t0)}{E_{0,28}} \qquad (1)$$

Where:

σ(to): It is the constant applied tension in to

E 0,t0 : Initial modulus of elasticity of concrete at the time of application to the load (defined in section 39.6 of the EHE (2003) (EHE)).

E0,28 : Initial modulus of elasticity of concrete at 28 days of age (defined in section 39.6 of the EHE (2003)).

φ (t-t0): Creep coefficient whose expression involves φo basic creep coefficient.

Then the evolution of creep for SCC-MSs occurs after 530 days of experiment. In Figures 8 and 9, the evolution of creep deformation occurs in SCC-MSs with cements CEM I 42.5 R and CEM II 32.5 R (Barcelona-Spain).

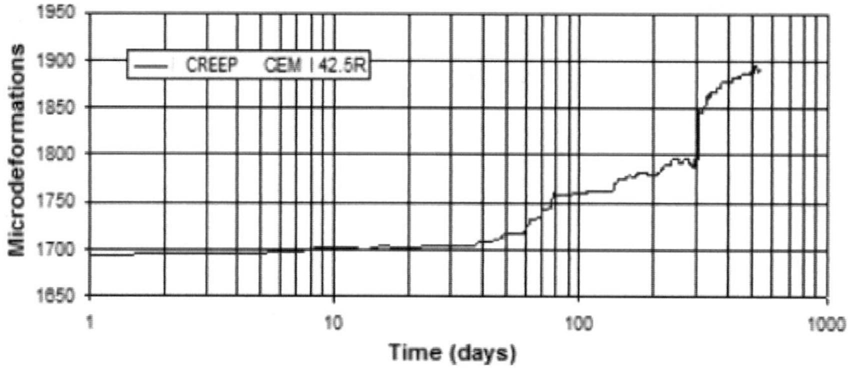

Figure 8. Evolution of the creep of SCC-MS with CEM I 42.5 R

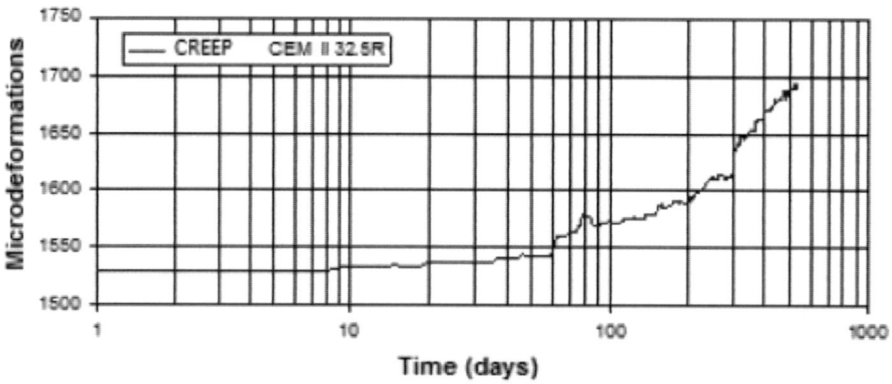

Figure 9. Evolution of the creep of SCC-MS with CEM II 32.5 R

To calculate the evolution of the flow coefficient ($) two (2) SCC-MSs with electronic reading values (gauge), the following Eq. (2) is applied according to Article 39 of EHE (2003):

$$\varphi = \frac{\varepsilon_{fluencia}}{\varepsilon\sigma} - 1 \quad (2)$$

With recent records deformation (microstrain) of the specimens prepared for determining shrinkage, creep coefficient ($) and creep of the SCC-MSs, it is shown in Figure 10. Next, the evolution of creep coefficients ($) for both concretes.

ϕvalue, for SCC-MS with CEM II 32.5 R 629 days to prepare specimens, is 1.67, as shown in Figure 11. Similarly, the $ results are presented for SCC-MSs concretes with CEM I 42.5 R, up to 629 days:

As shown in the previous figure, ϕ values are lower for concrete with CEM I 42.5 R 50% at 28 days and 61% at 42 days, compared to the other concrete CEM II 32.5 R.

According to the results of De La Cruz (2006), which are presented in Table 5, in the same specimens and 169 days, ϕ it had a value of 1.15 for SCC-MS with CEM I 42.5 R . Establishing compared to Cc and SCC (Roncero et al, 2001) ϕ has a value of 0.80 and 0.81 respectively; which according to the experimental data to 629 days, it is 1.24 (see Figure 11).

IOP Conf. Series: Materials Science and Engineering **175** (2017) 012060 doi:10.1088/1757-899X/175/1/012060

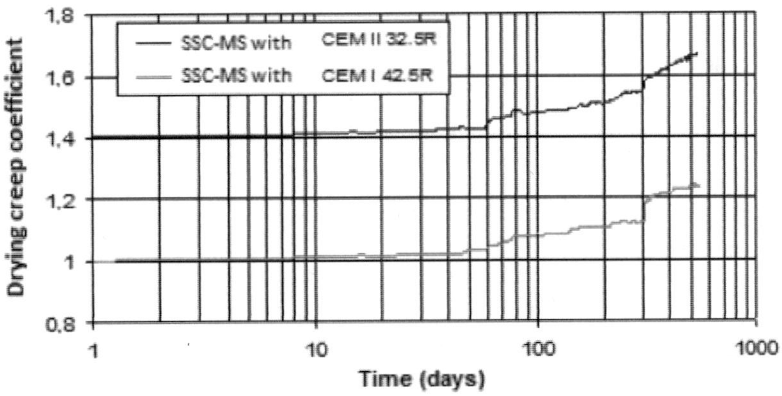

Figure 10. Evolution of creep coefficient (ϕ) for SCC-MSs with CEM II 32.5 R and CEM I 42.5 R

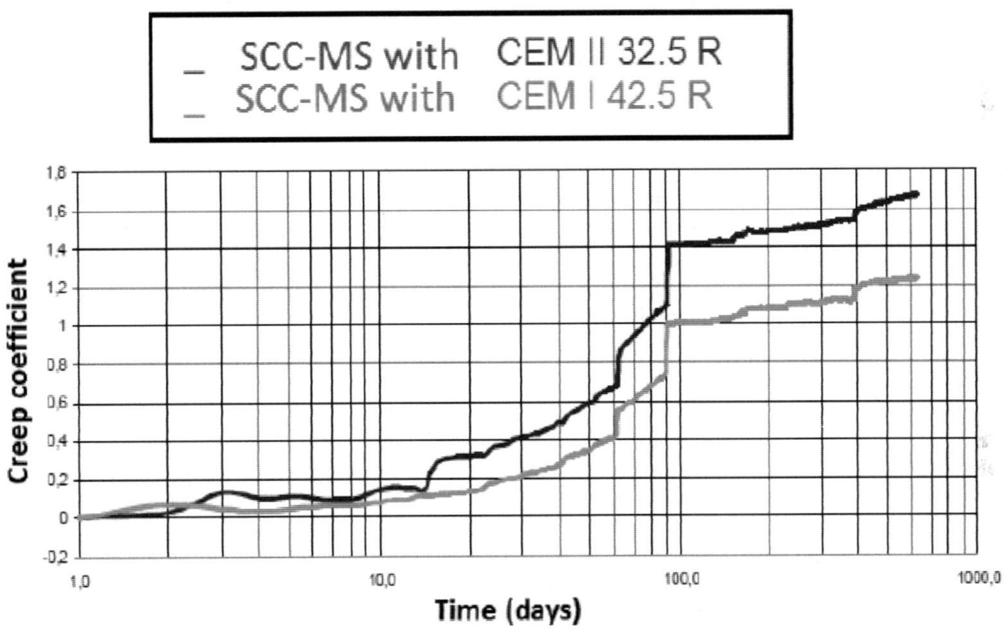

Figure 11. Evolution of creep coefficient (ϕ) for SCC-MSs with CEM II 32.5 R and CEM I 42.5 R

Table 5. Comparison of creep coefficients up to 169 days. (De La Cruz, 2006)

TYPE OF CONCRETE (y)(	169 DAYS)
SCC CEM I 42.5 R	0,81
SCC-MS CEM I 42.5 R	1,15
CC (Roncero et al. (2001))	0,80

IOP Conf. Series: Materials Science and Engineering **175** (2017) 012060 doi:10.1088/1757-899X/175/1/012060

In Table 6 below, the values of ϕ for concrete with the same mechanical properties similar environmental conditions while testing are presented, as proposed calculation Article 39 (EHE (2003) (Gettu et al, 2000)), Roncero et al. (2001) and Mari and Cladera, 2003 ("RH = 60%"); and it is observed that the value of ϕ with CEM I 42.5 R is less than for concrete with CEM II 32.5 R, by 75% and 62% respectively.

Table 6. Coefficients of creep ($) with points DEMEC and gauges in SCC-MS specimens (microdeformations) (Davies, 2006) (De La Cruz, 2006)

METHODOLOGY USED					
TYPE OF CONCRETE	**DEMEC**	**GAUGES**	**BY EHE**	**Roncero et al (2001)**	**Man y Cladera (2003)**
Days	28 42	28 42	42	42	42
SCC-MS CEM I 42.5 R SCC-MS CEM II	0,10 0,23 0,40 0,61	0,21 0,27 0,40 0,60	180	0,60	1,0

For purposes of analysis, comparing the electronic results (more data) with the calculations for CC 42 days of trial, as proposed by Article 39 the difference is 61% for SCC-MS - CEM I 42.5 R and to the other of only 14%. But if we make the same comparison with the research $ Roncero et al. (2001), the percentage difference is 55% for SCC-MS with CEM I 42.5 R and zero (0) for SCC-MS with CEM II 32.5 R. Finally, if we consider the proposal $ according to Mari and Cladera (2003), the SCC- MS with CEM I 42.5 R presents a percentage difference of 73% and 40% with the SCC-MS with CEM II 32.5 R, considering that this proposal is the most remote research (See Table 5.).

Another conclusion of De La Cruz (2006), regarding the creep coefficient ($) of the SCC-MSs with different types of cement, and according to estimates suggesting CCs Article 39 is applicable to SCC-MSs.

5. Conclusions

- As the percentage comparisons with the SCC-MS CEM II 32.5 R, it has a deformation similar to a CC yield the same mechanical properties. This situation differs in a high percentage to the other concrete CEM I 42.5 R. This does not preclude in any way that the latter has a strain similar to that of a CC creep, as it has been compared to concrete characteristic strength of 35 MPa, and its compressive strength at 28 days is 43 MPa.
- If the methodologies and approaches used only with SCC-MS with CEM II 32.5 compared, one can say that the results have not large percentage differences, suggesting verify the effectiveness of the measurements with both methods, whichever one it appropriate to determine the deformation creep of the SCC- MSs.
- It seems that the calculations suggesting the Article 39 to determine the evolution of the creep coefficient is applicable to SCC-MSs.
- The strain values obtained to 629 days, the specimens for shrinkage and creep tests show the same trend until 169 days (De La Cruz, 2006), but are increasing. Higher for SCC-MS with CEM II 32.5 R, which for SCC-MS with CEM I 42.5 R.
- As in the case of determining the retraction of this type of concrete, it is not excluded in any way, do a more thorough study of creep deformation and evolution of $ for SCC-MSs, continuing with measurements and calculations to help identify reliably, the deferred behavior of this type of SCCs.

References

[1] ASTM Designation: C 512 - 87 "Standard Test Method for Creep of Concrete in Compression" (Reapproved 1994).

[2] Bravo, M. (2004) "Caracterización y Utilización del Hormigón Autocompactable". Projecte o Tesina d'Especialitat. Enginyerie de la Construcció. Escola Técnica Superior d'Enginyers de Camins, Canals I Ports de Barcelona.

[3] De La Cruz, C. (2006), "Medium Strength self-compacting concrete for structural applications", *Doctoral Thesis.*

[4] Gettu, R., Aguado, A., Martín, M.A., y Roncero, J. (2000), El uso de aditivos reductores de retracción en el hormigón y sus implicaciones. El hormigón en las infraestructuras de transporte. CEMENTO- HORMIGÓN. N° **813**, pp 755-769.

[5] Marí, A. y Cladera, A. (2003), "Bases de cálculo para el hormigón de alta resistencia". Revista - Hormigón y Acero- núm. **228-229**, 2° y 3er trimestre 2003, pp. 18-23.

[6] Neville, A. M. (1996), "Properties of Concrete", Person Education Limited, Harlow, UK, **4**th Edition, 434 pp.

[7] Persson B. (2001), "A comparison between mechanical properties of self-compacting concrete and the corresponding properties of normal concrete", J Cement and Concrete Research., **31**, 193-198.

[8] R. Loser, A. Leemann. (2009), "Shrinkage and restrained shrinkage cracking of self compacting concrete compared to conventionally vibrated concrete", Materials and Structures, **42**, 71-82.

[9] Roncero, J., Gettu, R. y Martín, M. (2001), "Influencia de un Aditivo Reductor de Retracción en el Comportamiento Diferido de Hormigones Estructurales". II CONGRESO DE ACHE PUENTES Y ESTRUCTURAS Investigaciones y estudios. Materiales. Dept. Ingeniería de la construcción, Universitat Politécnica de Catalunya.

[10] Structural Concrete Instruction (EHE), European Standard.

[11] Xiao-jie Liu, Zhi-wu Yu and Lizhong Jiang (2008), "Long term behavior of self compacting reinforced concrete beams", *J. Cent. South Univ. Technol.*, **15**, 423-428.

IOP Conf. Series: Materials Science and Engineering **175** (2017) 012061 doi:10.1088/1757-899X/175/1/012061

Development of spraying agent for reducing drying shrinkage of mortar

Hiromi Fujiwara[1], Masanori Maruoka[2] and Lingling Liu[3]

[1]Professor, Dept. of Civil Engineering, Utsunomiya University, Japan
[2]Associate Professor, Dept. of Civil Engineering, Utsunomiya University, Japan
[3]Graduate School of Engineering, Utsunomiya University, Japan

E-mail: fhiromi@cc.utsunomiya-u.ac.jp

Abstract. Mortar used to repair is sometimes exposed to drying state in early ages after construction and a few days later water is sprayed frequently on the surface of the mortar in order to prevent cracks. This research studied on shrinkage characteristic of mortar subjected to drying conditions like this. The result showed that the water spraying on the mortar after initial drying did not have any effect to prevent shrinkage, but increased. And it also showed when various chemical agents are mixed and used in waterprayingit had the prevention effect on shrinkage. This report was to understand this kind of phenomenon and clarify the mechanism. In addition, based on the results, the new spraying agent was developed to reduce drying shrinkage.

1. Introduction

Concrete structures are prone to crack. And cracks have bad influence on structure members, and even can have a bad effect on the whole structure. Up to now, researches about preventing or reducing cracks especiallydrying shrinkage crack have been carried out. And some effective methods have been developed. Using chemical agent to reduce crack is one of those methods.

But for large buildings, using chemical agent is not affordable. Moreover, admixture is another effective method to prevent crack. It wasreported that as admixture, ureacould reduce drying shrinkage. When the urea is used 50kg/m^{3}in concrete, drying shrinkage decreases about 60%[1)].

Under this condition, this study is intended to develop new spraying agent to reduce drying shrinkage. And the new agent is mainly urea and some special chemical agent. Urea is inexpensive and can be got easily. What's more, the spraying is easier to operate than mixing admixture inalready mixed concrete plant. In this research, the length of test pieces had been measured at different days to calculate the drying shrinkage. And compressive strength test was also completed to find out the strength property of the test pieces under different spraying conditions.

2. Outline of Experiments

This research studied the drying shrinkage property of mortar and the effect that new chemical agent had to mortar compared with an existing agent.And this study also aimed to search out the best concentration of urea that can reduce the drying shrinkage to the greatest extent. And the experiment also had different series of water cement ratio and soaking day to find out the effect these conditions could have to drying shrinkage. This experiment soaked the test pieces to simulate spray. The length of the test pieces were measured at different days(1,3,7,14,21,28,42,56,70,91).

IOP Conf. Series: Materials Science and Engineering **175** (2017) 012061 doi:10.1088/1757-899X/175/1/012061

2.1. Materials
The materials used in the experiment are given in table 1. Urea was used when dissolved in water. The main ingredient of coating-type shrinkage reducing admixture is glycol ether-based derivatives.

2.2. Mixing Method
After the cement and the fine aggregate were mixed for 1 minute, water was added and mixed together one more minute by using Omni-type mixer.The mixer was with a nominal capacity of 10 litres. And thesize of mortar test piece was 40×40×160 mm.

2.3. Demoulding
Demoulding was done at one day after mixing. After demoulding, test pieces were put in the constant temperature room at 20°C with 60% relative humidity.

2.4. Test items

2.4.1. Mortar flow test. The test was conducted according to the JIS R5201-2015"Physical Testing Methods for Cement"[2]. And the results are given in table 2.

*2.4.2. Compressive Strength.*The test was conducted according to the JIS R1108-2006"Compressive Strength Testing Methods"[3].The conditions that compressive strength test was conducted are marked by ○ in table 2.

2.4.3. Drying shrinkage. The length of the test pieces were measured at different days(1,3,7,14,21,28,42,56,70,91).Gauge plug was used to measure the length of the test pieces.

2.5. Testing conditions
Experiment conditions are given in table 2.The soaking liquid were nothing, water, and urea solution. Urea30% means the mass of urea is 30% in the solution. Urea40% and urea50% are the same.And soaking time had three levels, 1 minute,10 minutes, and 30 minutes.Soaking was conducted at the 3rd day or 7th day after demoulding and strength tests were completed at different days (7 and 28).

Table 1. Materials

Materials	Symbol	Type	Density(g/cm^3)
Cement	C	Ordinary Portland cement	3.15
Water	W	Tap water	1.00
Fine aggregate	S	River sand	2.63
Chemical agent	U	Urea	1.32
	W	Tap water	1.00
	P	Coating-type shrinkage reducing admixture	0.99~1.05

Table 2. Experiment Conditions

W/C	S/C	soaking liquid	soaking time (minute)	soaking day (after demoulding)	Symbol	Mortar flow (mm)	Compressive Strength test
40%	1.2	nothing			40%N	191	
		water	1	3rd	40%3W		
				7th	40%7W		
		urea50%		3rd	40%3U50%		
				7th	40%7U50%		
50%	1.7	nothing			50%N	199	○
		water	1	3rd	50%3W		○
				7th	50%7W		
			10	3rd	50%3W10min		
			30	3rd	50%3W30min		
		urea30%	1	3rd	50%3U30%		○
		urea40%	1	3rd	50%3U40%		
		urea50%	1	3rd	50%3U50%		
				7th	50%7U50%		
			10	3rd	50%3U50%10min		
			30	3rd	50%3U50%30min		
		Coating-type shrinkage reducing admixture	1	3rd	50%3P		
				7th	50%7P		
60%	2.4	nothing			60%N	197	
		water	1	3rd	60%3W		
				7th	60%7W		
		urea50%		3rd	60%3U50%		
				7th	60%7U50%		

3. Test results and Discussion

3.1. Effects of water cement ratio and soaking day

For different water cement ratios, length changes are shown in figure 1, figure 2 and figure 3.

When the soaking liquid were nothing and water, drying shrinkages of test pieces were almost same and larger than other conditions.However, when the soaking liquid was urea, drying shrinkage of test pieces reduced. Compared with N, when using urea solution, drying shrinkage at most decreased 19% at 40% water cement ratio, 34% at 50% water cement ratio, and 33% at 60% water cement ratio.As for soaking day, at the same water cement ratio, no matter the soaking liquid was water or urea50%, the drying shrinkage was less of 3rd day then that of 7th day. Especially, at 50% water cement ratio, it was found that when using urea50% on 3rd day, the drying shrinkage decreased 19% compared with using urea50% on 7th day.

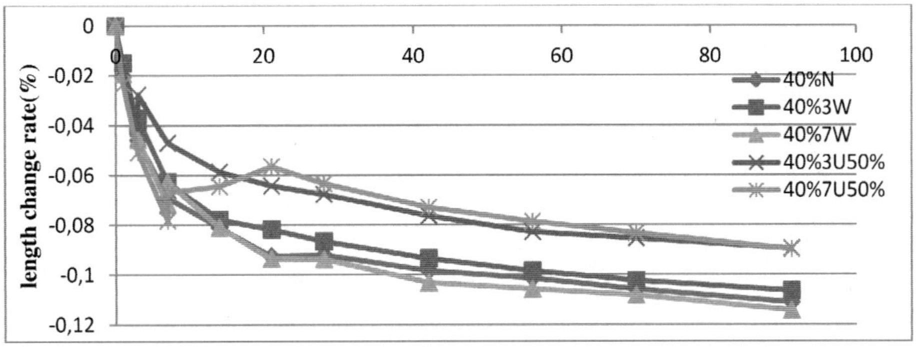

Figure 1. Length change rates(W/C=40%)

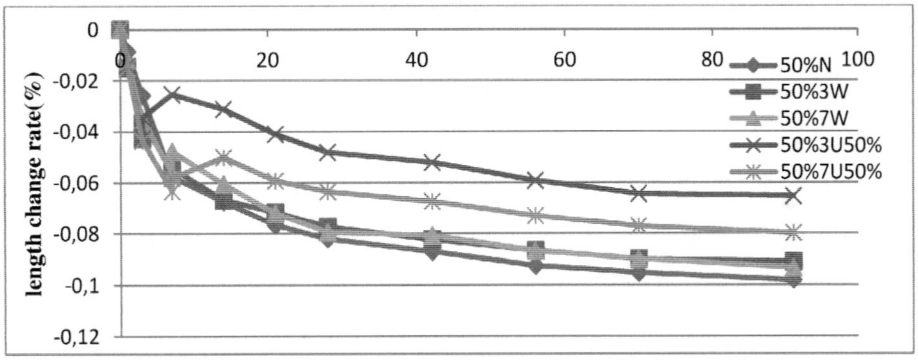

Figure 2. Lehgth change rates (W/C=50%)

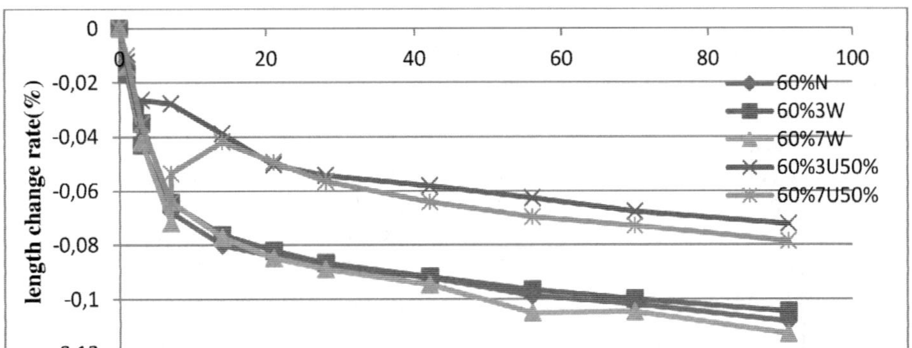

Figure 3. Length change rates (W/C=60%)

3.2. Effects of soaking time

Length changes of test pieces at 50% water cement ratio when the soaking time was different are shown in figure 4. From this figure, it was found that the longer the soaking time, the less the drying shrinkage. However, when using water, the longer the soaking time, the more the drying shrinkage. When soaking time was 30 minutes, the drying shrinkage was even only 23% as that of N. In addition, after soaking for 10 or 30 minutes, it was observed that urea crystallized on the surface of test piece out from the inside.

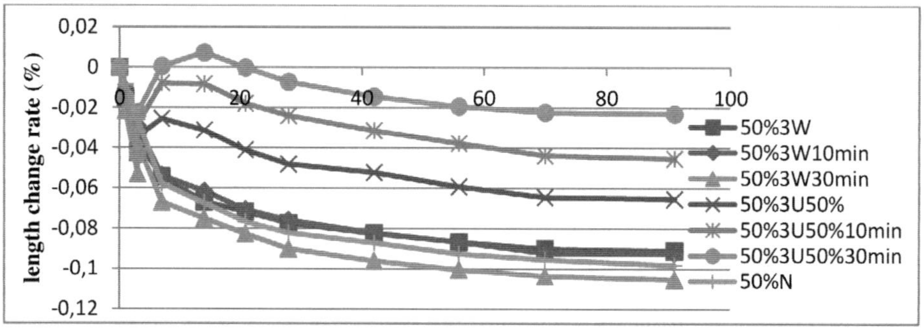

Figure 4. Length change rates (different soaking time)

3.3. Effects of concentration of Urea

Length changes of test pieces when the urea concentration was different are shown in figure 5at 50% water cement ratio.Using urea30%, urea40% and urea50%, the drying shrinkages were 76%, 75%, 64% in turn as that of N, and the drying shrinkage of test pieces soaked in water was less than that of N. In summary, the higher the urea concentration, the less the drying shrinkage. And all the results were better than the test pieces soaked in water or nothing.

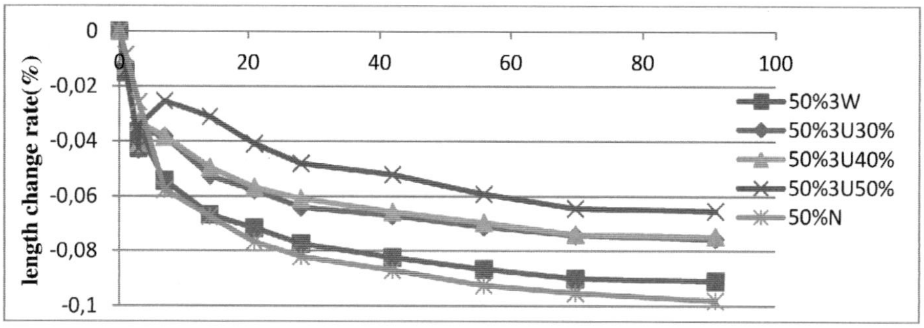

Figure 5. Length change rates (different urea concentration)

3.4. Compared to Coating type shrinkage reducing admixture

Length changes of test pieces when the soaking liquid were urea solution and coating type shrinkage reducing admixture are shown in figure6at 50% water cement ratio. When soaking day was 3rdday, the effect of coating-type shrinkage reducing admixture was not as good as urea solution.Nevertheless, when soaking day was 7th day, the effect of Coating-type shrinkage reducing admixture was better than urea solution.But all the results were better than using nothing to soak test pieces in.

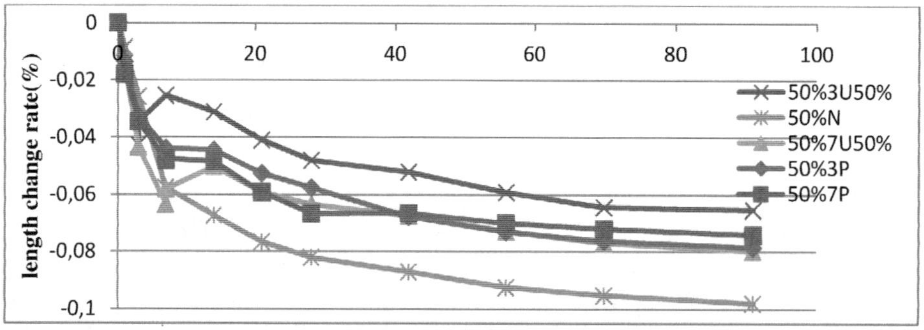

Figure 6. Length change rates (different soaking liquid)

3.5. Compressive Strength
The results are shown in figures 7.As the water cement ratio increased, the strength decreased. The compressive strengths of urea50%-type were higher than that of N-type at different water cement ratios.It was thought that after soaked in urea solution, the drying shrinkage of mortar became smaller, so the internal compressive stress inside mortar became smaller, so when being compressed, it can bear more external pressure. So the compressive strength increased.

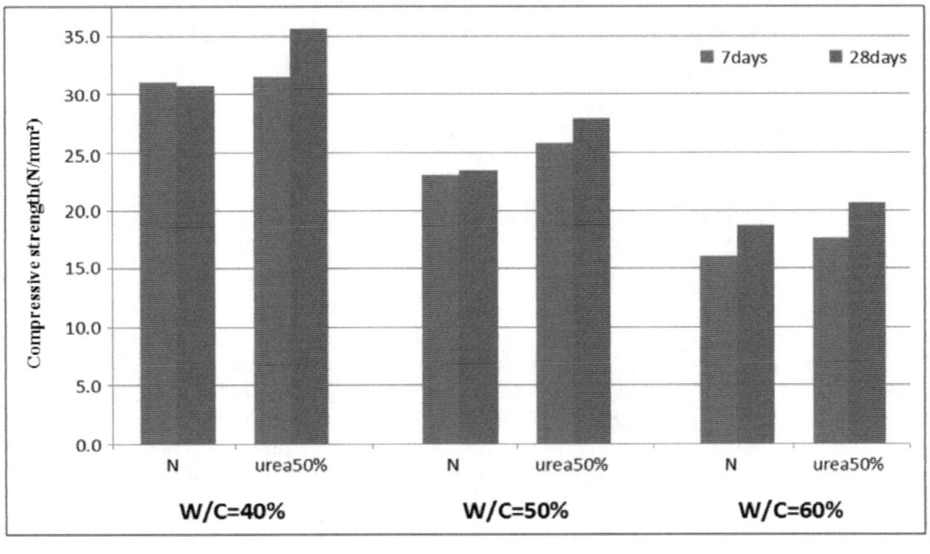

Figure 7. Compressive strength

4. Conclusions
Urea is easy to soluble in water, but also easy to precipitate from water. So it was probable that urea permeated mortar when soaking and crystallized in the gaps inside mortar when the water evaporates. So it was found that after soaked by urea solution, mortar's drying shrinkage decreased.
The results of this study can be summarized as following:
1) Using urea solution to soak mortar can reduce drying shrinkage.
2) When the urea solution concentration is higher, the effect is better.
3) When the soaking time is longer, the effect is better.
4) Compressive strength increases after using urea50%.
5) Some agents are better to be used at third day than seventh day.

References
[1] KawaiT andSakataK 2007 *Proceedings of the Japan Concrete Institute*.**29** 644
[2] Japan Standards Association 2015, *JIS R 5201 Physical test method of cement.*
[3] Japan Standards Association 2006, *JIS R 1108 Compressive strength test method.*

IOP Conf. Series: Materials Science and Engineering **175** (2017) 012062 doi:10.1088/1757-899X/175/1/012062

Numerical methods for estimating *J* integral in models with regular rectangular meshes

B Kozłowiec

Institute of Aeronautics and Applied Mechanics, Warsaw University of Technology

E-mail: bkozlowiec@meil.pw.edu.pl

Abstract. Cracks and delaminations are the common structural degradation mechanisms studied recently using numerous methods and techniques. Among them, numerical methods based on FEM analyses are in widespread commercial use. The scope of these methods has focused i.e. on energetic approach to linear elastic fracture mechanics (LEFM) theory, encompassing such quantities as the *J*-integral and the energy release rate *G*. This approach enables to introduce damage criteria of analyzed structures without dealing with the details of the physical singularities occurring at the crack tip. In this paper, two numerical methods based on LEFM are used to analyze both isotropic and orthotropic specimens and the results are compared with well-known analytical solutions as well as (in some cases) VCCT results. These methods are optimized for industrial use with simple, rectangular meshes. The verification is made based on two dimensional mode partitioning.

1. Introduction

It has been commonly known that in the scope of linear elastic fracture mechanics the *J* integral remains a measure of resistance to crack growth equivalent to the energy release rate (ERR) *G* [1, 2]. Moreover, in ideally elastic brittle, as well as ductile-brittle materials with plastic zone at the crack tip, the *J* integral is highly dependent on the loading mode. Calculating the different mode components of *J* integral with good precision is therefore vital for correct design against fracture of structures, including laminates. Despite the existence of analytic solutions in terms of both the *J* integral J [2, 3] and the ERR *G* [4, 5], the numerical calculations based on FEM are still considered as the primary efficient method for estimating values of *J* and *G* [6].

An important problem in the field of determining resistance to crack growth is estimation of mode mixity for a given body–crack configuration. Complex two-dimensional cases of loading modes are usually described using a relative percentage of the mode II (shearing mode) to the total ERR (G_{II} / G), the so-called mode mix, where:

$$G = G_I + G_{II} \tag{1.1}$$

The estimation of mode mix (the so-called decomposition or mode partitioning) has been the subject of numerous scientific works of analytic character [2–5]. Subsequently, many works of experimental nature have been conducted with the aim of verifying and comparing both methods, as well as discovering their limits of applicability [3, 5]. This paper is aimed at presenting and comparing numerical techniques which allow to calculate the *J* integral and its mode components. The results obtained using these techniques will be compared against those acquired using analytic methods. For detailed description of these methods, the readers are referred to [2, 3, 4, 6].

2. Numerical mode partitioning – the *J* integral method

Numerical methods have found application in fracture mechanics analyses since FEM software came into common usage. The *J* integral method uses the energetic contour path integral described as [1]:

$$J_1 = \int_K \left(W n_1 - \sigma_{ij} \frac{\partial u_i}{\partial x_1} n_j \right) ds \qquad (2.1)$$

where:

- $W = \int_0^{\varepsilon_{ij}} \sigma_{ij} d\varepsilon_{ij}$ – strain energy density function,
- n_j – component of the outward unit normal vector in the *j*-th direction,
- σ_{ij} – element of the stress tensor,
- u_i – component of the displacement vector in the *i*-th direction.

The *J* integral is a path integral over a scalar field along an arbitrary curve in *n*-dimensional space. In fracture mechanics problems the integral defined in two and three-dimensional Euclidean space is used. Another primary characteristic of the *J* integral is that it disappears when a path contour is closed. Consequently, *J* may be considered a measure of a crack's existence inside the analyzed contour. Additionally, *J* is an oriented integral – this last property together with the disappearance of *J* along a closed contour imply that the *J* integral is path independent.

Under the assumptions of LEFM, equality of the two quantities: *J* and *G* is proven. This fact together with the path independence property of the *J* integral cause *J* to be useful in analyzing non-homogeneous structures with interfacial cracks. The energetic approach with the use of the path independent *J* integral enables to bypass the singularity dominated zone in the vicinity of the crack tip and the issues related to its existence [5, 7], particularly in terms of mode partitioning accuracy.

J, similarly to *G*, may be partitioned into mode I and II (in a two-dimensional loading case), corresponding to opening and shearing fracture. The quantities J_I and J_{II} are obtained by partitioning the terms under the integral sign into symmetric and antisymmetric components respectively [8]. The equation (1.1) holds true also in case of the *J* integral.

3. Implementation of the *J* integral calculations

In the following section, two independent numerical techniques of calculating the *J* integral will be introduced. The details of computational implementation will be given in terms of the complete flowcharts of the methods' algorithms. Although the algorithms may already be used in commercial software environments, they have not yet been published to the best of the author's knowledge. The flowcharts are designed to be used as the basis for development of macros in typical parametric languages, such as ANSYS APDL, however general purpose programming languages may also be used if a FEM solution is known.

3.1. Direct integration method

The name *direct integration* was given to the algorithm in which an integration path is selected as a sorted set of nodes of finite element mesh from a rectangular contour of predefined thickness *t* (see Fig. 1), where $t \geq 0$. In practice, the cases with *t* equalling zero occur only when a mesh is regular.

Formulas describing the *J* integral and its components on separate respective segments of Γ can be obtained using the definition (2.1) and the conventions depicted in Fig. 1. The trapezoidal rule is used for calculating Riemann sums of the integrated functions between the nodes (endpoints of respective partitions) of the integration contour (the (*) box, Fig. 3):

$$J_M = J_M + \tfrac{1}{2} \Delta s_k (\tilde{J}_k^M + \tilde{J}_{k-1}^M) \qquad (3.1)$$

from which:

- $\Delta s_k = |P_i P_k| - |P_i P_{k-1}| = |P_{k-1} P_k| = \sqrt{(x_k - x_{k-1})^2 + (y_k - y_{k-1})^2}$ – length of a segment,
- $\tilde{J}_k^M = W_M(N_k)n_1 - \sigma_{ij}^M(N_k)u_{i,1}^M(N_k)n_j$ – value of the integrated function in k-th node.

While describing the direct integration algorithm it is worth noticing that, due to specifics of many FEM software packages, it is possible to switch to the nearest node neighbouring the selected node during the iteration process within the inner loop (which iterates over the nodes subscripted with the letter k). This action is described by the condition (**):

$$P_k : \underset{\varepsilon>0}{\exists} \underset{\delta<\varepsilon}{\forall} \left| P_k(x+\delta_x, y+\delta_y) - P_k(x) \right| = 0 \tag{3.2}$$

In general-purpose programming languages, this condition could be realized as a switch between two points (array elements) sorted in ascending order of distance from the starting point P_0.

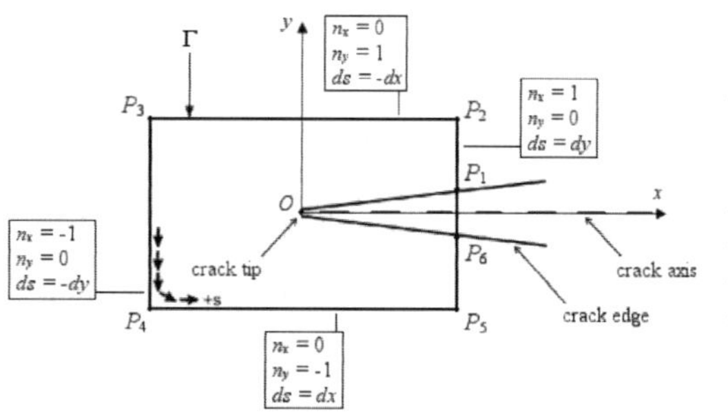

Figure 1. Integration domain in direct method: division and convections

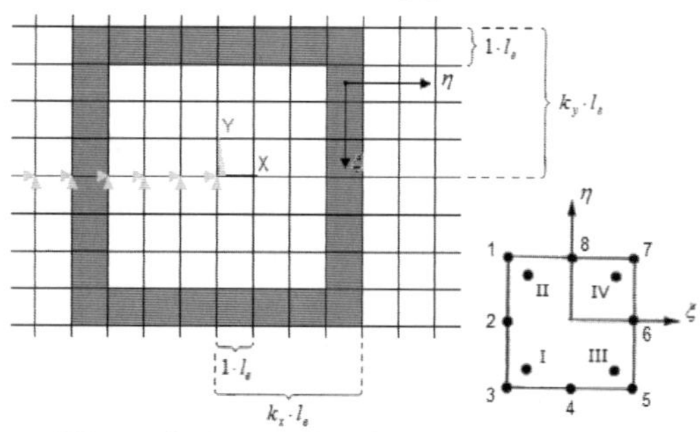

Figure 2. Integration domain in EDI method: coordinates

3.2. EDI integration method

Equivalent Domain Integral (EDI) is the name of a method, which invokes the divergence theorem to convert a line integral of a vector field around a simple closed curve to a domain integral of this field's divergence over the plane region bounded by this curve. By that means the J integral is calculated over a certain compact domain, which consists of finite elements which form a ring around the crack tip. The EDI method also introduces operations and symbols characteristic of a discrete space spread by the finite element mesh [8, 9].

Fig. 4 describes the EDI algorithm as implemented in MATLAB environment. The algorithm is executed as a post-processing phase after FEM calculations carried out in ANSYS Mechanical.

In Fig. 2 the use of a parent element coordinate system $O\xi\eta$ is depicted. This system is defined in general as curvilinear, with its origin at an element's geometric centre. The coordinates (ξ, η) are selected (box (*), Fig. 4) so that the corner nodes are positioned in the following way (see Fig. 2):

$$1:(\xi,\eta) = (-1,1) \quad 5:(\xi,\eta) = (1,-1)$$

$$3:(\xi,\eta) = (-1,-1) \quad 7:(\xi,\eta) = (1,1)$$

The coordinates of the even-numbered nodes are arithmetic means of the corresponding coordinates of two neighbouring nodes. The local system utilizes Gaussian points, which are the roots of a Gauss-Legendre polynomial, which in turn approximates the integrated function. The Gaussian points are numbered using Roman numerals according to the convention in Fig. 2.

The sum of the integrated function's values at the Gaussian points forms a Gaussian quadrature of the J integral's increment in an element [8, 9]. Using the Einstein notation this can be written as:

$$\Delta J_M = \sum_{m=1}^{m=4} \sum_{n=1}^{n=4} \left[W_M \cdot S_{,1} - u_{i,1}\sigma_{ij}S_{,j} + (W_{,1}^M - \sigma_{ij}\varepsilon_{ij})S \right] \det \mathbf{J} \tag{3.3}$$

The summation (3.3) takes place in the (***) box of the Fig. 4. The S function occurring in the expression (3.3) is a polynomial function of two variables:

$$S(\xi,\eta) = L_i S_i \tag{3.4}$$

where:

- L_i – shape function for the i-th node of a given element;
- S_i – value of S function in the i-th node.

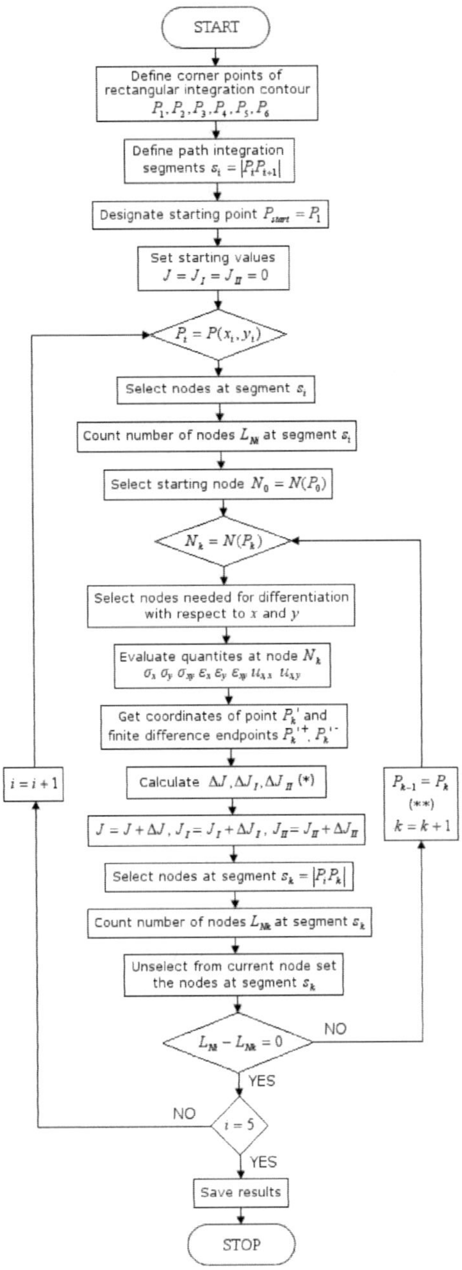

Figure 3. Flowchart of the direct integration algorithm

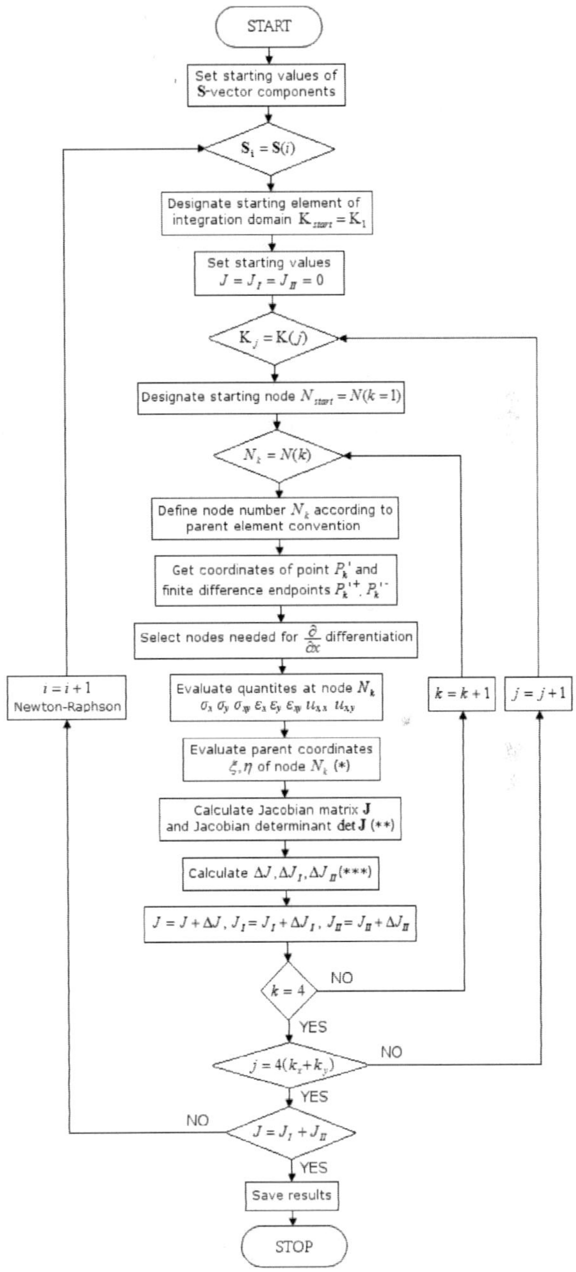

Figure 4. Flowchart of the EDI algorithm

The S_i values are set so that in the inner nodes of the contour ring (nodes numbered 3, 4, 5 according to Fig. 2) the value of S is 1, while in the outer nodes (numbered 1, 7, 8) S is 0. In nodes 2 and 6 the S function takes arbitrary values from the interval [0, 1]. These values are selected in the inner loop of the algorithm in Fig. 4 so that the condition (1.1) expressed in terms of J integral and its components is met with predefined accuracy.

For the purposes of the EDI algorithm and its application, a type of mesh with two geometrical parameters has been developed. Thanks to its simplicity, a derivative of any quantity in the direction x at every Gaussian point may be calculated using the central finite difference formula. Therefore, the values of derivatives are approximated with good accuracy and, at the same time, the distance between the contour ring and the crack tip, as well as the number of elements forming the ring, may be controlled. Obtaining the values of various quantities at Gaussian points may be achieved in most FEM environments by turning off the extrapolation of the Gaussian point results to the nodes.

4. Verification and application of the *J* integral analyses

Verification of results of numerical analyses was conducted basing on loading tests (specimens) used for evaluation of the critical ERRs in various modes of loading. All the specimens have geometries of slender beams and remain under plane stress condition (width equalling 25.4 mm). The specimens' models were built with the use of ANSYS Mechanical software. Material properties of the models correspond to isotropic (steel) and unidirectional orthotropic (graphite/epoxy) material as described in Table 1. Values of loads have been selected so that every test case remains in elastic strain regime.

Table 1.

Material type	Young's modulus [GPa]	Poisson's ratio [-]	Loading value [N]			Deflection [mm]			Number of plies	Ply thickness [mm]
			F_{DCB}	F_{ENF}	F_{SLB}	u_{DCB}	u_{ENF}	u_{SLB}		
isotropic (steel)	$E = 209.86$	$v = 0.33$	200	2000	1000	2.185	3.880	0.990	$n = 1$	$T = 4.8$ mm
orthotropic (CFRP)	$E_x = 146.86$ $E_y = 10.62$ $E_z = 10.62$	$v_{xy} = 0.33$ $v_{yz} = 0.33$ $v_{xz} = 0.33$	120	2000	800	1.806	2.470	1.009	$n = 16$	$T_i = 0.32$ mm

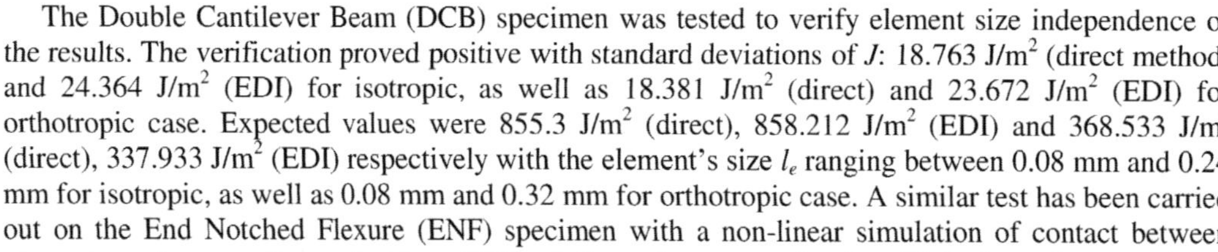

Figure 5. Geometry of specimen, a. DCB, b. ENF, c. SLB

The Double Cantilever Beam (DCB) specimen was tested to verify element size independence of the results. The verification proved positive with standard deviations of J: 18.763 J/m^2 (direct method) and 24.364 J/m^2 (EDI) for isotropic, as well as 18.381 J/m^2 (direct) and 23.672 J/m^2 (EDI) for orthotropic case. Expected values were 855.3 J/m^2 (direct), 858.212 J/m^2 (EDI) and 368.533 J/m^2 (direct), 337.933 J/m^2 (EDI) respectively with the element's size l_e ranging between 0.08 mm and 0.24 mm for isotropic, as well as 0.08 mm and 0.32 mm for orthotropic case. A similar test has been carried out on the End Notched Flexure (ENF) specimen with a non-linear simulation of contact between elements on the opposite cracked legs. ENF solutions also appeared to be of little dependence on l_e.

The results of the two tests displayed pure mode I (DCB) and pure mode II (ENF) loading as expected, with roughly 1% and 99% of mode mix respectively. The result J_{II}/J = 94% for the ENF specimen calculated using the direct method can be seen as underestimated.

The Single Leg Bending (SLB) test was performed to verify the mode prediction ability of the tested methods in complex loading configuration. A mode mix of approx. 40% was calculated using analytic methods. Fig. 7 illustrates changes of mode mix with increasing distance between the contour path and the crack tip (increment of the parameter k) for constant l_e. For both materials examined, the $J_{II}/J(k)$ curves appear to converge except for small oscillations. An exception is the EDI method, which fails when 4-node elements are used for meshing. A big improvement comes after using elements with mid-side nodes. The EDI solution for orthotropic specimen converges on the value of 41%, which lies between the two analytically predicted mode mixes. The solution obtained using direct integration is lower, amounting to approximately 38%. It may be concluded that both numerical methods estimate mode mixes accurately. Also, both numerical methods estimate the total value of J to be 7.5% ÷ 12% higher than analytic solutions. Similar results obtained using the well-known VCCT method suggest a possible influence of FEM modelling on the analytic-numerical discrepancies.

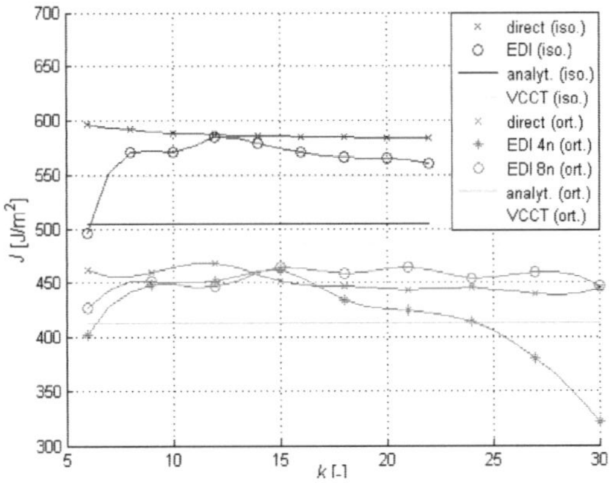

Figure 6. SLB test results: J total

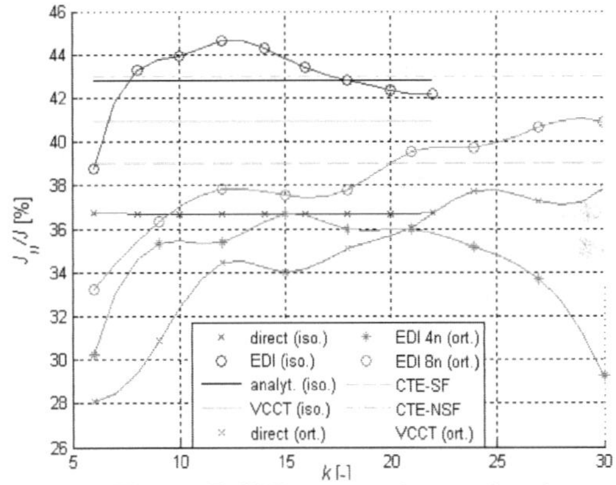

Figure 7. SLB test results: mode mix

5. Concluding remarks

In the present paper, the comparison between different methods which allow for the calculation of energy release rate (ERR) and its mode partitioning was presented. The methods used were of both analytic, semi-analytic and numerical origin. In the first part of the paper, the description of the two numerical techniques was provided, which use the J integral approach to calculate the ERR. The matter of numerical implementation in a typical FEM software of these techniques was also presented in detail. In the second part of the paper, the results obtained using the two numerical techniques, known as the direct integration method and the Equivalent Domain Integral method, were compared using three typical types of specimens: the DCB, ENF and SLB. The analytic, semi-analytic and (in some cases) VCCT solutions were considered as comparison criteria. The scope of the analysis (the loads applied, the maximum deflection of the specimens) never exceeded the linear elastic fracture mechanics (LEFM) criteria. The analysis allow for positive validation of the two numerical techniques compared. The direct integration method seems to diminish the J_{II}/J values in the whole scope of integration contours considered. The direct integration results, however, are of less deviation and converge faster than EDI counterparts. The matter of optimizing the integration algorithms for future use in simple rectangular meshes remains forthcoming.

References

[1] Anderson T L 2005 *Fracture Mechanics Fundamentals and Applications (3rd Edition)* (Boca Raton: CRC Press) pp 107–11
[2] Szekrenyes A 2012 *Period. Polytech. Mech.* **56** pp 63–71
[3] Conroy M, Kinloch A J, Williams J G and Ivankovic A 2015 *Eng. Fract. Mech.* **149** pp 351–67
[4] Williams J G 1988 *Int. J. Fracture* **36** pp 101–19

[5] Hutchinson J W and Suo Z 1992 *Adv. Appl. Mech.* **29** pp 63–191
[6] Rahman S 2001 *Eng. Fract. Mech.* **68** pp 107–25
[7] Wang S and Harvey C 2012 *Compos. Struct.* **94** pp 758–67
[8] Raju I S and Shivakumar K S 1989 An equivalent domain integral for analysis of two-dimensional mixed mode problems *30th Structures, Structural Dynamics and Materials Conf. (Mobile)* (Reston: AIAA Meeting Papers) pp 1231-1249
[9] Mohammadi S 2008 *Extended Finite Element Method for Fracture Analysis of Structures* (Oxford: Wiley-Blackwell) pp 56–59

IOP Conf. Series: Materials Science and Engineering **175** (2017) 012063 doi:10.1088/1757-899X/175/1/012063

Application of carbon FRP for fatigue strengthening of old steel structures

J Vůjtěch, P Ryjáček and M Vovesný

Faculty of Civil Engineering, Czech Technical University in Prague, 166 26 Prague, Czech Republic

E-mail: jan.vujtech@fsv.cvut.cz

Abstract. The traffic requirements on the existing infrastructure are rising still. This coupled with its age puts a strain on it. This is especially problematic for old steel bridges. Higher and more frequent loads will lead to development of fatigue damage to those structures. This causes an issue for the infrastructure owners as the existing methods of repair are difficult, time consuming and expensive. So there is a need to find some easier alternatives. One of such can be the use of carbon fibre reinforced polymers (CFRP). They are being successfully used for repairs and strengthening of concrete structures however their use with steel is still relatively new.

The purpose of this work is to establish how does a deteriorated steel reinforced with CFRP behave under fatigue loading. To test this a series of experiments was designed. With the help of a preliminary numerical study the dimensions of the specimens and the applied loading was established. There are two sets of specimens. With both we are using mild steel and each set has different level of surface deterioration (corrosion pits or corrosion holes). The specimens are reinforced using hand laid wet layup composites. They are subjected to fatigue loading and the difference between the fatigue life reinforced and unreinforced specimens is observed. Based on the preliminary study, it is expected, that the reinforcement will prolong the life expectancy by half.

1. Introduction

The current infrastructure in many countries in Europe is slowly reaching its life expectancy. This is particularly true for the steel bridges many of which was built at the turn of the19th century. That brings many issues to the infrastructure owners. The two main of those would be the corrosion and fatigue cracks. They are especially dangerous because one can feed the other. Thecorrosion increases the likelihood of cracks mainly by increasing the stress levels in the damaged surface. It was shown in experiments e.g. [1] the corrosion pits lead to localized plasticization in the defect even under small loads. And once existing the fatigue crack are extremely dangerous because of the possibility of sudden collapse.

The conventional ways to repair fatigue damage (replacement, welding) may not be always possible. Furthermore, the conventional method is usually quite expensive and time-consuming. It is also impossible to use it while the structure is under load, the subsequent closing of the repaired structure brings additional indirect costs. A possible way to find an easier method of repair is the use of new construction materials, such as using externally bonded FRP.Which could become a fast and easy solution for extending the service life of the structure.

IOP Conf. Series: Materials Science and Engineering **175** (2017) 012063 doi:10.1088/1757-899X/175/1/012063

The fibre-reinforced polymer (FRP) is a composite with of two main components, matrix and fibres. Together they form a new material of very high strength and durability. Matrix is usually polyester, vinyl-ester or epoxy glue. The fibres (glass, carbon, kevlar etc.) form the reinforcement. For steel members just carbon fibres are viable, as their modulus of elasticity is similar to the steel. However, steel and carbon create galvanic cell and they have to be separated by epoxy resin or other means.

2. The experiments

Using the carbon fibres for reinforcing against fatigue is of course not entirely new. There have been several studies into this matter. To mention just a few articles published [2], [3] and for a state of art review see [4]. There are also several national guidelines, such as the British [5], US [6] and the Italian [7]. However, the research into this topic is still relatively new and does not cover everything. Much more additional research is required to fully understand the behaviour of reinforced elements in all possible scenarios.

In our research we focus on the fatigue life of a corroded steel specimen reinforced with the CFRP. For the experiments there are two sets of specimen prepared (see Table 1). Each set has a different type of deterioration which corresponds to two most common types of damage found on old steel structures. The first is heavy surface corrosion, the corrosion pits. On our specimens the corrosion is idealised so the corrosion pits are symmetrical. The second is heavy corrosion in the form of corrosion hole. Which can develop either with corrosion going through the whole element or around a rivet.

To establish the dimensions of specimens and the faults a preliminary numerical study was performed. The study was also used to establish the amount and type of reinforcement and the loading force that is going to be used. The results of this study were presented at conference Esat2016 [8].

Table 1. List of specimens

Specimen Nr.	Fault	Surface treatment	Steel	Additional inf.
1 - 4	1	Sa 2 –sand blasting	mild	Unreinforced
5 - 8	1	Sa 2½ - sand blasting	mild	Reinforced
9 - 12	2	Sa 2 - sand blasting	mild	Unreinforced
13 - 16	2	St 2 - brushing	mild	Reinforced +100 freeze thaw cyc.
17 - 20	2	Sa 2½ - sand blasting	mild	Reinforced
21 - 24	2	Sa 2½ - sand blasting	mild	Reinforced +100 freeze thaw cyc.
33 - 36	1	Sa 2½ - sand blasting	S355	Reinforced
37 – 40	2	Sa 2½ sand blasting	S355	Reinforced
41 - 44	2	Sa 2½ sand blasting	S355	Reinforced+100 freeze thaw cyc.
45 - 48	2	Sa 2½ sand blasting	S355	Reinforced

2.1. Specimen

The geometry is identical for both types of specimens. The specimen is a steel plate 500 mm long, 14 mm thick and with the width of 60 mm in the tested part and 95 mm at the ends. Both faults are in the middle of the specimen. Fault 1 consists of 9 simulated circular corrosion pits 3 mm deep with diameter of 15 mm on both sides of the specimen (see Figure 1). Fault 2 is a single circular hole with surface diameter of 41 mm and inner diameter of 22mm (see Figure 2). The specimens are made of both old and new steel. The old steel is mild, pre 1890 steel and the new steel is S355. The behavior of the old steel is the main focus of this work and the new steel is used to check the results to exclude the influence of cyclic loads already introduced to the steel.

The FRP reinforcement is applied on both sides of the specimen. The wet layup process is used. The FRP patches are 300 mm long and 60 mm wide.There are 4 layers of heavy 600 g twill woven carbon textile on each side, made of the high strength carbon fibres, supplied by Havel Composity a.s.

IOP Conf. Series: Materials Science and Engineering **175** (2017) 012063 doi:10.1088/1757-899X/175/1/012063

They constitute the main reinforcement. One layer of glass textile is added between steel and carbon to prevent the creation of galvanic cell. The epoxy resin is SikaDur 300, supplied by Sika s.r.o.

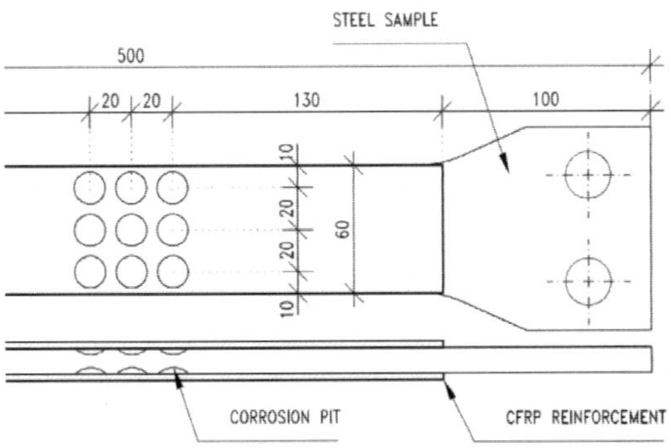

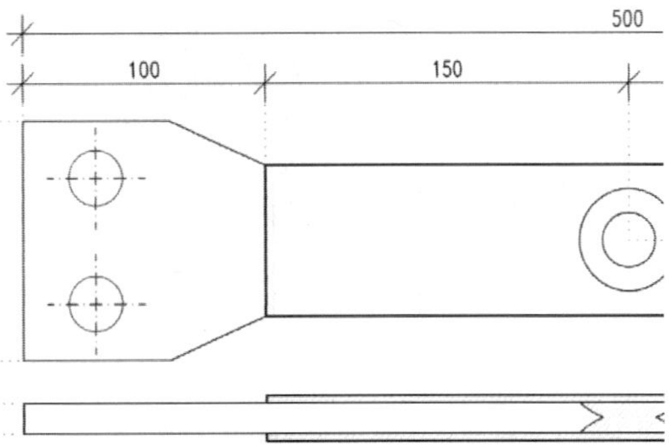

Figure 1. Geometry of specimen with fault type 1 **Figure 2**. Geometry of specimen with fault type 2.

Figure 3. Specimen fault 2 without reinforcement

Figure 4. Specimen fault 2 with reinforcement applied

2.2. The experiment set-up

In the test rig the specimen is connected to the fasteners by two prestressed bolts. The fasteners are made up of two parts for easy handling and they connect the specimen to the load cell on one side and to the ground on the other. To measure the stress distribution on the specimen two strain gauges are applied to the side of the specimen. The test rig can be seen on figure 5.

The specimens are loaded with cyclical loading force until the failure. The number of cycles needed is the main parameter measured. During the loading the specimens are always in tension. As mentioned above the force used was established in the numerical study. The magnitude was set to be between 3-66 kN for the specimen without the reinforcement and between 3-85 kN for the reinforced element at the frequency of 5 Hz. The different forces are used to produce similar level of stress inside the steel. The idea behind this was that the specimens will crack roughly at the same time.

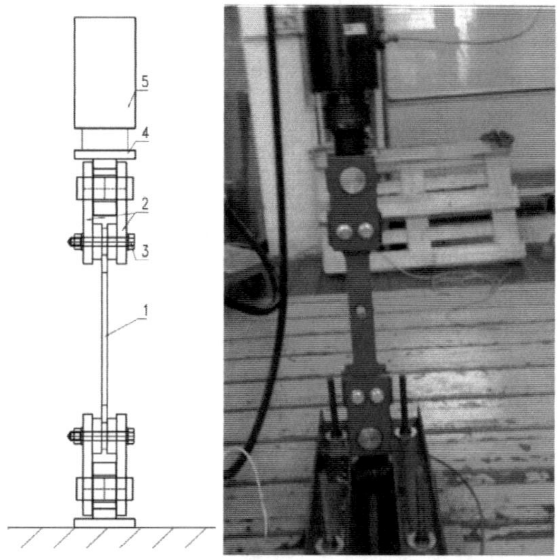

Figure 5. The test rig, schema (right) - (1) specimen, (2) the fastener – two parts, (3) prestressed bolts, (4) steel joint, (5) loading cell, and photo (left).

3. The results

There are 40 specimens in total. As the experiments started not long ago the results presented here are based on the first tested samples. So far five specimens were tested successfully and one specimen is currently being tested. They all have the fault type 2, the corrosion hole. Even with this number of cases at the beginning of the experiments, the results show increase in the number of cycles the reinforces specimen is able to withstand. Of course more data, that will be available after all 40 specimens are finished, is preferablefor making the definite conclusions.

3.1. Specimens without reinforcement

Three specimens without the reinforcement were tested. It was specimens number 9, 11 and 12. In the table 2 you can see the record for each experiment. Originally based on the numerical study the specimen was estimated to last about 250 000 cycles. This was determined using the UIC 778-2 R leaflet. The topic of this leaflet is the determination of remaining fatigue life for existing steel bridges. The adequate permissible stress range for this concrete fatigue prone detail is 85 MPa. Out of this the number of cycles was established using the standard procedure. The area of the damaged cross section is 371 mm^2 and the force applied is ca 63 kN, which corresponds to max stress range of 170 MPa.

Table 2. Unreinforced samples – test results

Specimen No	Total time	cycles	F_{MIN} [kN]	F_{MAX} [kN]	SG. 1[μm/m]	SG. 2[μm/m]
9	23:21	420218	3,53	53,2	749	329
11	23:18	419435	3,53	66,48	765	658
12	22:13	398959	3,53	66,52	704	719

The actual life expectancy during the experiment is however higher, as expected. All three of the specimens reached around 400 000 cycles before breaking. This is mainly due to the probabilistic nature of the S-N curves as they give 5% quantile of breaking.The applied force was modified for each sample. This was based on the real strains measured at the specimen, where optimal strain for all samples was app. 740 microstrains. That is why the specimen no. 9 was loaded only with maximal force of 50 kN. Due to the imperfections this specimen was loaded very asymmetrically.

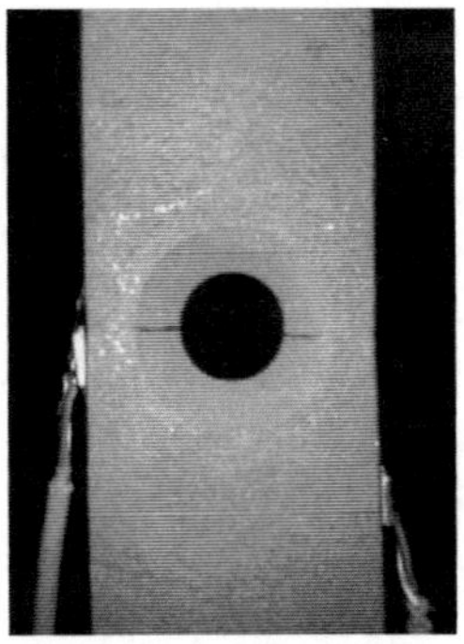

Figure 6. Specimen no 12 shortly before breaking

Figure 7. Specimen no 9 – asymmetrically loaded

3.2. Specimens with reinforcement

There are two reinforced specimens already tested (number 17 and 18). According to the preliminary study the reinforcement should under the same load lover the stress in the steel by one third. This was confirmed by statically loading of the reinforced specimen. From app. 750 microstrains the value measured on the reinforced element sunk to app. 550 microstrains. This reduction should effectively double the life expectancy (see [8]).

In order to test samples in the reasonable time range, a different approach was chosen to test the reinforced specimens. The specimens were loaded not with the same force as before but with force that would produce the same stress in steel. Same stress should mean the same time till the breaking of the steel. The loading force in the range between 3-85 kN was chosen.

Table 3. Reinforced samples – test results

Specimen No	**Total time**	cycles	F_{MIN} [kN]	F_{MAX} [kN]	SG. 1[μm/m]	SG. 2[μm/m]
17	17:21	312171	3,53	85,7	831	639
18	14:50	267695	3,53	85,7	744	688

The breaking of the specimen is still abrupt. The development of the initial crack leads to rise in the measured strains. This continues as the crack grows until the brittle fracture, which is denoted by sharp decrease of measured strain. The crack is now so wide that the remaining steel plasticises and finally breaks completely. The carbon lamella doesn't show any signs of delamination during this process. It breaks of the steel when the specimen breaks completely. The critical point in this is the shear strength of the epoxy resin.

Figure 8. Specimen no 18

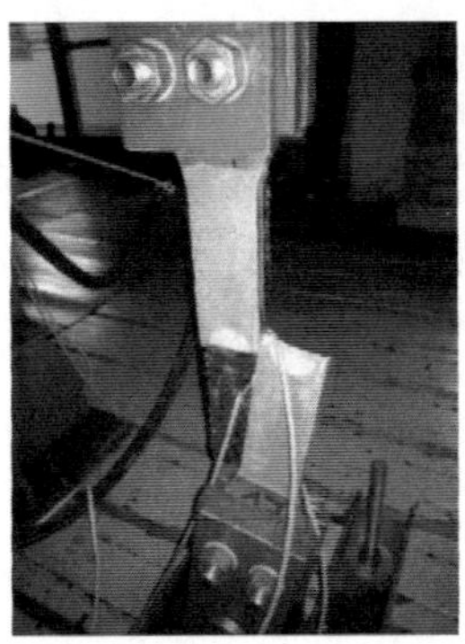

Figure 9. Specimen no 17 after breaking

Figure 10. Specimen no 18 after breaking

3.3. Comparison of reinforcement effectiveness

In reality the specimen did not resisted to the number of cycles, that were expected. Several reasons can be found, such as the complex stress behaviour in the corrosion pit, usage of older steel with worse material parameters etc. The impact of accumulated load was also verified, but the samples from the old bridge were taken from parts, which should have not been influenced by it.

To compare the two sets, we have to calculate the number of cycles that would occurif the stress range corresponds to that of unreinforced specimen. To do this we have to establish proper S-N curve for the reinforced detail. To do this we used Eurocode 1993-1-9 section 7.1. Out of the equation:

$$\Delta\sigma_R{}^m * N_R = \Delta\sigma_C{}^m * 2 * 10^6, m = 3\, for\; N \leq 5 * 10^6 \qquad (1),$$

where we know the number of cycles N (300 000 app. the mean value) and the real stress range $\Delta\sigma_R$ (221 MPa in the steel without calculating the influence of ideal cross-section). From here we can calculate the fatigue prone detail to have fatigue limit at 117 MPa with $2 * 10^6$cycles. Then we use this value again in the same equation (1) to calculate the number of cycles needed while using the stress range of the unreinforced specimen (170 MPa).

$$170^3 * N_R = 117^3 * 2 * 10^6 \rightarrow N_R = 651\,990\; cycle \qquad (1.1),$$

If we would be using the same force for the reinforced element as was used for the unreinforced one, we should expect the failure after approximately 650 000 cycles. The life expectancy will grow only 1.6 times and not the predicted 2 times. However, this is still quite substantial rise.

4. Conclusions

The carbon fibre reinforcement is shaping to be a viable option for strengthening of old steel structures. As was shown in this article the reinforcement can increase the life expectancy of such structures. Furthermore, the materials are very easy to handle and they are quite versatile to use. And also in terms of costs the carbon reinforcement available is reasonably cheap.

On the other hand,these assumptions are based only on the experiments that have happened so far. To make the final conclusions on effectiveness more data is needed and we have to wait to finish all of planed experiments. The method used to strengthen the specimens has also some drawbacks. As it is a hand layout the quality of the reinforcement depends heavily on the person doing it. It does not prevent cancelation of traffic. In order to activate the reinforcement, there can´t be any load applied. This can be however remedied by prestressing the reinforcement.

Acknowledgements

Research in this paper was supported by COST CZ (LD) project of the Ministry of Education, youth and sports LD (No. LD15131) "Fatigue behaviour of FRP reinforced steel members under severe environment".

References

[1] Macho M and Ryjáček P2016*Proc. Int. Conf. on Engineering Sciences and Technologies* ESaT 2015 123-28
[2] Colombi P, Bassetti A and Nussbaumer A2003*Fatigue Fract. Eng. Mater. Struct.***26**59-66
[3] Tavakkolizadeh M and Saadatmanesh H 2003 Feb *J. Struct. Eng.* 186-96
[4] Tavakkolizadeh M and Saadatmanesh H 2001 *J. Compos. Constr.* **5** 200-10
[5] Pipinato A, Pellegrino C and Modena C 2012 *Mod. App. Sci.* **6** no 10
[6] Xiao-Ling Zhao X and Zhang L 2007 *Eng. Struct***29** 1808-23
[7] CNR Italian Advisory Committee on Technical Recommendations for Construction*2005 Guide for the designand construction of externally bonded FRP systems for strengthening existing structures. FRP systems for strengthening existing structures*
[8] Vůjtěch J, Ryjáček P and Vovesný M2016 *Proc.2nd Int. Conf. on Engineering Sciences and Technologies* ESaT 2016art. no 126

IOP Conf. Series: Materials Science and Engineering **175** (2017) 012064 doi:10.1088/1757-899X/175/1/012064

Investigations on new carbon-based nanohybrids combining carbon nanotubes, HfO_2 and ZnO nanoparticles

P Rauwel[1], A Galeckas[2], M Salumaa[1], A Aasna[1], F Ducroquet[3] and E Rauwel[1]

[1]Tartu College, Tallinn University of Technology, 51008 Tartu, Estonia
[2]Department of Physics, University of Oslo, 0316 Oslo, Norway
[3]IMEP-LAHC, CNRS, Université de Grenoble-Alpes, Minatec campus, 38016 Grenoble, France

Email: protima.rauwel@ttu.ee

Abstract: In this work, we present two types of hybrid materials. The first hybrid material is a combination of carbon nanotubes (CNT) with HfO_2 nanoparticles. The latter constituent on its own exhibits unusual visible photoluminescence, which is in stark contrast to non-luminescent HfO_2 in the bulk form. The small size of HfO_2 nanoparticles, 2.6 nm in average, suggests surface-defect related origin of the observed photoluminescence. The other hybrid material is CNT-ZnO with the embedded ZnO nanoparticles ranging from 50nm to 100nm in size. ZnO represents a direct bandgap semiconductor renowned as highly-luminescent in a broad spectral range. The visible region is attributed to luminescence involving deep-level defects, and hence depends on the synthesis conditions. In this study we compare the morphology of the two hybrid materials with transmission electron microscopy. We further compare the photoluminescence properties and the influence of the CNT coupling on enhancing or suppressing defect related emissions. Finally, we present a novel hybrid material CNT-HfO_2 capable of producing a photocurrent under zero bias.

1. Introduction

Hybrid materials are gaining interest as they combine different classes of materials, commonly involving both inorganic and organic components. By combining these two materials, it is possible to obtain a composite material that displays properties of the individual constituents as well as entirely new ones as a result of hybridizing. At present, carbon based materials have gained momentum owing to their superior electrical, mechanical, physical and optical properties [1]. Moreover, carbon based materials are abundant and are known to be non-toxic and bio-compatible thereby enlarging their sphere of application [2]. HfO_2 is a dielectric material and produces no visible luminescence once in the bulk form [3], however, ultrasmall HfO_2 nanoparticles were recently demonstrated to display strong photoluminescence originating from surface defects viz., Hf and O vacancies [4, 5]; the signature of these defects is an emission in the blue-green part of the visible spectrum. On the other hand, ZnO is a semiconductor material with a direct bandgap of 3.2eV. The spectral range of highly-luminescent ZnO covers the UV and visible regions, the latter strongly dependent on the morphology, synthesis route and size [6]. For instance, the characteristic green luminescence dominates in oxygen deficient ZnO, whereas hydrothermal or sol gel methods produce ZnO emitting in the yellow-orange region [7]. Although ZnO has shown photovoltaic activity on its own, more efficient charge transfer to an external load can be obtained by hybridizing them with carbon nanotubes [8]. Moreover, from an industrial point of view, ZnO is already used in sun screens and water purification owing to its nontoxic character.

In our previous study, we have demonstrated that photocurrent generation is possible at zero bias in the hybrid CNT-HfO_2 structure [9]. In the present work, we compare the photoluminescence properties of these two hybrid structures viz., CNT-HfO_2 and CNT-ZnO, and demonstrate how the size, morphology and synthesis routes affect the optical properties of the material. We also present proof-of-concept results on photocurrent generation of a novel, non-conventional optical material such as CNT-HfO_2 hybrid.

2. Experimental

2.1 Synthesis

2.1.1 Synthesis of CNT-ZnO hybrid: The procedure for synthesizing ZnO NPs was carried out under air. In a typical synthesis, zinc acetate (3.41 mmol) (99.99%, Aldrich) was added to 20mL (183mmol) benzyl amine (>99.0%, Aldrich). The reaction mixture was transferred into a stainless steel autoclave and carefully sealed and heated in a furnace at temperature ranging from 200°C to 300°C for 2 days. The resulting milky suspensions were centrifuged, the precipitates thoroughly washed with ethanol and dichloromethane and subsequently dried in air at 60°C. In the case of the carbon nanohybrid synthesis, NANOCYL™ NC7000 MWCNT with an average diameter and length of 10 nm and 1.5 µm, respectively were used in the synthesis. For the synthesis of carbon based ZnO nanocomposite, the MWCNTs were directly homogeneously dispersed into the solution of zinc acetate and benzylamine before transferring the solution into autoclave for the reaction synthesis.

2.1.2 Synthesis of HfO_2-CNT hybrid: The procedure for synthesizing cubic HfO_2 NPs was carried out in a glove box (O_2 and H_2O, <1 ppm). In a typical synthesis, hafnium tert-butoxide (($Hf(O^tBu)_4$) precursor (STREM 99.9%) (0.87 mmol) was added to 20 mL (183 mmol) of benzylamine (purified by redistillation (99.5%), Aldrich). The reaction mixture was transferred into a stainless steel autoclave and carefully sealed. Thereafter, the autoclave was taken out of the glove box and heated in a furnace at 300°C for 2 days. The resulting milky suspensions were centrifuged; the precipitates were thoroughly washed with ethanol and dichloromethane and subsequently dried in air at 60°C [5]. NANOCYL NC7000 MWCNTs with an average diameter and length of 10 nm and 1.5 µm, respectively, were used in the synthesis. The nanoparticles were then dispersed in pure ethanol along with the CNT and sonicated for a total of 2h.

2.2 Characterization:

Transmission electron microscopy studies were carried out on a Titan G2 80-200 working in TEM mode at 200kV. PL was investigated at a room temperature by employing a 325 nm wavelength CW He-Cd laser with an output power of 6 mW as an excitation source. The emission was collected by a microscope and directed to a fiber optic spectrometer (Ocean Optics, USB4000) with a spectral resolution of 2 nm. The electrical measurements were carried out using a source measure unit (Agilent 4156). The sample was illuminated by a 125 W Hg lamp emanating a wavelength of 365 nm.

3. Results and discussions:

3.1 Morphological characterizations

TEM images of CNT decorated with HfO_2 and ZnO are presented in Figures 1(a) and (b) respectively. In Figure 1(a) the HfO_2 nanoparticles with average sizes of 2.6 nm appear to be agglomerated. It was previously shown that they attach themselves to topological defect sites of the CNT which were already present or were created after sonication [9]. In the case of CNT decorated with ZnO nanoparticles in Figure 1(b), ZnO nanoparticles are larger with sizes ranging from 50-100 nm. Moreover, they decorate the CNT randomly as compared to HfO_2. In Figure 1(c) a HRTEM image of the CNT-HfO_2 hybrid structure is presented clearly showing the atomic resolution of HfO_2 and its high crystalline quality along with the CNT walls where the outermost walls on one hand appear to be damaged but on the other hand facilitate the decoration of the HfO_2 nanoparticles on them and create a strong contact at the interface of CNT and HfO_2. The HRTEM image in Figure 1(d) of CNT used with ZnO not only show wall damage but also amorphous carbon on the side walls due to the synthesis and sonication processes. The damaged side walls contain dangling bonds which are either functional groups or carbon and act as anchorage points for ZnO.

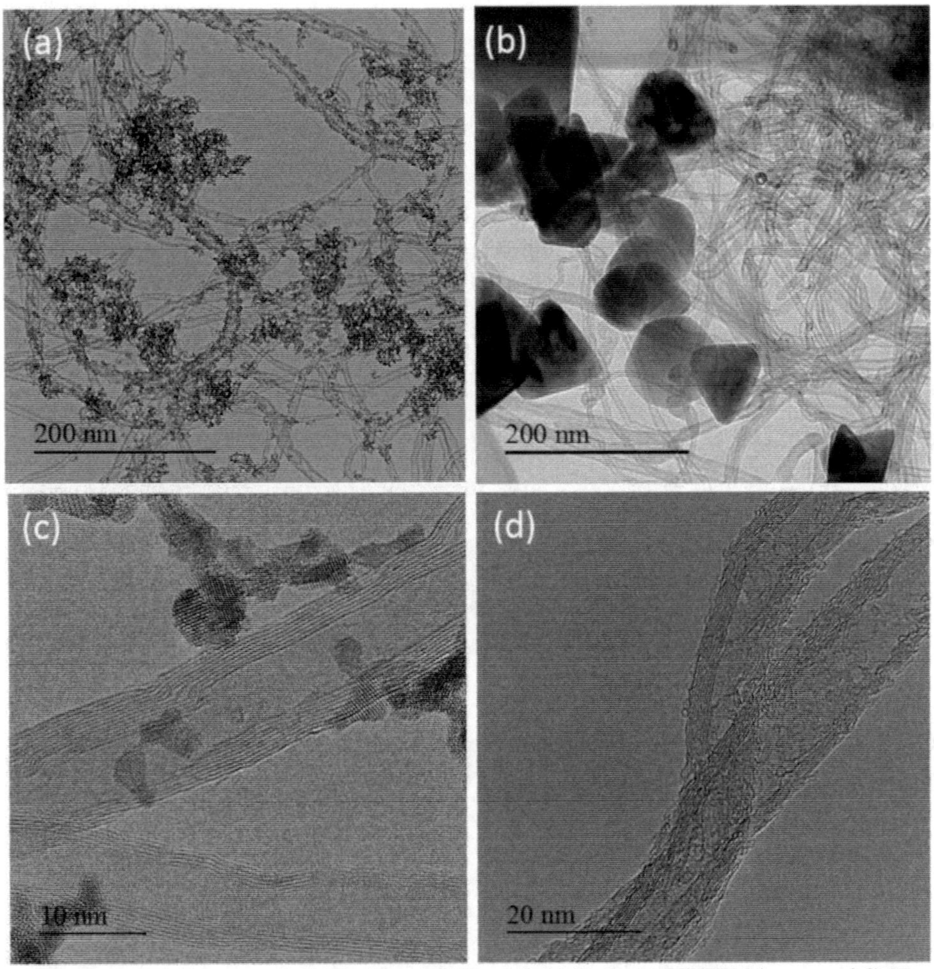

Figure 1. TEM images of CNT with (a) HfO_2 and (b) ZnO, (c) HRTEM images of CNT with HfO_2 (d) part of a CNT decorated with ZnO

3.2 Photoluminescence characterization

The room-temperature PL spectra of the two hybrid materials, i.e. CNT decorated with HfO_2 and ZnO, are represented in Figure 2(a) and (b), respectively. The PL spectrum of the HfO_2 decorated hybrid material consists of several components which can be clearly discriminated using Gaussian deconvolution algorithm. The emission components centered at photon energies of 2.2 eV, 2.5 eV, 2.8 eV and 3.1 eV emerge as a result of the synthesis method applied as well as particle size reduction to nanoscale, which accordingly yield extrinsic luminescent centers and enhance the role of surface defects. The emission around 2.5 eV is attributed to luminescent surface-defects associated with oxygen vacancy, while the emission component at 3.1 eV corresponds to $Hf^{3}+$ centers on the surface of the HfO_2 linked to oxygen deficiency [4]. In a recent report comparing PL of the hybrid CNT-HfO_2

nanocomposites dispersed in colloidal suspension in ethanol and of the free-standing HfO_2 nanoparticles [9], the observed stronger emission at 3.1eV in nanocomposites is attributed to an enhancement of Hf^{3+} defects at CNT-HfO_2 interface, which in turn indicates an increase in oxygen vacancies. The CNT-ZnO hybrid material also presents several emission components at room temperature. The emission at 3.2 eV corresponds to the near band edge emission of ZnO or its band gap and its intensity is negligible compared to the intensity of the visible emissions located at 2.5 eV, 2.2 eV and 2.1 eV corresponding to the green-orange emission in the visible spectrum. The enhanced deep level defect at 2.5 eV indicates an increase in the oxygen vacancy related defects at the interface of CNT-ZnO viz., O_{Zn} or neutral Oi. The latter is all the more pronounced considering the surface to volume ratio increase in the nanostructured material.

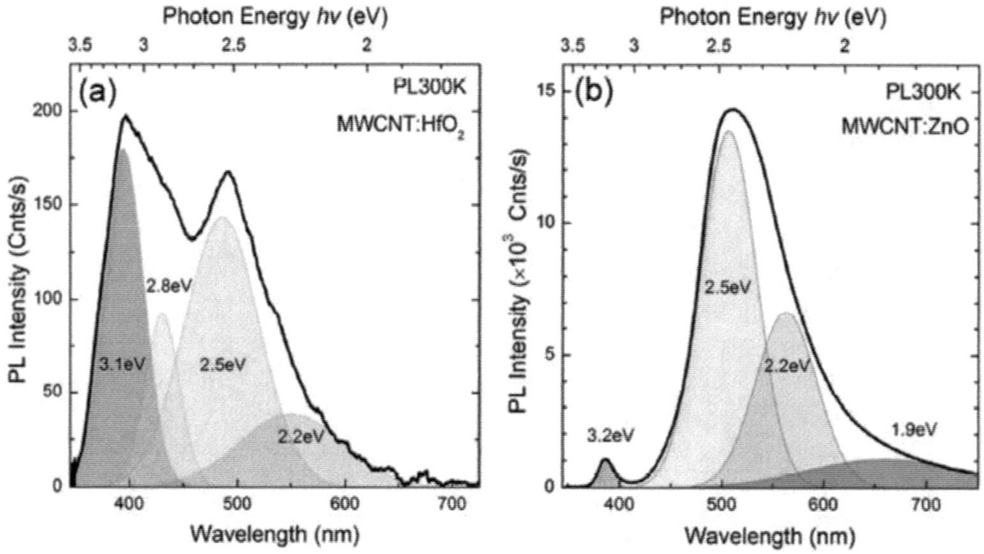

Figure 2: Room-temperature PL spectra of CNT decorated with (a) HfO_2 and (b) ZnO

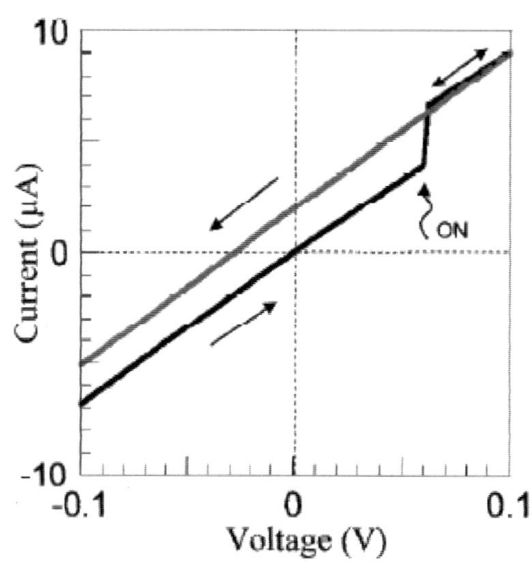

Figure 3. Photocurrent generation in CNT-HfO_2 hybrid materials

3.3Photocurrent generation

For the electrical characterization, the nanocomposite is deposited on a glass slide and contacted using a micromanipulator manual probe station. The dark current-voltage characteristic illustrated by the black graph is linear, indicating an ohmic type conduction via the CNTs. On illumination, the I-V curve is shifted to a higher current value. This extra current is not depending on the voltage and is maintained at 0V, indicating the generation of a photocurrent.

4. Conclusion

Here we present a novel hybrid material CNT-HfO_2 as a possible candidate for photocurrent generation. We have compared its morphology and photoluminescence to CNT-ZnO hybrid. In terms of morphology, the HfO_2 nanoparticles have an average size of 2.6 nm. Moreover, TEM observations illustrate that these nanoparticles are decorated on defect sites of the CNT and are usually in the form of agglomerates. ZnO on the other hand are much larger in the range of 50-100nm. Both materials show similarities with orange- yellow emissions originating from synthesis via sol-gel methods. Also, on hybridizing, at the interface of the hybrid and the oxide nanoparticle, the oxygen vacancies increase whereby an intense emission in both cases is observed at 2.5eV. In the case of HfO_2, these oxygen vacancies at the interface are also responsible for the increase in Hf^{3+} vacancies thereby emanating light in the UV-blue region of the spectrum. A similar but a less intense emission is observed for ZnO however, the origin is completely different. The emission at 3.1eV for ZnO is the near band edge due to the band gap of the material and indicates the purity of ZnO when compared to the defect emission in the visible region. In both cases, we observe an increase in defect related emission. We have also studied photocurrent generation in the CNT- HfO_2 material. Considering the large amount of defects which act as traps and accumulate charges in the free-standing HfO_2, the dissipation of these charges on contacting them with CNTs seems to be the most probable explanation for photocurrent generation. The photocurrent generation in CNT-ZnO hybrid structure is being presently studied to understand explicitly the origin of photocurrent generation in the two hybrid materials

Acknowledgements

Financial support from through the Estonian Research Council (grant number PUT431), MENESR and MAEDI French ministries (Parrot program n°33787YJ),, and the Estonian Road Map infrastructure and the NAMUR projects for the use of the Titan G2 80-200 Transmission electron microscope is acknowledged. The research has been supported by the European Regional Development Fund project TK134 (TAR19016).

References

[1] O. Kozák, *et al.* 2016 *Chemistry of Materials* **28** 4085-128
[2] J. K. Jaiswal, *et al.* 2004 *Nat Meth* **1** 73-8
[3] P. Rauwel et al. 2014 Stabilization of Higher Symmetry HfO_2 Polymorphs as Thin Films and Nanoparticles in: *Hafnium: Chemical Characteristics, Production and Applications,* ed. H. Yu: Nova, pp 77-112
[4] E. Rauwel, et al. 2014 *Materials Research Express* **1** 015035
[5] P. Rauwel, et al. 2012 *Journal of Applied Physics* **112** 104107
[6] E. Rauwel, et al. 2011 *The Journal of Physical Chemistry C* **115** 25227-33
[7] A. B. Djurisic, et al. 2006 *Applied Physics Letters* **88** 103107
[8] P. Rauwel, et al. *2016 Journal of Nanomaterials In Press*
[9] *P. Rauwel, et al.* 2016 *Beilstein Journal of Nanotechnology* **7** 1075-85

IOP Conf. Series: Materials Science and Engineering **175** (2017) 012065 doi:10.1088/1757-899X/175/1/012065

Preparation of SiC–$MgAl_2O_4$–$Y_3Al_5O_{12}$-MWCNTs nanocomposites by spark plasma sintering

E V Zharikov[1], V V Kapustin[1], P P Faikov[1], N A Popova[1], A A Barmin[2], A V Ivanov[2], R N Rizakhanov[2]

[1] D. Mendeleev University of Chemical Technology of Russia, 125047 Moscow, Russia
[2] State Scientific Centre "Keldysh Research Center", 125438 Moscow, Russia

E-mail: evzh@mail.ru

Abstract. Fabrication a composite materials based on silicon carbide (SiC) reinforced with multi-walled carbon nanotubes (MWCNTs) with addition of magnesium alumina spinel $MgAl_2O_4$, and yttrium aluminum garnet $Y_3Al_5O_{12}$ by spark plasma sintering are presented.
Two series of composites differing by the particle size of starting SiC were prepared. Mechanical characteristics of composites including microhardness, fracture toughness and flexural strength are determined.

1. Introduction

In recent years, in aerospace engineering there is a trend for the use of ceramic composite materials at sites exposed to high temperatures (1500 K) and aggressive environment. This is because, ceramics and composite materials in these conditions have higher characteristics compared to metals and alloys, particularly flexural strength [1]. However, ceramics is a more brittle material, therefore, the most promising are materials combining strength and thermal stability of ceramics and a sufficiently high fracture toughness and elasticity [2, 3].

Silicon carbide is one of the most promising oxygen-free refractory compounds for obtaining construction materials for use at high temperatures. Reinforcement of ceramics with various additives (powders, fibers, nanotubes) leads to a significant increase in strength and fracture toughness, which place the ceramic composite materials on one of the first places among the promising structural materials for aerospace engineering [4, 5, 6].SiC-based composites are also promising for use as protective armor [7]. Popular sintering additives for silicon carbide ceramics are oxides (such as Al_2O_3, Y_2O_3, RE_2O_3, MgO, etc.) and their mixtures that create a liquid-phase sintering of SiC composites providing the notable lowering of synthesis temperature. It was found that oxide additives facilitate densification of sinters and significantly improve mechanical properties of SiC ceramics [8, 9]. The microstructure of such materials is characterized by a framework of crystalline grains surrounded by a layer of inter-granular phases [10, 11].

Among common reinforcing constituents, the most preferred care are the carbon nanotubes demonstrating excellent mechanical properties that can improve the performance of a polycrystalline ceramic matrix, such as hardness, strength, Young's modulus , fracture toughness, wear resistance [12, 13].

There are many routes for the fabrication of silicon carbide. Spark Plasma Sintering (SPS) stands out among them. It is a recent development in sintering technology that combines axial pressure with heating by an electrical current passing through the die that contains the powder body. The basic idea of SPS is to heat a powder material by passing a pulsed current with simultaneous pressure application [14]. It provides very fast heating and lower sintering temperatures allowing preparation of materials with a high compactness in a few minutes [15].

The important advantage of this process is the possibility to obtain dense material with a fine microstructure due to the very low sintering time at low temperature. There are some reports on the SPS synthesis of SiC composites containing separately either oxide additives [16, 17] or multi-walled carbon nanotubes [18, 19]. It is possible that simultaneous introduction of two different reinforcing agents as oxide dopants and MWCNTs could give a synergetic effect in improvement of mechanical properties.

In this report we present the results on fabrication of SiC nanocomposites containing together the multi-walled carbon nanotubes and complex oxides if form of magnesium alumina spinel $MgAl_2O_4$, and yttrium alumina garnet $Y_3Al_5O_{12}$ by the Spark Plasma Sintering.

2. Experimental

As the starting materials the commercially available industrial products, namely SiC powder and multi-walled carbon nanotubes were used. We used two different batches of silicon carbide: first one with an average size of 2-3 μm, and second one with the particle size up to 30 microns. Based on these batches two corresponding series of composites were fabricated (series #1 from batch #1 and series #2 from batch #2). Both kinds of SiC powders were subjected to additional treatment in hydrofluoric acid HF for removing residual SiO_2 impurity.

The additives of magnesium alumina spinel $MgAl_2O_4$, and yttrium alumina garnet $Y_3Al_5O_{12}$ were prepared in nanodisperse form by specific version of sol–gel technology [20, 21], the fundamental essence of which is to obtain a gel of high molecular weight polymer and realize the distribution in it the homogenous true solution of several oxide components derived from the nitrates and chlorides of the metals. The dried xerogel is calcined at relatively low temperatures (from 600 to 1100°C depending on the type and nature of the oxide), turning the porous mass into granular agglomerated powder consisting of flakes with a size of 50-70 microns that in turn are consisting of nanoparticles of the resulting oxide.

Commercial multi-walled carbon nanotubes produced by Bayer Material Science AG were used as main reinforcing component for SiC-based composites. The outer diameter and inner diameter of the MWCNT are around 13 nm and 4 nm respectively and MWCNT length is less than 1 micron. The bulk density of nanotubes was 0.13-0.15 g/cm^3. The nature as well as the way of special treatment of carbon nanotubes strongly influence on the structure and strength of ceramic composites [22]. MWCNTs have previously been calcined at 300°C for removing a preservative agent. Further the nanotubes were subjected to sonication in polyvinyl alcohol solution. The use of ultrasound allows dispersing MWCNTs in a liquid by the destruction of their agglomerates, particularly the large bundles and aggregates.

In the course of development we tried different surfactants as the dispersion medium, including methyl, ethyl, isopropyl alcohol solutions and water. This has been studied in detail in work [23]. The comparative study of destruction efficiency and stability of solution shoved that polyvinyl alcohol (PVA) is the most suitable dispersion surfactant. We have developed a method of CNTs dispersing in PVA solution (concentration 0.1%). Dispersion of MWCNT performed using ultrasound dispersant Sonoplus HD 3100, with power 100 W, dispersing time 10-20 minutes, frequency of 22 kHz, the pulse duration 1-1.5 seconds and duty cycle of 1 sec. As a result we obtained the homogenized suspension, stable up to 48 hours.

After the dispersion, the suspension containing MWCNTs was mixed with SiC powder. The mixing of all the components of the composite material was performed using a planetary mill for 20–40 min. Then, after separation of the suspension from the grinding media, it was placed in a heat resistant bowl and exposed to the drying. After a series of investigations it was found that the best results of drying are obtained by simultaneous heating from bottom (using an electric hotplate) and from top (using an infrared lamp) under constant stirring. In this case, it is used a low-temperature drying from IR lamp source. The distance between the lamp and the surface of the heated object is 5 - 12 cm. This simultaneous deep and surface heating from two sources under continuous stirring results in a quick drying of the slurry; no peeling MWNTs or sedimentation of individual particles occurs during the procedure. Thus it is achieved a uniform distribution of all components of the ceramic composite material. The resulting powder was further homogenized by rubbing through a sieve with a mesh size of 0.25 mm to increase the homogeneity, flowability and ease filling of the mold by press-powder.

The composite samples were prepared by spark plasma sintering on installation 25 D HP (FCT Systeme, Germany) at a temperature in the range of 2000-2200 °C with heating rate 450 °C/min under pressure 19-22 kN and exposure at maximal temperature 10-15 minutes. Resulting composite samples appeared as dark gray discs of 20 mm in diameter and 4.5 mm thick. Polishing of material carried on grinding and polishing machine EcoMet 250 (Buehler, Germany).

The measurements of the composite micro-hardness were performed on apparatus Micromet 5114 by Vickers pyramid indentation. Determination of the average micro-hardness of the sample was performed at loads in the range of 200-300 g., the loading time for all tests was 15 seconds (5 seconds-load, 10 seconds – exposure under load). 5 prints were done on the each sample.

The micro-hardness indentation method was using also to measure fracture toughness (K_{1C}). For calculation used the following relation [24]:

$$K_{1C} = 0{,}016\ (E/H)^{1/2}\ (P/c^{3/2}), \tag{1}$$

where E and H are Young's modulus and hardness, respectively, P is the applied load, c is the radial crack length. The load on the indenter in the measurements was 19.6 N.

Investigation of the thermo-mechanical properties of the samples was performed on a high-temperature electromechanical universal machine UTS 110M with a nominal load of 100kN and furnace VE-324-RM. A study of the thermo-mechanical properties of SiC-based composites was carried out by three-point bending at a temperature of 1500 °C under vacuum (10^{-2}Torr). The heating rate was 10 °C per minute. Calculation of flexural strength was performed according to relation:

$$\sigma = \frac{3 \cdot P \cdot l}{2 \cdot b \cdot h^2}\ (\text{MPa}), \tag{2}$$

where P - breaking load in N, l - span in mm, b - width of the specimen in mm,
h - the thickness of the sample in mm.

3. Results and Discussion

The microstructure of samples belonging to series #1 is formed from dense equi-crystallite material with grain size a few microns. Nanotubes are distributed rather uniformly over the entire volume. The nanotubes form in the SiC ceramic composite a reticular framework along the facets of the crystals of the ceramic matrix, like it is in alumina-CNT composites [25], that impedes the growth of SiC grains and promotes the removal of the closed porosity. These reticular carcasses strongly contribute to the increase in the mechanical properties of composite. The microstructure of SiC + 6% MWCNTs + 1% $MgAl_2O_4$ composite shown on Fig. 1a presents rather uniform distribution of silicon carbide crystals with size 2-3 microns that covered by reticular-carcass framework of carbon nanotubes. As can be seen from Table 1, such representative mechanical properties as micro-hardness and fracture toughness are increasing after addition of magnesium alumina spinel and then yttrium alumina garnet.

It is believed [26] that oxides, like YAG, spinel, Al_2O_3 and some others, either due to reaction with passivating silica layer coating the SiC grains, or due to reaction among themselves, form liquid phases during sintering. On one hand they activate the densification processes, and on the other they improve some material's properties. We assume that appearance of fine films formed by magnesium alumina spinel and YAG around grains of silicon carbide provides the slowing down migration of their boundaries, and together with MWCNTs inhibits the SiC grain growth. Moreover, due to the

IOP Conf. Series: Materials Science and Engineering **175** (2017) 012065 doi:10.1088/1757-899X/175/1/012065

chemical interaction between $MgAl_2O_4$ and $Y_3Al_5O_{12}$ and formation of eutectics, as well as due to the difference of the melting points (2135°C for spinel and 1950°C for YAG) their joint action takes place in a broader temperature range that contributes to a more fine-grained and homogeneous structure.

(a) **(b)**

Figure 1. SEM image of a cleaved samples:SiC + 6% MWCNTs + 1% $MgAl_2O_4$ (series #1) **(a)** SiC + 1% MWCNTs + 1% $MgAl_2O_4$ + 1% $Y_3Al_5O_{12}$ (series #2) **(b)**

The following to spinel addition of yttrium-aluminum garnet leads to further suppression of growth of silicon carbide crystals in the composite during sintering. The reduction in the grain size of SiC-MWCNTs ceramic matrix composite containing both additives $MgAl_2O_4$ and $Y_3Al_5O_{12}$ is notable compared to single doping by spinel. The grain size of SiC becomes only 0.5-2 microns. This behavior results in further enhancement of mechanical properties of composite including high-temperature characteristics (see Tables 1 and 2).

Table 1. Averaged values of the mechanical properties of the samples of series #1

Composite	**Load (g)**	**Microhardness (GPa)**	**Fracturetoughness (MPa/m^2)**
SiC+6%MWCNT	300	28,3	4,3
SiC+6%MWCNT +1% $MgAl_2O_4$	300	31,4	5,1
SiC+6%MWCNT +1% $MgAl_2O_4$ +1% $Y_3Al_5O_{12}$	300	34,8	5,7

Note: The contents of MWCNT and oxide additives here and below are given in vol. %.

Composites of series #2 demonstrate the double-fractional grain structure shown on Fig.1b. Large difference in size of crystals and presence of small grain fraction results in the better densification of structure due to the distribution of fine grains of silicon carbide around the majorones. Nevertheless such dense packing doesn't result in full-dense structure and demonstrate the presence of certain amount of closed pores, as it can see from Fig.5. However, this structure and the presence of large crystals surrounded by small grains, leads to uneven and hindered cracks propagation. Fracture toughness measured by the method of indentation of Vickers pyramid corroborates that (see Table 3).

Table 2. The thermo-mechanical properties of composites of series #1

Composite	Load (g)	Microhardness (GPa)	Fracturetoughness (MPa/m^2)
SiC+6%MWCNT	300	28,3	4,3
SiC+6%MWCNT +1% $MgAl_2O_4$	300	31,4	5,1
SiC+6%MWCNT +1% $MgAl_2O_4$ +1% $Y_3Al_5O_{12}$	300	34,8	5,7

Table 3. The mechanical and thermo-mechanical properties of composites of series #2

Composite	Microhardness (GPa)	Fracturetoughness (MPa/m^2)	Flexural strength at 25°C, (MPa)	Flexural strength at 1500°C (MPa)
SiC	32,8	3,77	240	130
SiC+1% MWCNT	31,3	4,05	280	160
SiC+6% MWCNT +1% $Y_3Al_5O_{12}$	33,3	5,62	270	170
SiC+1% MWCNT +1% $Y_3Al_5O_{12}$	33,7	8,01	320	210
SiC+1% MWCNT+1% $MgAl_2O_4$	32,7	5,69	310	200
SiC+1% MWCNT +1% $Y_3Al_5O_{12}$ +1% $MgAl_2O_4$	32,2	8,67	390	280

Comparison of data presented in Tables 1-3 shows that other mechanical properties, such as micro-hardness and flexural strength of composites of series #1 and #2 are rather close to each other. Only difference is in fracture toughness, which reaches to 8 MPa/m^2 in case of composites belonging to series #2. We believe that increased value of fracture toughness caused by double-fractional grain structure of the composites. The influence of YAG and spinel addition on mechanical properties of series #2 composites is similar to that for series #1.

4. Conclusions

The composite materials based on silicon carbide reinforced with multi-walled carbon nanotubes with addition of magnesium alumina spinel $MgAl_2O_4$, and yttrium aluminum garnet $Y_3Al_5O_{12}$ fabricated by spark plasma sintering. As starting materials two different SiC commercial powders with particle sizes 2-3 μm and up to 30 μm used for composites fabrication.

The commercial multi-walled carbon nanotubes produced by Bayer MaterialScience AG, and the nanodisperse additives $MgAl_2O_4$and $Y_3Al_5O_{12}$ prepared by specific version of sol–gel technology were used as additives.

Introduction MWCNT increases the mechanical properties (strength and crack resistance) of composite material compared to silicon carbide ceramics.

Additives $MgAl_2O_4$ and $Y_3Al_5O_{12}$not only increase the strength characteristics of the material, but also lead to the suppression of SiC crystal growth during sintering, resulting in increased strength and reduced porosity of the composite.

Acknowledgments
This work was supported by the Russian Science Foundation, grant number 14-19-00522

References

[1] Chanthapan S, RapeA, Gephart S, et al 2011 *Advanced materials & processes,* **169** 21
[2] Izhevskyi V A, Genova L A, Bressiani J C, Bressiani A H A 2000 *Cerâmica* **46** 297
[3] Fan J, Chu P K 2014 *Silicon Carbide Nanostrucures, Fabrication, Structure,and Properties* (Cham: Springer)
[4] Kablov E N, Grashchenkov D V, Isayev N V, Solntsev S S, Sevastyanov V G 2012 *Glass and ceramics* **69** 109
[5] Ersova N I, Kelina I Y 2009 *Építőanyag-JSBCM* **61** 34 http://dx.doi.org/10.14382/epitoanyag-jsbcm.2009.6
[6] Gömze L A, Gömze L N 2010 *Építőanyag-JSBCM* **62** 98 http://dx.doi.org/10.14382/epitoanyag-jsbcm.2010.18
[7] Zhitnyuk S V, Makarov N A, Guseva T V 2014 *Glass and ceramics* **71** 6
[8] Cutler R A, Ikar A V, Hurford A C 1989 *US Patent 4829027*
[9] Gubernat A, Stobierski L, Labaj P 2007 *J. Europ. Ceram. Soc.* **27** 781
[10] Zhou Y, Hirao K, Toriyama M, Yamauchi Y, Kanzaki S 2001 *J. Am. Ceram. Soc.* **84** 1642
[11] Perevislov S N 2013 *Glass and Ceramics* **70** 265
[12] Samal S S, Bal S 2008 *J. Minerals & Materials Characterization &Engineering* 73 55
[13] Cho J, Boccaccini A R, Shaffer M S P 2009 *J. Mater. Sci.* **44** 1934
[14] Suárez M, Fernández A, Menéndez JL, et al. 2013 In: *Sintering Applications*, B. Ertug (Ed.), InTechPubl.319-42
[15] Fedosova N A, Koltsova E M, Popova N A, Zharikov E V, Lukin E S 2015 *Novye Ogneupory* **1** 13
[16] Zhan G D, Mitomo M, Xie R J, Mukherjee A K 2001 *J. Am. Ceram. Soc.* **84** 2448
[17] Perevislov S N, Nesmelov D D, Tomkovich M V 2013 *Proc. Lobachevsky Nizhegorodsky University* **2** 107
[18] Hayun S, Paris V, Mitrani R, S. Kalabukhov S, et al. 2012 *Ceramics Intern.* **38** 6335
[19] Sarkar K, Sarkar S, Das P K 2016 *J. Mater. Sci.* **51** 6697
[20] Zharikov E V, Zaramenskikh K S, Iskhakova L D, et al. 2011 *Chem. Engineering* **12** 76
[21] Zaramenskikh K S, Faikov P P, Zharikov E V, McRae E, Devaux X 2012 *Fullerenes, Nanotubes, and Carbon Nanostructures* **20** 424
[22] Fedosova N A, Faikov P P, Popova N A, et al. 2014 *Glass and Ceramics* **71** 128
[23] Zharikov E V, Duong T T T, Faikov P P, Popova N A, Sovyk D M 2016 *Inorganic Materials: Applied Research* **7** 163
[24] Anstis G R, Chantikul P, Lawn B R, Marshall DB 1981 *J. Am. Ceram. Soc.* **64** 533
[25] Zharikov E V, Zaramenskikh K S, Popova N A, et al. 2011 *Glass and Ceramics* **68** 80
[26] Gubernat A, Stobierski L, Labaj P 2007 *J. Europ. Ceram. Soc.***27** 781